Salinity Responses and Tolerance in Plants, Volume 2

Vinay Kumar • Shabir Hussain Wani
Penna Suprasanna • Lam-Son Phan Tran
Editors

Salinity Responses and Tolerance in Plants, Volume 2

Exploring RNAi, Genome Editing and Systems Biology

Editors
Vinay Kumar
Department of Biotechnology
Modern College of Arts Science
and Commerce
Savitribai Phule Pune University
Pune, India

Department of Environmental Science
Savitribai Phule Pune University
Pune, India

Penna Suprasanna
Nuclear Agriculture and Biotechnology
Division
Bhabha Atomic Research Centre (BARC)
Mumbai, India

Shabir Hussain Wani
Genetics and Plant Breeding
Mountain Research Center for Field Crops
Khudwani, Anantnag
Sher-e-Kashmir University of Agricultural
Sciences and Technology of Kashmir
Jammu and Kashmir, India

Lam-Son Phan Tran
Stress Adaptation Research Unit
RIKEN Center for Sustainable
Resource Science
Yokohama, Japan

ISBN 978-3-030-13265-1 ISBN 978-3-319-90318-7 (eBook)
https://doi.org/10.1007/978-3-319-90318-7

Softcover re-print of the Hardcover 1st edition 2018

Printed on acid-free paper

This Springer imprint is published by the registered company Springer International Publishing AG part of Springer Nature.
The registered company address is: Gewerbestrasse 11, 6330 Cham, Switzerland

Preface

Hyper soil salinity has emerged as a key abiotic stress factor and poses serious threats to crop yields and quality of produce. Owing to the underlying complexity, conventional breeding programs have met with limited success. The conventional genetic engineering approaches via transferring/overexpressing a single "direct action gene" per event did not yield optimal results as expected. Nevertheless, the biotechnological advents in the last decade coupled with the availability of genomic sequences of major crops and model plants have opened new vistas for understanding salinity responses and improving salinity tolerance in important glycophytic crops. Through this two-volume book series entitled "Salinity Responses and Tolerance in Plants," we are presenting critical assessments of the potent avenues to be targeted for imparting salt stress tolerance in major crops in the postgenomic era.

Specifically, this book series is an attempt to update the state-of-the-art on intrinsic mechanisms underlying salinity responses and adaptive mechanisms in plants, as well as novel approaches to target them with assistance of sound biotechnological tools and platforms for developing salt-tolerant crops to feed the ever-increasing global population. Volume 1 deals with plant response to salinity stress and perspectives for improving crop salinity tolerance via targeting the sensory, ion transport, and signaling mechanisms.

Volume 2 aims at discussing the potencies of postgenomic era tools like DNA helicases, RNAi, genomics intervention, and systems biology and other potent and novel approaches enabling the breeders and biotechnologists to develop salt-tolerant crops in the era of climate change.

This book series is an excellent and comprehensive reference material for plant breeders, geneticists, biotechnologists, under/postgraduate students of agricultural biotechnology as well as for agricultural companies, providing novel and powerful options for understanding and targeting molecular mechanisms for producing salt-tolerant and high-yielding crops. The chapters are written by internationally reputed scientists, and all the chapters underwent review process for assuring scientific accuracy and high-quality presentation.

We express our sincere thanks and gratitude to our esteemed authors to come together on this important task and contribute their excellent work. We are grateful to Springer-Nature for giving us the opportunity to complete this book project.

Pune, India — Vinay Kumar
Srinagar, India — Shabir Hussain Wani
Mumbai, India — Suprasanna Penna
Yokohama, Japan — Lam-Son Phan Tran

Contents

Contributors

Chedly Abdely Laboratory of Extremophile Plants, Centre of Biotechnology of Borj Cedria, Hammam Lif, Tunisia

Ahmad Arzani Department of Agronomy and Plant Breeding, College of Agriculture, Isfahan University of Technology, Isfahan, Iran

Sudip Biswas Department of Biochemistry and Molecular Biology, Dhaka University, Dhaka, Bangladesh

Marian Brestic Department of Plant Physiology, Slovak University of Agriculture in Nitra, Nitra, Slovakia

Petronia Carillo Dipartimento di Scienze e Tecnologie Ambientali, Biologiche e Farmaceutiche, Università degli Studi della Campania "Luigi Vanvitelli", Caserta, Italy

Huatao Chen Division of Plant Sciences, University of Missouri, Columbia, MO, USA

Institute of Industrial Crops, Jiangsu Academy of Agricultural Sciences, Nanjing, China

Pengyin Chen Division of Plant Sciences, University of Missouri, Delta Research Center, Portageville, MO, USA

Xin Chen Institute of Industrial Crops, Jiangsu Academy of Agricultural Sciences, Nanjing, China

Loredana F. Ciarmiello Dipartimento di Scienze e Tecnologie Ambientali, Biologiche e Farmaceutiche, Università degli Studi della Campania "Luigi Vanvitelli", Caserta, Italy

Luisa D'Amelia Dipartimento di Scienze e Tecnologie Ambientali, Biologiche e Farmaceutiche, Università degli Studi della Campania "Luigi Vanvitelli", Caserta, Italy

Amira Dabbous Laboratory of Extremophile Plants, Centre of Biotechnology of Borj Cedria, Hammam Lif, Tunisia

Emilia dell'Aversana Dipartimento di Scienze e Tecnologie Ambientali, Biologiche e Farmaceutiche, Università degli Studi della Campania "Luigi Vanvitelli", Caserta, Italy

Tuyen D. Do Division of Plant Sciences, University of Missouri, Columbia, MO, USA

Hassan El Shaer Desert Research Centre, Cairo, Egypt

Sabrina M. Elias Department of Biochemistry and Molecular Biology, Dhaka University, Dhaka, Bangladesh

S. A. Ghuge Division of Biochemical Sciences, National Chemical Laboratory (NCL), Pune, India

Karim Ben Hamed Laboratory of Extremophile Plants, Centre of Biotechnology of Borj Cedria, Hammam Lif, Tunisia

Tushar Khare Department of Biotechnology, Modern College of Arts, Science and Commerce, Savitribai Phule Pune University, Pune, India

Pannaga Krishnamurthy Department of Biological Sciences, and NUS Environmental Research Institute (NERI), National University of Singapore, Singapore, Singapore

Prakash P. Kumar Department of Biological Sciences, and NUS Environmental Research Institute (NERI), National University of Singapore, Singapore, Singapore

Ratanesh Kumar Plant RNAi Biology Group, International Centre for Genetic Engineering and Biotechnology, New Delhi, India

Sudhir Kumar Plant RNAi Biology Group, International Centre for Genetic Engineering and Biotechnology, New Delhi, India

Vinay Kumar Department of Biotechnology, Modern College of Arts, Science and Commerce, Savitribai Phule Pune University, Pune, India

Department of Environmental Science, Savitribai Phule Pune University, Pune, India

S. J. Mirajkar Division of Vegetable Science, ICAR-Indian Agricultural Research Institute (IARI), New Delhi, India

Pragati Misra Department of Molecular and Cellular Engineering, Jacob Institute of Biotechnology and Bioengineering, Sam Higginbottom University of Agriculture, Technology and Sciences, Allahabad, Uttar Pradesh, India

Henry T. Nguyen Division of Plant Sciences, University of Missouri, Columbia, MO, USA

G. C. Nikalje Department of Botany, R.K. Talreja College of Arts, Science and Commerce, Ulhasnagar, Thane, India

V. Y. Patade Defence Research & Development Organisation (DRDO), Defence Institute of Bio-Energy Research (DIBER) Field Station, Pithoragarh, Uttarakhand, India

Lin Qingsong Department of Biological Sciences, and NUS Environmental Research Institute (NERI), National University of Singapore, Singapore, Singapore

Preeti Rajoriya Department of Molecular and Cellular Engineering, Jacob Institute of Biotechnology and Bioengineering, Sam Higginbottom University of Agriculture, Technology and Sciences, Allahabad, Uttar Pradesh, India

Kaori Sako Plant Genomic Network Research Team, RIKEN Center for Sustainable Resource Science, Yokohama, Kanagawa, Japan

Core Research for Evolutional Science and Technology (CREST), Japan Science and Technology Agency (JST), Kawaguchi, Saitama, Japan

Neeti Sanan-Mishra Plant RNAi Biology Group, International Centre for Genetic Engineering and Biotechnology, New Delhi, India

Motoaki Seki Plant Genomic Network Research Team, RIKEN Center for Sustainable Resource Science, Yokohama, Kanagawa, Japan

Plant Epigenome Regulation Laboratory, RIKEN Cluster for Pioneering Research, Wako, Saitama, Japan

Core Research for Evolutional Science and Technology (CREST), Japan Science and Technology Agency (JST), Kawaguchi, Saitama, Japan

Kihara Institute for Biological Research, Yokohama City University, Yokohama, Kanagawa, Japan

Zeba I. Seraj Department of Biochemistry and Molecular Biology, Dhaka University, Dhaka, Bangladesh

Grover J. Shannon Division of Plant Sciences, University of Missouri, Columbia, MO, USA

Pradeep Kumar Shukla Department of Biological Sciences, Faculty of Science, Sam Higginbottom University of Agriculture, Technology and Sciences, Allahabad, Uttar Pradesh, India

Saumya Shukla Department of Biological Sciences, Faculty of Science, Sam Higginbottom University of Agriculture, Technology and Sciences, Allahabad, Uttar Pradesh, India

Sonia Mbarki Laboratory of Extremophile Plants, Centre de Biotechnologie de Borj Cedria, Hamam Lif, Tunisia

Amrita Srivastav Department of Biotechnology, Modern College of Arts, Science and Commerce, Savitribai Phule Pune University, Pune, India

P. Suprasanna Nuclear Agriculture and Biotechnology Division, Bhabha Atomic Research Centre (BARC), Mumbai, India

Oksana Sytar Plant Physiology and Ecology Department, Institute of Biology, Taras Shevchenko National University of Kyiv, Kyiv, Ukraine

Department of Plant Physiology, Slovak University of Agriculture in Nitra, Nitra, Slovakia

Neveen B. Talaat Department of Plant Physiology, Faculty of Agriculture, Cairo University, Giza, Egypt

Narendra Tuteja Amity Institute of Microbial Technology, Amity University, Noida, Uttar Pradesh, India

Minoru Ueda Plant Genomic Network Research Team, RIKEN Center for Sustainable Resource Science, Yokohama, Kanagawa, Japan

Core Research for Evolutional Science and Technology (CREST), Japan Science and Technology Agency (JST), Kawaguchi, Saitama, Japan

Babu Valliyodan Division of Plant Sciences, University of Missouri, Columbia, MO, USA

Pasqualina Woodrow Dipartimento di Scienze e Tecnologie Ambientali, Biologiche e Farmaceutiche, Università degli Studi della Campania "Luigi Vanvitelli", Caserta, Italy

Heng Ye Division of Plant Sciences, University of Missouri, Columbia, MO, USA

Jianfeng Zhou Division of Agricultural Systems Management, University of Missouri, Columbia, MO, USA

Marek Zivcak Department of Plant Physiology, Slovak University of Agriculture in Nitra, Nitra, Slovakia

About the Editors

Dr. Vinay Kumar is an Assistant Professor at the Post-Graduate Department of Biotechnology, Progressive Education Society's Modern College of Arts, Science and Commerce, Ganeshkhind, Pune, India, and a Visiting Faculty at the Department of Environmental Sciences, Savitribai Phule University, Pune, India. He earned his PhD in Biotechnology from Savitribai Phule Pune University (Formerly University of Pune) in 2009. For his PhD, he worked on metabolic engineering of rice for improved salinity tolerance. He has published 32 peer-reviewed research/review articles and contributed 13 book chapters in edited books published by Springer, CRC Press, and Elsevier. He is a recipient of Government of India's Science and Engineering Board, Department of Science and Technology (SERB-DST) Young Scientist Award in 2011. His current research interests include elucidating the molecular mechanisms underlying salinity stress responses and tolerance in plants. His research group is engaged in assessing the individual roles and relative importance of sodium and chloride ions under salinity stress in rice and has made significant contributions in elucidating the individual and additive (under NaCl) effects of sodium and chloride ions.

Dr. Shabir Hussain Wani is an Assistant Professor cum Scientist at the Mountain Research Centre for Field Crops, Khudwani Sher-e-Kashmir University of Agricultural Sciences and Technology of Kashmir, Srinagar, Jammu and Kashmir, India. He has published more than 100 papers/chapters in peer-reviewed journals, and books of international and national repute. He was Review Editor of *Frontiers in Plant Sciences* (2015–2018), Switzerland. He is editor of *SKUAST Journal of Research* and *LS: An International Journal of Life Sciences*. He has also edited ten books on current topics in crop improvement published by CRC press, Taylor and Francis Group, USA, and Springer in 2015 and 2016. His PhD research fetched the first prize in north zone at national-level competition in India. He is the fellow of the Linnean Society of London and Society for Plant Research, India. He received various awards including Young Scientist Award (Agriculture) 2015, Young Scientist Award 2016, and Young Achiever Award 2016 by various prestigious scientific societies. He has also worked as a Visiting Scientist in the Department of Plant Soil and Microbial Sciences, Michigan State University, USA, for the year 2016–2017 under the Raman Post-Doctoral Research Fellowship programme sponsored by University Grants Commission, Government of India, New Delhi. He is a member of the Crop Science Society of America.

Dr. Penna Suprasanna (PhD Genetics, Osmania University, Hyderabad) is a Senior Scientist and Head of Plant Stress Physiology and Biotechnology Group in the Nuclear Agriculture and Biotechnology Division, Bhabha Atomic Research Centre, Mumbai, India. Dr. Suprasanna made significant contributions to crop biotechnology research through radiation-induced mutagenesis, plant cell and tissue culture, genomics, and abiotic stress tolerance. His research on radiation-induced mutagenesis and *in vitro* selection in sugarcane yielded several agronomically superior mutants for sugar yield and stress tolerance. He has made intensive efforts to apply radiation mutagenesis techniques in vegetatively propagated plants through collaborative research projects. He is actively associated with several national and international bodies (IAEA, Vienna) in the areas of radiation mutagenesis, plant biotechnology, and biosafety. He is the recipient of the "Award of Scientific and Technical Excellence" by the Department of Atomic Energy, Government of India; is the Fellow of Maharashtra Academy of Sciences, Andhra Pradesh Academy of Sciences, Telangana Academy of Sciences, and Association of Biotechnology; and is the Faculty Professor, Homi Bhabha National Institute, DAE. Dr. Suprasanna has published more than 225 research papers/articles in national and international journals and books. His research centered on molecular understanding of abiotic stress (salinity, drought, and arsenic) tolerance, and salt-stress adaptive mechanism in halophytic plants. The research group led by him on crop genomics has successfully identified novel microRNAs, early responsive genes besides validating the concept of redox regulation toward abiotic stress tolerance and plant productivity.

Dr. Lam-Son Phan Tran is Head of the Stress Adaptation Research Unit at RIKEN Center for Sustainable Resource Science, Japan. He obtained his MSc in Biotechnology in 1994 and PhD in Biological Sciences in 1997, from Szent Istvan University, Hungary. After doing his postdoctoral research at the National Food Research Institute (1999–2000) and the Nara Institute of Science and Technology of Japan (2001), in October 2001, he joined the Japan International Research Center for Agricultural Sciences to work on the functional analyses of transcription factors and osmosensors in *Arabidopsis* plants under environmental stresses. In August 2007, he moved to the University of Missouri-Columbia, USA, as a Senior Research Scientist to coordinate a research team working to discover soybean genes to be used for genetic engineering of drought-tolerant soybean plants. His current research interests are elucidation of the roles of phytohormones and their interactions in abiotic stress responses, as well as translational genomics of legume crops with the aim to enhance crop productivity under adverse environmental conditions. He has published over 125 peer-reviewed papers with more than 90 research and 35 review articles in *American Society of Agronomy*, *Crop Science Society of America*, and *Soil Science Society of America*, and contributed 8 book chapters to various book editions published by Springer and Wiley-Blackwell. He has also edited 11 book volumes, including this one, for Springer and Elsevier.

Chapter 1
Salinity Responses and Adaptive Mechanisms in Halophytes and Their Exploitation for Producing Salinity Tolerant Crops

Karim Ben Hamed, Amira Dabbous, Hassan El Shaer, and Chedly Abdely

Abstract The increasing salinization of cultivated lands and associated annual losses in agricultural production require a better understanding of key physiological mechanisms conferring salinity tolerance in crops. The effective way of gaining such knowledge comes from studying halophytes. Halophytes have the advantages of tolerating and even benefitting from salt concentrations that kill most crop species. This review summarized the main strategies of resistance of halophytes to salt stress and the specificities in their responses that distinguish them from glycophytes. Many studies showed that the superior salinity tolerance in halophytes is mainly attributed to a set of complementary and well-orchestrated mechanisms for ion, osmotic and reactive oxygen species (ROS) homeostasis. We also gave special attention to the acclimation in halophytes that allows plants to improve stress tolerance to salt at a later period of plant growth. Armed with such information on halophytes, it will be possible to produce salt tolerant crops through genetic modification, priming and effective breeding strategies.

Keywords Halophytes · Signalling · Salinity · Priming · Osmolytes · ROS homeostasis · Acclimation · Crops · Vacuolar compartmentation · Gene expression · H_2O_2

Abbreviations

CAT Catalases
HKT High potassium transporter

K. Ben Hamed (✉) · A. Dabbous · C. Abdely
Laboratory of Extremophile Plants, Centre of Biotechnology of Borj Cedria, Hammam Lif, Tunisia
e-mail: karimbenhamed2016@gmail.com

H. El Shaer
Desert Research Center, Cairo, Egypt

V. Kumar et al. (eds.), *Salinity Responses and Tolerance in Plants, Volume 2*,
https://doi.org/10.1007/978-3-319-90318-7_1

NHX Na/H+ exchanger
P5CS Pyrroline-5-carboxylate synthetase
PDH Proline dehydrogenase
PLC Phospholipase C
PLD Phospholipase D
POD Peroxidases
ROS Reactive oxygen species
SOD Superoxide dismutase
SOS Salt overly sensitive
TF Transcription factor
VHA Vacuolar H^+ ATPase

1.1 Breeding for Salinity Tolerance: Current Stand

Worldwide, salt stress is one of the main environmental constraints that decrease plant growth and crop productivity. Alarmingly, salinity is increasing particularly in the most populous and least developed countries, in Central Asia and Africa, where salinization affects up to 50% of irrigated areas. Every minute, 3 ha are lost for agriculture due to inappropriate irrigation practices and between 10 and 20 M ha of irrigated land deteriorates to zero productivity each year (Panta et al. 2014). Most crop plants like cereals, forages and legumes are very sensitive to the presence of sodium in the soil. Interestingly some plants called halophytes have developed original adaptation to cope with the presence of salt.

The biological approach that consists in improving the performance of cultivated plants has been proposed as one of the effective means to overcome salinity issue. For example, the overexpression of genes encoding sodium and potassium transporters like Na/H antiporter or HKT1 genes (Zhang and Blumwald 2001; Ruiz 2002; Shi et al. 2003; Apse and Blumwald 2002; Munns et al. 2012), improved the growth of some salt sensitive species like rice, tomato, wheat rapeseed and Arabidopsis under salt stress. Nevertheless, the progress towards improving the salt tolerance of these crops remains disappointingly slow. Thus, in spite of the fact that it is not a single crop that can be used seawater salinity (Flowers 2004). The main reason for this is the multigenic nature of the salt response, which also involves simultaneous changes at different levels (whole plant, tissues, cells and even organelles). On the other hand, when a gene is over-expressed for example, the effects of the modification will be attenuated by dilution at whole plant level. Thus, the analysis of several bibliographic data shows that the improvement of tolerance to salinity is only possible through coupling of genetic approach and functional genomics, biochemical and physiological approaches of biophysics, ecophysiology taking into account the interaction of the genome with its environment. The main aim is to look for mechanisms which extend beyond the cellular scale and which are part of the whole plant, which amounts to adopting an approach of integrative biology.

Moreover, *Thellungiella salsuginea* is a halophyte from the eastern coast of China, very close to *Arabidopsis thaliana* since the degree of homology of the sequences between the two species is about 90%. *Thellungiella* becomes the halophyte model for functional genomics. Curiously, very few genes are induced by salt in this plant. On the other hand, many genes are constitutively expressed at a level higher than that of *Arabidopsis* under salinity. Some of these genes are involved in Na^+ excretion, antioxidant systems, osmolyte biosynthesis, protection against desiccation etc. (Taji et al. 2004; Gong et al. 2005).

From the comparison of *Arabidopsis* and *Thellungiella* emerges the hypothesis that halophyte is more salt tolerance is not associated with the acquisition of new genes but with a change in the regulation of many pre-existing genes, thus the interest of the investigation and the comparative analysis of these regulatory pathways in plants with the same ability to tolerate salt stress.

In this chapter, we have summarized the main strategies of salt resistance in halophytes and focus on the special hallmarks that are better activated or pre-activated to elicit quick and better responses in halophytes than in glycophytes, resulting in adaptation to saline environment. We continue with some proper approaches that used halophyte specificities to develop salt tolerant crops.

1.2 Strategies of Salt Resistance in Halophytes

In salty environments, halophytes are experiencing three types of constraints, (1) Osmotic constraint: the salt causes a lowering of the water potential of the medium thus disrupting the water supply to the plant. (2) Ion stress associated with toxic accumulation of Na^+ and Cl^- in tissues. (3) Nutritional stress resulting from the competition of Na^+ with the other essential cations K^+, Mg^{2+}, Ca^{2+} and Cl^- with NO_3^-, $H_2PO_4^-$ and SO_4^{2-}. These three stresses have a common denominator is that they may be responsible for an accumulation of reactive oxygen species (ROS). Halophytes, plants living in a salt environment, had to develop adaptive mechanisms for these stresses. These adaptations are located at different levels of the plant, in organs, cells, membranes and at the molecular level (Pitman and Laüchli 2002). Two main strategies are adopted by halophytes (Fig. 1.1), depending on whether they are dicotyledons or monocotyledons, the former adopt an osmotic regulation strategy (osmotic or tolerance strategy) and the latter adopt an avoidance strategy. These strategies could not be mutually exclusive, and a particular plant may use several of these different strategies depending on particular circumstances (Souid et al. 2016; Shabala and McKay 2011) Fig. 1.2.

1.2.1 Tolerance Strategy or Osmotic Strategy

This strategy is based on a high ability to absorb and transport Na^+ and Cl^- from the roots to the aerial parts before their accumulation in the vacuoles of the halophytes (Flowers 1985). This salt accumulation may be beneficial to these plants

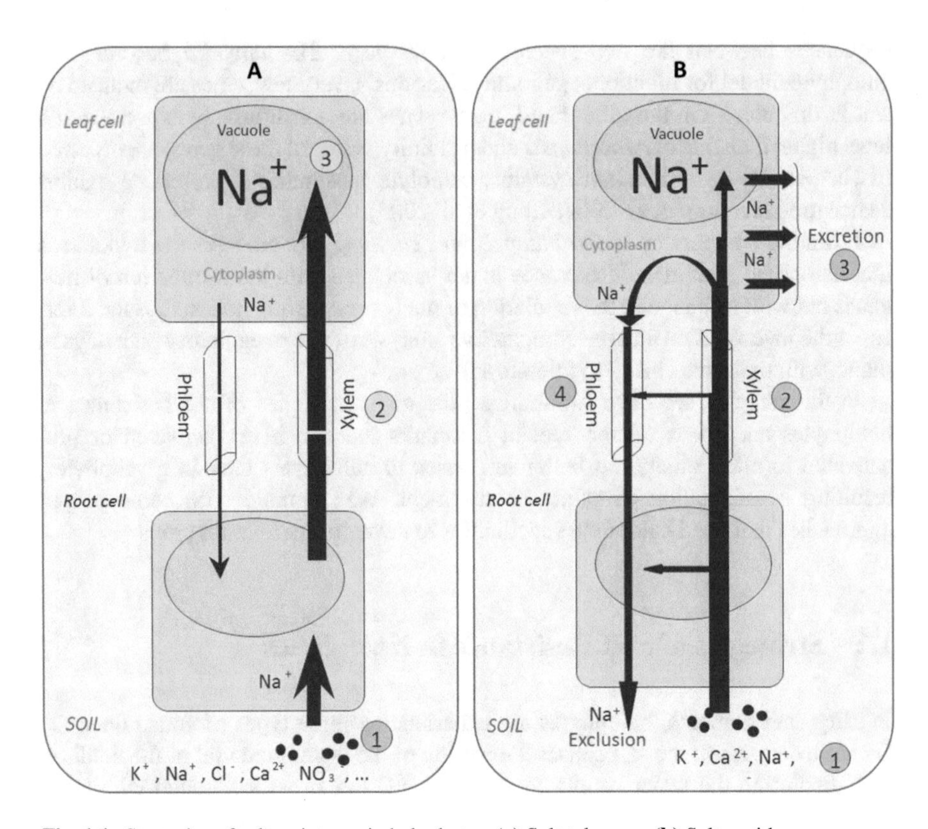

Fig. 1.1 Strategies of salt resistance in halophytes. (**a**) Salt tolerance (**b**) Salt avoidance

compared to glycophytes since it could lead to a lowering of the water potential of the cells, which could increase their water supply and reduce sweating. In most dicotyledonous halophytes, "inclusive" behavior is effective because it is associated with high growth rates and sequestration of large quantities of inorganic ions in the vacuole (Flowers and Colmer 2008). Most halophytes used Na^+ and Cl^- for osmotic adjustment, some others used K^+ and SO_4^{2-}. However, levels of accumulation depend on species and environmental conditions. The accumulation of Na^+ and Cl^-in the vacuole also increases stem or leaf succulence. The succulence in leaves or stems characteristic of numerous halophytes, is associated with an increase in cell size, a decrease in surface area per tissue volume and a high water content per unit surface area (Short and Colmer 1999). The increased succulence being correlated with the water status of the leaves could be interpreted as an adaptation in terms of conservation of internal water and dilution of accumulated salts (Debez et al. 2004).

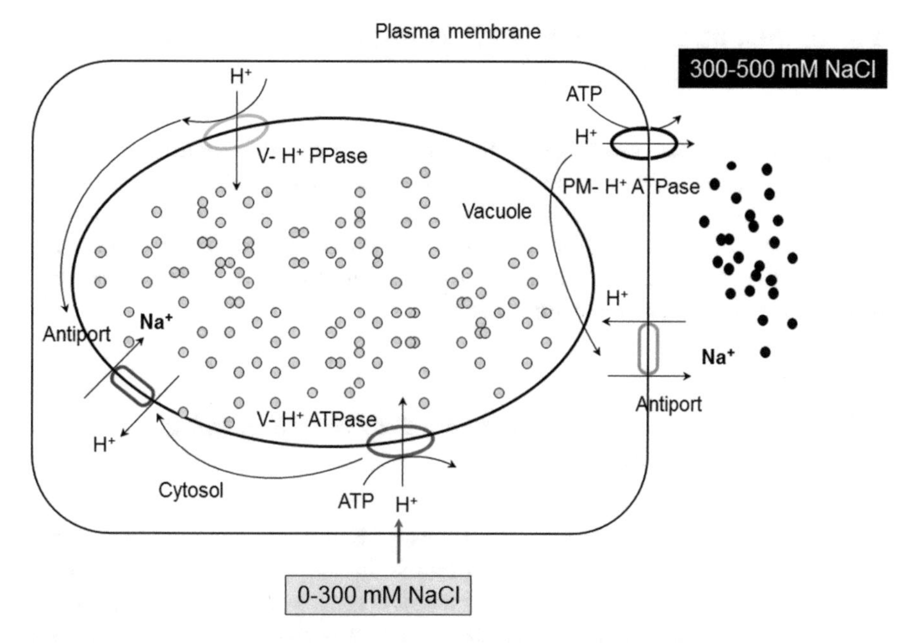

Fig. 1.2 Complementary mechanisms for ion homeostasis in the halophyte *Cakile maritima* involving V H+-ATPase (at moderate salinity, up to 300 mM NaCl) and PM H+ ATPase (at high salinity, 300–500 mM NaCl)

1.2.2 Avoidance Strategy

The avoidance of salt requires certain physiological and structural adaptations. There are two main saline avoidance methods for halophytes. (1) Root exclusion: Na^+ is transported in the xylem, transported to the leaves by transpiration, and then partially recirculated by the phloem to the roots and excreted in the medium (Bradley and Morris 1991; Berthomieu et al. 2003). (2) Excretion: Some halophytes are capable to deposit high (molar) amounts of NaCl in specialized external (epidermal) structures like salt glands and salt bladders (Dassanayake and Larkin 2017; Shabala et al. 2014). About 50% of all halophyte plants contain salt bladders. With the diameter of the bladder being often 10 times bigger than epidermal cells, each epidermal bladder cell has about 1000 times larger volume, and hence, could sequester 1000-fold more Na^+ compared with leaf cell vacuoles (Shabala et al. 2014). The mechanism of excretion has been described particularly in the case of salt glands. Thus, the movement of the liquid from the basal cell to the apical cell is done through the connections of the plasmodesm between the two cells. A solution enriched with salt is stored in the collecting chamber between the wall of the apical cell and the cuticle which has peeled off. The latter stretches by bringing the saline solution to the surface of the leaf via small open spaces of the cuticle (Gorham 1996).

1.3 Learning from Signaling Pathways in Halophytes in Response to Salinity

The general consensus is that there is nothing really unique to halophytes, neither in their anatomical features, nor in their physiological mechanisms. The major difference to glycophytes is how efficiently these mechanisms are controlled in these two plant groups (Shabala and Mackay 2011; Shabala 2013). The key to improving salinity stress tolerance in crops may lie in understanding how salt stress-responsive mechanisms are regulated in halophytes.

1.3.1 Regulation of Ion Homeostasis by SOS Signaling Pathway in Halophytes

Several studies have established the Salt Overly Sensitive (SOS) pathway as the canonical model for the mechanism responsible for salt tolerance. The SOS pathway involves interplay among Na^+-H^+ antiporters for transporting sodium, and the activation of the kinase that phosphorylates the transporter (Ji et al. 2013). Among them, SOS1, a plasma membrane Na^+-H^+ antiporter, has been shown to be a critical component for maintaining salt homeostasis by pumping sodium out of cells upon activation. Considering the characterization of sodium extrusion based on SOS1 in the highly sodium-sensitive, glycophytic Arabidopsis, it remained to be studied whether the SOS pathway, and particularly SOS1, had any function and relevance for the sodium tolerance of halophytic species. The function of SOS1 in *Thellungiella salsuginea* has now been characterized (Bressan et al. 2001; Amtmann 2009) with special emphasis on intracellular processes that might be altered by the inhibition of SOS1 expression. The study identified SOS1 as an intrinsic part of the halophytic nature of this species because the down-regulation of SOS1 transcript expression converted *Thellungiella* into a glycophytic species (Oh et al. 2009a, b). A computational analysis of the primary and secondary structures of halophytes and glycophytes showed that no major differences in SOS1 between both groups (Kim and Bressan 2016). The major difference in Na^+ tolerance between *A. thaliana* and *T. salsuginea* was due to a much higher Na^+ influx in *A. thaliana*. This observation is consistent with others' findings that SOS1 is required for Na^+ tolerance but overexpression of SOS1 from either *A. thaliana* or *T. salsuginea* does not result in a major improvement in Na^+ tolerance (Shi et al. 2003).

1.3.2 Vacuolar Compartmentalization System

The vital control of intracellular sodium concentration for plants coping with salinity was achieved by intracellular accumulation and compartmentalization of Na^+ predominantly in vacuoles, and thereby the cytoplasm keeps tolerable concentrations

(Shabala 2013; Flowers and Colmer 2008).The sequestration of Na^+ in the vacuole is possible thanks to Na^+/H^+ antiporters located in the tonoplast. The great benefit of the halophytes is that they could have a constitutive expression of this tonoplast antiporter and can stimulate their activity under a salt stress (Zhang et al. 2008). The activity of NHX Na^+/H^+ antiporters which use the pH gradient generated by vacuolar V H^+-ATPases to actively transport Na^+ against its electrochemical gradient toward the vacuole (Zhu 2001; Hasegawa 2013). The combined stimulation of these both transporters seems to be important in the adaptation of halophytes to salt stress (Qiu et al. 2007). The compartmentalization of Na^+ in the vacuole contributes not only to ion homeostasis and cell turgor, it also protects the metabolic enzymes from salt toxicity (Aharon et al. 2003). The accumulation of Na^+ and Cl^- in the vacuole also increases stem or leaf succulence. The succulence in leaves or stems characteristic of numerous halophytes, is associated with an increase in cell size, a decrease in surface area per tissue volume and a high water content per unit surface area (Short and Colmer 1999). In *C. maritima* the Na^+ concentration of the leaves increased with increasing NaCl concentration in the medium (Debez et al. 2004) and no salt-related toxicity symptoms were noted. Moreover high leaf Na^+ concentration was associated with increased leaf thickness and succulence (Debez et al. 2006) suggesting the existence of mechanisms for compartmentalization of Na^+ in these organs (Fig. 1.2). The increased succulence of the leaves could be interpreted as an adaptation in terms of conservation of internal water and dilution of accumulated salts (Koyro and Lieth 2008). Moreover, the increased activity of V H^+-ATPases since 300 mM NaCl in *C. maritima* leaves has been reported (Debez et al. 2006). This activity could provide the necessary proton driving force triggering sodium transport towards the vacuole (Fig. 1.2).

1.3.3 Osmolyte Biosynthesis Pathways: Opposite Regulation in Halophytes and Glycophytes

In halophytes, the accumulation of glycine betaine and proline two important organic solutes is involved in osmotic adjustment in response to salt stress (Flowers and Colmer 2008). Halophytes are more efficient in controlling the catabolism of the proline and greatly accumulate it in salt stressed cells than glycophytes (Slama et al. 2015). The leaves and roots of *C. maritima* present a proline pool similar to *A. thaliana* in control situation but after 72 h of salt stress, the leaves of *C. maritima* accumulate twofold more proline than *A. thaliana* and threefold in the roots (Ellouzi et al. 2014). They have also the ability to accumulate 18-fold more betaine than the glycophytes (Jdey et al. 2014). Thus, *C. maritima* could use organic osmolytes in addition to Na^+ for osmotic adjustment. It is further noteworthy that glycine-betaine as proline accumulation could also play a role in ROS homeostasis in halophytes (Katschnig et al. 2013) and serve in preservation of the structural and functional integrity at the cellular level (Jdey et al. 2014).The beneficial effect of proline accumulation on salt tolerance has been demonstrated in a range of halophyte species (Szabados and Savoure 2010; Slama et al. 2015). *Thellungiella salsuginea* and

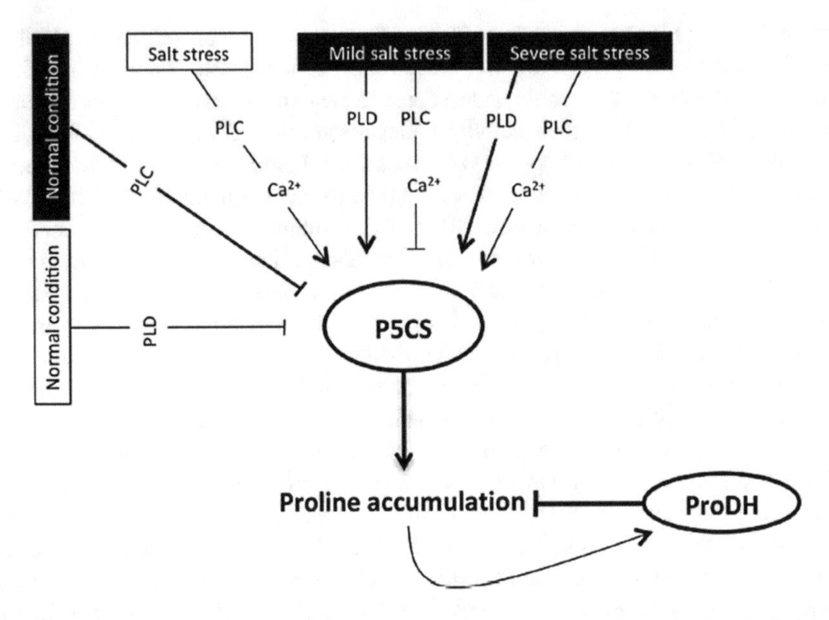

Fig. 1.3 Regulation of proline metabolism in plants. Most data were obtained using *Arabidopsis thaliana* and *Thellungiella salsuginea*. Circles designate enzymes and rectangles the environmental context; open and black boxes correspond to the signalling pathways demonstrated in A. thaliana and T. salsuginea, respectively. Abbreviations: *P5CS* pyrroline-5-carboxylate synthetase, *ProDH* proline dehydrogenase, *PLC* phospholipase C, PLD, phospholipase D, *ROS* reactive oxygen species. (Modified from Ghars et al. 2012)

Lepidium crassifolium, two halophytic wild relatives of *Arabidopsis*, accumulated more proline under control and salt-stressed conditions (Murakeozy et al. 2003; Taji et al. 2004; Ghars et al. 2008). Proline hyperaccumulation in *T. salsuginea* was shown to be a result of enhanced proline synthesis via pyrroline-5-carboxylate synthetase (P5CS) and reduced proline catabolism by proline dehydrogenase PDH (Taji et al. 2004; Kant et al. 2006). Interestingly, in *Arabidopsis*, phospholipase D (PLD) functions as a negative regulator of proline accumulation under control conditions (Thiery et al. 2004) whereas phospholipase C (PLC) acts as a positive regulator of proline accumulation during salt stress (Parre et al. 2007). However, this regulation is opposite in *Thellungiella*, where PLD functions as a positive regulator and PLC acts as a negative regulator (Ghars et al. 2012). This opposite regulatory function exerted by *T. salsuginea* on proline accumulation (Fig. 1.3) suggests that halophytes have a 'stress-anticipatory preparedness' strategy (Taji et al. 2004).

1.3.4 ROS Homeostasis

Reactive oxygen species (ROS) are important in the cell for all organisms in their signaling pathway. But, large accumulation of ROS in the cell triggered an oxidative stress. This stress is a source of damages as lipid peroxidation, DNA damage, and protein denaturation for examples. The sensitivity to ROS seems to be different between the glycophytes and halophytes. Actually, halophytes endure a higher level of ROS since high salinity can lead to the formation of ROS. Scavenging of ROS is thus critical in plant salt tolerance, and accordingly halophytes are equipped with powerful antioxidant systems, including enzymatic and non-enzymatic components to regulate the level of ROS in the cell (Bose et al. 2014). Halophytes have the capacity to use the superoxide dismutase (SOD) at its better yield (Jithesh et al. 2006; Ozgur et al. 2013). This enzyme converts the $O_2^{\cdot -}$ in H_2O_2, a more stable ROS that can be next transformed into H_2O_2 thanks to the catalase (CAT) and the peroxidase (POX).

The kinetics of H_2O_2 signalling appears to be much faster in halophytes than in glycophytes (Bose et al. 2014). For example, the salt-induced H_2O_2 accumulation peaked at 4 h upon salt stress onset in the leaves of a halophyte *Cakile maritime* and declined rapidly afterwards. In the glycophyte *A. thaliana*, however, H_2O_2 continued to accumulate even after 72 h after salt application (Ellouzi et al. 2011) (Fig. 1.4). Similarly, salt stress-induced H_2O_2 production was higher in the leaves of a halophyte (*Populus euphratica*) in comparison with the leaves of a glycophyte (*P. popularis*) after 24 h of salinity treatment (Sun et al. 2010), suggesting that elevated H_2O_2 levels upon stress exposure are *essential* to confer salt stress signalling and adaptation to stress in halophytes. Enhanced H_2O_2 production in the chloroplasts of *Thellungiella salsuginea* is pivotal for the "salt stress preparedness" mechanism in this halophyte species (Wiciarz et al. 2015), consistent with this notion. While (transiently) elevated H_2O_2 appears to be essential for salt-stress signaling in halophytes, interaction between H_2O_2 and transition metals (Fe^{2+} or Cu^{2+}) present in cell walls may result in the generation of highly reactive hydroxyl radicals ($OH^{\cdot}$) via the Fenton reaction. Hydroxyl radicals damage cellular structures, decrease cytosolic K^+/Na^+ ratio by directly activating a

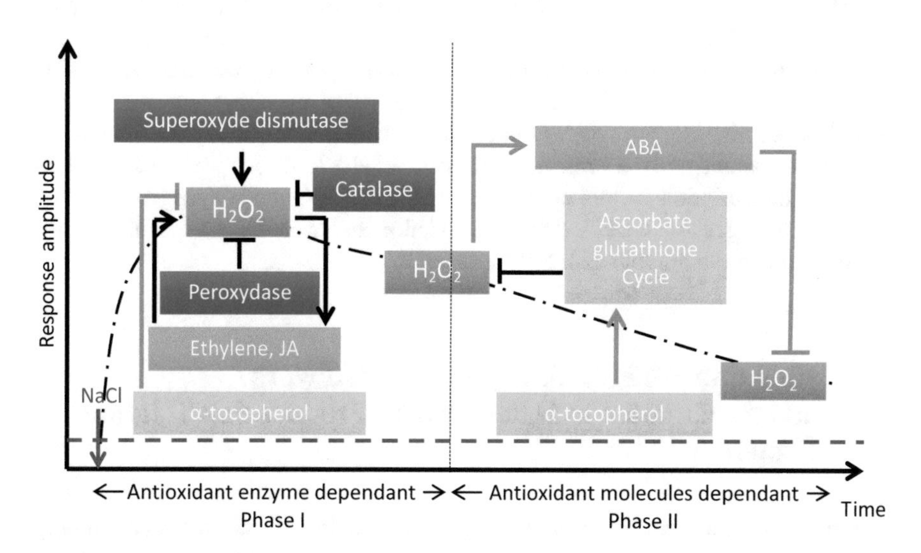

Fig. 1.4 Early oxidative burst and coordination of the antioxidant response of *C. maritima* to salt stress. Red dotted line: response in absence of salt. *ABA* abscissic acid, *JA* jasmonic acid

range of Na^+, K^+ and Ca^{2+}-permeable cation channels and eventually cause cell death (Demidchik et al. 2010). The abundance of the metal binding proteins (e.g. ferritin) in specific tissues or cellular compartments controls the concentration of the transition metal ions available for the Fenton reaction, hence regulates the H_2O_2-mediated signal propagation in plants. Salt stress increases ferritin deposits in the leaves of the halophytic *Mesembryanthemum crystallinum* (Paramonova et al. 2004). Similarly, a green micro-alga (*Dunaliella salina*) that grows only under highly saline conditions accumulates large quantities of a triplicated transferrin-like protein (Liska et al. 2004). Furthermore, there is a transient increase in the transcripts of ferritin gene (*Fer1*) in the leaves of a mangrove *Avicennia marina* 12 h after salt stress (Jithesh et al. 2006). These observations suggest that halophytes can indeed regulate the propagation of H_2O_2-mediated signalling by increasing the abundance of metal binding proteins in all the tissue/cellular organelles where H_2O_2 accumulates, preventing it from being converted into hydroxyl radicals.

1.4 Learning from the Acclimation of Halophytes to Their Natural Biotopes

In saline environments, salinity is heterogenous spatially and temporally. For example, halophytes naturally adapted to seasonal salinity fluctuations or unpredictable changes in salinity, owing to their ability to keep a stress imprint or 'memory' that improved plant responses when challenged by future salt stress events (Ben Hamed et al. 2013). Our recent studies on the halophyte *Cakile maritima*, native to Tunisian coasts, suggest a relatively long-term stress memory in plants pre-exposed to salinity which resulted in a lower oxidative stress when subsequently exposed to salinity (Ellouzi et al. 2014). This suggests that among the set of defence mechanisms triggered by stress pre-exposure there are imprints which persist for several weeks after relief of a first stress that allowed stress pre-exposed plants either to prevent and/or scavenge reactive oxygen species more efficiently than non-pre-exposed plants. As possible mechanisms for a stress imprint, Bruce et al. (2007) suggested the accumulation of transcription factors or signalling proteins, epigenetic changes including chemical changes at the DNA (DNA methylation and acetylation), histone modification or accumulation of small RNAs.

1.5 How Can We Exploit Regulation Pathways and Acclimation in Halophytes to Produce Salt Tolerant Crops?

There are several possible practical ways of increasing the salt tolerance in crop species. These include: (1) over-expression of genes from halophytes and (2) priming involving prior exposure to a biotic or abiotic stress factor making a plant more resistant to future exposure. Priming can also be achieved by applying natural or synthetic compounds which act as signalling transducers, 'activating' the plant's defense system. It appears that each of these approaches targets intracellular K^+/Na^+ and ROS homeostasis and ion transport across cell membranes, improving the cytosolic K^+/Na^+ ratio and assisting cell osmotic adjustment (Fig. 1.4).

1.5.1 Over-Expression of Genes from Halophytes in Salt Sensitive Species

The recent isolation and characterization of salt tolerance genes encoding signalling components from halophytes has provided tools for the development of transgenic crop plants with improved salt tolerance and economically beneficial traits (Fig. 1.5). These genes encode protein implicated in Na^+ sequestration (H^+ ATPase, NHX transporters) and exclusion (SOS1), synthesis of osmolytes, ROS detoxification, signal perception and regulating factors and other unknown functions (Himabindu et al. 2016).

Halophytic genes involved in salt tolerance mechanisms have been extensively studied in various model species (approximately 80 genes), including E. coli, yeast, Arabidopsis, tobacco and few crop species (only 20 genes) (Himabindu et al. 2016). Few reports used these genes from monocot halophytes for crop improvement, in comparison to the studies that used dicot halophytes genes (Himabindu et al. 2016). For example, vacuolar H^+-ATPases encoding genes from dicots halophytes were proved to be potential key players of salt tolerance in glycophytes. We recently showed that this gene (*Lm*VHA-E1) is up-regulated in the halophyte *Lobularia maritima* under different stress conditions (Dabbous et al. 2017), similar to what was reported by *Mesembryanthemum crystallinum* (Kluge et al. 2009), *Suaeda salsa* (Ratajczak 2000; Wang et al. 2001) and *Tamarix hispida* (Gao et al. 2011). The overexpression of this gene in *A. thaliana* led to improved stress tolerance to

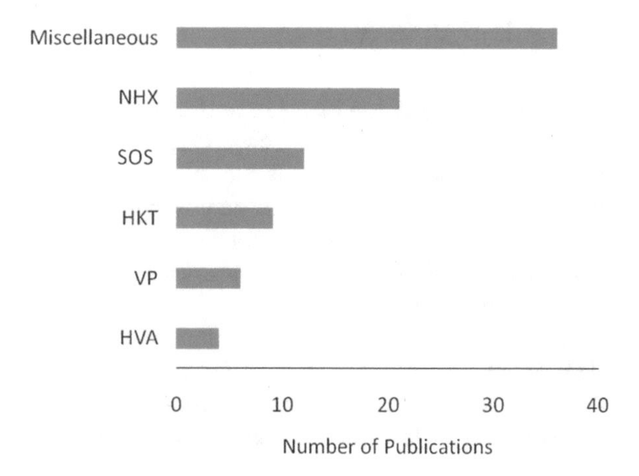

Fig. 1.5 Salt tolerance studies using halophyte genes in different crop plants

salinity and osmotic stress in transgenic plants, mainly associated with reduced relative water loss and oxidative damages and increased levels of sodium possibly due to higher H+-ATPase activity compared to the wild-type plants (WT). Accordingly, the higher levels of expression of different stress related genes such as AtNHX1 AtP5CS, AtCAT, AtSOD, AtPOD and AtLEA, indicated higher levels of activities in sodium sequestration into vacuoles, in osmotic regulation and in ROS scavenging of transgenic plants. Altogether, our results suggest an essential role for V-H^+-ATPase subunit E1 in improving the tolerance of plants not only to to salinity but also osmotic stress. In another study, Hu et al. (2012) found that tomato plants overexpressing an MdVHA-B from apple (*Malus domestica*) exhibited improved drought-stress tolerance. These results indicate that enhanced vacuolar H+ pumping in transgenic plants could be an effective approach to improve the drought tolerance of crops. In addition to their role in ion transport, V-H^+-ATPases energize the transport of solutes such as amino acids (proline), betaine, polyols and sugars, across the tonoplast (Chen and Murata 2002). There is also evidence that the V-ATPase subunits may have additional functions independent of proton pumping, thus adding another layer of complexity to the biology of V-ATPases. For example, VHA-B1 has been identified in nuclear complexes with hexokinase, and a series of phenotypes of a vha-B1 mutant indicate that this subunit is involved in glucose signalling (Cho et al. 2006). According to our results, *Lm*VHA-E1 improved the tolerance of *A. thaliana* to mannitol-induced osmotic stress probably through an increased accumulation of compatible solutes like proline, as indicated by an increased expression of P5CS gene in transgenic lines. Further experiments are needed if a clear conclusion on the relationship between V-H^+-ATPases and osmotic stress tolerance is to be drawn. It is evident from our study and several other reports that the over-expression of a single gene could impart salt tolerance to plants. But, it is well-known that salinity tolerance is a multi-gene trait and transformation with more than one gene will be necessary to meet the requirements to obtain insights into the mechanisms of tolerance improvement.

1.5.2 H_2O_2 Priming

Several recent studies on plants have demonstrated that pre- treatment with exogenous H_2O_+ can induce salt tolerance (Table 1.1). These findings demonstrate that H_2O_2 priming can induce tolerance to salinity in plants by modulating physiological and metabolic processes such as photosynthesis, proline accumulation and ROS detoxification, and that this ultimately leads to better growth and development. Importantly, ROS metabolism also plays a pivotal role in the development of stress and cross stress tolerance.

H_2O_2 priming represents a fruitful area of future research, which should help plant scientists explore the molecular mechanisms associated with abiotic stress tolerance and promote a more environmental friendly sustainable agriculture (Hossain et al. 2015). As shown in previous section, halophytes have mechanisms

Table 1.1 List of some successful studies showing the ameliorative effect of H_2O_2 priming on the response of crop species to salinity

Plant	Method of application	Response to salt stress after pre-treatment	Reference
Rice	Low concentrations (<10 μM) of H_2O_2 or NO	Greener leaves and a higher photosynthetic activity increases in ROS scavenging enzyme activities and increased expression of genes encoding pyrroline-5-carboxylatesynthase, sucrose- phosphate synthase, and the small heat shock protein 26	Uchida et al. (2002)
Maize	1 μM $H_2O_2$for 2 days, 25 mM for 1, 5 and 10 days	By enhancing antioxidant metabolism and reducing lipid peroxidation in both leaves and roots	Azevedo-Neto et al. (2005)
Triticum aestivum	Seeds were soaked in H_2O_2 (1–120 μM,8 h) and subsequently grown in saline conditions (150 mM NaCl)	Lower H2O2 in seedlings from primed seeds, better photosynthesis, maintained turgor, under salt stress	Wahid et al. (2007)
Hordeum vulgare	Seedlings pre-treated with H_2O_2 (1 and 5 mM), 150 mM NaCl for 4 and 7 d	Lower Cl^- leaf content, Higher rates of CO_2 fixation and lower malondialdehyde (MDA) and H_2O_2 contents under salt stress	Fedina et al. (2009)
Wheat	H_2O_2 0.05 μM	Enhanced GSH content and increased the activities of APX, CAT, SOD, and POD	Li et al. (2011)
Avena sativa	H_2O_2 0.5 mM	Up-regulation of the activities of CAT and SOD	Hossain et al. (2015)
Maize	Foliar H_2O_2 priming	Less H_2O_2 accumulation and maintenance of the leaf RWC and chlorophyll contents	Gondim et al.(2012, 2013)
Triticum aestivum	H_2O_2: 50, 100 μM	Reductions in both Na^+ and Cl^- ion levels and an increase in proline content and in N assimilation. Improved water relations, increased levels of photosynthetic pigments and greater growth rates	Ashfaque et al. (2014)
Panax ginseng	100 μM H_2O_2, for 2 days	Better seedling growth, and chlorophyll and carotenoid contents, lower oxidative stress increased activities of APX, CAT, and guaiacol peroxidase,	Sathiyaraj et al. (2014)

to utilize ROS, especially H_2O_2, for signaling purposes that confer acclamatory stress tolerance through the modulation of osmotic adjustment, ROS detoxification, photosynthetic C fixation and hormonal regulation (Ben Hamed et al. 2016). A large number of studies have suggested that H_2O_2 treatment is capable of inducing salt stress tolerance through the induction of a small oxidative burst. This burst subsequently activates a ROS-dependent signalling network, thereby enhancing the accumulation of latent defense proteins, such as ROS-scavenging enzymes and transcription factors (TFs), resulting in a primed state and an enhanced stress response (Borges et al. 2014). Many researchers have suggested a central role for H_2O_2 in intracellular and systemic signaling routes that increase tolerance and acclimation to abiotic stresses. Recent findings have shown that effective ROS signaling may require an increased flux of thiol-dependent antioxidants. With respect to signal transduction, ROS can interact with other signaling pathways, such as the activation of NADPH oxidase dependent or monomeric G protein; lipid-derived signals; induction of MAPK; redox sensitive TFs; regulation of Ca^{2+}; and plant hormone signal transduction. An understanding of the H_2O_2 physiology of plants, particularly H_2O_2 sensing and the identification of the components of H_2O_2 signaling network and H_2O_2 cross-talk with other growth factors, is of practical importance if we aim at improving the performance of crop plants growing under salt stress conditions.

1.5.3 Salinity Pre-treatment

Plants exposure to low level salinity activates an array of processes leading to an improvement of plant stress tolerance. This has already been demonstrated for different crop species such as soybean, rice, sorghum, pea and maize (Amzallag et al. 1990; Umezawa et al. 2000; Djanaguiraman et al. 2006, Pandolfi et al. 2012, 2016). For example, soybean pretreated for 23 days showed a higher survival rate under severe stress conditions (Umezawa et al., 2000); in rice, 1 week of pretreatment decreased leaf area and total dry matter production, but improved growth rate and shoot and root length after 1 week of severe salt treatment (Djanaguiraman et al. 2006); and in sorghum pre-treated plants maintained the same growth rate before and after the exposure to high level of salt, and they could stand a concentration much higher than non acclimated plants (Amzallag et al. 1990). In *Pisum staivum* and *Zea mays*, 7 days were enough to activate an array of processes leading to an improvement of plant stress tolerance (Pandolfi et al. 2012, 2016).

In many studies, acclimation to salinity was reported mainly to be related to ion-specific rather than osmotic component of salt stress. For example, in *Pisum sativum*, although acclimation took place primary in the root tissues, the control of xylem ion loading and efficient Na^+ sequestration in mesophyll cells are found to be important components of this process (Pandolfi et al. 2012). In maize, acclimation to salinity is not attributed to better ability of roots to exclude Na^+, given the lack of any significant difference in net Na^+ fluxes between acclimated and non-acclimated roots (Pandolfi et al. 2016). These findings are in a full agreement that acclimated plants accumulated more Na^+ in the shoot compared with non-acclimated ones, although further investigations are needed in order to unravel a clear picture of the ionic component of the acclimation mechanisms at the molecular level. Reported results allow us to suggest that the contribution of the root SOS1 plasma membrane transporters in this process is relatively minor, and instead points out at the important role of vacuolar compartmentation of Na^+ as a component of acclimation mechanism.

1.6 Conclusions and Perspectives

There is an urgent need to obtain high salt tolerant crops, because salinity is continuously spreading, productivity of conventional crop plants is significantly decreasing and world population is growing. This challenge can be achieved through the exploitation of halophytes, either through the engineering of their salt tolerance or miming acclamatory processes in salt sensitive crops. Most studies of halophytes concentrate on changes in plant physiology and gene expression to understand the mechanisms of salt tolerance. However adaptation to salinity is also connected with complex ecological processes within the rhizosphere induced by microorganisms (procaryotes and fungi) inhabiting roots and leaves of halophytes. It is therefore necessary to also learn from such microorganisms in order to increase the salt tolerance in crop species.

References

Aharon R, Shahak Y, Wininger S, Bendov R, Kapulnik Y, Galili G (2003) Over expression of a plasma membrane aquaporin in transgenic tobacco improves plant vigor under favorable growth conditions but not under drought or salt stress. Plant Cell 15:439–447

Amtmann A (2009) Learning from evolution: *Thellungiella* generates new knowledge on essential and critical components of abiotic stress tolerance in plants. Mol Plant 2:3–12

Amzallag N, Lerner HR, Poljakoff Mayber A (1990) Induction of increased salt tolerance in *Sorghum bicolor* by NaCl pretreatment. J Exp Bot 41:29–34

Apse MP, Blumwald E (2002) Engineering salt tolerance in plants. Nat Biotech 21:81–85

Ashfaque F, Khan MIR, Khan NA (2014) Exogenously applied H_2O_2 promotes proline accumulation, water relations, photosynthetic efficiency and growth of wheat (*Triticum aestivum* L.) under salt stress. Annu Res Rev Biol 14:105–120

Azevedo-Neto AD, Prisco JT, Eneas-Filho J, Medeiros JVR, Gomes-Filho E (2005) Hydrogen peroxide pre-treatment induces salt stress acclimation in maize plants. J Plant Physiol 162:1114–1122

Ben Hamed K, Ellouzi H, Talbi OZ, Hessini K, Slama I, Ghnaya T, Abdelly C (2013) Physiological response of halophytes to multiple stresses. Funct Plant Biol 40:883–896

Ben Hamed K, Ben Hamad I, Bouteau F, Abdelly C (2016) Insights into the ecology and the salt tolerance of the halophyte *Cakile maritima* using mutidisciplinary approaches. In: Khan MA, Ozturk M, Gul B, Ahmed MZ (eds) Halophytes for food security in dry lands. Elsevier, Oxford, pp 197–211

Berthomieu P, Conéjéro G, Nublat A, Brackenbury WJ, Lambert C, Savio C, Uozumi N, Oiki S, Yamada K, Cellier F et al (2003) Functional analysis of AtHKT1 in *Arabidopsis* shows that Na^+ recirculation by the phloem is crucial for salt tolerance. EMBO J 22:2004–2014

Borges AA, Jiménez-Arias D, Expósito-Rodríguez M, Sandalio LM, Pérez JA (2014) Priming crops against biotic and abiotic stresses: MSB as a tool for studying mechanisms. Front Plant Sci 5:642. https://doi.org/10.3389/fpls.2014.00642

Bose J, Rodrigo-Moreno A, Shabala S (2014) ROS homeostasis in halophytes in the context of salinity stress tolerance. J Exp Bot 65:1241–1257

Bradley PM, Morris JT (1991) Relative importance of ion exclusion, secretion and accumulation in *Spartina alterniflora* Loisel. J Exp Bot 42:1525–1532

Bressan RA, Zhang C, Zhang H, Hasegawa PM, Bohnert HJ, Zhu JK (2001) Learning from the *Arabidopsis* experience. The next gene search paradigm. Plant Physiol 127:1354–1360

Bruce TJA, Matthes M, Napier JA, Pickett JA (2007) Stressful 'memories' of plants: evidence and possible mechanisms. Plant Sci 173:603–608

Chen THH, Murata N (2002) Enhancement of tolerance of abiotic stress by metabolic engineering of betaines and other compatible solutes. Curr Opin Plant Biol 5:250–257

Dabbous A, Ben Saad R, Brini F, Farhat-Khemekhem A, Zorrig A, Abdely C, Ben Hamed K (2017) Over-expression of a subunit E1 of a vacuolar H^+-ATPase gene (*Lm* VHA-E1) cloned from the halophyte *Lobularia maritima* improves the tolerance of *Arabidopsis thaliana* to salt and osmotic stresses. Environ Exp Bot 127:128–141

Dassanayake M, Larkin JC (2017) Making plants break a sweat: the structure, function, and evolution of plant salt glands. Front Plant Sci 8:406. https://doi.org/10.3389/fpls.2017.00406

Debez A, BenHamed K, Grignon C, Abdelly C (2004) Salinity effects on germination, growth, and seed production of the halophyte *Cakile maritima*. Plant Soil 262:179–189

Debez A, Saadaoui D, Ramani B, Ouerghi Z, Koyro HW, Huchzermeyer B, Abdelly C (2006) Leaf H+-ATPase activity and photosynthetic capacity of Cakile maritima under increasing salinity. Environ Exp Bot 57:285–295

Demidchik V et al (2010) Arabidopsis root K^+-efflux conductance activated by hydroxyl radicals: single-channel properties, genetic basis and involvement in stress-induced cell death. J Cell Sci 123:1468–1479

Djanaguiraman M, Sheeba JA, Shanker AK, Devi DD, Bangarusamy U (2006) Rice can acclimate to lethal level of salinity by pretreatment with sublethal level of salinity through osmotic adjustment. Plant Soil 284:363–373

Ellouzi H, Ben Hamed K, Cela J, Munné-Bosch S, Abdelly C (2011) Early effects of salt stress on the physiological and oxidative status of *Cakile maritima* (halophyte) and *Arabidopsis thaliana* (glycophyte). Physiol Plant 142:128–143

Ellouzi H, Ben Hamed K, Hernández I, Cela J, Müller M, Magné C, Abdelly C, Munné-Bosch S (2014) A comparative study of the early osmotic, ionic, redox and hormonal signalling response in leaves and roots of two halophytes and a glycophyte to salinity. Planta 240:1299–1317

Fedina IS, Nedeva D, Çiçek N (2009) Pre-treatment with H2O2 induces salt tolerance in barley seedlings. Biol Plant 53:321–324

Flowers TJ (1985) Physiology of halophytes. Plant Soil 89:41–56

Flowers TJ (2004) Improving crop salt tolerance. J Exp Bot 396:307–319

Flowers TJ, Colmer TD (2008) Salinity tolerance in halophytes. New Phytol 179:945–963

Gao C, Wang Y, Jiang B, Liu G, Yu L, Wei Z, Yang C (2011) A novel vacuolar membrane H+-ATPase c subunit gene (ThVHAc1) from Tamarix hispida confers tolerance to several abiotic stresses in Saccharomyces cerevisiae. Mol Biol Rep 38:957–963

Ghars MA, Parre E, Debez A, Bordenave M, Richard L, Leport L, Bouchereau A, Savouré A, Abdelly C (2008) Comparative salt tolerance analysis between *Arabidopsis thaliana* and *Thellungiella halophila*, with special emphasis on K^+/Na^+ selectivity and proline accumulation. J Plant Physiol 165:588–599

Ghars MA, Richard L, Lefebvre-De Vos D, Leprince AS, Parre E, Bordenave M, Abdelly C, Savouré A (2012) Phospholipases C and D modulate proline accumulation in *Thellungiella halophila/salsuginea* differently according to the severity of salt or hyperosmotic stress. Plant Cell Physiol 53:183–192

Gondim FA, Gomes-Filho E, Costa JH, Alencar NLM, Priso JT (2012) Catalase plays a key role in salt stress acclimation induced by hydrogen peroxide pre treatment in maize. J Plant Physiol Biochem 56:62–71

Gondim FA, Miranda RS, Gomes-Filho E, Prisco JT (2013) Enhanced salt tolerance in maize plants induced by H_2O_2 leaf spraying is associated with improved gas exchange rather than with non-enzymatic antioxidant system. Theor Exp Plant Physiol 25:251–260

Gong Q, Li P, Ma S, Rupassara SI, Bohnert H (2005) Salinity stress adaptation competence in the extremophile *T. halophila* in comparison with its relative *A. thaliana*. Plant J 31:826–839

Gorham J (1996) Mechanisms of salt tolerance of halophytes. In: Choukrallah R et al (eds) Halophytes and biosaline agriculture. Marcel Dekker, Inc, New York, pp 31–53

Hasegawa PM (2013) Sodium (Na^+) homeostasis and salt tolerance of plants. Environ Exp Bot 92:19–31

Himabindu Y, Chakradhar T, ReddyMC KA, Redding KE, Chandrasekhar T (2016) Salt-tolerant genes from halophytes are potential key players of salt tolerance in glycophytes. Environ Exp Bot 124:39–63

Hossain MA, Bhattacharjee S, Armin S-M, Qian P, Xin W, Li H-Y, Burritt DJ, Fujita M, Tran L-SP (2015) Hydrogen peroxide priming modulates abiotic oxidative stress tolerance: insights from3 ROS detoxification and scavenging. Front Plant Sci 6:420. https://doi.org/10.3389/fpls.2015.00420

Hu DG, Wang SH, Luo H, Ma QJ, Yao YX, You CX, Hao YJ (2012) Overexpression of MdVHA-B, a V–ATPase gene from apple, confers tolerance to drought in transgenic tomato. Sci Hortic 145:94–101

Jdey A, Slama I, Rouached A, Abdelly C (2014) Growth, Na^+, K^+, osmolyte accumulation and lipid membrane peroxidation of two provenances of *Cakile maritima* during water deficit stress and subsequent recovery. Flora 209:54–62

Ji H, Pardo JM, Batelli G, Van Oosten MJ, Bressan RA, Li X (2013) The salt overly sensitive (SOS) pathway: established and emerging roles. Mol Plant 6:275–286

Jithesh MN, Prashanth SR, Sivaprakash KR, Parida AK (2006) Antioxidative response mechanisms in halophytes: their role in stress defence. J Genet 85:237–254

Kant S, Kant P, Raveh E, Barak S (2006) Evidence that differential gene expression between the halophyte, *Thellungiella halophila*, and Arabidopsis thaliana is responsible for higher levels of the compatible osmolyte proline and tight control of Na+ uptake in *T. halophila*. Plant Cell Environ 29:1220–1234

Katschnig D, Broekman R, Rozema J (2013) Salt tolerance in the halophyte *salicorniadolichostachya* moss: growth, morphology and physiology. Environ Exp Bot 92:32–42

Kim C, Bressan R (2016) A computational analysis of salt overly sensitive 1 homologs in halophytes and glycophytes. PeerJ PrePrints 4:e1668v1 https://doi.org/10.7287/peerj.preprints.1668v1

Kluge C, Lamkemeyer P, Tavakoli N, Golldack D, Kandlbinder A, Dietz KJ (2009) cDNA cloning of 12 subunits of the V-type ATPase from Mesembryanthemum crystallinum and their expression under stress. Mol Membr Biol 20:171–183

Koyro HW, Lieth H (2008) Global water crisis: the potential of cash crop halophytes to reduce the dilemma. In: Lieth H, Sucre MG, Herzog B (eds) Mangroves and halophytes: restoration and utilisation. Tasks for vegetation sciences, vol 43. Springer, Dordrecht, pp 7–19

Li JT, Qiu ZB, Zhang XW, Wang LS (2011) Exogenous hydrogen peroxide can enhance tolerance of wheat seedlings to salt stress. Acta Physiol Plant 33:835–842

Liska AJ, Shevchenko A, Pick U, Katz A (2004) Enhanced photosynthesis and redox energy production contribute to salinity tolerance in *Dunaliella* as revealed by homology-based proteomics. Plant Physiol 136:2806–2817

Munns R, James RA, Xu B, Athman A, Conn SJ, Jordans C, Byrt CS, Hare RA, Tyerman SD, Tester M, Plett D, Gilliham M (2012) Wheat grain yield on saline soils is improved by an ancestral Na^+ transporter gene. Nat Biotechnol 30:360–364

Murakeözy EP, Nagy Z, Duhazé C, Bouchereau A, Tuba Z (2003) Seasonal changes in the levels of compatible osmolytes in three halophytic species of inland saline vegetation in Hungary. J Plant Physiol 160:395–401

Oh DH, Leidi E, Zhang Q, Hwang SM, Li Y, Quintero FJ et al (2009a) Loss of halophytism by interference with SOS1 expression. Plant Physiol 151:210–222

Oh DO, Zahir A, Yun DJ, Bressan RA, Bohnert HJ (2009b) SOS1 and halophytism. Plant Signal Behav 4(11):1081–1083

Ozgur R, Uzilday B, Sekmen AH, Turkan I (2013) Reactive oxygen species regulation and antioxidant defence in halophytes. Funct Plant Biol 40:832–847

Pandolfi C, Mancuso S, Shabala S (2012) Physiology of acclimation to salinity stress in pea (*Pisum sativum*). Environ Exp Bot 84:44–51

Pandolfi C, Azzarello E, Mancuso S, Shabala S (2016) Acclimation improves salt stress tolerance in Zea mays plants. J Plant Physiol 201:1–8

Panta S, Flowers T, Lane P, Doyle R, Haros G, Shabala S (2014) Halophyte agriculture: success stories. Environ Exp Bot 107:71–83

Paramonova NV, Shevyakova NI, Kuznetsov VV (2004) Ultrastructure of chloroplasts and their storage inclusions in the primary leaves of *Mesembryanthemum crystallinum* affected by putrescine and NaCl. Russ J Plant Physiol 51:86–96

Parre E, Ghars MA, Leprince AS, Thiery L, Lefebvre D, Bordenave M, Luc R, Mazars C, Abdelly C, Savouré A (2007) Calcium signaling via phospholipase C is essential for proline accumulation upon ionic but not nonionic hyperosmotic stresses in Arabidopsis. Plant Physiol 144:503–512

Pitman MG, Läuchli A (2002) Global impact of salinity and agricultural ecosystems. In: Läuchli A, Lüttge U (eds) Salinity: environment – plants – molecules. Springer Netherlands, Dordrecht, pp 3–20

Qiu NW, Chen M, Guo JR, Bao HY, Ma XL, Wang BS (2007) Coordinate up-regulation of V-H^+-ATPase and vacuolar Na+/H+ antiporter as a response to NaCl treatment in a C-3 halophyte *Suaeda salsa*. Plant Sci 172:1218–1225

Ratajczak R (2000) Structure, function and regulation of the plant vacuolar H+ translocating ATPase. Biochim Biophys Acta 1465:17–36

Ruiz JM (2002) Engineering salt tolerance in crop plants. Curr Opin Biotechnol 13:146–150

Sathiyaraj G, Srinivasan S, Kim YJ, Lee OR, Balusamy SDR, Khorolaragchaa A et al (2014) Acclimation of hydrogen peroxide enhances salt tolerance by activating defense-related proteins in *Panax ginseng* CA. Meyer. Mol Biol Rep 41:3761–3771

Shabala S (2013) Learning from halophytes: physiological basis and strategies to improve abiotic stress tolerance in crops. Ann Bot 112:1209–1221

Shabala S, Mackay A (2011) Ion transport in halophytes. Adv Bot Res 57:151–199

Shabala S, Bose J, Hedrich R (2014) Salt bladders: do they matter? Trends Plant Sci 19:687–691

Shi H, Lee BH, Wu SJ, Zhu JK (2003) Overexpression of a plasma membrane Na^+/H^+ antiporter gene improves salt tolerance in *Arabidopsis thaliana*. Nat Biotechnol 21:81–85

Short DC, Colmer TD (1999) Salt tolerance in the halophyte *Halosarcia pergranulata* subsp. *Pergranulata*. Ann Bot 83:207–213

Slama I, Abdelly C, Bouchereau A, Flowers T, Savouré A (2015) Diversity, distribution and roles of osmoprotective compounds accumulated in halophytes under abiotic stress. Ann Bot 115:433–447

Souid A, Gabriele M, Longo V, Pucci L, Bellani L, Smaoui A, Abdelly C, Ben Hamed K (2016) Salt tolerance of the halophyte *Limonium delicatulum* is more associated with antioxidant enzyme activities than phenolic compounds. Funct Plant Biol 43:607–619

Sun J, Wang MJ, Ding MQ, Deng SR, Liu MQ, Lu CF, Zhou XY, Shen X, Zheng XJ, Zhang ZK, Song J, Hu ZM, Xu Y, Chen SL (2010) H_2O_2 and cytosolic Ca^{2+} signals triggered by the PM H^+-coupled transport system mediate K^+/Na^+ homeostasis in NaCl-stressed *Populus euphratica* cells. Plant Cell Environ 33:943–958

Szabados L, Savouré A (2010) Proline: a mulitfunctional amino acid. Trends Plant Sci 15:89–97

Taji T, Seki M, Satou M, Sakurai T, Kobayashi M, Ishiyama K, Narusaka Y, Narusaka M, Zhu JK, Shinozaki K (2004) Comparative genomics in salt tolerance between *Arabidopsis* and *Arabdopsis*-related halophyte salt cress using *Arabidopsis* microarray. Plant Physiol 135:1697–1709

Thiery L, Leprince A-S, Lefebvre D, Ghars MA, Debarbieux E, Savouré A (2004) Phospholipase D is a negative regulator of proline biosynthesis in *Arabidopsis thaliana*. J Biol Chem 279:14812–14818

Uchida A, Jagendorf AT, Hibino T, Takabe T, Takabe T (2002) Effects of hydrogen peroxide and nitric oxide on both salt and heat stress tolerance in rice. Plant Sci1 63:515–523

Umezawa T, Shimizu K, Kato M, Ueda T (2000) Enhancement of salt tolerance in soybean with NaCl pretreatment. Physiol Plant 110:59–63

Wahid A, Perveen M, Gelani S, Basra SMA (2007) Pre treatment of seed with H_2O_2 improves salt tolerance of wheat seedlings by alleviation of oxidative damage and expression of stress proteins. J Plant Physiol1 64:283–294

Wang BS, Lüttge U, Ratajczak R (2001) Effects of salt treatment and osmotic stress on V-ATPase and V-PPase in leaves of the halophyte Suaeda salsa. J Exp Bot 52:2355–2365

Wiciarz M, Gubernator B, Kruk J, Niewiadomska E (2015) Enhanced chloroplastic generation of H_2O_2 in stress-resistant *Thellungiella salsuginea* in comparison to *Arabidopsis thaliana*. Physiol Plant 153:467–476

Zhang HX, Blumwald E (2001) Transgenic salt-tolerant tomato plants accumulate salt in foliage but not in fruit. Nat Biotechnol 19:765–768

Zhang GH, Su Q, An LJ, Wu S (2008) Characterization and expression of a vacuolar Na(+)/H(+) antiporter gene from the monocot halophyte Aeluropus littoralis. Plant Physiol Biochem 46:117–126

Zhu JK (2001) Plant salt tolerance. Trends Plant Sci 6:66–71

Chapter 2
The Involvement of Different Secondary Metabolites in Salinity Tolerance of Crops

Oksana Sytar, Sonia Mbarki, Marek Zivcak, and Marian Brestic

Abstract Salt stress decreased plant growth and development; affects carbon metabolism, ion toxicity, nutritional status, and oxidative metabolism; and modulates the levels of secondary metabolites which are important physiological parameters in salt stress tolerance. Recent progress has been made in the identification and characterization of the mechanisms that allow plants to tolerate high salt concentrations and drought stress. Accumulation of secondary metabolites often occurs in plants subjected to stresses including various elicitors or signal molecules. The focus of the present chapter is the influence of salt stress on secondary metabolite production and some of important plant pharmaceuticals. Enhanced synthesis in the cytosol of determined secondary metabolites (anthocyanins, flavones, phenolics, and specific phenolic acids) under stress condition may protect cells from ion-induced oxidative damage by binding the ions and thereby showing reduced toxicity on cytoplasmic structures. The aim of this study was to determine the physiological implication of secondary metabolites in salt-tolerant crops.

Keywords Secondary metabolites · Salt stress · Phenolic acids · Anthocyanins · Phenolics · Salt-tolerant plants · Non-salt-tolerant plants

---

O. Sytar (✉)
Plant Physiology and Ecology Department, Institute of Biology, Taras Shevchenko National University of Kyiv, Kyiv, Ukraine

Department of Plant Physiology, Slovak University of Agriculture in Nitra, Nitra, Slovakia

S. Mbarki
Laboratory of Extremophile Plants, Centre de Biotechnologie de Borj Cedria, Hamam Lif, Tunisia

M. Zivcak · M. Brestic
Department of Plant Physiology, Slovak University of Agriculture in Nitra, Nitra, Slovakia

V. Kumar et al. (eds.), *Salinity Responses and Tolerance in Plants, Volume 2*,
https://doi.org/10.1007/978-3-319-90318-7_2

Abbreviation

CAT	Catalase
GPX	Glutathione peroxidase
GSTs	Glutathione-S-transferase
PAL	Phenylalanine ammonium lyase
ROS	Reactive oxygen species

2.1 Introduction

Drought, high salinity, and freezing temperatures are environmental conditions that cause adverse effects on the growth of plants and the productivity of crops. Environmental factors, viz., temperature, humidity, and light intensity, the supply of water, minerals, and CO_2 influence the growth of a plant and secondary metabolite production. Salt environment leads to cellular dehydration, which causes osmotic stress and removal of water from the cytoplasm resulting in a reduction of the cytosolic and vacuolar volumes. Salt stress often creates both ionic and osmotic stress in plants, resulting in accumulation or decrease of specific secondary metabolites in plants (Mahajan and Tuteja 2005).

It is expected that exposure of plants to salinity will also result in a wide range of metabolic responses. The changes in levels of several metabolites following salt stress are well documented. One of the best described metabolic responses is the increase in intracellular concentration of a range of soluble, neutral organic compounds that are collectively termed "compatible solutes" (Bohnert et al. 1995; Nelson et al. 1998).

The most common compatible solutes in plants are polyhydroxy compounds (sucrose, oligosaccharides, and polyhydric alcohols) and nitrogen-containing compounds (proline, other amino acids, quaternary ammonium compounds, and polyamines) (Bohnert et al. 1995; Hare et al. 1998). However, it is noteworthy that increases in amino acid concentrations have been attributed simply to arising from the decrease in growth rate that results from stress (Munns 2002). Such accumulation may not, therefore, be an adaptive response to salinity but simply a secondary consequence of reduced growth.

Anthocyanins are reported to increase in response to salt stress (Parida and Das 2005; Slama et al. 2017). In contrast to this, salt stress decreased anthocyanin level in the salt-sensitive species (Daneshmand et al. 2010).

Polyphenol synthesis and accumulation is generally stimulated in response to biotic or abiotic stresses (Muthukumarasamy et al. 2000). Increase in polyphenol content in different tissues under increasing salinity has also been reported in a number of plants (Parida and Das 2005; Slama et al. 2017). Increasing total phenolic content with moderately saline level has been observed in red peppers (Navarro et al. 2006). Exposure *S. portulacastrum* as halophyte under 800 mM NaCl impaired significantly photosynthesis, proline, polyphenol, antocyanins and carotenoids accumulation and strong antiradical activity (DPPH) observed at this extreme salinity and exhibits important antioxidant potentialities.might partly explain the plant survival which be used in the rehabilitation and the stabilisation of saline or saline arid land (Slama et al. 2017).

Total chlorophyll (CHL) and CAR involved in photochemical reaction of leaves are the most important secondary metabolites. Decrease in CHL content is associated with enhanced expression of chlorophyllase activity under stress (Majumdar et al. 1991).

Plants synthesise a variety of organic solutes such as proline, glycine betaine, soluble sugars, polyamines, etc., which are collectively called as osmolytes or compatible solutes. The osmolyte accumulation is frequently reported in plants including halophytes exposed to salt stress and has been correlated with a plant's capacity to tolerate and adapt to salinity conditions (Mssedi et al. 2000; Slama et al. 2008). When compared to other plants, *S. portulacastrum* appeared as a high proline accumulator, with proline levels reaching 300 µmol g−1 leaf dry matter (Slama et al. 2006).

Besides proline, higher salinity also causes increased accumulation of glycine betaine (Lokhande et al. 2010). The accumulation of glycine betaine was assumed to have positive functions in relation to maintenance of membrane integrity and stability of other cellular structures under salt and drought stress conditions (Martinez et al. 2005).

Studies of plant responses to salinity stress and the differences between salt-sensitive and salt-tolerant species using involvement of secondary metabolites have revealed an interesting diversity of patterns. First, there are differences in constitutive levels of metabolites between tolerant and sensitive varieties/species, and, second, different species show conserved as well as divergent metabolite responses in response to salinity.

Many major crops such as pepper, eggplant, potato, lettuce, and cabbage are salt-sensitive (Shannon and Grieve 1999). In addition, important cereals such as rice and maize are also sensitive to hyperosmotic stresses, and their production seriously decreases in saline soils (Ngara et al. 2012). Therefore increasing soil salinization as well as the growing world population shows the increasing need to develop crops which are able to adapt to salt stress.

Some plants grow and complete their life cycle in the habitats with a high salt content. They are known as salt plants or halophytes adapted to saline environments that are able to complete their life cycles in the presence of salt concentrations in the soil equivalent to at least 200 mM NaCl (Katschnig et al. 2013; Flowers and Colmer 2008; Santos et al. 2015). The halophytes distribution and their responses to salinity habitats is used for development of halophytes classification into the following two main categories. First category is salt-enduring halophytes which show optimum development in non-saline habitats but can tolerate salts (Abdell et al. 2006). Second category is salt-resistant halophytes which show optimum development in saline habitats (Van Eijk 1939; Slama et al. 2017).

The high structural diversity of osmolytes combined with their multifunctionality and the seasonal flexibility of the metabolism in halophytes has been shown (Murakeözy et al. 2003; Slama et al. 2015). Depending on the accumulated osmolyte, halophyte plants can be classified into three physiotypes (Briens and Larher 1982): species that produce high levels of soluble carbohydrates and/or polyols but in which nitrogenous water-soluble compounds are present at very low concentrations, plants accumulating nitrogenous water-soluble compounds at higher concentrations than nonstructural carbohydrates, and species that accumulate both carbohydrates and nitrogenous solutes, the first remaining quantitatively predominant.

Accumulation of carbohydrates may require a great deal of carbon, which is less readily available under stress conditions when net assimilation is decreased (Jefferies and Rudmik 1984). Moreover, a saline environment is generally deficient in nitrogen. The nitrogen-containing compounds able to accumulate in the plant, despite NO_3 reduction in many species in response to salt stress. Nitrogen in salt-stressed plants may not be a limiting factor (Mansour 2000). Salt tolerance requires the coordinated metabolic adjustment of different organs and tissues, with inherent reallocation processes of both C and N (Adams et al. 1990); therefore, investigation of whole-plant responses of primary metabolites is studied nowadays. In many cases known in salt response, primary metabolites can influence or can be affected by changes in secondary metabolites which can take part in detoxification of reactive oxygen species (ROS) as antioxidants. It is not fully available data regarding effect of salt stress on level, accumulation etc. of specific secondary metabolites which present in specific halophytes and crop plants.

Therefore, the aim of this chapter is to describe the presence of specific secondary metabolites in halophyte plants. To discuss possible role of these secondary metabolites in response reactions under salt stress is main aim of proposed chapter.

2.1.1 Divisions of Plant Secondary Metabolites in Known Salt-Tolerant Plants

Secondary metabolites are organic compounds produced by plants originating of primary metabolism, have no direct relation to its growth and development, and show no functions hitherto recognized in vital processes of plants, such as photosynthesis, cellular respiration, protein synthesis, solute transport, and nutrient assimilation, unlike the primary metabolites (Olivoto et al. 2017). The synthesis of specific metabolites being restricted to a few species (Taiz and Zeiger 2010) is not observed in all species of the plant kingdom.

Most of the secondary metabolites are classified based on their biosynthetic origin. Plant secondary metabolites considered as end products of primary metabolism and not involved in metabolic activity can be divided into chemically distinct main groups: phenolic compounds, glycosides (saponins, cyanogens), alkaloids, steroids, essential oils, lignins, tannins, lectins (hemagglutinins), etc. (Salah 2015).

The environmental stress factors such as salt, dryness, temperature, and soil pH are essential factors for release of secondary metabolites with potent antioxidant activity (Selvam et al. 2013). For example, there is a confirmed correlation between antioxidant activity and total phenolic content in plants of different origin (Piluzza and Bullitta 2011; Sadeghi et al. 2015). The total antioxidant activity is an integrated parameter of antioxidants present in a complex. Other secondary metabolites which can have antioxidant capacity are participating in creating parameters of total antioxidant activity. These are flavonoids, tannins, anthocyanins, and phenolic acids - representatives of phenols group (Dyduch-Siemińska et al. 2015). Secondary metabolites containing nitrogen are alkaloids which also have antioxidant activity (Velioglu et al. 1998) and can be connected with changes of metabolites containing nitrogen in primary metabolism under salt stress.

The most important characteristic of halophyte plants is their capacity to grow under high concentrations of NaCl. The biochemical mechanisms leading to salt tolerance in these plants are regulated in such a way that allows a more successful response to salt stress than in other plants (Hasegawa et al. 2000). We suppose that response mechanism of halophyte plants to salinity can be connected with secondary metabolite composition and specific gene expression under salt stress (Panich et al. 2010; Ksouri et al. 2012; Falleh et al. 2013).

Many of secondary metabolites participate in the development of salt stress response as antioxidants (Fig. 2.1).

Some physiological and biochemical mechanisms of tolerance to salinity are common to many halophytes when plants are subjected to salinity, whereas others

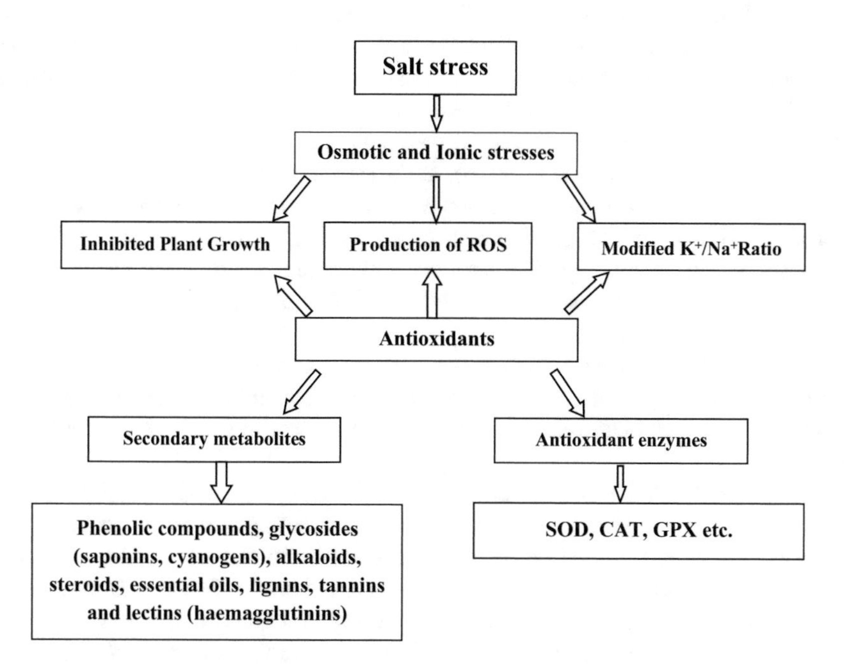

Fig. 2.1 Participation of secondary metabolites in development of salt stress response

are specifically amplified under a combination of stresses. Therefore, the response of halophytes to multiple stressors may not reflect an additive effect of these constraints but rather constitute specific response to a new situation (Visser et al. 2006) where many constraints are operating simultaneously. The effects of stressors can lead to deficiencies in growth, crop yields, permanent damage, or death if the stress exceeds the plant tolerance limits. Comparative studies between halophytes and glycophytes have shown that halophytes are better equipped with the mechanisms of cross-stress tolerance and are constitutively prepared for stress. Moreover, other data has shown that the pretreatment of halophytes with salinity or other constraints in the early stages of development improves their subsequent response to salinity, which suggests the capacity of these plants to "memorize" a previous stress allows them to respond positively to subsequent stress (Hamed et al. 2013). Halophytes can be sources for production of antioxidant compounds, and their accession-dependent capacity to induce antioxidative mechanisms in response to salt may result in a corresponding variability for growth sustainability (Ksouri et al. 2007; Flowers et al. 2015; Jdey et al. 2017).

It is supposed that halophytes vary regarding the composition of secondary metabolites but possible that some secondary metabolites especially which present in high content can be responsible for the salt tolerance capacities of halophyte plants. A large number of medicinally important plants are halophytes (Qasim et al. 2016; Jdey et al. 2017). Most halophytes produce secondary metabolites that affect animal performance when consumed, even when palatable to the animals (Salah 2015). Plant secondary metabolites are produced in plants for protective purposes for the plants themselves and to adapt to environmental stresses and are unique sources for pharmaceuticals, food additives, flavors, and industrially important biochemicals. The most of halophytes contains moderate or low levels of anti-nutritional factors (secondary metabolites), which lies under the safe limits. The halophytes can be used as a potential feed source to raise animals, particularly in arid/semiarid areas (Ehsen et al. 2016). At the same time these plants are also worldwide used in several areas of industry used in traditional medicine. It was done screening of halophytes regarding pharmacological properties to set out new promising sources of natural ingredients for cosmetic or nutraceutical applications (used as antioxidant, antibiotic as well as new natural skin lightening agents) (Jdey et al. 2017).

In previous botanical researches, the most well-known salt-tolerant plants were chosen (Armitage 2000; Darke 1999; Dirr 1997; Gerhold et. al. 2001). Table 2.1 shows endogenous plant secondary metabolites present in some halophytic fodder crops.

Plant metabolomics reveals conserved and divergent metabolic responses to salinity. Few studies have utilized metabolic fingerprinting or profiling technologies to discover changes in higher plants upon exposure to salt stress. The salt stress impact on secondary metabolites of crops such as tomato, *Solanum lycopersicum* cv. Edkawy and cv. Simge F1; rice, *Oryza sativa*; and grapevine, *Vitis vinifera* cv. Cabernet and the models *Arabidopsis thaliana* eco. Col-0 as well as *A. thaliana* T87 cell cultures has been investigated (Cramer et al. 2007). Comparisons to halophytic

Table 2.1 Endogenous plant secondary metabolites present in some halophytic fodder crops

Name of plant species	Presented secondary metabolites			Reference
Deciduous trees that tolerate salt				
Horse chestnut (*Aesculus hippocastanum*)	Flavonoids (tamarixetin 3-*O*- [β-D-glucopyranosyl(1→ 3)]-*O*-β-D-xylopyranosyl-(1→2)-*O*-β-D-glucopyranoside)		Aescin	Kapusta et al. (2007)
Cherry birch (*Betula lenta*)	Isoprenoids and phenolics			Palo (1984)
Maidenhair tree (*Ginkgo biloba*)	Phenolics, flavonoid glycosides	Terpenes (terpene trilactones)		He et al. (2009), Hao et al. (2010), and Singh et al. (2009)
Eastern red cedar (*Juniperus virginiana*)	Essential oils (α-pinene, β-pinenes, and β-phellandrene)	Lignans (podophyllotoxin and deoxypodophyllotoxin)		Stewart et al. (2014), Kusari et al. (2011)
Magnolia (*Magnolia* spp.)	Terpenes (sesquiterpene)	Alkaloid (kachirachiranine)	Phenolic lignans ((+)-de-*O*-methylmagnolin, (+)-phillygenin and (+)-pinoresinol)	Chang et al. (2011) and Mário et al. (2001)
White oak (*Quercus alba*)	Condensed tannins (ellagitannins)	Polyphenols		Forkner et al. (2004) and Scalbert and Haslam (1987)
Red oak (*Quercus rubra*)	Condensed tannins (ellagitannins)			Herve du Penhoat et al. (1991)
Japanese Pagoda tree (*Sophora japonica*)	Flavonoids (rutin)	Phenolic compounds	Alkaloids (sophoridine, matrine)	Paniwnyk et al. (2001) and Küçükboyaci et al. (2011)
Tamarix sp.	Sulfated flavonoids	Phenolics (phenolic acids)	Flavonoids (isoquercitin)	Ksouri et al. (2009) and Karker et al. (2016)
Laurus nobilis	Condensed tannins	Flavonoids	Total phenolics	Ouchikh et al. (2011)

(continued)

Table 2.1 (continued)

Name of plant species	Presented secondary metabolites			Reference
Shrubs and ground covers that tolerate salt				
Bearberry (*Arctostaphylos uva-ursi*)	Tannins			Haslam (1996)
Chokeberry (*Aronia* spp.)	Phenolics (phenolic acids)			Szopa et al. (2013)
European sea rocket (*Cakile maritima*)	Phenolics			Ksouri et al. (2007)
Cardoon (*Cynara cardunculus*)	Phenolics (phenolic acids)	Flavonoids	Tannins	Falleh et al. (2008)
Siberian pea shrub (*Caragana* spp.)	Flavonoids	Stilbenoids (carasinol B, kobophenol A, etc.)	Phytosterols (β-sitosterol), lectins	Meng et al. (2009)
Rock cotoneaster (*Cotoneaster horizontalis*)	Phenolics (phenolic acids)			Sytar et al. (2015, 2016)
Rose of Sharon (*Hibiscus syriacus*)	Alpha- and beta-carotenes	Phenolics	Flavonoids	Hanny and Hedin (1972) and Gomes Maganha et al. (2010)
St. John's wort (*Hypericum* spp.)	Naphtodianthrones (hyperforin, hypericin, pseudohypericin)	Flavonoids (rutin, quercetrin, etc.)		Sytar et al. (2013), Smelcerovic et al. (2006), and Maffi et al. (2001)
Bayberry (*Myrica pensylvanica*)	Anthocyanins (cyanidin-3,5-O-diglucoside), flavonol (quercetin-3-*o*-glucoside)	Phenolics (Phenolic acids)		Shao-huan Zhou et al. (2009) and Niu et al. (2010)
Potentilla (*Potentilla fruticosa*)	Tannins	Triterpenes		Tomczyk and Latté (2009)
Phragmites karka (Retz.)	Flavonoids	Phenolics		Abideen et al. (2015)
Saltspray rose (*Rosa rugosa*)	Hydrolyzable tannins	Phenolics (catechin derivatives), flavonoids	Monoterpenes, sesquiterpenes, triterpenes	Hashidoko (1996)

Lilacs (*Syringa* spp.)	Phenolics, flavonoids	Phenylpropanoids secoiridoids (oleuropein and nuzhenide)		Tóth et al. (2016)
Blueberry/Cranberry (*Vaccinium* spp.)	P-sitosterol	Phenolics (phenolic acids)	Flavonoids	Eichholz et al. (2011) and Vattem et al. (2005)
Conifers that tolerate salt				
Larch (*Larix* spp.)	Phenolics (phenolic acids, tannins)	Flavonoids	Lignans, labdane-type diterpene	Nabeta (1994) and Xue et al. (2004)
White spruce (*Picea glauca*)	Terpenoids (camphor)	Phenolics		Sinclair et al. (1988) and Warren et al. (2015)
Japanese white pine (*Pinus strobus*)	Phenolics (phenolic acids)	Diterpene resin acids		Shadkami et al. (2007)
Ponderosa pine (*Pinus ponderosa*)	Phenolics	Piperidine alkaloids		Gerson et al. (2009) and Tingey et al. (1976)
Mugo pine (*Pinus mugo*)	Phenolics (diacylated flavonol, kaempferol, shikimic acid)	Diterpenes	Essential oils (α-pinene, β-pinene, etc.)	Venditti et al. (2013) and Hajdari et al. (2015)
Perennials that tolerate salt				
Sea thrift (*Armeria maritima*)	Phenolics (phenolics acids)			Gourguillon et al. (2016)
Cheddar pink (*Dianthus gratianopolitanus*)	Flavonoids, anthocyanins	Flavanone (dianthalexin)	Phenolics	Matern (1994)
Lenten rose (*Helleborus orientalis*)	Phenolics	Bufadienolide glycoside	Spirostanol saponin	Watanabe et al. (2003) and Shahri et al. (2011)
Daylily (*Hemerocallis* spp.)	Phenolics (phenolic acids, especially chlorogenic acid)	Tannins (oxypinnatanine)		Clifford et al. (2006) and Nguyen et al. (2015)

(continued)

Table 2.1 (continued)

Name of plant species	Presented secondary metabolites			Reference
Sea lavender (*Limonium latifolium*, *Limonium delicatulum*)	Phenolics	Alkaloids	Tannins	Chandna et al. (2013), Salah (2015), and Faten et al. (2014)
Lilyturf (*Liriope spicata*)	Phenolics	Saponins		Chang et al. (2006) and Song et al. (1996)
Fountain grass (*Pennisetum alopecuroides*)	Phenolics			Mane et al. (2010)
Little bluestem (*Schizachyrium scoparium*)	Phenolics	Terpenoids	Allelochemicals	Fischer et al. (1994) and Hierro and Callaway (2003)
Barren strawberry (*Waldsteinia fragarioides*)	Phenolics	Flavonoids (isoquercitrin, etc.)		Matkowski et al. (2006) and Chattopadhyay et al. (2010)

species, such as the tree *Populus euphratica* or the shrubs *Thellungiella halophila* and *Limonium latifolium*, have also been performed (Gagneul et al. 2007). Thus, applications range from ecological metabolomic approaches to artificial in vitro systems.

An increase in the photosynthetic pigments at lower salt levels might be due to the osmotic adjustment mechanism developed by the grass species incumbent with salt stress. The increased levels of polyphenols at elevated salinity might be due to the accumulation of secondary metabolites (Mane et al. 2010).

Medicinal plants are rich in secondary metabolites like alkaloids, glycosides, steroids, and flavonoids which are potential sources of drugs. The question of subjecting medicinal herbs to modern scientific test has often been raised. Biosynthesis of secondary metabolites is affected strongly by salt stress resulting in considerable fluctuations in quality and quantity. There can be variation in the pattern of secondary product composition within the same plant. It is possible that the efficacy of the herb depends on the total effect of plant contents rather than the few chemical fractions separated from the herb (Muthulakshmi Santhi et al. 2015).

2.1.2 Role of Alkaloids in Salt Stress

Alkaloids are a group of naturally occurring chemical compounds that mostly contain basic nitrogen atoms. Primarily alkaloids are found in plants and are especially common in original families of flowering plants. More than 3000 different types of alkaloids have been identified in a total of more than 4000 plant species. The plant families which are rich in alkaloids are Papaveraceae, Ranunculaceae, Solanaceae, and Amaryllidaceae families. Among presented in Table 2.1, halophytic fodder crop alkaloids were found and present in *Magnolia* sp., *Sophora japonica*, and *Pinus ponderosa*.

Effect of salinity and different nitrogen sources on the activity of antioxidant enzymes and indole alkaloid content in *Catharanthus roseus* seedlings shows increased accumulation of alkaloid in all leaf pairs, as well as in roots of *C. roseus* of NO_3^--fed plants as compared to NH_4^+-fed plants (Misra and Gupta 2006). The obtained results show the possibility to increase content of steroidal alkaloid solsasodine level production under salinity stress in *Solanum nigrum*. Positive dose-dependent correlations were observed between the NaCl levels and solasodine content accumulation, proline content, and solasodine accumulation in calli of *Solanum nigrum* (Šutković et al. 2011). Increasing level of plant alkaloid vincristine under drought stress was found (Amirjani 2013). It was observed that the level of ricinine alkaloids in roots of *Ricinus communis* was significantly lowered by salt stress, but it was increased in shoots (Ali et al. 2008).

Salinity in soil presents a stress condition for growth of the plants. Under natural conditions of growth and development, plants are inevitably exposed to different types of stress, which may cause increased production of reactive oxygen species (ROS) (Smirnoff 1993). Under highly suppressive conditions, these alkaloids play the role as ROS scavengers. Some experimental studies have also shown that ROS production also regulates the alkaloid pathway occurring in undifferentiated cells (Sachan et al. 2010).

2.1.3 Role of Phenolic Compounds in Salt Stress

Phenolics are aromatic benzene ring compounds with one or more hydroxyl groups produced by plants mainly for protection against stress. The functions of phenolic compounds in plant physiology are difficult to overestimate, especially their interactions with biotic and abiotic environments. Phenolics play important roles in plant development, particularly in biosynthesis of pigments and lignin (Bhattacharya et al. 2010).

However, phenolic compounds are potential antioxidants because there is a relation between antioxidant activity and presence of phenols in common vegetables and fruits (Cai et al. 2004; Fu et al. 2011). A positive linear correlation between antioxidant capacities and total phenolic contents implied that phenolic compounds in tested 50 medicinal plants could be the main components contributing to the observed activities. The results showed that *Geranium wilfordii*, *Loranthus parasiticus*, *Polygonum aviculare*, *Pyrrosia sheareri*, *Sinomenium acutum*, and *Tripterygium wilfordii* possessed the highest antioxidant capacities and total phenolic content among 50 plants tested and could be rich potential sources of natural antioxidants (Yi et al. 2010). Among presented in Table 2.1, halophytic fodder crops, mostly all of them, have phenolic comounds and specific flavonoid composition.

Phenylalanine ammonium lyase (PAL) and glutathione-S-transferase (GSTs) also get induced from unfavorable effects of stresses (Marrs 1996). PAL, in action with cinamates 4-hydroxylase, forms an essential group of enzymes that helps in biosynthesis of several important secondary metabolites from phenylalanine (Singh et al. 2009). Involvement of phenylpropanoid and particularly flavonoid pathways in safflower plants during wounding and especially salinity stress was estimated (Dehghan et al. 2014).

Increasing phenolic compound content in the leaves of buckwheat sprouts compared with the control has been observed under treatment with various concentrations of NaCl. Moreover, the accumulation of phenolic compounds was primarily caused by an increase in the levels of four compounds: isoorientin, orientin, rutin, and vitexin (Lim et al. 2012). Salt stress in strawberry fruits increased the antioxidant capacity, antioxidant pools (ascorbic acid, anthocyanins, superoxide dismutase), and selected minerals such as Na^+, Cl^-, K^+, N, P, and Zn^{2+}, as well as lipid peroxidation (Keutgen and Pawelzik 2008). There was also an observed connection between production of polyphenolic antioxidants and salt stress in halophyte *P. karka* which indicates salinity as an effective tool to produce antioxidant-rich biomass for industrial purposes (Abideen et al. 2015).

It was estimated that salt stress in some concentrations can improve the nutritional value of radish sprouts too. The germination of sprouts under adequate salt stress can be one useful way to enhance health-promoting compounds of plant food. Antioxidant activity of radish sprouts was not affected by experimental concentrations of NaCl. But lower concentrations of NaCl have decreased total glucosinolate and total phenolic contents, and higher concentrations of NaCl have increased their content (Yuan et al. 2010).

Salt-induced lipid peroxidation was observed only in old leaves, whereas the accumulation of the major phenolic compounds under saline conditions was higher in young leaves of *Carthamus tinctorius* L (Abdallah et al. 2013). During stress treatment, the accumulation of the major polyphenolic compounds is found higher in young leaves of maize plants too. At the same time a difference in phenolic response under salt stress between different plant cultivars has been observed (Ghosh et al. 2011; Ashraf et al. 2010; Hajlaoui et al. 2009). A high polyphenol accumulation, especially anthocyanins can show their efficient participation in restriction of oxidative damages caused by the H_2O_2 generation (Hajlaoui et al. 2009).

It was suggested that accession-dependent capacity to induce antioxidative mechanisms in response to salt may result in a corresponding variability for growth sustainability of some plants. In this case halophytes may be interesting for production of antioxidant compounds. The high leaf polyphenol content and antioxidant activity have been estimated in leaves of the halophyte *Cakile maritima* (Ksouri et al. 2007). It was observed that soil salinity also causes higher chlorogenic acid concentrations in halophyte honeysuckle flower buds, with significant increases in total chlorogenic acid concentration (Yan et al. 2016).

Flavonoids are one of the largest classes of plant phenolics that play a role in plant defense (Kondo et al. 1992). Isoflavonoids are derivatives of flavonone intermediate, e.g., naringenin, secreted by legumes which play a critical role in plant development and defense response and also in supporting the configuration of nitrogen- fixing nodules by symbiotic rhizobia (Sreevidya et al. 2006). Posmyk et al. (2009) reported that synthesis of these flavonoids is an effective strategy against ROS. It was found that under salt stress, *Azospirillum brasilense* promoted root branching in bean seedling roots and increased secretion of *nod* gene-inducing flavonoid species (Dardanelli et al. 2008). Anthocyanins are reported to increase in response to salt stress (Parida and Das 2005). In contrast to this, salt stress decreased anthocyanin level in the salt-sensitive species (Daneshmand et al. 2010).

2.2 Role Lignins in Salt Stress

The one representative of phenols group is important to the structure of the plant and is called lignin. Lignin is one of the major characteristics of plant secondary cell wall that provides structural rigidity for the cells and tissues and hydrophobicity to tracheary elements. It was estimated that salt stress exerts a synergistic effect in the lignification of both protoxylem and metaxylem vessels and induces an earlier and stronger lignification of the secondary thickenings of the phloem fiber cell walls. Salt exposure influenced cell wall composition involving increases in the lignin-carbohydrate ratio (Janz et al. 2012). The lignin deposition is a factor which inhibits root growth and which contributes to the structural integrity of xylem vessels as an adaptation mechanism in resisting the stress imposed by salinity (Cachorro et al. 1993).

In the primary roots of maize, the Casparian strip matured closer to the root tip under conditions of salt stress. It has been shown that formation of the strip at positions closer to the root tip under salt stress was due to a decrease in the number of cells and in the lengths of cells in the endodermis between the root tip and the lowest position of the strip. The age of a cell at the lowest position of the strip is equivalent to the time required for the cell to complete the formation of the strip since the cell itself had been produced. It was indicated that salinity did not have a major effect on the time-dependent formation of the strip itself in individual cells (Karahara et al. 2004).

The average number of lignified cells in vascular bundles was significantly greater together with increased *S*-adenosyl-L-methionine synthase activity in plants under salt stress, with a maximal expansion of the lignified area found in the root vasculature. It was indicated that increased *S*-adenosyl-L-methionine synthase activity correlated with a greater deposition of lignin in the vascular tissues of plants under salinity stress (Sánchez-Aguayo et al. 2004).

The effects caused by NaCl may be owing to the enhanced lignin production that solidifies the cell wall and restricts root growth. The total phenolic and lignin contents and p-hydroxyphenyl (H) and syringyl (S) monomers of lignin increased with PAL activity decreased in NaCl-treated roots (Neves et al. 2010). Activation of lignin biosynthetic enzymes during internodal development of *Aeluropus littoralis* under salt stress has been observed. It induced activities of phenylalanine ammonia lyase and cinnamyl alcohol dehydrogenase and increased the lignin content (Kelij et al. 2015).

2.3 Role of Glucosinolates in Salt Stress

Glucosinolates are a large group of sulfur- and nitrogen-containing glucosides. Glucosinolates, a class of secondary metabolites, mainly found in *Brassicaceae*, are affected by the changing environment. In the plant, glucosinolates coexist with an endogenous enzyme myrosinase. The glucosinolates are stored in the vacuoles of so-called S-cells and myrosinase in separate but adjacent cells. Upon plant tissue disruption, glucosinolates are released at the damage site and become hydrolyzed by myrosinase. The chemical nature of the hydrolysis products depends on the structure of the glucosinolate side chain, plant species, and reaction conditions (Redovnikovi et al. 2008). The mechanism of glucosinolate turnover regulation under salinity still is not completely clarified. However, the different physiological stage of the plant or the level of tolerance to salinity and the individual glucosinolate response (in relation to glucosinolate-myrosinase system) needs a detailed analysis to discuss the response of glucosinolates to salt stress. It is found that lower concentrations of

NaCl have decreased glucoraphasatin (4-methylthio-3-butenyl-glucosinolate) and total glucosinolate in radish seedlings and higher concentrations of NaCl have increased their content (Yuan et al. 2010; López-Pérez et al. 2009). In broccoli plant under influence of salinity is determined decreasing of glucosinolates due to stress conditions (Sarikamiş and Çakir 2017). These results could be related to the strong metabolism alteration focused in turgor adjusting and leading to a high growth reduction (López-Pérez et al. 2009; López-Berenguer et al. 2009). The fact that glucosinolates were accumulated under low water potential when the leaves have to maintain turgor suggested that during salinity stress, the primary metabolism and growth were restricted, but not the secondary metabolism and the production of glucosinolates. It was suggested that the increase in glucosinolates was related to the synthesis of osmoprotective compounds (Martínez-Ballesta et al. 2013).

Leaf water balance under salt stress can be supported by aquaporins and glucosinolate synthesis. NaCl addition occurs decrease in SO_4^{2-}–, and an increase in glucosinolates what can indicate about involvement of these compounds in the water-balance response of plants to salt stress (López-Berenguer et al. 2008).

Glucosinolate degradation and turnover are carried out by the activity of myrosinases (Halkier and Gershenzon 2006; Yan and Chen 2007). Tissue damage brings myrosinase (cytoplasm) in active contact with glucosinolates (vacuole) (Galletti et al. 2006; Andersson et al. 2009). The complexity of the regulation of the glucosinolate-myrosinase system has been pointed out, since no relationship has been observed between the glucosinolate level as a result of altered myrosinase activity in plants under salt stress (Pang et al. 2012). In this study, an increase in glucosinolate degradation was expected to be observed as a consequence of membrane damage by salt stress indicated by high relative electrolyte leakage. Therefore, the lack of relation between myrosinase activity and glucosinolate levels supported the hypothesis that salinity results in an alteration of metabolic activity producing an increase in glucosinolate content.

2.4 The Essential Oils Under Salt Stress

Essential oils are volatile, natural, complex compounds characterized by a strong odor and are formed by aromatic plants as secondary metabolites. They can be synthesized by all plant organs, i.e., buds, leaves, stems, flowers, twigs, fruits, seeds, roots, bark, or wood, and are stored in secretory cells, cavities, canals, or epidermal cells. Essential oils can be also produced and stored in trichomes and are involved in plant-animal, plant-microorganism, and plant-plant interactions that aid plant maintenance, survival, and adaptation to environmental conditions (Castro et al. 2008). Because of the mode of extraction, mostly by distillation from aromatic plants, they contain a variety of volatile molecules such as terpenes and terpenoids, phenol-derived aromatic components, and aliphatic components (Bakkali et al. 2008).

In general, terpenoids are a predominant constituent of plant essential oils, but many of these oils are also composed of other chemicals like phenylpropanoids. The terpenoids are a large and diverse class, derived from five-carbon isoprene units assembled and modified in thousands of ways. In fact, almost all essential oils are extremely complex in composition, with respect to the presence of a large variety of highly functionalized chemical entities, belonging to different chemical classes (monoterpenoids, sesquiterpenoids, phenylpropanoids, etc.). Essential oil productivity is ecophysiologically and environmentally friendly (Sangwan et al. 2001).

The cultivation of essential oils containing species from different families has been initiated in a number of countries to respond to food and cosmetic industry needs. Developing ways to manipulate the concentration and yield of some essential oils under different stresses is important nowadays. In fact, soil salinity may favor a directional production of particular components of interest and can be used for studying their possible role under salt stress for some particular plants.

One of the most important characteristics of oil accumulation is its dependence on the developmental stage/phase of the plant per se as well as its concerned part/organ, tissue, and cells (Sangwan et al. 2001). Comparing the performance of two *Cymbopogons C. winterianus* and *C. flexuosus* under increasing salinity and sodic conditions, it has been estimated that increasing salinity and sodicity stresses caused a reduction, both in shoots and root yield of citronella, lemongrass, and vetiver (Kumar and Gill 1995). Ansari et al. (1998) has recorded the decline in oil yield in *C. winterianus* and least in *C. flexuosus* too. Compositionally, proportions of citral and geraniol increased under salt stress, respectively, in the oils of lemongrass and palmarosa. A salt-tolerant line of palmarosa, developed by an in vitro procedure (Patnaik and Debata 1997), has a regeneration potential, even under high salt concentrations (up to 200 mM).

Irrigation with a solution containing 100 mM NaCl for 4 weeks increased considerably 1.8-cineole, camphor, and β-thujone concentrations in common sage, whereas lower concentrations had no effects (Tounekti and Khemira 2015). Within the essential oil, the relative level for various constituents increased, decreased, or did not change in peppermint (*Mentha* x *piperita* L.), pennyroyal (*Mentha pulegium* L.), and apple mint (*Mentha suaveolens Ehrh.*) under salt stress as compared with nonstressed control plants. Reduction of the essential oil yields and induced marked quantitative changes in the chemical composition of the essential oils in safflower (*Carthamus tinctorius* L.) have been observed (Harrathi et al. 2012). Salt stress did not affect the yield or the concentrations of the constituents of the essential oil of *Lippia gracilis* Schauer, carvacrol and thymol showing the highest concentrations in all treatments (Ragagnin et al. 2014). In experimental plants, oil concentration in the plant tissue under salt stress increased as compared with untreated controls, suggesting that oil synthesis and/or oil degradation processes were less sensitive to salt stress than similar processes in plants with decreased oil concentration content (Aziz et al. 2008).

2.5 Tannins Under Salt Stress

Tannins is a term widely used to refer to any large polyphenolic compound containing sufficient hydroxyls and other suitable groups (such as carboxyls) to form strong complexes with various macromolecules. The tannin compounds are widely distributed in many species of plants, where they play a role in protection from such biotic stress as predation and in plant growth regulation (Ferrell and Thorington 2006). Among presented in Table 2.1, halophytic fodder crops tannins were found and present in bearberry (*Arctostaphylos uva-ursi*), red oak (*Quercus rubra*), white oak (*Quercus alba*), potentilla (*Potentilla fruticosa*), larch (*Larix* spp.), daylily (*Hemerocallis* spp.), and sea lavender (*Limonium latifolium*).

Condensed tannins are polymeric flavanols formed by condensation of monomeric units such as flavan-3-ols and flavan-3-4-diols (Foo et al. 1996). When condensed tannins get depolymerized, they produce mainly cyanidin or delphinidin and therefore have been further classified as procyanidins or prodelphinidins (Bruneton 1999). Only a low degree of absorption of condensed tannins by the digestive tract of herbivores has been reported. One of their most important chemical properties is the ability to form soluble and insoluble complexes with macromolecules, such as protein, fiber, and starch. Condensed tannins have been proposed to play a role in the interactions between plants and microorganisms, either pathogenic or mutualistic, as well as in plant responses to abiotic stresses (Gebrehiwot et al. 2002; Reinoso et al. 2004; Paolocci et al. 2005). Lotus species exhibit significant variations in shoot and root condensed tannins (Escaray et al. 2007).

Leaf anatomical changes like plant height, fresh weight, and leaf area were found to have negative correlation with salinities. With increment of salinity, the thickness of upper and lower cuticle and upper and lower epidermis and the densities of tannin cell increased, but the thickness of upper and lower endodermis and the intercellular spaces of spongy tissue of leaves decreased (Liu et al.2009).

The salt-sensitive and salt-tolerant poplar species reveals evolutionary adaption of stress tolerance mechanisms. An increased respiration, greater tannin and soluble phenolic contents, as well as, higher glucose and fructose level in the two different phylogenetically poplar species has been observed under salt stress. Increasing tannin content is connected to pathways of phenylpropanoid and flavonoid biosynthesis which got higher transcript abundances of enzymes (Janz et al. 2010).

2.6 Manipulations/Genetic Engineering of Secondary Metabolites for Conferring Salinity Tolerance

Nowadays mostly studied salt-tolerant genes encode a putative Na+/H+ antiporter (Shi et al., 2000) (Table 2.2).

Table 2.2 Genetic engineering of secondary metabolism for conferring salinity tolerance

Species	Gene name	Gene functions	References
Oryza sativa	OsHsp17.0, OsHsp23.7	Heat-shock proteins, molecular chaperones, and folding, assembling, and transporting proteins	Zou et al. (2012)
Artemisia annua	CYP71AV1	Increase of artemisinin production	Sheludko (2010)
Tobacco (*Nicotiana tabacum* L.)	Constitutively silenced for homogentisate phytyltransferase (HPT) and γ-tocopherol methyltransferase (γ-TMT) activity	98% reduction of total tocopherol accumulation in transgenic line and hence decreased degree of salt tolerance	Abbasi et al. (2007)
Common buckwheat (*Fagopyrum esculentum*)	*AtNHX1*, a vacuolar Na^+/H^+ antiporter gene	Increase of rutin accumulation	Chen et al. (2008)
Arabidopsis thaliana	MYB112 transcription factor	Increase of anthocyanin accumulation	Lotkowska et al. (2015)
Tobacco (*Nicotiana tabacum* L.)	NtMYC2a-recognized G-box motifs	Increase of nicotine content	Chen et al. (2016)
Arabidopsis thaliana	Expression of genes *F3H*, *F3′H*, and *LDOX*	Increase of anthocyanin accumulation	Van Oosten et al. (2013)
Arabidopsis thaliana	*VvbHLH1* gene from grape	Increase of anthocyanin accumulation	Wang et al. (2016)
Lotus japonicus	Gene of plastidic glutamine synthetase (GS_2) deficiency	Changes in phenolic metabolism	García-Calderón et al. (2015)
Aeluropus littoralis (Willd) Parl	Cell wall CWPRX gene expression	Increase of cell wall phenolic acids (ferulic acid, *p*-coumaric acid, and sinapic acid)	Haghighi et al. (2014)
Sacharum sp.	Expression patterns of miRNAs 156, 159, and 166	Overexpression of SPL5, GA-Myb, and Glass III HD-Zip protein 4 which enhanced salt tolerance	Shriram et al. (2016)
Rauvolfia serpentina *Arabidopsis thaliana*	miRNAs	Activation of flavonoid biosynthetic	Gupta et al. 2017
Diospyros kaki			

2.7 Conclusions

In the presented chapter, some literature data and analysis about possible roles of plant secondary metabolites under conditions of salt stress are described. Participation of primary metabolites in osmotic adjustment which can induce development of oxidative stress with increasing ROS is also known. Secondary metabolites as antioxidants are able to participate in balancing oxidative status of plant organism. Phenolic compounds, tannins, essential oils, glucosinolates, and lignins are present in halophytes and their possible role in salt tolerance has been discussed. Secondary metabolites also have significant practical applications in medicine, serve as nutritive unique sources for food additives and cosmetics, and has importance in plant stress physiology for adaptation.

Acknowledgment This work was supported by the research project of the Scientific Grant Agency of Slovak Republic VEGA- 613 1-0923-16, VEGA-1-0831-17, and APVV-15-0721.

References

Abbasi A, Hajirezaei M, Hofius D, Sonnewald U, Voll LM (2007) Specific roles of α- and γ-tocopherol in abiotic stress responses of transgenic tobacco. Plant Physiol 143:1720–1738

Abdallah SB, Rabhi M, Harbaoui F et al (2013) Distribution of phenolic compounds and antioxidant activity between young and old leaves of *Carthamus tinctorius* L. and their induction by salt stress. Acta Physiol Plant 35:1161

Abdell C, Barhoumi Z, Ghnaya T, Debez A, Hamed KB, Ksouri R, Talbi O, Zribi F, Ouerghi Z, Smaoui A (2006) Potential utilisation of halophytes for the rehabilitation and valorisation of salt-affected areas in Tunisia. In: Öztürk M, Waisel Y, Khan MA, Görk G (eds) Biosaline agriculture and salinity tolerance in plants. Springer, London, pp 163–172

Abideen Z, Qasim M, Rasheed A, Yousuf AM, Gul B, Khan M (2015) A. Antioxidant activity and polyphenolic content of *Phragmites karka* under saline conditions. Pak J Bot 47(3):813–818

Adams PR, Kendall E, Kartha KK (1990) Comparison of free sugars in growing and desiccated plants of Selaginella lepidophylla. Biochem Syst Ecol 18:107–110

Ali RM, Elfeky SS, Abbas H (2008) Response of salt stressed *Ricinus communis* L. to exogenous application of glycerol and/or aspartic acid. J Biol Sci 8(1):171–175

Amirjani MR (2013) Effects of drought stress on the alkaloid contents and growth parameters of *Catharanthus roseus*. J Agric Biol Sci 8(11):745–750

Andersson D, Chakrabarty R, Bejai S, Zhang J, Rask L, Meijer J (2009) Myrosinases from root and leaves of *Arabidopsis thaliana* have different catalytic properties. Phytochemistry 70:1345–1354

Ansari SR, Frooqi AHA, Sharma S (1998) Interspecific variation in sodium and potassium ion accumulation and essential oil metabolism in three Cymbopogon species raised under sodium chloride stress. J Essent Oil Res 10:413–418

Armitage AM (2000) Armitages garden perennials. Timber Press. ISBN-10: 0881924350

Ashraf MA, Ashraf M, Ali Q (2010) Response of two genetically diverse wheat cultivars to salt stress at different growth stages: leaf lipid peroxidation and phenolic contents. Pak J Bot. 2010 42(1):559–565

Aziz EE, Hussein A-A, Lyle EC (2008) Influence of salt stress on growth and essential oil production in peppermint, pennyroyal, and apple mint. J Herbs Spices Med Plants 14(1–2):77–87

Bakkali F, Averbeck S, Averbeck D, Idaomar M (2008) Biological effects of essential oils – a review. Food Chem Toxicol 46(2):446–475

Bhattacharya A, Sood P, Citovsky V (2010) The roles of plant phenolics in defence and communication during *Agrobacterium* and *Rhizobium* infection. Mol Plant Pathol 11(5):705–719

Bohnert HJ, Nelson DE, Jensen RG (1995) Adaptations to environmental stresses. Plant Cell 7:1099–1111

Briens M, Larher F (1982) Osmoregulation in halophytic higher plants: a comparative study of soluble carbohydrates, polyols, betaines and free proline. Plant Cell Environ 5:287–292

Bruneton J (1999) Tanins. In: Pharmacognosie, phytochimie, plantes médicinales, 3rd edn. Lavoisier, Paris, France, pp 370–404

Cachorro P, Ortiz A, Barcelö AR, Cerdä A (1993) Lignin deposition in vascular tissues of *Phaseolus vulgaris* roots in response to salt stress and Ca^{2+} ions. Phyton (Horn, Austria) 33:33–40

Cai Y, Luo Q, Sun M, Corke H (2004) Antioxidant activity and phenolic compounds of 112 traditional Chinese medicinal plants associated with anticancer. Life Sci 74(17):2157–2184

Castro NEA, Carvalho GJ, Cardoso MG, Pimentel FA, Correa RM, Guimarães LGL (2008) Avaliação de rendimento e dos constituintes químicos do óleo essencial de folhas de *Eucalyptus citriodora* Hook. Colhidas em diferentes épocas do ano em municípios de Minas Gerais. Revista Brasileira de Plantas Medicinais 10(1):70–75

Chang HL, Jiny L, Ki SK (2006) Changes of phenolic compounds and abscisic acid in *Liriope spicata* seeds according to cold stratification and seed harvesting date and their relationships to germination. Hortic Environ Biotechnol 47(1):34–40

Chang C-Y, Chiang JCH, Wehner MF, Friedman A, Ruedy R (2011) Sulfate aerosol control of tropical Atlantic climate over the 20th century. J Clim 24:2540–2555. https://doi.org/10.1175/2010JCLI4065.1

Chattopadhyay S, Marques JT, Yamashita M, Peters KL, Smith K, Desai A, Williams BR, Sen GC (2010) Viral apoptosis is induced by IRF-3-mediated activation of Bax. EMBO J 29(10):1762–1773. https://doi.org/10.1038/emboj.2010.50

Chandna R, Azooz MM, Ahmad P (2013) Recent advances of metabolomics to reveal plant response during salt stress. In: Ahmad P, Azooz MM, Prasad MNV (eds) Salt stress in plants: signalling, omics and adaptations. Springer, New York, pp 1–14

Chen L-H, Zhang B, Xu Z-Q (2008) Salt tolerance conferred by overexpression of *Arabidopsis* vacuolar Na^{+}/H^{+} antiporter gene *AtNHX1* in common buckwheat (*Fagopyrum esculentum*). Transgenic Res 17:121

Chen X, Zhang X, Jia A, Xu G, Hu H, Hu X, Hu L (2016) Jasmonate mediates salt-induced nicotine biosynthesis in tobacco (*Nicotiana tabacum* L.). Plant Divers 38:118–123

Clifford MN, Wu W, Kuhnert N (2006) The chlorogenic acids of *Hemerocallis*. Food Chem 95(4):574–578

Cramer GR, Ergul A, Grimplet J, Tillett RL, Tattersall EAR, Bohlman MC, Vincent D, Sonderegger J, Evans J, Osborne C, Quilici D, Schlauch KA, Schooley DA, Cushman JC (2007) Water and salinity stress in grapevines: early and late changes in transcript and metabolite profiles. Funct Integr Genomics 7:111–134

Daneshmand F, Arvin MJ, Kalantari KM (2010) Physiological responses to NaCl stress in three wild species of potato in vitro. Acta Physiol Plant 32:91–101

Dardanelli MS, Fernandez de Cordoba FJ, Espuny MR, Rodrıguez Carvajal MA, Mes D, Antonio M, Serrano G, Yaacov O, Manuel M (2008) Effect of *Azospirillum brasilense* coinoculated with *Rhizobium* on *Phaseolus vulgaris* flavonoids and Nod factor production under salt stress. Soil Biol Biochem 40:2713–2721

Darke R (1999) The color encyclopedia of ornamental grasses. Timber Press, Portland. isbn: 10: 0881924644

Dehghan N, de Mestral C, McKee MD, Schemitsch EH, Nathens A (2014) Flail chest injuries: a review of outcomes and treatment practices from the National Trauma Data Bank. J Trauma Acute Care Surg 76(2):462–468. https://doi.org/10.1097/TA.0000000000000086

Dirr MA (1997) Dirrs hardy trees. Timber Press, Portland. isbn: 0881924040

Dyduch-Siemińska M, Najda A, Dyduch J, Gantner M, Klimek K (2015) The content of secondary metabolites and antioxidant activity of wild strawberry fruit (*Fragaria vesca* L.). J Anal Methods Chem 2015:8. Article ID 831238

Ehsen S, Qasim M, Abideen Z, Rizvi RF, Gul B, Ansari R, Ajmal Khan M (2016) Secondary metabolites as anti-nutritional factors in locally used halophytic forage/fodder. Pak J Bot 48(2):629–636

Eichholz I, Huyskens-Keil S, Keller A, Ulrich D, Kroh LW, Rohn S (2011) UV-B-induced changes of volatile metabolites and phenolic compounds in blueberries (*Vaccinium corymbosum* L.). Food Chem 126(1):60–64

Escaray F, Pesqueira J, Damiani F, Paolocci F, Pedro CS, Ruiz OA (2007) Condensed tannins in *Lotus* species under salt stress. Lotus Newsl 37(2):81–83

Falleh H, Ksouri R, Chaieb K, Karray-Bouraoui N, Trabelsi N, Boulaaba M, Abdelly C (2008) Phenolic composition of *Cynara cardunculus* L. organs, and their biological activities. C R Biologies 331:372–379

Falleh H, Msilini N, Oueslati S, Ksouri R, Magné C, Lachaâl M (2013) Diplotaxis harra and diplotaxis simplex organs: assessment of phenolics and biological activities before and after fractionation. Industr Crops Prod 45:141–147

Faten M, Hanen F, Riadh K, Chedly A (2014) Total phenolic, flavonoid and tannin contents and antioxidant andantimicrobial activities of organic extracts of shoots of the plant *Limonium delicatulum*. J Taibah Univ Sci 8:216–224

Ferrell KE, Thorington RW (2006) Squirrels: the animal answer guide. Johns Hopkins University Press, Baltimore, p 91

Fischer NH, Williamson GB, Weidenhamer JD et al (1994) In search of allelopathy in the Florida scrub: the role of terpenoids. J Chem Ecol 20:1355

Flowers TJ, Colmer TD (2008) Salinity tolerance in halophytes. New Phytol 179:945–963

Flowers TJ, Munns R, Colmer TD (2015) Sodium chloride toxicity and the cellular basis of salt tolerance in halophytes. Ann Bot 115:419–431

Foo LY, Newman R, Waghorn G, Mcnabb WC, Ulyatt MJ (1996) Proanthocyanidins from Lotus corniculatus. Phytochemistry 41:617–624

Forkner RE, Marquis RJ, Lill JT (2004) Feeny revisited: condensed tannins as anti-herbivore defences in leaf-chewing herbivore communities of *Quercus*. Ecol Entomol 29:174–187

Fu XZ, Ullah Khan E, Hu SS, Fan QJ, Liu JH (2011) Overexpression of the betaine aldehyde dehydrogenase gene from Atriplex hortensis enhances salt tolerance in the transgenic trifoliate orange (Poncirus trifoliata L. Raf.). Environ Exp Bot 74:106–113

Gagneul D, Ainouche A, Duhaze C, Lugan R, Lahrer FR, Bouchereau A (2007) A reassessment of the function of the so-called compatible solutes in the halophytic plumbaginaceae *Limonium latifolium*. Plant Physiol 144:1598–1611

Galletti S, Barillari J, Iori R, Venturi G (2006) Glucobrassicin enhancement in woad (*Isatis tinctoria*) leaves by chemical and physical treatments. J Sci Food Agric 86:1833–1838

García-Calderón M, Pons-Ferrer T, Mrázova A et al (2015) Modulation of phenolic metabolism under stress conditions in a *Lotus japonicus* mutant lacking plastidic glutamine synthetase. Front Plant Sci 6:760

Gebrehiwot L, Beuselinck PR, Roberts CA (2002) Seasonal variations in condensed tannin concentration of three Lotus species. Agron J 94:1059–1065

Gerhold HD et al (2001) Landscape tree factsheets. Penn State University, State College

Gerson EA, Kelsey RG, St Clair JB (2009) Genetic variation of piperidine alkaloids in Pinus ponderosa: a common garden study. Ann Bot 103:447–457

Ghosh N, Adak MK, Ghosh PD et al (2011) Differential responses of two rice varieties to salt stress. Plant Biotechnol Rep 5:89

Gourguillon L, Lobstein A, Gondet L (2016) Effects of explant type, culture media and growth regulators for callus induction of a potential bioactive halophyte: *Armeria maritima* (*Plumbaginaceae*). Planta Med 81(S 01):S1–S381

Gupta OP, Karkute SG, Banerjee S, Meena NL, Anil D (2017) Contemporary understanding of miRNA-based regulation of secondary metabolites biosynthesis in plants. Front Plant Sci 8:374

Haghighi L, Majd A, Zadeh GN, Shokri M, Kelij S, Irian S (2014) Salt-induced changes in cell wall peroxidase (CWPRX) and phenolic content of *Aeluropus littoralis* (Willd) Parl. Aust J Crop Sci 8(2):296–300

Hajdari A, Behxhet M, Gresa A, Bledar P, Brigitte L, Alban I, Gjoshe S, Quave CL, Novak J (2015) Essential oil composition variability among natural populations of *Pinus mugo* Turra in Kosovo. Springerplus 4:828

Hajlaoui H, Denden M, El Ayeb N (2009) Differential responses of two maize (*Zea mays* L.) varieties to salt stress: changes on polyphenols composition of foliage and oxidative damages. Ind Crop Prod 30(1):144–151

Halkier BA, Gershenzon J (2006) Biology and biochemistry of glucosinolates. Annu Rev Plant Biol 57:303–333

Hamed KB, Ellouzi H, Talbi OZ, Hessini K, Slama I, Ghnaya T, Bosch SM, Savouré A, Abdelly C (2013) Physiological response of halophytes to multiple stresses. Funct Plant Biol 40:883–896

Hanny BW, Hedin PA (1972) Phytochemical studies in the family *Malvaceae*. II. Analysis of some chemical constituents of four *Hibiscus* species. Diss Abstr Int B 33:1424a

Hao G, Du X, Zhao F et al (2010) Fungal endophytes-induced abscisic acid is required for flavonoid accumulation in suspension cells of *Ginkgo biloba*. Biotechnol Lett 32:305

Hare PD, Cress WA, Staden JV (1998) Dissecting the roles of osmolyte accumulation during stress. Plant Cell Environ 21:535–553. https://doi.org/10.1046/j.1365-3040.1998.00309.x

Harrathi J, Hosni K, Karray-Bouraoui N, Attia H, Marzouk B, Magné C, Lachaâl M (2012) Effect of salt stress on growth, fatty acids and essential oils in safflower (*Carthamus tinctorius* L.). Acta Physiol Plant 34:129

Hasegawa M, Bressan R, Pardo JM (2000) The dawn of plant salt tolerance genetics. Trends Plant Sci 5:317–319

Hashidoko Y (1996) The phytochemistry of *Rosa rugosa*. Phytochemistry 43(3):535–549

Haslam E (1996) Natural polyphenols (vegetable tannins) as drugs: possible modes of action. J Nat Prod 59(2):205–215

He X, Huang W, Chen W, Dong T, Liu C, Chen Z, Xu S, Ruan Y (2009) Changes of main secondary metabolites in leaves of *Ginkgo biloba* in response to ozone fumigation. J Environ Sci (China) 21(2):199–203

Herve du Penhoat CLM, Michon VMF, Peng S, Viriot C, Scalbert A, Gage D, Roburins A-E (1991) Structural elucidation of new dimeric ellagitannins from *Quercus robur* L. J Chem Soc Perkin Trans 7:1653–1660

Hierro JL, Callaway RM (2003) Allelopathy and exotic plant invasion. Plant Soil 256:29

Janz D, Behnke K, Schnitzler JP, Kanawati B, Schmitt-Kopplin P, Polle A (2010) Pathway analysis of the transcriptome and metabolome of salt sensitive and tolerant poplar species reveals evolutionary adaption of stress tolerance mechanisms. BMC Plant Biol 10:150

Janz D, Lautner S, Wildhagen H, Behnke K, Schnitzler JP, Rennenberg H, Fromm J, Polle A (2012) Salt stress induces the formation of a novel type of 'pressure wood' in two *Populus* species. New Phytol 194(1):129–141

Jdey A, Falleh H, Jannet SB, Hammi KM, Dauvergne X, Magné C, Ksouri R (2017) Anti-aging activities of extracts from Tunisian medicinal halophytes and their aromatic constituents. EXCLI J 16:755

Jefferies RL, Rudmik T (1984) The responses of halophytes to salinity: an ecological perspective. In: Staples RC, Toenniessen GH (eds) Salinity tolerance in plants: strategies for crop improvement. Wiley, New York, pp 213–227

Kapusta I, Bogdan J, Barbara S, Stochmal A, Piacente S, Pizza C, Franceschi F, Chlodwig F, Wieslaw O (2007) Flavonoids in horse chestnut (*Aesculus hippocastanum*) seeds and powdered waste water byproducts. J Agric Food Chem 55(21):8485–8490

Karahara I, Ikeda A, Kondo T, Uetake Y (2004) Development of the casparian strip in primary roots of maize under salt stress. Planta 219:41

Karker M, De Tommasi N, Smaoui A, Abdelly C, Ksouri R, Braca A (2016) New sulphated flavonoids from *Tamarix africana* and biological activities of its polar extract. Planta Med 82:1374–1380

Katschnig D, Broekman R, Rozema J (2013) Salt tolerance in the halophyte *Salicornia dolichostachya* Moss: growth, morphology and physiology. Environ Exp Bot 92:32–42

Kelij S, Majd A, Nematzade G, Jounobi P (2015) Activation of lignin biosynthetic enzymes during internodal development of *Aeluropus littoralis* exposed to NaCl. J Genet Resour 1(1):19–24

Keutgen AJ, Pawelzik E (2008) Quality and nutritional value of strawberry fruit under long term salt stress. Food Chem 107(4):1413–1420

Kondo T, Yoshida K, Nakagawa A, Kawai T, Tamura H, Goto T (1992) Commelinin, a highly associated metalloanthocyanin present in the blue flower petals of *Commelina communis*. Nature 358:515–517

Ksouri R, Megdiche W, Debez A, Falleh H, Grignon C, Abdelly C (2007) Salinity effects on polyphenol content and antioxidant activities in leaves of the halophyte *Cakile maritima*. Plant Physiol Biochem 45(3–4):244–249

Ksouri R, Falleh H, Megdiche W, Trabelsi N, Mhamdi B, Chaieb K, Bakrouf A, Magné C, Abdelly C (2009) Antioxidant and antimicrobial activities of the edible medicinal halophyte *Tamarix gallica* L. and related polyphenolic constituents. Food Chem Toxicol 47:2083–2091

Ksouri R, Ksouri WM, Jallali I, Debez A, Magné C, Hiroko I, Abdelly C (2012) Review Medicinal halophytes: potent source of health promoting biomolecules with medical, nutraceutical and food applications. Crit Rev Biotechnol 32(4):289–326

Küçükboyaci N, Özkan S, Adigüzel N, Tosun F (2011) Characterisation and antimicrobial activity of *Sophora alopecuroides* L. var. alopecuroides alkaloid extracts. Turk J Biol 35:379–385

Kumar AA, Gill KS (1995) Performance of aromatic grasses under saline and sodic stress condition. Salt tolerance of aromatic grasses. Indian Perfumer 39:39–44

Kusari S, Zühlke S, Spiteller M (2011) Chemometric evaluation of the anti-cancer pro-drug podophyllotoxin and potential therapeutic analogues in *Juniperusand podophyllum* species. Phytochem Anal 22:128–143

Lim JH, Park KJ, Kim BK, Jeong JW, Kim HJ (2012) Effect of salinity stress on phenolic compounds and carotenoids in buckwheat (*Fagopyrum esculentum* M.) sprout. Food Chem 135(3):1065–1070

Liu R, Sun W, Chao MX, Ji CJ, Wang M, Ye BP (2009) Leaf anatomical changes of *Bruguiera gymnorrhizaseedlings* under salt stress. J Trop Subtropical Bot 17(2):169–175

Lokhande VH, Nikam TD, Suprasanna P (2010) Biochemical, physiological and growth changes in response to salinity in callus cultures of Sesuvium portulacastrum L. Plant Cell Tissue Organ Cult 102:17–25

López-Berenguer C, Martínez-Ballesta CM, García-Viguera C, Carvajal M (2008) Leaf water balance mediated by aquaporins under salt stress and associated glucosinolate synthesis in broccoli. Plant Sci 174(3):321–328

López-Berenguer C, Martínez-Ballesta MC, Moreno DA, Carvajal M, García-Viguera C (2009) Growing hardier crops for better health: salinity tolerance and the nutritional value of broccoli. J Agric Food Chem 57:572–578

López-Pérez L, Martínez Ballesta MC, Maurel C, Carvajal M (2009) Changes in plasma membrane composition of broccoli roots as an adaptation to increase water transport under salinity. Phytochemistry 70:492–500

Lotkowska ME, Tohge T, Fernie AR, Xue G-P, Balazadeh S, Mueller-Roeber B (2015) The *Arabidopsis* transcription factor MYB112 promotes anthocyanin formation during salinity and under high light stress. Plant Physiol 169(3):1862–1880

Maffi L, Benvenuti S, Fornasiero RB, Bianchi A, Melegari M (2001) Inter-population variability of secondary metabolites in *Hypericum* spp. (*Hypericaceae*) of the northern apennines, Italy. Nord J Bot 21:585–593

Maganha EG, da Costa Halmenschlager R, Moreira Rosa R, Henriques JAP, de Paula Ramos ALL, Saffi J (2010) Pharmacological evidences for the extracts and secondary metabolites from plants of the genus *Hibiscus*. Food Chem 118(1):1–10

Mahajan S, Tuteja N (2005) Cold, salinity and drought stresses: an overview. Arch Biochem Biophys 444(2):139–158

Majumdar S, Ghosh S, Glick BR, Dumbroff EB (1991) Activities of chlorophyllase, phosphoenolpyruvate carboxyllase and ribulose-1,5-bisphosphate carboxylase in the primary leaves of soybean drying senescence and drought. Physiol Plant 81:473–480

Mane AV, Karadge BA, Samant JS (2010) Salt stress induced alteration in photosynthetic pigments and polyphenols of *Pennisetum alopecuroides* (L.). J Ecophysiol Occup Health 10:177–182

Mansour MMF (2000) Nitrogen containing compounds and adaptation of plants to salinity stress. Biol Plant 43:491–500

Marrs KA (1996) The function and regulation of glutathione-S-transferase in plants. Ann Rev Plant Physiol Plant Mol Biol 47:127–158

Mário M, Pavan Marcos A, Ziglio Cláudio O, Franchini Júlio C (2001) Reduction of exchangeable calcium and magnesium in soil with increasing pH. Braz Arch Biol Technol 44(2):149–153. https://doi.org/10.1590/S1516-89132001000200007

Martinez JP, Kinet JM, Bajji M, Lutts S (2005) NaCl alleviates polyethylene glycol-induced water stress in the halophyte species Atriplex halimus L. J Expt Bot 56:2421–2431

Martínez-Ballesta MdC, Moreno DA, Carvajal M (2013) The physiological importance of glucosinolates on plant response to abiotic stress in *Brassica*. Int J Mol Sci 14(6):11607–11625

Matern U (1994) Dianthus species (Carnation): in vitro culture and the biosynthesis of dianthalexin and other secondary metabolites. In: Medicinal and aromatic plants VII. Volume 28 of the series Biotechnology in agriculture and forestry, pp 170–184

Matkowski LA, Świąder K, Ślusarczyk S, Jezierska-Domaradzka A, Oszmiański J (2006) Free radical scavenging activity of extracts obtained from cultivated plants of *Potentilla alba* L. and *Waldsteinia geoides*. Herba Pol 52(4):91–97

Meng Q, Niu Y, Niu X, Roubin RH, Hanrahan JR (2009) Ethnobotany, phytochemistry and pharmacology of the genus Caragana used in traditional Chinese medicine. J Ethnopharmacol 124(3):350–368

Misra N, Gupta AK (2006) Effect of salinity and different nitrogen sources on the activity of antioxidant enzymes and indole alkaloid content in *Catharanthus roseus* seedlings. J Plant Physiol 163(1):11–18

Mssedi D, Sleimi N, Abdelly C (2000) Some plants: the origin of diacylglycerol moiety. Arch. Biophysiological and biochemical aspects of salt tolerchem. Biophys. 240, 851D858. ance of Sesuvium portulacastrum. In: Cash Crop Halophytes: Potentials, Pilot Projects, Basic and Applied Research on Halophytes and Saline Irrigabrane (Lieth H. and Moschenko M., eds.). EU conmembranes action (IC18CT96-0055) symposium, Germany, p. 25.

Munns R (2002) Comparative physiology of salt and water stress. Plant Cell Environ 25(2):239–250

Murakeözy ÉP, Nagy Z, Duhazé C, Bouchereau A, Tuba Z (2003) Seasonal changes in the levels of compatible osmolytes in three halophytic species of inland saline vegetation in Hungary. J Plant Physiol 160:395–401

Muthukumarasamy M, Gupta SD, Pannerselvam R (2000) Enhancement of peroxidase, polyphenol oxidase and superoxide dismutase activities by triadimefon in NaCl stressed *Raphanus sativus* L. Biol Plant 43:317–320

Muthulakshmi Santhi M, Gurulakshmi SG, Rajathi S (2015) Effect of salt stress on physiological and biochemical characteristics in *Solanum nigrum* L. Int J Sci Res 4(3):567–571

Nabeta K (1994) *Larix leptolepis* (Japanese Larch): In vitro culture and the production of secondary metabolites. In: Medicinal and aromatic plants VII. Volume 28 of the series Biotechnology in agriculture and forestry, pp 271–288

Navarro JM, Flores P, Garrido C, Martinez V (2006) Changes in the contents of antioxidant compounds in pepper fruits at ripening stages, as affected by salinity. Food Chem 96:66–73

Nelson DE, Shen B, Bohnert HJ (1998) The regulation of cell-specific inostol metabolism and transport in plant salinity tolerance. In: Setlow JK (ed) Genetic engineering, principles and methods. Plenum Publication, New York, pp 153–176

Neves GYS, Marchiosi R, Ferrarese MLL, Siqueira-Soares RC, Ferrarese-Filho O (2010) Root growth inhibition and lignification induced by salt stress in soybean. J Agron Crop Sci 196(6):467–473

Ngara R, Ndimba R, Borch-Jensen J et al (2012) Identification and profiling of salinity stress-responsive proteins in Sorghum bicolor seedlings. J Proteome 75:4139–4150. https://doi.org/10.1016/j.jprot.2012.05.038

Nguyen NTH, Arima S, Konishi T, Ogawa Y, Adaniya S, Keiji M (2015) Variation of oxypinnatanine concentration in daylily (*Hemerocallis* spp.) influenced by ploidy levels, growth stages, and environmental factors. Trop Agric Dev 59(4):179–189

Niu SS, Xu CJ, Zhang WS, Zhang B, Li X, Lin-Wang K, Ferguson IB, Allan AC, Chen KS (2010) Coordinated regulation of anthocyanin biosynthesis in Chinese bayberry (*Myrica rubra*) fruit by a R2R3 MYB transcription factor. Planta 231:887

Olivoto T, Nardino M, Carvalho IR, Follmann DN, Szareski VJ, Pelegrin AJ, Souza VQ (2017) Plant secondary metabolites and its dynamical systems induction in response to environmental factors. Afr J Agric Res 12(2):71–84. 2017

Ouchikh O, Thouraya C, Riadh K, Mouna Ben T, Hanen F, Chedly A, Mohamed EK, Brahim M (2011) The effects of extraction method on the measured tocopherol level and antioxidant activity of *L. nobilis* vegetative organs. J Food Compos Anal 24:103–110

Palo RT (1984) Distribution of birch (*Betula* SPP.), willow (*Salix* SPP.), and poplar (*Populus* SPP.) secondary metabolites and their potential role as chemical defense against herbivores. J Chem Ecol 10:499

Pang Q, Guo J, Chen S, Chen Y, Zhang L, Fei M, Jin S, Li M, Wang Y, Yan X (2012) Effect of salt treatment on the glucosinolate-myrosinase system in *Thellungiella salsuginea*. Plant Soil 355:363–374

Panich U, Kongtaphan K, Onkoksoong T, Jaemsak K, Phadungrakwittaya R, Thaworn A (2010) Modulation of antioxidant defense by *Alpinia galanga* and *Curcuma aromatica* extracts correlates with their inhibition of UVA-induced melanogenesis. Cell Biol Toxicol 26:103–116

Paniwnyk L, Beaufoy E, Lorimer JP, Mason TJ (2001) The extraction of rutin from flower buds of *Sophora japonica*. Ultrason Sonochem 8(3):299–301

Paolocci F, Bovone T, Tosti N, Arcioni S, Damiani F (2005) Light and an exogenous transcription factor qualitatively and quantitatively affect the biosynthetic pathway of condensed tannins in *Lotus corniculatus* leaves. J Exp Bot 56:1093–1103

Parida AK, Das AB (2005) Salt tolerance and salinity effects on plants: a review. Ecotoxicol Environ Saf 60(3):324–349

Patnaik J, Debata BK (1997) Regeneration of plantlets from NaCl tolerant callus lines of *C. martinii* (Roxb). Wats Plant Sci 128:67–74

Piluzza G, Bullitta S (2011) Correlations between phenolic content and antioxidant properties in twenty-four plant species of traditional ethno veterinary use in the Mediterranean area. Pharm Biol 49(3):240–247

Posmyk MM, Kontek R, Janas KM (2009) Antioxidant enzymes activity and phenolic compounds content in red cabbage seedlings exposed to copper stress. Ecotoxicol Environ Saf 72:596–602

Qasim SE, Jacobs J (2016) Human hippocampal theta oscillations during movement without visual cues. Neuron 89(6):1121–1123. https://doi.org/10.1016/j.neuron.2016.03.003

Ragagnin RCG, Albuquerque CC, Oliveira FFM, Santos RG, Gurgel EP, Diniz JC, Rocha SAS, Viana FA (2014) Effect of salt stress on the growth of *Lippia gracilis* Schauer and on the quality of its essential oil. Acta Bot Brasilica 28(3):346–351

Redovnikovi IR, Gliveti T, Delonga K, Vorkapi-Fura J (2008) Glucosinolates and their potential role in plant. Period Biol 110(4):297–309

Reinoso H, Sosa L, Ramírez L, Luna V (2004) Salt-induced changes in the vegetative anatomy of *Prosopis strombulifera (Leguminosa*e). Can J Bot 82(5):618–628

Sachan N, Rogers DT, Yun KY, Littleton JM, Falcone DL (2010) Reactive oxygen species regulate alkaloid metabolism in undifferentiated *N. tabacum* cells. Plant Cell Rep 29:437

Sadeghi Z, Valizadeh J, Azyzian Shermeh O, Akaberi M (2015) Antioxidant activity and total phenolic content of *Boerhavia elegans* (choisy) grown in Baluchestan, Iran. Avicenna J Phytomedicine 1:1–9

Salah A-I (2015) Chapter 8: Plant secondary metabolites of halophytes and salt tolerant plants. In: El Shaer HM, Squires VR (eds) Halophytic and salt-tolerant feedstuffs, impacts on nutrition, physiology and reproduction of livestock, 1st edn. CRC Press, Boca Raton, pp 127–142

Sánchez-Aguayo I, Rodríguez-Galán JM, García R, Torreblanca J, Pardo JM (2004) Salt stress enhances xylem development and expression of *S*-adenosyl-L-methionine synthase in lignifying tissues of tomato plants. Planta 220:278

Sangwan N, Farooqi A, Shabih F, Sangwan RS (2001) Regulation of essential oil production in plants. Plant Growth Regul 34:3

Santos J, Al-Azzawi M, Aronson J, Flowers TJ (2015) eHALOPH a data base of salt-tolerant plants: helping put halophytes to work. Plant Cell Physiol 57:e10

Sarikamiş G, Çakir A (2017) Influence of salinity on aliphatic and indole glucosinolates in broccoli (*Brassica oleracea* var. *italica*). Appl Ecol Environ Res 15(3):1781–1788

Scalbert A, Haslam E (1987) Polyphenols and chemical defence of the leaves of *Quercus robur*. Phytochemistry 26(12):3191–3195

Selvam R, Jurkevich A, Kang SW, Mikhailova MV, Cornett LE, Kuenzel WJ (2013) Distribution of the vasotocin subtype four receptor (VT4R) in the anterior pituitary gland of the chicken, Gallus gallus, and its possible role in the avian stress response. J Neuroendocrinol 25:56–66. https://doi.org/10.1111/j.1365-2826.2012.02370.x

Shadkami F, Helleur RJ, Cox RM (2007) Profiling secondary metabolites of needles of ozone-fumigated white pine (*Pinus strobus*) clones by thermally assisted hydrolysis/methylation GC/MS. J Chem Ecol 33:1467

Shahri W, Tahir I, Islam ST, Bhat MA (2011) Physiological and biochemical changes associated with flower development and senescence in so far unexplored *Helleborus orientalis* Lam. cv. Olympicus. Physiol Mol Biol Plants 17(1):33–39

Shannon MC, Grieve CM (1999) Tolerance of vegetable crops to salinity. Sci Hortic 78:5–38

Sheludko YV (2010) Recent advances in plant biotechnology and genetic engineering for production of secondary metabolites. Cytol Genet 44(1):52–60

Shi H, Ishitani M, Kim C, Zhu J-K (2000) The *Arabidopsis thaliana* salt tolerance gene SOS1 encodes a putative Na+/H+ antiporter. Proc Natl Acad Sci USA 97(12):6896–6901

Shriram V, Kumar V, Devarumath RM, Khare TS, Wani SH (2016) MicroRNAs as potential targets for abiotic stress tolerance in plants. Front Plant Sci 7:817

Sinclair ARE, Jogia MK, Andersen RJ (1988) Camphor from juvenile white spruce as an antifeedant for snowshoe hares. J Chem Ecol 14:1505

Singh K, Kumar S, Rani A, Gulati A, Ahuja PS (2009) Phenylalanine ammonia-lyase (PAL) and cinnamate 4-hydroxylase (C4H) and catechins (fl avan-3-ols) accumulation in tea. Funct Integr Genomics 9:125–134

Slama I, Messedi D, Ghnaya T, Savoufe A, Abdelly C (2006) Effects of water-deficit on growth and proline metabolism in Sesuvium portulacastrum. Environ Exp Bot 56:231–238

Slama I, Ghnaya T, Savoufe A, Abdelly C (2008) Combined effects of long-term salinity and soil drying on growth, water relations, nutrient status and proline accumulation of Sesuvium portulacastrum. C R Biol 331:442–451

Slama I, Abdelly C, Bouchereau A, Flowers T, Savouré A (2015) Diversity, distribution and roles of osmoprotective compounds accumulated in halophytes under abiotic stress. Ann Bot 115(3):433–447. https://doi.org/10.1093/aob/mcu239

Slama I, M'Rabet R, Ksouri R, Talbi O, Debez A, Abdelly C (2017) Effects of salt treatment on growth, lipid membrane peroxidation, polyphenol content, and antioxidant activities in leaves of Sesuvium portulacastrum L. Arid Land Res Manag 31:1–14

Smelcerovic A, Verma V, Spiteller M, Mudasir AS, Satish CP, Ghulam NQ (2006) Phytochemical analysis and genetic characterization of six Hypericum species from Serbia. Phytochemistry 67(2):171–177

Smirnoff N (1993) The role of active oxygen in the response of plants to water deficit and desiccation. New Phytol 125:27–58

Song X-L, Gao G-Y, Ye L-H (1996) Effects of total saponin of liriope spicata lour on experimental myocardial ischemia. Chin Pharmacol Bull 12:329–332

Sreevidya VS, Srinivasa RC, Rao C, Sullia SB, Ladha JK, Reddy PM (2006) Metabolic engineering of rice with soyabean iso fl avone synthase for promoting nodulation gene expression in rhizobia. J Exp Bot 57:1957–1969

Stewart CD, Jones CD, Setzer WN (2014) Essential oil compositions of *Juniperus virginiana* and *Pinus virginiana*, two important trees in Cherokee traditional medicine. Am J Essent Oils Nat Prod 2(2):17–24

Šutković J, Lerl D, Ragab MGA (2011) In vitro production of solasodine alkaloid in *Solanum nigrum* under salinity stress. J Phytology 3(1):43–49

Sytar O, Zhenzhen C, Marian B, Prasad MNV, Taran N, Smetanska I (2013) Foliar applied nickel on buckwheat (*Fagopyrum esculentum*) induced phenolic compounds as potential antioxidants. Clean (Weinh) 41(11):1129–1137

Sytar O, Bruckova K, Hunkova E, Zivcak M, Kiessoun K, Brestic M (2015) The application of muliplex flourimetric sensor for analysis flavonoids content in the medical herbs family *Asteraceae, Lamiaceae, Rosaceae*. Biol Res. 2015 48(5):1–9. https://doi.org/10.1186/0717-6287-48-5

Sytar O, Hemmerich I, Zivcak M, Rauh C, Brestic M (2016) Comparative analysis of bioactive phenolic compounds composition from 26 medicinal plants. Saudi J Biol Sci. https://doi.org/10.1016/j.sjbs.2016.01.036

Szopa A, Ekiert H, Muszyńska B (2013) Accumulation of hydroxybenzoic acids and other biologically active phenolic acids in shoot and callus cultures of *Aronia melanocarpa* (Michx.) Elliott (black chokeberry). Plant Cell Tissue Organ Cult 113:323

Taiz L, Zeiger E (2010) Plant physiology, 5th edn. Sinauer Associates, Massachusetts

Tingey DT, Wilhour RG, Standley C (1976) The Effect of chronic ozone exposures on the metabolite content of ponderosa pine seedlings. For Sci 22(3, 1):234–241

Tomczyk M, Latté KP (2009) Potentilla – a review of its phytochemical and pharmacological profile. J Ethnopharmacol 122(2):184–204

Tóth G, Barabás C, Tóth A, Kéry Á, Béni S, Boldizsár I, Varga E, Noszál B (2016) Characterization of antioxidant phenolics in *Syringa vulgaris* L. flowers and fruits by HPLC-DAD-ESI-MS. Biomed Chromatogr 30(6):923–932

Tounekti T, Khemira H (2015) NaCl stress-induced changes in the essential oil quality and abietane diterpene yield and composition in common sage. J Intercult Ethnopharmacol 4(3):208–216

Van Eijk M (1939) Analyse der Wirkung des NaCl auf die entwicklung sukkulenze und transpiration bei *Salicornia herbacea*, sowie untersuchungen über den einfluss der salzaufnahme auf die wurzelatmung bei *Aster tripolium*. Rec Trav Bot Neerl 36:559–657

Van Oosten MJ, Sharkhuu A, Batelli G, Bressan Ray A, Maggio A (2013) The *Arabidopsis thaliana* mutant *air1* implicates SOS3 in the regulation of anthocyanins under salt stress. Plant Mol Biol 83(4–5):405–415

Vattem DA, Ghaedian R, Shetty K (2005) Enhancing health benefits of berries through phenolic antioxidant enrichment: focus on cranberry. Asia Pac J Clin Nutr 14(2):120–130

Velioglu YS, Mazza G, Gao L, Oomah BD (1998) Antioxidant activity and total phenolics in selected fruits, vegetables, and grain products. J Agric Food Chem 46(10):4113–4117

Venditti A, Serrilli AM, Vittori S, Papa F, Maggi F, Di Cecco M, Ciaschetti G, Bruno M, Rosselli S, Bianco A (2013) Secondary metabolites from *Pinus mugo* Turra subsp. *Mugo* growing in the Majella National Park (Central Apennines, Italy). Chem Biodivers 10:2091–2100

Visser JM, Sasser CE, Cade BS (2006) The effect of multiple stressors on salt marsh end-of-season biomass. Estuaries Coasts 29(2):328–339

Wang F, Zhu H, Chen D, Li Z, Peng R, Yao Q (2016) A grape bHLH transcription factor gene, *VvbHLH1*, increases the accumulation of flavonoids and enhances salt and drought tolerance in transgenic *Arabidopsis thaliana*. Plant Cell, Tissue Organ Cult (PCTOC) 125(2):387–398

Warren RL, Keeling CI, Yuen MM, Raymond A, Taylor GA, Vandervalk BP, Mohamadi H, Paulino D, Chiu R, Jackman SD, Robertson G, Yang C, Boyle B, Hoffmann M, Weigel D, Nelson DR, Ritland C, Isabel N, Jaquish B, Yanchuk A, Bousquet J, Jones SJ, MacKay J, Birol I, Bohlmann

J (2015) Improved white spruce (*Picea glauca*) genome assemblies and annotation of large gene families of conifer terpenoid and phenolic defense metabolism. Plant J 83(2):189–212
Watanabe K, Mimaki Y, Sakagami H, Sashida Y (2003) Bufadienolide and spirostanol glycosides from the rhizomes of helleborus orientalis. J Nat Prod 66(2):236–241
Xue J-J, Fan C-Q, Dong L, Yang S-P, Yue J-M (2004) Novel antibacterial diterpenoids from *Larix chinensis* Beissn. Chem Biodivers 1:1702
Yan X, Chen S (2007) Regulation of plant glucosinolate metabolism. Planta 226:1343–1352
Yan K, Cui M, Zhao S, Chen X, Tang X (2016) Salinity stress is beneficial to the accumulation of chlorogenic acids in honeysuckle (*Lonicera japonica* Thunb.). Front. Plant Sci 7:1563
Yi G, Lei Z, Zhong-Ji S, Zhu-Xia S, Yi-Qiong Z, Gen-Xuan W (2010) Stomatal clustering, a new marker for environmental perception and adaptation in terrestrial plants. Bot Stud 51:325–336
Yuan G, Wang X, Guo R, Wang Q (2010) Effect of salt stress on phenolic compounds, glucosinolates, myrosinase and antioxidant activity in radish sprouts. Food Chem 121(4):1014–1019
Zhou S-H, Fang Z-X, Lü Y, Chen J-C, Liu D-H, Ye X-Q (2009) Phenolics and antioxidant properties of bayberry (*Myrica rubra* Sieb. et Zucc.) pomace. Food Chem 112(2):394–399
Zou J, Liu C, Liu A, Zou D, Chen X (2012) Overexpression of OsHsp17.0 and OsHsp23.7 enhances drought and salt tolerance in rice. J Plant Physiol 169(6):628–635

Chapter 3
Exploring Halotolerant Rhizomicrobes as a Pool of Potent Genes for Engineering Salt Stress Tolerance in Crops

Neveen B. Talaat

Abstract Soil salinization is a constant threat to crop productivity and ecology worldwide. The conventional approach, breeding salt-tolerant plant cultivars, has often failed to efficiently alleviate this devastating environmental stress factor. In contrast, the use of a diverse array of microorganisms harbored by plants has attracted increasing attention because of the remarkable beneficial effects of them on plants. Among these microorganisms, halophilic and halotolerant rhizomicrobes is one of the most important extremophilic microorganisms. They can be found in saline or hypersaline ecosystems and have developed different adaptations to survive in salty environments. Their proteins and encoding genes are magnificently engineered to function in a milieu containing 2–5 M salt and represent a valuable repository and resource for reconstruction and visualizing processes of habitat selection and adaptive evolution. Indeed, the natural occurrence of these microorganisms in saline soils opens up a possible important role of them in increasing the salt tolerance in crops. They are capable of eliciting physical, chemical, and molecular changes in plants which enhanced their tolerance and promoted their growth, and thus they can refine agricultural practices and production under saline conditions. Likewise, their ability to serve as bioinoculants could be a more ready utilizable and sustainable solution to ameliorate the deleterious salt effects on plants. However, the ecology of their interactions with plants is still under investigation and not fully understood. This chapter aims to introduce the halotolerant rhizomicrobes and shed light on their special mechanisms to adapt to salinity conditions. A special section would be dedicated for their potential to be exploited in engineering salt tolerance in crops.

Keywords Halotolerant rhizomicrobes · Plant halotolerant-microbe interaction · Plant salt tolerance · Salt stress

---

N. B. Talaat (✉)
Department of Plant Physiology, Faculty of Agriculture, Cairo University, Giza, Egypt

V. Kumar et al. (eds.), *Salinity Responses and Tolerance in Plants, Volume 2*,
https://doi.org/10.1007/978-3-319-90318-7_3

Abbreviations

ACC	1-Aminocyclopropane-1-carboxylic acid
AMF	Arbuscular mycorrhizal fungi
APX	Ascorbate peroxidase
CAT	Catalase
DHAR	Dehydroascorbate reductase
EPS	Exopolysaccharides
GPX	Guaiacol peroxidase
GR	Glutathione reductase
IAA	Indole acetic acid
LCO	Lipo-chitooligosaccharide
MDA	Malondialdehyde
MDHAR	Monodehydroascorbate reductase
*mtl*D	Mannitol 1-phosphate dehydrogenase
PGDH	3-Phosphoglycerate dehydrogenase
PGPR	Plant growth-promoting rhizobacteria
QACs	Quaternary ammonium compounds
ROS	Reactive oxygen species
SHMT	Serine hydroxymethyltransferase
SOD	Superoxide dismutase
SOS	Salt overly sensitive
VOCs	Volatile organic compounds
VSP2	Vegetative storage protein 2

3.1 Introduction

Soil salinization is one of the most widespread soil degradation processes on the earth endangering the potential use of soils. It is a major environmental stress negatively affecting plant growth and severely limiting the agricultural productivity worldwide. Globally, it is estimated that 20% of irrigated land and almost 5% of the world's cultivable land are salt-affected (Selvakumar et al. 2014). Exposure of plants to excess salt induces osmotic, ionic, and oxidative stresses as well as nutritional imbalances. Osmotic stress is induced by limited water adsorption from soil, while ionic stress resulted from high concentration of Na^+ and Cl^- inside the plant cells (Talaat and Shawky 2011, 2015). Oxidative stress is usually caused by generation of reactive oxygen species (ROS) and hydroxyl radicals, which are detrimental to plant survival under salt stress (Talaat 2014, 2015a). Reduced physiological activity is due to nutritional imbalance under salt stress, which consequently suppresses the growth and development of plants (Talaat 2015b; Talaat et al. 2015a). Furthermore, plants evolved an array of genetic and epigenetic regulatory systems to respond to salinity stress (Vannier et al. 2015). Indeed, the integration of genomics approach with traditional breeding is useful for improving crop salinity tolerance. Transgenic plants overexpressing a wide range of genes adapt well to high saline condition (Roy et al. 2014). However, this approach is time-consuming, expensive, and labor-intensive and frequently results in unstable mutants due to the simultaneous manipulation of numerous genes involved in abiotic stress responses. It is also still uncertain whether transgenic crops will become generally publicly acceptable (Jewell et al. 2010). Moreover, molecular techniques are not widely applicable to important *Brassica* and *Triticum* species that are tetraploid or hexaploid (Kumar et al. 2015). Finally, transgenic salt-tolerant crops, especially those whose transcription factors have been genetically modified, sometimes suffer yield penalties (Roy et al. 2014). Thus, care should be taken to maintain yield stability in transgenic crops. To bypass these limitations, it is necessary to investigate additional alternative strategy. Indeed, it is worthwhile to explore the question of how to mitigate the adverse effects of salt stress, enhance plant salt tolerance, and eventually increase crop yields in high-salinity soils by using sustainable and eco-friendly biological solution.

Microbes are a key component of all ecosystems on earth, playing major roles in the biogeochemical cycles. Compatible plant-microbe interactions favor plant growth and development and help plants deal with different environmental challenges (Talaat and Shawky 2015; Subramanian et al. 2016). Although many microbes are hard to grow under high salt conditions, some halotolerant or halophilic rhizomicrobes are identified (Liu et al. 2016; Subramanian et al. 2016). Halophiles, as a member of extremophiles, are salt-loving organisms that can live in hypersaline environments such as hypersaline soils, salt lakes, marine sediments, salted food, and brines. They adapt with their saline surroundings by their capability to balance the osmotic pressure of the environment, via either producing compatible organic solutes or accumulating large salt concentrations in their cytoplasm. Their low nutritional requirements and resistance to high concentrations of salt make them a potent candidate in a wide range of biotechnological applications (Edbeib et al. 2016). Halotolerant rhizomicrobes not only tolerate salt-stress conditions but also have the ability to promote plant growth and development under this harsh environment (Essghaier et al. 2014; Singh and Jha 2016). Using them as bioinoculants could be a more readily utilizable and sustainable way to mitigate the detrimental salt effects on crops (Bharti et al. 2016). Thus, there is high potential for bioremediation applications for salt-affected soils using halophilic microbes. This biotechnology method has proven to be more efficient than plant breeding and genetic modification approaches.

Application of bioinoculants such as plant growth-promoting rhizobacteria (PGPR), arbuscular mycorrhizal fungi (AMF), and *Rhizobium* has been known as an efficient, eco-friendly, and cost-effective approach for ameliorating saline soils and increasing plant growth via several mechanisms (Munns and Gilliham 2015). PGPR are a group of microorganisms that can be found in the rhizoplane and rhizosphere, phyllosphere, or inside of plant tissues as endophytes. Halotolerant PGPR provides plants with their activities to challenge salinity stress. They impart benefit to the growth of plants under high salinity by using different mechanisms such as synthesis of phytohormones (auxins, cytokinins, gibberellins, and abscisic acid);

production of essential enzymes, 1-aminocyclopropane-1-carboxylate (ACC) deaminase, to reduce ethylene level in the root of developing plants; fixation of atmospheric nitrogen; solubilization of insoluble phosphate; mobilization of nutrient in the rhizosphere; synthesis and excretion of siderophores that enhance the bioavailability of iron; as well as production of volatile organic compounds (VOCs), antioxidants, and exopolysaccharides (EPS). Moreover, some halotolerant PGPB colonize plants endophytically, produce various antimicrobial metabolites against pathogenic fungi and bacteria, support plant health by improving systemic resistance, and contribute to soil fertility and remediation (Orhan 2016; Singh and Jha 2016; Sorty et al. 2016; Bhise et al. 2017). In addition, some salt-tolerant *Rhizobium* strains can grow at NaCl concentration up to 500 mM. These salt-tolerant rhizobia go through some morphological, metabolical, and structural modifications to muddle through the salt stress (Guasch-Vidal et al. 2013; Liu et al. 2014). Interestingly, combined application of halotolerant *Rhizobium* and halotolerant PGPR *Pseudomonas* strains could be explored as an effective strategy to induce salt tolerance in mung bean (Ahmad et al. 2013).

AMF form beneficial symbiotic associations with most plants and play a vital role in plant growth under various stressed conditions by modifying the root system, enhancing mobilization and the uptake of several essential elements, defending roots against soilborne pathogens, stimulating phytohormone synthesis, inducing osmoregulator accumulation, controlling ROS accumulation by enhancing antioxidant enzyme activity and antioxidant molecule content, improving photosynthesis process, enhancing protein synthesis, and changing transcript levels of genes involved in signaling pathway or stress response as well as structural adaptations (Talaat and Shawky 2013, 2014a, b; Shabani et al. 2016). Halotolerant AMF colonization of maize roots improved plant salt tolerance by increasing water acquisition and shoot K^+ while decreasing shoot Na^+ concentration (Estrada et al. 2013). Furthermore, endophytes can colonize the internal tissues of their host plants and can promote growth, stress tolerance, and nutrient uptake. Plants living under stressful conditions may enhance their resistance to stress by associating with endophytes that may be useful in mitigating impacts of climate change and expanding agricultural production into stressful environments (Soares et al. 2016). In this concern, endophytic salt-tolerant bacteria promote plant growth directly or indirectly through production of phytohormones, biocontrol of host plant diseases, and/or improvement of plant nutritional status (Arora et al. 2014; Piernik et al. 2017). Therefore, treatment with halotolerant rhizomicrobes is an attractive option to mitigate the negative impact of soil salinization and improve crop yields under this harsh condition.

The present chapter illustrates the different mechanisms used by plants to adapt to saline condition. It also highlights the distribution of the halophiles and appraises the various strategies adopted by them to adjust with their saline surroundings. Moreover, it explores the possible mechanisms involved in halotolerant rhizomicrobe-mediated salinity tolerance in plants. It demonstrates the changes in the expression patterns of genes participating in inherent salt tolerance in plants when inoculated with this kind of microorganisms. A special section dedicates for mitigate salinity tolerance in crops by introducing halotolerant microbial genes. Finally, major aspects for future work in the current direction have also been highlighted.

3.2 Plant Salt Adaptation Mechanisms to the Salt-Stressed Conditions

Plants basically counteract the negative effects of salinity by activation of biochemical and genetic responses. They deploy a variety of traits to combat salt like osmolyte synthesis, reducing ROS level by following antioxidant defense mechanism and ion transport and their compartmentalization (Talaat and Shawky 2011, 2015; Talaat et al. 2015a). Indeed, the most essential trait is osmotic adjustment – cells accumulate sufficient solutes to balance extra osmotic pressure in the soil solution to maintain turgor. Osmotic balance can be achieved either by solute uptake from the soil or alternatively by compatible solute synthesis. Failure of this balance results in loss of turgidity, cell dehydration, and ultimately, death of cells (Munns and Gilliham 2015). The synthesis and accumulation of low molecular weight compatible solutes termed as "osmolytes" such as proline, sugars, polyols, trehalose, and quaternary ammonium compounds (QACs) like glycine betaine, alanine betaine, proline betaine, hydroxyproline betaine, choline *o*-sulfate, and pipecolate betaine are one of these mechanisms that are able to stabilize proteins and cellular structures, maintain cell turgor pressure by osmotic adjustment, and detoxify ROS protection of membrane integrity (Yokoi et al. 2002). Accumulation of compatible solutes is often regarded as basic strategy for the protection and survival of plants under salt stress (Talaat and Shawky 2014a; Talaat 2015a).

Salt stress in plants is a cumulative effect of osmotic and ionic stress which negatively affects the plant growth and yield. Ionic stress generated by salinity is mitigated through the participation of salt overly sensitive (SOS) pathway and ion transporters via ion homeostasis (Bharti et al. 2016). Excess salt disrupts ion homeostasis in plant cells, and thus plants utilize three strategies to prevent high Na^+ concentrations: active Na^+ efflux, Na^+ influx prevention, and Na^+ compartmentalization in vacuoles. Indeed, ion homeostasis is controlled by the cell via various ion transporters (*NHX1* and *SOS1, 2, 3, 4*) that restrict Na^+ entry into the cytoplasm and regulate its accumulation in the vacuoles and simultaneously selectively import K^+ ions. Na^+/H^+ antiporters (*SOS1* and *NHX1*) minimize the cytotoxicity by maintaining the optimal cytosolic ion concentration. *NHX1*, located in tonoplast, reduces cytosolic Na^+ concentration by pumping Na^+ into the vacuole, whereas plasma membrane-localized *SOS1* regulates the long-distance Na^+ transport from root to shoot, and both of these processes are driven by proton-motive force generated by the H^+-ATPase (Blumwald 2000). *SOS2* encodes for a serine/threonine protein kinase, and *SOS3* encodes for a myristoylated calcium-binding protein and senses salt-specific cytosolic Ca^{2+} concentration. *SOS3* interacts with *SOS2* using calcium as the second messenger and targets vegetative storage protein 2 (*VSP2*) to impart salt tolerance, simultaneously controlling the Na^+/

H^+ antiporter system (Qiu et al. 2002). Differences in Ca^{2+} concentration trigger protein phosphorylation cascades that provoke mitogen-activated protein kinases, which in turn, regulate the stress response (Chinnusamy et al. 2004). *SOS4* encodes a pyridoxal kinase, involved in the biosynthesis of pyridoxal-5-phosphate, which acts as an essential cofactor for numerous cellular enzymes. *SOS4* regulates Na^+ and K^+ homeostasis by modulating the activities of ion transporters (Bharti et al. 2016). Another ion carrier channel, *AtHKT1*, has been shown to function as a selective Na^+ transporter. *AtHKT1* plays a role in long-distance Na^+ transport and Na^+ circulation in the plant, with *AtHKT1* mediating Na^+ loading into the leaf phloem and Na^+ unloading from the root phloem sap (Berthomieu et al. 2003).

Furthermore, salinity alters the normal homeostasis of the cell and causes an increased production of ROS such as the superoxide radical, hydrogen peroxide, and hydroxyl radical. Under optimal growth conditions, ROS are mainly produced at low levels in organelles such as chloroplasts, mitochondria, and peroxisomes. The enhanced production of ROS during salt stress can pose a threat to cells, but it is thought that ROS also act as signals for the activation of stress response and defense pathways (Miller et al. 2010). Hence, ROS play two divergent roles as both deleterious and beneficial species depending on their concentration in plants. At high concentration they cause damage to biomolecules and trigger genetically programmed cell suicide events, whereas at low concentration it acts as second messenger in intracellular signaling cascades that mediate several responses in plant cells (Mittler et al. 2011). At high levels, ROS are unwelcome harmful by-products of normal cellular metabolism. It causes oxidative damage to lipid, protein, and DNA leading to altered intrinsic membrane properties like fluidity, ion transport, loss of enzyme activity, protein cross-linking, inhibition of protein synthesis, and DNA damage, ultimately resulting in cell death. In order to avoid the oxidative damage, higher plants possess an elaborate and highly redundant plant ROS network, composed of antioxidant enzymes and antioxidant molecules, and are responsible for keeping ROS levels under control (Talaat and Shawky 2014b; Talaat 2014, 2015a). The enzymatic components of the antioxidative defense system comprise of several antioxidant enzymes such as superoxide dismutase (SOD, EC 1.15.1.1), catalase (CAT, EC 1.11.1.6), guaiacol peroxidase (GPX), and enzymes of ascorbate-glutathione (AsA-GSH) cycle, ascorbate peroxidase (APX, EC 1.11.1.11), monodehydroascorbate reductase (MDHAR, EC 1.6.5.4), dehydroascorbate reductase (DHAR, EC 1.8.5.1), and glutathione reductase (GR, EC 1.6.4.2). Nonenzymatic mechanisms include compounds, such as ascorbic acid, glutathione, and α-tocopherol, capable of directly scavenging several ROS (Noctor and Foyer 2011). On the other hand, spatial and temporal fluctuations of ROS levels are interpreted as signals required for numerous biological processes such as growth, development, tolerance to abiotic stress factors, proper response to pathogens, and cell death. The molecular language associated with ROS-mediated signal transduction, leading to modulation in gene expression, is one of the specific early stress responses in the acclamatory performance of the plant (Bhattacharjee 2012). Moreover, overexpression of genes involved in ROS scavenging has resulted in lower cellular damage, the maintenance of photosynthetic energy capture, and an improvement in shoot and root growth under saline conditions (Roy et al. 2014).

3.3 Presence of Halotolerant or Halophilic Rhizomicrobes

Halophilic microbes are salt-loving organisms that flourish in hypersaline ecosystems such as the Dead Sea, saltern crystallizer ponds, natural inland salt lakes, brines, alkaline saline habitats, salt-contaminated soils, salt flats, evaporated ponds, subsurface salt formations, deep-sea hypersaline basins, salted foods, cold saline environments, and decayed monuments which may be the consequence of seawater evaporation (Edbeib et al. 2016). Hypersaline ecosystems are divided into two groups: thalassohaline, which is from seawater like oceans and contains sodium chloride such as the predominant salt, and athalassohaline, which is from non-seawater sources like dead seas, alkaline soda lakes, carbonate springs, saltern brines, and alkaline soil and contains different ion ratios. They vary in ionic composition and pH. Alkaline soda lakes have a high concentration of carbonate/bicarbonate ions, while the Mg^{+2} and Ca^{+2} are abundant in the Dead Sea (Oren 2008). Halophiles were isolated from salty soils in a variety of regions throughout the world, viz., *Oceanobacillus iheyensis*, *Halassobacillus devorans*, *Bacillus halmapalus*, and *Virgibacillus chiguensis* from Death Valley, CA, USA (Piubeli et al. 2015); *Kocuria* sp., *Nocardiopsis* sp., and *Micromonospora* sp. from Ribandar (Goa, India); as well as *Haloterrigena dagingensis G3-1*, *Natrinema altunense SD3*, and *Halostagnicola larsenii OS2* from the Algerian Sahara (Quadri et al. 2016).

Halophiles can be divided into slightly, moderately, or extremely halophilic depending on their sodium chloride requirements. Extreme halophiles are those which can grow optimally in media with 15–30% (2.5–5.2 M) NaCl, moderate halophiles are growing optimally in media with 3–15% (0.5–2.5 M) NaCl, and slight halophiles are able to grow optimally between 1% and 3% (0.2–0.5 M) NaCl (DasSarma and DasSarma 2015). Halophilic microbes can survive and thrive in saline conditions containing up to (>300 g/l) NaCl concentration and are found all over the small subunit rRNA-based tree of life. They are found within the three domains of life, *Archaea*, *Bacteria*, and *Eukarya* (Fig. 3.1; Ciccarelli et al. 2006). There are aerobic as well as anaerobic halophiles, heterotrophic, phototrophic, and chemoautotrophic types (Oren 2008). The bacterial sequences were assigned into 5784 operational taxonomic units (OTUs, based on ≥97% sequence identity), representing 24 known bacterial phyla, with *Proteobacteria* (44.9%), *Actinobacteria* (12.3%), *Firmicutes* (10.4%), *Acidobacteria* (9.0%), *Bacteroidetes* (6.8%), and *Chloroflexi* (5.9%) being predominant. *Lysobacter* (12.8%) was the dominant bacterial genus in saline soils, followed by *Sphingomonas* (4.5%), *Halomonas* (2.5%), and *Gemmatimonas* (2.5%). Archaeal sequences were assigned to 602 OTUs, primarily from the phyla *Euryarchaeota* (88.7%) and *Crenarchaeota* (11.3%). *Halorubrum* and *Thermofilum* were the dominant archaeal genera in saline soils. Rarefaction analysis indicated less than 25% of bacterial diversity and approximately 50% of archaeal diversity in saline soil (Trivedi 2017).

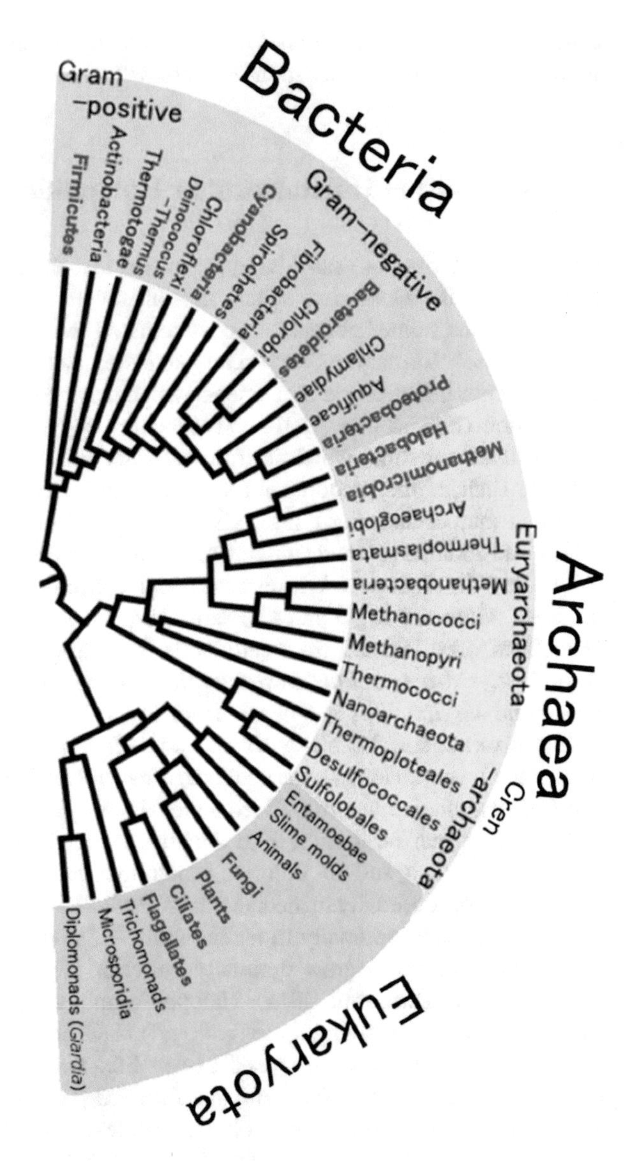

Fig. 3.1 Distribution of halophiles within the three domains of life, i.e., Archaea, Bacteria, and Eukarya. The tree was built using small subunit rRNA gene sequences from Ciccarelli et al. (2006)

3.4 Halophilic Rhizomicrobe Adaptation Mechanisms to the Hypersaline Environments

The population of microorganisms generally decreases with increasing salinity, and diversity is exceptionally low in hypersaline water masses; only halotolerant microbes can survive and thrive at higher salinity conditions. Indeed, the survival of microorganisms in hypersaline conditions requires specialized cellular adaptation mechanisms to preserve the osmotic balance with the highly saline environments. Halotolerant microbes cope with this harsh environmental condition by two strategies, namely, "salt-in" and "compatible solutes" (Oren 2008). Halophiles accumulate inorganic ions such as potassium and chloride inside their cells to balance the salt concentration in their environment. They import Cl^- from the environment into the cytoplasm via the Cl^- pumps. Arginines and/or lysines are positioned at both ends of the channel to facilitate Cl^- uptake and release (Lanyi 1990). Extreme halophiles of the archaeal Halobacteriaceae family such as *H. salinarum*, *Haloarcula marismortui*, and *Halococcus morrhuae* and the bacterial Halanaerbiales family such as *Haloanaerobium praevalens*, *Haloanaerobium acetoethylicum*, and *H. halobius* accumulate intracellular K^+ to maintain their osmotic balance with the environment (Oren 2008). This is achieved by the concerted action of the membrane-bound proton-pump bacteriorhodopsin, the ATP synthase, and the Na^+/H^+ antiporter and results in an electrical potential that drives the uptake of K^+ into cells via a K^+-uniport mechanism (Kixmuller and Greie 2012).

Furthermore, halophilic microbes respond to salt stress by regulating the cytosolic pools of organic solutes to achieve osmotic equilibrium. Indeed, halophiles can survive in a highly saline environment by maintaining osmotic balance and establishing the proper turgor pressure, via producing organic compatible solutes which include polyols such as glycerol, sugars and their derivatives, amino acids and their derivatives, and quaternary amines such as glycine, betaine, ectoines, etc. inside their cells. Osmoprotectant accumulation in microorganism's cells either by de novo synthesis or by uptake from the surroundings is regulated according to the salt concentration outside the cell (Edbeib et al. 2016). In addition, the types of compatible solutes accumulated inside halophilic strains can determine the degree of halotolerance. Low salt-tolerant strains generally accumulate simple sugars such as sucrose and/or trehalose. Sugar accumulation in low salt-tolerant strains serves as a compatible solute in response to salt stress and can play crucial roles as a signaling molecule and involved in glycogen synthesis as well as insoluble polysaccharide synthesis (Kolman et al. 2015). Moderately salt-tolerant strains also accumulate glucosylglycerol (Fulda et al, 2006). Moreover, the moderately halophilic bacterial strains *Halomonas hydrothermalis* VITP9, *Bacillus aquimaris* VITP4, *Planococcus maritimus* VITP21, and *Virgibacillus dokdonensis* VITP14 accumulated glutamic acid under salt stress, indicating the significance of glutamic acid as the anionic counterpart of K^+/Na^+ ions and precursor for other synthesized nitrogenous osmolytes. Major other solutes were also accumulated by these strains under saline conditions, such as proline (VITP4), ectoine (VITP14 and VITP9), sugars (VITP21), and glycine betaine (by all strains under study), referring that the total intracellular organic solutes increased significantly with increasing NaCl concentration in the growth medium, and the compositions of the solutes were dependent on the type of strain (Joghee and Jayaraman 2016). Likewise, in the presence of high salinities, the moderately halophilic *Halobacillus halophilus* produces proline as the main organic compatible solute (Saum and Muller 2008). Indeed, the frequency of proline and glycine is higher in case of extreme halophiles. Highly salt-tolerant strains generally accumulate ectoine and the quaternary ammonium compounds such as glycine betaine and glutamate betaine. Significant increase in the frequency of proline among extreme halophiles is observed in the genes reported as regulators of ion transporters (*aca4*, *esi47*, *hal3*, *hal*), transporters (*ema1*, *gbuA*), osmotic tolerance proteins (*gpd1*, *relA*), and salt toxicity target (*hal2*). On the other hand, high frequency of glycine among extreme halophile is observed in the genes reported to regulators of ion transporters (esi47, hal3, hal5), transporters (*ena1*, *ena2*, *gbuA*, *hal11*), osmotic tolerance protein (*cysK*, *gdp1*, *relaA*), and chaperones (*dnaK*, *groEL*) (Anwar and Chauhan 2012). These compatible solutes exert their effect through changes in solvent structure and/or elusive changes in the dynamic properties of the protein rather than by changing the structure of the protein itself and also help in protein-DNA interaction (Kurz 2008). They can also serve as stabilizers of biomolecules, i.e., enzymes, DNA, membranes and whole cells, salt antagonists, and stress-protective agents (Detkova and Boltyanskaya 2007). In this concern, ectoine can increase shelf life and activity of enzyme preparations by protecting many unstable enzymes and nucleic acids against the detrimental action of high salinity. Such compatible solutes are sometimes termed "molecular chaperones" (Kolp et al. 2006).

The unique internal microbial protein structure of halophilic microbes allows them to thrive in extreme saline environments. The difference between halophile proteins and non-halophile proteins that helps halophiles to prosper in hypersaline environments is the presence of a larger proportion of acidic amino acids like glutamate and aspartate on their protein surfaces and thus increases the negative protein surface potential and improves its ability to compete with ions for water molecules and, thereby, preferentially solubilizes the protein (Zhang and Yi 2013). This hypothesis is supported by the observations that crystal structures of halophilic proteins include water molecules bound to surface acidic residues (Karan et al. 2012). Glutamate has a water-binding nature higher than any other amino acid. In addition, halophilic enzymes are more stable than their non-halophilic counterparts owing to their polyextremophilic characteristics (Edbeib et al. 2016). The surface of halophilic enzymes is highly negatively charged, and spheres of hydration are clustered to the protein surface to prevent their degradation (Paul et al. 2008). Halophilic enzymes are divided into three categories: (1) intracellular enzymes, which are not directly in contact with the ionic concentration of the surroundings; (2) membrane-bound enzymes (carrier proteins), which are in direct contact with the cytoplasmic content as well as the outside medium; and (3) extracellular enzymes, which are directly exposed to the saline medium. Halophilic bacteria belonging to *Bacillus* group including *Solibacillus*, *Halobacillus*, *Oceanobacillus*, *Gracilibacillus*, *Virgibacillus*, *Thalassobacillus*, and *Piscibacillus* and Gram-negative bacterial species of *Salinivibrio*, *Chromohalobacter*, and *Halomonas* can secrete extracellular hydrolytic enzymes such as amylases, proteases, lipases, DNases, pullulanases, and xylanases (Rohban et al. 2009).

Most importantly, halophiles can use hydrocarbons as their sole carbon and energy sources. Therefore, they may prove to be valuable bioremediation agents for the treatment of saline effluents and hypersaline waters contaminated with toxic compounds that are resistant to degradation (Edbeib et al. 2016). Among the bacteria domain, members of the genera *Micrococcus*, *Pseudomonas*, *Alcaligenes*, *Rhodococcus*, *Arthrobacter*, *Bacillus*, *Halomonas*, *Chromohalobacter*, *Alcanivorax*, *Marinobacter*, *Idiomarina*, *Thalassospira*, *Halomonas*, and *Arhodomonas* were able to degrade aliphatic and/or aromatic compounds as the sole sources of carbon at salinity ranging from 5% to 15% (Moreno et al. 2011; Bonfá et al. 2013; Zhou et al. 2016). Halophilic *Thalassospira* sp. strain TSL5-1 effectively degrades high molecular weight polycyclic aromatic hydrocarbons, namely, pyrene (41.4%) in hypersaline environments at salinity ranging from 0.5% to 19.5% in 25 days (Zhou et al. 2016). Besides, Moreno et al. (2011) found the presence of common genes, which are responsible for metabolism of phenol and benzoate in *Halomonas organivorans*, viz., *catA*, *catB*, *catC*, and *catR* encode 1,2- CAT, cis-,cis-muconate cycloisomerase, muconolactone delta-isomerase, and a transcriptional regulator, respectively. These genes were flanked downstream by the benzoate catabolic genes *benA* and *benB*, which code for the large and small subunits of benzoate 1,2-dioxygenase, respectively.

3.5 Root Halotolerant Microbial Inoculation and Plant Salt Tolerance

Halotolerant microbes isolated from saline environments have potential to alleviate the toxic effects of soil salinity on plant growth and improve salt tolerant through a number of direct and/or indirect mechanisms and would be most appropriate as bioinoculants under such condition. Salt-tolerant microbes can reduce Na^+ content of shoots, increase the expression of stress-responsive transcription factors, induce greater osmolyte synthesis, induce phytohormone production, enhance ROS scavenging, and thus improve plant biomass under salinity stress.

3.5.1 Halotolerant Plant Growth-Promoting Rhizobacteria (PGPR)

Mechanisms of Salt-Stress Amelioration in Halotolerant PGPR-Inoculated Plants

3.5.1.1 Halotolerant PGPR-Induced Morphophysiological Alterations

To address salt-stress management, plants' morphological and physiological characteristics have great impact. Inoculation with *Azospirillum* strains isolated from saline soil increased salinity tolerance and improved shoot dry weight, grain productivity, and N concentrations of wheat plants (Nia et al. 2012). Likewise, inoculation with halotolerant bacterial strains *Halobacillus* sp. and *Bacillus halodenitrificans* ameliorated salt stress by enhancing root length, dry weight, and grain yield of wheat (Ramadoss et al. 2013). *Exiguobacterium oxidotolerans*, a halotolerant PGP rhizobacteria, improved yield and content of secondary metabolites in *Bacopa monnieri* under salt stress (Bharti et al. 2013). Halotolerant PGPB ameliorated the negative impact of salinity on wheat plant by increasing the leaf's relative water content and enhancing photosynthetic pigment production (Saghafi et al. 2013). Inoculation of two halophilic bacteria strain J31 of *Terribacillus halophilus* and strain M3-23 of *Virgibacillus marismortui* producing halotolerant and thermotolerant chitinases that help in decomposing chitin-based organic matters to tomato seeds improved the stem growth compared to the uninoculated control (Essghaier et al. 2014). Inoculation of halotolerant bacterium *Bacillus licheniformis* HSW-16 protected wheat plants from growth inhibition caused by NaCl and increased plant growth (6–38%) in terms of root length, shoot length, fresh weight, and dry weight (Singh and Jha 2016). Similarly, halotolerant PGPR *Dietzia natronolimnaea* STR1 inoculation promoted the growth of salt-stressed wheat plants by increasing the plant height, root length, and dry weight (Bharti et al. 2016). However, inoculated salt-stressed rice plants with salt-tolerant bacteria *Pseudomonas* sp. PDMZnCd2003 had a negative impact on the plant growth. This unexpected effect was a case that should be concerned before the application of this kind of bacteria as a PGPB (Nakbanpote et al. 2014).

3.5.1.2 Halotolerant PGPR-Induced Phytohormone Production

Noteworthy, inoculated salt-sensitive plants with halotolerant PGP bacteria can ameliorate the salinity negative impact on plant growth and development. Indeed, halotolerant PGPB stimulate plant growth under high salinity by using several mechanisms, such as synthesis of IAA, GA, CK, and ABA, solubilization of insoluble phosphate, synthesis and excretion of siderophores, and production of ACC deaminase. In this respect, inoculated salt-stressed soybean plants with halotolerant phytohormones producing PGP rhizobacteria induced salt tolerance to plants by improving proline production, shoot/root length, and dry weight (Naz et al. 2009). Salt-tolerant IAA-producing bacterial strains *P. aureantiaca* TSAU22 and *P. extremorientalis* TSAU20 alleviated quite successfully the reductive effect of salt stress on percentage of germination (up to 79%). They may supply additional phytohormone to the plants and thus may help stimulate root growth and reverse the growth-inhibiting effect of salt stress to a certain extent in both the shoot and the root (Egamberdieva 2009). Regarding phytohormone quantification, lipo-chitooligosaccharide (LCO)-treated *A. thaliana* rosettes had increased levels of ABA and free SA, while thuricin 17 (Th17)-treated rosettes showed increased levels of IAA and SA. ABA regulation was observed in salt-stress tolerance and IAA-regulated protein degradation using the ubiquitin proteasome pathway, which decreased the toxic effects of ROS (Subramanian et al. 2016). Moreover, halotolerant PGPR improves salt-stress tolerance in different plant species by producing ACC deaminase and thus keeps ethylene levels low, which is helpful to root growth and survival of plants under this harsh condition. Canola seeds inoculated with halotolerant bacteria containing ACC deaminase activity under 150 mM NaCl exhibited high biomass production (Siddikee et al. 2010). Inoculated tomato plants with salt-tolerant isolates *Pseudomonas stutzeri* (C4) and *Pseudomonas aeruginosa* (T15), having different PGP traits like ACC deaminase, phytohormones, siderophore production, and P solubilization, enhanced plant salinity tolerance and ameliorated the deleterious effect of stress by increasing root and shoot length as well as biomass production (Tank and Saraf 2010). Halotolerant bacteria with ACC deaminase activity conferred salt tolerance in red pepper seedlings by showing higher biomass production and lower ethylene content (Siddikee et al. 2011). Inoculated wheat plants with salt-tolerant isolates *Bacillus pumilus*, *Pseudomonas mendocina*, *Arthrobacter* sp., *Halomonas* sp., and *Nitrinicola lacisaponensis*, isolated from high saline habitats and exhibited plant growth-promoting traits like P solubilization as well as IAA, siderophore, and ammonia production, increased root and shoot length, biomass, and biochemical levels such as chlorophyll, carotenoids, protein, and phenolics (Tiwari et al. 2011). The halophilic bacterium *Planococcus rifietoensis* has PGP activities like IAA and ACC deaminase production, and phosphate-solubilizing activity enhanced the growth and yield of *T. aestivum* under salinity stress (Rajput et al. 2013). Salt-stressed *Cucumis sativus* plants showed an upregulation of stress-responsive abscisic acid which is not occurring when osmotolerant PGPB were inoculated, whereas salicylic acid and GA were highly produced in PGPB-inoculated plants (Kang et al. 2014). Halotolerant PGPB strains *Bacillus*, *Pantoea*, *Marinobacterium*, *Acinetobacter*, *Enterobacter*, *Pseudomonas*, *Rhizobium*, and *Sinorhizobium* (LC027447-53; LC027455; LC027457, LC027459, and LC128410) associated with the weed *Psoralea corylifolia* L. had IAA-producing activity and promoted the seed germination and seedling growth of wheat under saline conditions (Sorty et al. 2016). The plant growth reduction caused by NaCl was ameliorated with the application of halotolerant bacterial strains, *Thalassobacillus* sp., *Bacillus* sp., *Halomonas* sp., *Oceanobacillus* sp., *Bacillus* sp., *Zhihengliuella* sp., and *S. succinus*, isolated from salt-affected soils of the East Anatolian region and possessed PGP activities such as IAA, ACC deaminase, and ammonia production, phosphate solubilization, and atmospheric nitrogen fixation. They significantly increased the root and shoot length and total fresh weight of the wheat plants. The growth rates of the plants inoculated with bacterial strains ranged from 62.2% to 78.1% (Orhan 2016). Inoculation with *Enterobacter cloacae* strain KBPD, a salt-tolerant PGP rhizobacteria with ACC deaminase activity, phosphate solubilization, and indole acetic acid, siderophore, ammonia, hydrogen cyanide, and exopolysaccharide production, alleviated salt toxicity in *Vigna radiata* L. by increasing shoot length, root length, fresh and dry weights, as well as total chlorophyll content. Salt-affected plants had higher proline content, while inoculation with *E. cloacae* KBPD reduced its content (Bhise et al. 2017).

3.5.1.3 Halotolerant PGPR-Induced Osmolyte Accumulation

Halotolerant rhizomicrobes are the potential tools for sustainable agriculture and trend for the future. One of the mechanisms by which these bacteria mediated salinity tolerance in host plant is the enhancement of the organic solute accumulation (Singh and Jha 2016). Indeed, plants accumulate a variety of osmoprotectants that improve their ability to combat abiotic stresses. Among them, betaine appears to play an important role in conferring resistance to stresses (Talaat and Shawky 2014a; Talaat 2015a; Talaat et al. 2015b). Halotolerant cyanobacterium, *Aphanothece halophytica*, has a unique biosynthetic pathway of betaine consisting of three-step methylation of glycine, which is catalyzed by two *N*-methyltransferases (ApGSMT and ApDMT). *Arabidopsis* plants that were transformed with ApGSMT and ApDMT accumulated substantial amounts of betaine and increased tolerance to salt stress (Waditee et al. 2005). Halotolerant *A. halophytica*, isolated from the Dead Sea, is known to accumulate significant amounts of betaine and can grow in media of up to 3.0 M NaCl. The gene encoding 3-phosphoglycerate dehydrogenase (PGDH), which catalyzes the first step of the phosphorylated pathway of serine biosynthesis, was isolated from *A. halophytica* and transferred into *Arabidopsis* plants, in which the betaine synthetic pathway was introduced via glycine methylation and further increased betaine levels and improved the stress tolerance. Thus, PGDH enhanced the levels of betaine by providing the precursor serine for both choline oxidation and glycine methylation pathways. In addition, transgenic *Arabidopsis* plants expressing ApPGDH, ApGSMT, and ApDMT exhibited an enhanced tolerance for salt stress compared with those of the wild-type or transgenic plants expressing ApGSMT and ApDMT (Waditee et al. 2007). These results revealed that introducing the betaine synthetic pathway into betaine non-accumulating plants improved salt tolerance of plants. Moreover, increasing accumulation of serine as a precursor of betaine in the transgenic plants expressing ApPGDH enhanced betaine accumulation and subsequently conferred salt-stress tolerance. Indeed, serine is an essential amino acid that plays important roles in a variety of biological processes including metabolism, purine and pyrimidine biosynthesis, and generation of activated one-carbon (C-1) unit and is also a precursor of phosphatidylcholine (phospholipid) and cysteine (antioxidant compounds) and therefore involves in adaptive responses to abiotic stresses in plants (Tasseva et al. 2004). Thus, serine may play important roles either directly and/or indirectly in the responses of plants to various environmental stresses. Additionally, through serine hydroxymethyltransferase (SHMT), serine associates with glycine metabolism via the glycine decarboxylase complex. In halotolerant cyanobacterium, *A.* halophytica, the *SHMT* gene, was suggested to be essential for cell survival. *ApSHMT* is a salt-inducible gene, and overexpression of it increases the levels of not only glycine and serine but also choline and glycine betaine and conferred tolerance to salinity stress (Waditee et al. 2012). As many crop plants do not have a glycine betaine synthetic pathway, genetic engineering of glycine betaine biosynthesis pathways represents a potential way to improve the tolerance of crop plant to stress (Chen and Murata 2002). Therefore, attempt to express PGDH, SHMT, and glycine betaine synthesis gene together would be worthwhile to test for the improvement of salinity stress in crop plants via boosting the levels of glycine betaine. Furthermore, moderately halophilic bacterial strains (*S. haemolyticus* and *B. subtilis*) isolated from saline rhizosphere of chickpea can accumulate endogenous osmolytes such as proline, glycinebetaine, and choline. These osmolytes improve the growth of bacterial strains and plants by alleviating salt stress (Qurashi and Sabri 2013). Salt-stressed maize plants inoculated with halotolerant PGPRs, *Bacillus* sp., and *Arthrobacter pascens* sp. recorded higher sugar and proline accumulation (Ullah and Bano 2015). A halotolerant ACC deaminase bacterium *Bacillus licheniformis* HSW-16 induced systemic tolerance to salt stress in wheat plant, enhanced plant growth, and increased certain osmolytes such as total soluble sugar, total protein content, and decreased malondialdehyde content (Singh and Jha 2016).

3.5.1.4 Halotolerant PGPR-Induced Ion Homeostasis and Nutrient Acquisition

Halotolerant microbes mediate salinity tolerance in host plant by altering the selectivity of Na^+, K^+, and Ca^{2+} and sustain a higher K^+/Na^+ ratio in plant tissues. All halophilic microorganisms contain potent transport mechanisms, generally based on Na^+/H^+ antiporters, to expel sodium ions from the interior of the cell. A crucial function of ion transporters is to maintain and retain favorable cytosolic K^+/Na^+ ratios in the face of low K^+/Na^+ ratios in the environment. K^+ is an important monovalent cation inside the cell, where it is not only crucial for salt or turgor acclimation but is also involved in membrane energetic, pH regulation, enzyme activities, and gene expression (Hagemann 2011). In this concern, *A. halophytica* contains Na^+/H^+ antiporters such as Ap-NhaP1, Ap-NapA1–1, Ap-NapA1–2, and Ap-Mrp, as well as a putative F_1F_0-type Na^+-ATP synthase (ApNa$^+$-ATPase) operon (*ApNa*$^+$-*atp*) in the cytoplasmic membrane that is playing a potential role in salt tolerance (Soontharapirakkul et al. 2011).

Inoculation of salt-tolerant diazotrophic PGPR to rice increased nutrient uptake (N, P, and K), and in parallel, Na^+ level in plant tissues was reduced. In addition to these characters, these rhizobacteria showed the potential to fix nitrogen and solubilize phosphate under high salinity conditions (Sarathambal and Ilamurugu 2013). Inoculation of halotolerant bacteria containing ACC deaminase *Bacillus mojavensis* to salt-stressed wheat plants increased the root and shoot weight, chlorophyll content, as well as K^+, Ca^{2+}, and Mg^{2+} uptake in comparison with the uninoculated stressed plants (Pourbabaee et al. 2016). The halotolerant rhizobacteria that belong to the genera *Klebsiella*, *Pseudomonas*, *Agrobacterium*, and *Ochrobactrum* have PGP activities like IAA and ACC deaminase production and phosphate-solubilizing activity which promoted growth and enhanced salinity tolerance in peanut by maintaining ion homeostasis. Maintenance of high N, K^+, and Ca^{2+} levels as well as K^+/Na^+ ratio is a potential mechanism to reduce the damage caused by salt stress (Sharma et al. 2016). Inoculation of halotolerant ACC deaminase bacterium *Bacillus licheniformis* HSW-16 protected wheat plants from growth inhibition caused by

NaCl by regulating ion transporters, favoring K^+/Na^+ ratio, decreasing Na^+ accumulation (51%), and increasing K^+ (68%) and Ca^{2+} content (32%) in plants. Production of exopolysaccharide by the *B. licheniformis* HSW-16 can also protect from sodium by binding this ion (Singh and Jha 2016).

3.5.1.5 Halotolerant PGPR-Induced Antioxidative System

Some halotolerant PGPRs are capable of inducing tolerance against salt stress in plants by elevating ROS-scavenging system. They can activate plant antioxidant defense machinery by upregulating the activity of the antioxidant enzymes that scavenges overproducing ROS and protect the plants from salt toxicity. PGPRs, *Bacillus* sp., and *Arthrobacter pascens* sp. isolated from rhizospheric soil of halophyte regions showed reliability in growth promotion of salt-stressed maize plants and elevation in the activity of antioxidant enzymes including SOD, POX, CAT, and APX (Ullah and Bano 2015). Wheat plants inoculated with halotolerant rhizobacteria *Dietzia natronolimnaea* STR1 recorded higher proline accumulation and lower malondialdehyde (MDA) level under salt-stressed conditions in comparison to the uninoculated salt-affected plants. MDA content, a result of lipid peroxidation, is a marker of the extent of membrane damage owed to oxidative damage due to the salt stress (Bharti et al. 2016). Inoculation of halotolerant rhizobacteria that belong to the genera *Klebsiella*, *Pseudomonas*, *Agrobacterium*, and *Ochrobactrum* exhibited PGP traits like P solubilization as well as IAA and ACC deaminase production which enhanced peanut growth under saline conditions by decreasing ROS accumulation (Sharma et al. 2016).

3.5.1.6 Halotolerant PGPR-Induced Gene Expression

Halotolerant PGPRs mediate salinity tolerance in host plant by inducing the expression of different plant genes that involved in salt tolerance. Indeed, salinity adaptation process in inoculated plants is genetically and physiologically very complex. It is based on genes whose effects not only limit the rate of salt uptake from the soil but also limit the transport of salt through the plant and thus adjust the ionic and osmotic balance of the cells in roots and shoots as well as regulate the plant development and the onset of senescence. Carotenoid producing halotolerant PGPR *Dietzia natronolimnaea* STR1 protected wheat plants from salt-stress damaging effect by modulating the transcriptional machinery responsible for salinity tolerance in plants such as ABA signaling, SOS pathway, ion transporters, and antioxidant machinery. ABA-signaling cascades, such as *TaABARE* and *TaOPR1*, were upregulated in PGPR-inoculated plants leading to induction of *TaMYB* and *TaWRKY* expression followed by stimulation of expression of a plethora of stress-related genes. Enhanced expression of *TaST*, a salt stress-induced gene, associated with promoting salinity tolerance was observed in PGPR-inoculated plants in comparison to uninoculated control plants. Expression of SOS pathway-related genes (*SOS1* and *SOS4*) was also modulated in PGPR-applied wheat shoot and root systems. Improved *SOS4* expression in PGPR-inoculated wheat roots could be correlated with improvement in root length and overall plant growth in salt-stressed plants. Likewise, higher expression of ion transporters *TaNHX1*, *TaHAK*, and *TaHKT1* in STR1-inoculated plants was observed in both shoots and roots in comparison to their uninoculated counterparts suggesting a likely role of halotolerant rhizobacteria in modulating ion transport mechanisms under saline conditions. STR1-inoculated wheat plant tolerance to salinity stress was also correlated with the increased expression of various antioxidant enzymes such as *CAT*, *APX*, *MnSOD*, *POD*, *GPX*, and *GR*, suggesting that PGPR triggered the abiotic stress-related defense pathways. Thus, halophilic PGPR can mediate salt tolerance in plants through modulation of ROS-scavenging enzyme expression (Bharti et al. 2016). Inoculation with halotolerant PGPRs which belong to the genera *Klebsiella*, *Pseudomonas*, *Agrobacterium*, and *Ochrobactrum* induced salinity tolerance in peanut plants by affecting the transcript levels of antioxidant genes *APX*, *CAT*, and *SOD* and by regulating their expression (Sharma et al. 2016).

3.5.2 Halotolerant Endophytic Bacteria

Endophytes are the microorganisms that thrive inside the plants. They face less competition for nutrients and are more protected from adverse changes in the environment than bacteria in the rhizosphere and phyllosphere as they interact closely with the host plant. Halotolerant endophytic bacteria can positively enhance plant growth and productivity under saline conditions by involving several mechanisms such as production of phytohormones (auxins and cytokinins), increasing the amount of available nutrients by a number of biochemical processes (e.g., N_2-fixation, phosphate solubilization, siderophore release increasing Fe availability), suppression of ethylene production by ACC deaminase, or induction of plant defense mechanisms (Weyens et al. 2009). These properties of salt-tolerant endophytic bacteria make them suitable candidate for application in salt-stressed soils.

Several functional gene groups in endophytic salt-tolerant bacterium *Stenotrophomonas rhizophila* DSM14405^T were upregulated as a result of salt shock including those responsible for the synthesis and transport of cell wall, outer membrane, and cytoplasmic membrane; the metabolism and transport of amino acids, nucleotide, and secondary metabolites; and the energy production. In contrast, genes responsible for cell motility, secretion, and intracellular trafficking and the transport and metabolism of inorganic ions were downregulated. In addition, *S. rhizophila* is equipped with several other genes, which play a role in root colonization, such as those that encode the O-antigen, capsule polysaccharide biosynthesis pathways, hemagglutinin, and outer membrane adhesion proteins (Alavi et al. 2013). Endophytic salt-tolerant bacteria promote plant growth directly or indirectly through production of phytohormones, biocontrol of host plant diseases, and improvement of plant nutritional status (Arora et al. 2014). Inoculation of

Beta vulgaris L. seeds with the two halotolerant endophytic bacteria strains, *Pseudomonas* sp. ISE-12 (B1) and *Xanthomonadales* sp. CSE-34 (B2), under saline conditions positively affected germination percentage and germination index as well as shortened mean germination time, which led to a quickening of the growth stages of seedlings. Moreover, salt-stressed inoculated plants had a greater root length, higher dry biomass, lower tissue water content, and lower specific leaf area compared with the control (Piernik et al. 2017).

3.5.3 Halotolerant Rhizobia

The nitrogen-fixing symbiosis formed between rhizobia and legumes can decrease the damaging effect caused by soil salinity on stressed plants. In this respect, several studies focus on salt-tolerant rhizobia and their mechanisms of salt resistance. Selection of salt-tolerant strains of symbiotic bacteria able to efficiently nodulate plants under saline conditions is considered as a great challenge to improve the productivity of N_2-fixing plants. Among them, a *Frankia* Ceq1 was able to grow at NaCl concentration of more than 300 mM (Tani and Sasakawa 2003). Some species of halotolerant *Rhizobium* are capable to nodulate under high NaCl concentration, which could be attributed to the induction of the *nod* genes in the absence of flavonoid inducers (Guasch-Vidal et al. 2013). *Mesorhizobium alhagi* CCNWXJ12-2 is a highly salt-tolerant and alkali-tolerant rhizobium which can form nodules with the desert plant *Alhagi sparsifolia*. It contains a putative PrkA-family serine protein kinase, *PrkA*. The expression of *prkA* was found to be downregulated in high salt conditions (Liu et al. 2014). Recent report showed that PrkA could reduce the survival of *M. alhagi* under environmental stress and deletion of *prkA* dramatically improved the salt and alkaline tolerance as well as the antioxidant capacity of *M. alhagi*. Moreover, it is possible to increase the salt and alkali tolerance of a bacterium by constructing specific mutants in genes highlighted by RNA-Seq data (Liu et al. 2016).

Interestingly, inoculation/co-inoculation with salt-tolerant *Rhizobium* and halotolerant, auxin-producing *Pseudomonas* containing ACC deaminase improved the total dry matter and salt tolerance index of mung bean plants. Indeed, combined application of salt-tolerant *Rhizobium* and halotolerant *Pseudomonas* strains could be explored as an effective strategy to improve salt tolerance in mung bean (Ahmad et al. 2013).

3.5.4 Halotolerant Fungi

The halotolerant and halophilic fungi use polyols including glycerol, erythritol, arabitol, and mannitol as osmotic solutes and retain low salt concentrations in their cytoplasm. Na^+-sensitive 3-phosphoadenosine-5-phosphatase HwHal2 is one of the putative determinants of halotolerance in *Hortaea werneckii* and is promising transgene to improve halotolerance in crops (Plemenitas et al. 2014). Furthermore, halotolerant endophytic fungi can help plants adapt to extreme environments. Fungal endophytes isolated from hosts on saline soils have potential application to increase salt tolerance of important plants via habitat-adapted symbiosis. Endophytic fungi isolated from *Leymus mollis* and *Dichanthelium lanuginosum* promoted plant growth in rice seedlings under salt stress (Redman et al. 2011). Endophytic fungi play a role in increasing the capacity of *Phragmites australis* to grow in high salinity soils, probably contributing to invasion in saline environments. Moreover, endophytes found at the high salinity site increased tolerance of rice seedlings to elevated levels of salinity (Soares et al. 2016).

Accumulation of Na^+ and impairment of K^+ nutrition are a typical characteristic of plant subjected to salt stress (Talaat and Shawky 2011, 2013; Talaat 2015b). Plants have evolved molecular mechanisms in order to cope with the negative effects of salinity. These include the regulation of genes with a role in the uptake, transport, or compartmentation of Na^+ and/or K^+, which are responsible of the adequate ionic homeostasis in the plant. Hence, regulation of plant transporter genes involved in ion homeostasis by the halotolerant AMF under saline conditions is of great importance. Salt-stressed maize plants inoculated with three native AMF from a Mediterranean saline area showed significant increase in K^+ and reduction in Na^+ accumulation as compared to non-mycorrhizal-stressed plants, concomitantly with higher K^+/Na^+ ratios in their tissues, and these effects correlated with the regulation of *ZmAKT2*, *ZmSOS1*, and *ZmSKOR* genes in roots of maize colonized by these native AMF. Indeed, the protective effect of the halotolerant AMF on maize plants under salinity was mediated by improved K^+ retention in the plant tissues due to the upregulation of *ZmAKT2* and *ZmSKOR* as well as by enhanced Na^+ extrusion from the root to the external medium and/or prevented Na^+ from reaching to the photosynthetic tissues via redistribution of Na^+ between roots and shoots due to the upregulation of *ZmSOS1* (Estrada et al. 2013).

3.6 Plant Salt Mitigation by Introducing Halotolerant Microbial Genes

Salt stress is one of the major environmental factors that is severely affecting plant growth and productively resulting in significant losses worldwide. Improving plant salt tolerance is an important issue in plant molecular biology. Indeed, engineering salt tolerance in crops by introducing salt-resistant microbial gene to salt-stressed plants is of great importance and could be a sustainable solution to ameliorate the deleterious salt effects on plants. Hence, halotolerant microbial genes can contribute significantly to plant performance under this harsh condition. In this concern, the Na^+/H^+ antiporter encoding gene *AaNha*D from halotolerant bacterium *Alkalimonas amylolytica* increased salt tolerance in transgenic tobacco BY-2 cells.

Indeed, *AaNha*D even in plant cells functions as a pH-dependent tonoplast Na^+/H^+ antiporter, thus presenting a new avenue for the genetic improvement of salinity tolerance (Zhong et al. 2012). Overexpression of a stress-responsive gene FcSISP from *Fortunella crassifolia* in tobacco enhanced its salt tolerance. The transgenic plants accumulated lower Na^+ concentrations, which led to reduced Na^+/K^+ ratio, while accumulating more proline than the WT. Steady-state mRNA levels of genes involved in Na^+ exchange (three SOS genes and three NHX genes) and proline synthesis (P5CS and P5CR) were higher in the transgenic plants in comparison with WT (Gong et al. 2014). Moreover, Metwali et al. (2015) isolated AgNHX1, a vacuolar Na^+/H^+ antiporter from a halophytic species *Atriplex gmelini*, and introduced it into fig plants. Overexpression of the AgNHX1 gene in salt-treated transgenic plants enhanced their tolerance to salt stress, improved plant growth, increased proline and K^+ content, and decreased Na^+ concentration compared to non-transgenic control plants.

Moreover, accumulation of compatible solutes in salt-stressed plants is useful as they can act as osmoprotectants, stabilizers of biomolecules and whole cells, salt antagonists, and/or stress-protective agents. Glycine betaine as one of these organic solutes is the most effective cellular osmoprotectant against abiotic stresses. It also plays a role in stabilizing the structures of complex proteins, protecting the transcriptional machinery, and/or maintaining the integrity of membranes (Yang et al. 2008). *Arabidopsis* plants transformed with ApGSMT and ApDMT from a halotolerant cyanobacterium, *Aphanothece halophytica*, accumulated betaine, which improved their stress tolerance (Waditee et al. 2005). Introduced PGDH gene encoding 3-phosphoglycerate dehydrogenase from *A. halophytica* to *Arabidopsis* plants induced betaine accumulation and increased tolerance to salt stress (Waditee et al. 2007). Additionally, introduced codA gene encoding choline oxidase from *Arthrobacter globiformis* to tomato plants induced the synthesis of glycine betaine and improved the tolerance of plants to salt stress. The codA-transgenic plants revealed higher tolerance to salt stress during seed germination and subsequent growth of seedlings than WT plants. Mature salt-stressed transgenic plants showed higher contents of the relative water, chlorophyll, and proline than those of WT plants (Goel et al. 2011). Likewise, halophilic methanoarchaeon *Methanohalophilus portucalensis* can de novo synthesize betaine. Introduced genes encoding betaine biosynthesizing enzymes, Mpgsmt and Mpsdmt, into salt-stressed *Arabidopsis* plants enhanced their growth by increasing relative water content, photosystem II activity, and quaternary ammonium compound accumulation (Lai et al. 2014).

Bacterial mannitol 1-phosphate dehydrogenase (*mtl*D) gene introduced to potato plants enhanced NaCl tolerance in transgenic plants. The improved tolerance of this transgenic line could be due to the induction and progressive accumulation of mannitol in plant shoots and roots, which act as an osmoprotectant (Rahnama et al. 2011). Moreover, better growth and productivity were observed in mtlD peanut lines under saline conditions, which might be attributed to increasing mannitol accumulation; improving osmotic adjustments; enhancing nutrient uptake; decreasing H_2O_2 and MDA contents; increasing antioxidant enzymes activity such as SOD,

CAT, APX, and GR; enhancing ascorbic acid and relative water content; and stabilizing macromolecules such as membrane proteins and enzymes (Patel et al. 2016).

The proteome of *A. thaliana* rosettes is altered by two bacterial signal compounds: lipo-chitooligosaccharide (LCO) from *Bradyrhizobium japonicum* strain 532C and thuricin 17 (Th17), a bacteriocin from *Bacillus thuringiensis* strain NEB17, which isolated from bacteria that reside in the soybean rhizosphere and more so under salt stress. Carbon and energy metabolic pathways were affected under both unstressed and salt-stressed conditions when treated with these signals. PEP carboxylase, Rubisco-oxygenase large subunit, and pyruvate kinase were some of the noteworthy proteins enhanced by the signals. Upregulation of other stress-related proteins such as *A. thaliana* membrane-associated progesterone-binding protein, chloroplast proteins, photosystems I and II proteins, LEA proteins, and nodulin-related proteins was observed in the LCO and Th17 treatments with NaCl stress. These findings suggested that the proteome of *A. thaliana* rosettes is altered by the bacterial signals tested, and more so under salt stress, thereby imparting a positive effect on plant growth under high saline conditions (Subramanian et al. 2016).

3.7 Conclusion and Future Perspectives

Salt stress is a global problem that extends all over the world. It is a major threat to the agricultural production, leading to adverse implications for food security, environmental health, and economic welfare. Salinization is increasing worldwide, and plants growing under this worse condition are facing several detrimental effects. Indeed, crop cultivation in saline soils is one of the major challenges facing agriculture today. As human population is growing fast and food requirement is a global problem, then an environmental-friendly, sustainable, and efficient method is required to save the world from pollution-producing technologies, improve agriculture in salty soils, and help solve hunger in the world. Exploitation of plant-microbe interactions is one of the most important eco-friendly biological solutions that can promote crop productivity worldwide. In addition, the ability of microbes to confer stress resistance to plants may provide a novel strategy for mitigating the impacts of global climate change on agricultural and native plant communities. Application and exploration of halotolerant rhizomicrobes for sustainable improvement in growth and production under salt stress hold promise. These organisms have evolved many cellular adaptations, which allow them to survive and grow under extreme saline conditions. They adapt with their saline surroundings by their capability to balance the osmotic pressure of the environment, via either producing compatible organic solutes or accumulating large salt concentrations in their cytoplasm. The applications of these microorganisms trigger recovery of salt-affected soils by directly supporting the establishment and growth of vegetation in soils stressed with salts and thus opening a new and emerging application of microorganisms.

Halotolerant rhizomicrobes can play an effective role in mitigating the unfavorable effect of salt stress on plant development through one or more mechanisms. They can alter the selectivity of Na^+, K^+, and Ca^{2+} and sustain a higher K^+/Na^+ ratio in plant tissues, increase the expression of stress-responsive transcription factors, induce greater osmolyte synthesis, induce phytohormone production, enhance ROS scavenging, and thus improve plant biomass under this harsh condition. Therefore, it requires the exploration of suitable halotolerant microbial strain(s) which can ameliorate salt stress in plants through one or more properties. Moreover, the presence of plant growth-promoting properties in halotolerant PGPB which includes nitrogen fixation, phosphate solubilization, phytohormone production, siderophore production, and biocontrol activities can help the plant grow in a sustainable manner. These microbes could play a significant role in stress management, once their unique properties of tolerance to extremes, their ubiquity, and their genetic diversity are understood, and methods for their successful deployment in agriculture production have been developed. These microorganisms also provide excellent models for understanding stress tolerance mechanisms that can be subsequently engineered into crop plants. Hence, the challenge to develop and popularize an inoculant formulation for saline environment with long shelf life and high efficacy is considered as a promising approach in order to take advantage of halophilic microbe's unique molecules in sustainable agriculture. Intensive selection of halotolerant bioinoculants could lead to development of salt-tolerant crops which could advance the research toward bioengineering of susceptible plant lines. In fact, a lot of effort is still required to make halotolerant rhizomicrobes an efficient technique in sustainable agriculture.

Despite all advantages of halophilic microbes, there are few commercially developed examples so far. Implementation of new approaches such as genomics can be considered as an even more powerful tool in discovering the potentials and applications of halophiles. Indeed, in focusing commercial market of halotolerant rhizomicrobes as biofertilizers, a lot of hard work is still to be done. Also, isolation of new halophilic and halotolerant rhizomicrobes may lead to novel molecules that could be used for applications in different fields.

Selection of the appropriate halophilic microbial inoculants (PGPB, *rhizobia*, or mycorrhizae) is one of the most important technical traits. In addition, using compatible multiple microbial consortia consisting of halotolerant bacterial symbionts and halotolerant fungal symbionts acting synergistically, providing various beneficial effects, is also a powerful strategic tool. Indeed, investigations on interaction of halotolerant PGPR with other halotolerant microbes and their effect on the physiological response of plants under saline conditions are still in incipient stage. Inoculations with selected halotolerant PGPR and other selected halotolerant microbes could serve as the potential tool for alleviating the detrimental effects of salt stress on salt-sensitive crops. Therefore, an extensive investigation is needed in this area, and the use of PGPR and other symbiotic microorganisms can be useful in developing strategies to facilitate sustainable agriculture in saline soils. Future research has to be focused on the application of multi-halotolerant microbial inoculation, which could be an effective approach to reduce harmful impact of salt stress on plant development, but prerequisites for effective combinations need to be established. Furthermore, the challenge in the twenty-first century lays on developing stable multiple stress tolerance traits and thus improving yields particularly in areas with adverse environmental conditions and contributing to global food security.

Genetic techniques may point out to new insight in the alleviating role of halophilic microbial inoculates under salinity stress. Using these microorganisms as an elicitor to increase plant stress tolerance and to incorporate microbial genes into stressed plants is now being addressed and getting the interest of scientists in such studies. Indeed, their use as bioinoculants could help to emerge a new dimension into the microbial inoculant application to plants under abiotic stress conditions. Future research should be focused on genetic techniques, and molecular approaches may indicate new insight in the alleviating role of halotolerant rhizomicrobes under saline conditions, to identify target genes for promoting growth under this harsh environment and to transfer target genes into plants through biotechnology.

References

Ahmad M, Zahir ZA, Nazli F, Akram F, Arshad M, Khalid M (2013) Effectiveness of halo-tolerant, auxin producing *Pseudomonas* and *Rhizobium* strains to improve osmotic stress tolerance in mung bean (*Vigna radiata* L.). Braz J Microbiol 44:1341–1348

Alavi P, Starcher MR, Zachow C, Müller H, Berg G (2013) Root- microbe systems: the effect and mode of interaction of Stress Protecting Agent (SPA) *Stenotrophomonas rhizophila* DSM14405^T. Front Plant Sci 4:141

Anwar T, Chauhan RS (2012) Computational analysis of halotolerance gene from halophilic prokaryotes to infer their signature sequences. Int J Adv Biotechnol Bioinforma 1(1):69–78

Arora S, Patel P, Vanza M, Pao GG (2014) Isolation and characterization of endophytic bacteria colonizing halophyte and other salt tolerant plant species from Coastal Gujarat. Afr J Microbiol Res 8(17):1779–1788

Berthomieu P, Conejero G, Nublat A, Brackenbury WJ, Lambert C, Savio C, Uozumi N, Oiki S, Yamada K, Cellier F, Gosti F, Simonneau T, Essah PA, Tester M, Very A-A, Sentenac H, Casse F (2003) Functional analysis of AtHKT1 in Arabidopsis shows that Na^+ recirculation by the phloem is crucial for salt tolerance. EMBO J 22:2004–2014

Bharti N, Yadav D, Barnawal D, Maji D, Kalra A (2013) *Exiguobacterium oxidotolerans*, a halotolerant plant growth promoting rhizobacteria, improves yield and content of secondary metabolites in *Bacopa monnieri* (L.) Pennell under primary and secondary salt stress. World J Microbiol Biotechnol 29:379–387

Bharti N, Pandey SS, Barnawal D, Patel VK, Kalra A (2016) Plant growth promoting rhizobacteria *Dietzia natronolimnaea* modulates the expression of stress responsive genes providing protection of wheat from salinity stress. Sci Rep 6:34768

Bhattacharjee S (2012) The language of reactive oxygen species signaling in plants. J Bot 2012:1–22

Bhise KK, Bhagwat PK, Dandge PB (2017) Plant growth-promoting characteristics of salt tolerant *Enterobacter cloacae* strain KBPD and its efficacy in amelioration of salt stress in *Vigna radiata* L. J Plant Growth Regul 39:215–226

Blumwald E (2000) Sodium transport and salt tolerance in plants. Curr Opin Cell Biol 12:431–434

Bonfá MRL, Grossman MJ, Piubeli F, Mellado E, Durrant LR (2013) Phenol degradation by halophilic bacteria isolated from hypersaline environments. Biodegradation 24:699–709

Chen TH, Murata N (2002) Enhancement of tolerance of abiotic stress by metabolic engineering of betaines and other compatible solutes. Curr Opin Plant Biol 5:250–257

Chinnusamy V, Schumaker K, Zhu JK (2004) Molecular genetic perspectives on cross-talk and specificity in abiotic stress signaling in plants. J Exp Bot 55:225–236

Ciccarelli FD, Doerks T, von Mering C, Creevey CJ, Snel B, Bork P (2006) Toward automatic reconstruction of a highly resolved tree of life. Science 311:1283–1287

DasSarma S, DasSarma P (2015) Halophiles and their enzymes: negativity put to good use. Curr Opin Microbiol 25:120–126

Detkova EN, Boltyanskaya YV (2007) Osmoadaptation of haloalkaliphilic bacteria: role of osmoregulators and their possible practical application. Microbiology 76:511–522

Edbeib MF, Wahab RA, Huyop F (2016) Halophiles: biology, adaptation, and their role in decontamination of hypersaline environments. World J Microbiol Biotechnol 32:135

Egamberdieva D (2009) Alleviation of salt stress by plant growth regulators and IAA producing bacteria in wheat. Acta Physiol Plant 31:861–864

Essghaier B, Dhieb C, Rebib H, Ayari S, Boudabous ARA, Sadfi-Zouaoui N (2014) Antimicrobial behavior of intracellular proteins from two moderately halophilic bacteria: strain J31 of *Terribacillus halophilus* and strain M3-23 of *Virgibacillus marismortui*. J Plant Pathol Microbiol 5:214

Estrada B, Aroca R, Maathuis FJM, Barea JM, Ruiz-Lozano JM (2013) Arbuscular mycorrhizal fungi native from a Mediterranean saline area enhance maize tolerance to salinity through improved ion homeostasis. Plant Cell Environ 36:1771–1782

Fulda S, Mikkat S, Huang F et al (2006) Proteome analysis of salt stress response in the cyanobacterium *Synechocystis* sp. strain PCC 6803. Proteomics 6:2733–2745

Goel D, Singh AK, Yadav V, Babbar SB, Murata N, Bansal KC (2011) Transformation of tomato with a bacterial codA gene enhances tolerance to salt and water stresses. J Plant Physiol 168(11):1286–1294

Gong X, Zhang J, Liu JH (2014) A stress responsive gene of *Fortunella crassifolia* FcSISP functions in salt stress resistance. Plant Physiol Biochem 83:10–19

Guasch-Vidal B, Estévez J, Dardanelli MS, Soria-Díaz ME, Fernández de Córdoba F, Balog CIA, Manyani H, Gil-Serrano A, Thomas-Oates J, Hensbergen PJ, Deelder AM, Megías M, Van Brussel AAN (2013) High NaCl concentrations induce the nod genes of *Rhizobium tropici* CIAT899 in the absence of flavonoid inducers. Mol Plant-Microbe Interact 26:451–460

Hagemann M (2011) Molecular biology of cyanobacterial salt acclimation. FEMS Microbiol Rev 35(1):87–123

Jewell MC, Campbell BC, Godwin ID (2010) Transgenic plants for abiotic stress resistance. In: Kole C, Michler C, Abbott AG, Hall TC (eds) Transgenic crop plants. Springer, Berlin, pp 67–132

Joghee NN, Jayaraman G (2016) Biochemical changes induced by salt stress in halotolerant bacterial isolates are media dependent as well as species specific. Prep Biochem Biotechnol 46(1):8–14

Kang S-M, Khan AL, Waqas M, You Y-H, Kim J-H, Kim J-G, Hamayun M, Lee I-J (2014) Plant growth-promoting rhizobacteria reduce adverse effects of salinity and osmotic stress by regulating phytohormones and antioxidants in *Cucumis sativus*. J Plant Interact 9:673–682

Karan R, Capes MD, DasSarma S (2012) Function and biotechnology of extremophilic enzymes in low water activity. Aquat Biosyst 8(1):4

Kixmuller D, Greie JC (2012) An ATP-driven potassium pump promotes long-term survival of Halobacterium salinarum within salt crystals. Environ Microbiol Rep 4:234–241

Kolman MA, Nishi CN, Perez-Cenci M et al (2015) Sucrose in cyanobacteria: from a salt-response molecule to play a key role in nitrogen fixation. Life (Basel) 5:102–126

Kolp S, Pietsch M, Galinski EA, Gutschow M (2006) Compatible solutes as protectants for zymogens against proteolysis. Biochim Biophys Acta 1764:1234–1242

Kumar M, Choi JY, Kumari N, Pareek A, Kim SR (2015) Molecular breeding in *Brassica* for salt tolerance: importance of microsatellite (SSR) markers for molecular breeding in *Brassica*. Front Plant Sci 6:688

Kurz M (2008) Compatible solute influence on nucleic acids: many questions but few answers. Saline Syst 4:6

Lai SJ, Lai MC, Lee RJ, Chen YH, Yen HE (2014) Transgenic Arabidopsis expressing osmolyte glycine betaine synthesizing enzymes from halophilic methanogen promote tolerance to drought and salt stress. Plant Mol Biol 85:429

Lanyi JK (1990) Halorhodopsin, a light-driven electrogenic chloride transport system. Physiol Rev 70:319–330

Liu XD, Luo YT, Mohamed OA, Liu DY, Wei GH (2014) Global transcriptome analysis of *Mesorhizobium alhagi* CCNWXJ12-2 under salt stress. BMC Microbiol 14:1

Liu X, Luo Y, Li Z, Wei G (2016) Functional analysis of PrkA – a putative serine protein kinase from *Mesorhizobium alhagi* CCNWXJ12-2 – in stress resistance. BMC Microbiol 16:227

Metwali EMR, Soliman HIA, Fuller MP, Al-Zahrani HS, Howladar SM (2015) Molecular cloning and expression of a vacuolar Na^+/H^+ antiporter gene (AgNHX1) in fig (*Ficus carica* L.) under salt stress. Plant Cell Tissue Organ Cult 123(2):377–387

Miller G, Susuki N, Ciftci-Yilmaz S, Mittler R (2010) Reactive oxygen species homeostasis and signalling during drought and salinity stresses. Plant Cell Environ 33:453–467

Mittler R, Vanderauwera S, Suzuki N, Miller G, Tognetti VB, Vandepoele K, Gollery M, Shulaev V, Van Breusegem F (2011) ROS signaling: the new wave? Trends Plant Sci 16:300–309

Moreno ML, Sanchez-Porro C, Piubeli F, Frias L, Garcia MT, Mellado E (2011) Cloning, characterization and analysis of *cat* and *ben* genes from the phenol degrading halophilic bacterium *Halomonas organivorans*. PLoS One 6:e21049

Munns R, Gilliham M (2015) Salinity tolerance of crops – what is the cost? New Phytol 208:668–673

Nakbanpote W, Panitlurtumpai N, Sangdee A, Sakulpone N, Sirisom P, Pimthong A (2014) Salt-tolerant and plant growth-promoting bacteria isolated from Zn/Cd contaminated soil: identification and effect on rice under saline conditions. J Plant Interact 9:379–387

Naz I, Bano A, Ul-Hassan T (2009) Isolation of phytohormones producing plant growth promoting rhizobacteria from weeds growing in Khewra salt range, Pakistan and their implication in providing salt tolerance to *Glycine max* L. Afr J Biotechnol 8(21):5762–5766

Nia SH, Zarea MJ, Rejali F, Varma A (2012) Yield and yield components of wheat as affected by salinity and inoculation with *Azospirillum* strains from saline or non-saline soil. J Saudi Soc Agric Sci 11:113–121

Noctor G, Foyer CH (2011) Ascorbate and glutathione: the heart of the redox hub. Plant Physiol 155:2–18

Oren A (2008) Microbial life at high salt concentrations: phylogenetic and metabolic diversity. Saline Syst 4:1–13

Orhan F (2016) Alleviation of salt stress by halotolerant and halophilic plant growth-promoting bacteria in wheat (*Triticum aestivum*). Braz J Microbiol 47:621–627

Patel KG, Mandaliya VB, Mishra GP, Dobaria JR, Thankppan R (2016) Transgenic peanut overexpressing mtlD gene confers enhanced salinity stress tolerance via mannitol accumulation and differential antioxidative responses. Acta Physiol Plant 38:181

Paul S, Bag SK, Das S, Harvill ET, Dutta C (2008) Molecular signature of hypersalin adaptation: insights from genome and proteome composition of halophilic prokaryotes. Genome Biol 9:R70

Piernik A, Hrynkiewicz K, Wojciechowska A, Szymańska A, Lis MI, Muscolo A (2017) Effect of halotolerant endophytic bacteria isolated from *Salicornia europaea* L. on the growth of fodder beet (*Beta vulgaris* L.) under salt stress. Arch Agron Soil Sci 63:1–15

Piubeli F, Lourdes Moreno M, Kishi LT, Henrique-Silva F, Garcia MT, Mellado E (2015) Phylogenetic profiling and diversity of bacterial communities in the Death Valley, an extreme habitat in the Atacama Desert. Indian J Microbiol 55:392–399

Plemenitas A, Lenassi M, Konte T, Kejzar A, Zajc J, Gostincar C, Gunde-Cimerman N (2014) Adaptation to high salt concentrations in halotolerant/halophilic fungi: a molecular perspective. Front Microbiol 5:199

Pourbabaee AA, Bahmani E, Alikhani HA, Emami S (2016) Promotion of wheat growth under salt stress by halotolerant bacteria containing ACC deaminase. J Agric Sci Technol 18:855–864

Qiu Q, Guo Y, Dietrich M, Schumaker KS, Zhu JK (2002) Regulation of SOS1, a plasma membrane Na^+/H^+ exchanger in *Arabidopsis thaliana*, by SOS2 and SOS3. Proc Natl Acad Sci USA 99:8436–8441

Quadri I, Hassani II, l'Haridon S, Chalopin M, Hacene H, Jebbar M (2016) Characterization and antimicrobial potential of extremely halophilic archaea isolated from hypersaline environments of the Algerian Sahara. Microbiol Res 186–187:119–131

Qurashi AW, Sabri AN (2013) Osmolyte accumulation in moderately halophilic bacteria improves salt tolerance of chickpea. Pak J Bot 45:1011–1016

Rahnama H, Vakilian H, Fahimi H, Ghareyazie B (2011) Enhanced salt stress tolerance in transgenic potato plants (*Solanum tuberosum* L.) expressing a bacterial mtlD gene. Acta Physiol Plant 33:1521–1532

Rajput L, Imran A, Mubeen F, Hafeez FY (2013) Salt-tolerant PGPR strain *Planococcus rifietoensis* promotes the growth and yield of wheat (*Triticum aestivum*) cultivated in saline soil. Pak J Bot 45:1955–1962

Ramadoss D, Lakkineni VK, Bose P, Ali S, Annapurna K (2013) Mitigation of salt stress in wheat seedlings by halotolerant bacteria isolated from saline habitats. Springer Plus 2(6):1–7

Redman RS, Kim YO, Woodward CJDA, Greer C, Espino L, Doty SL, Rodriguez RJ (2011) Increased fitness of rice plants to abiotic stress via habitat adapted symbiosis: a strategy for mitigating impacts of climate change. PLoS One 6(7):e14823

Rohban R, Amoozegar MA, Ventosa A (2009) Screening and isolation of halophilic bacteria producing extracellular hydrolyses from Howz Soltan Lake, Iran. J Ind Microbiol Biotechnol 36:333–340

Roy SJ, Negrão S, Tester M (2014) Salt resistant crop plant. Curr Opin Biotechnol 26:115–124

Saghafi K, Ahmadi J, Asgharzadeh A, Bakhtiari S (2013) The effect of microbial inoculants on physiological responses of two wheat cultivars under salt stress. Int J Adv Biol Biomed Res 4:421–431

Sarathambal C, Ilamurugu K (2013) Saline tolerant plant growth promoting diazotrophs from rhizosphere of Bermuda grass and their effect on rice. Indian J Weed Sci 45:80–85

Saum SH, Muller V (2008) Regulation of osmoadaptation in the moderate halophile *Halobacillus halophilus*: chloride, glutamate and switching osmolyte strategies. Saline Syst 4:2014

Selvakumar G, Kim K, Hu S, Sa T (2014) Effect of salinity on plants and the role of arbuscular mycorrhizal fungi and plant growth-promoting rhizobacteria in alleviation of salt stress. In: Ahmad P, Wani MR (eds) Physiological mechanisms and adaptation strategies in plants under changing environment. Springer, New York, pp 115–144

Shabani L, Sabzalian MR, Pour SM (2016) Arbuscular mycorrhiza affects nickel translocation and expression of ABC transporter and metallothionein genes in *Festuca arundinacea*. Mycorrhiza 26:67–76

Sharma S, Kulkarni J, Jha B (2016) Halotolerant rhizobacteria promote growth and enhance salinity tolerance in peanut. Front Microbiol 9:1600

Siddikee MA, Chauhan PS, Anandham R, Han GH, Sa T (2010) Isolation, characterization, and use for plant growth promotion under salt stress, of ACC deaminase-producing halotolerant bacteria derived from coastal soil. J Microbiol Biotechnol 20(11):1577–1584

Siddikee MA, Glick BR, Chauhan PS, Yim WJ, Sa T (2011) Enhancement of growth and salt tolerance of red pepper seedlings (*Capsicum annuum* L.) by regulating stress ethylene synthesis with halotolerant bacteria containing 1-aminocyclopropane-1-carboxylic acid deaminase activity. Plant Physiol Biochem 49(4):427–434

Singh RP, Jha PN (2016) A halotolerant bacterium *Bacillus licheniformis* HSW-16 augments induced systemic tolerance to salt stress in wheat plant (*Triticum aestivum*). Front Plant Sci 7:1890

Soares MA, Li H, Kowalski KP, Bergen M, Torres MS, White JF (2016) Evaluation of the functional roles of fungal endophytes of *Phragmites australis* from high saline and low saline habitats. Biol Invasions 18:2689–2702

Soontharapirakkul K, Promden W, Yamada N, Kageyama H, Incharoensakdi A, Iwamoto-Kihara A, Takabe T (2011) Halotolerant cyanobacterium *Aphanothece halophytica* contains an Na^+-dependent F_1F_0-ATP synthase with a potential role in salt-stress tolerance. J Biol Chem 286(12):10169–10176

Sorty AM, Meena KK, Choudhary K, Bitla UM, Minhas PS, Krishnani KK (2016) Effect of plant growth promoting bacteria associated with halophytic weed (*Psoralea corylifolia* L) on germination and seedling growth of wheat under saline conditions. Appl Biochem Biotechnol 180:872–882

Subramanian S, Souleimanov A, Smith DL (2016) Proteomic studies on the effects of lipochitooligosaccharide and thuricin 17 under unstressed and salt stressed conditions in *Arabidopsis thaliana*. Front Plant Sci 7:1314

Talaat NB (2014) Effective microorganisms enhance the scavenging capacity of the ascorbate-glutathione cycle in common bean (*Phaseolus vulgaris* L.) plants grown in salty soils. Plant Physiol Biochem 80:136–143

Talaat NB (2015a) Effective microorganisms improve growth performance and modulate the ROS-scavenging system in common bean (*Phaseolus vulgaris* L.) plants exposed to salinity stress. J Plant Growth Regul 34:35–46

Talaat NB (2015b) Effective microorganisms modify protein and polyamine pools in common bean (*Phaseolus vulgaris* L.) plants grown under saline conditions. Sci Hortic 190:1–10

Talaat NB, Shawky BT (2011) Influence of arbuscular mycorrhizae on yield, nutrients, organic solutes, and antioxidant enzymes of two wheat cultivars under salt stress. J Plant Nutr Soil Sci 174:283–291

Talaat NB, Shawky BT (2013) Modulation of nutrient acquisition and polyamine pool in salt-stressed wheat (*Triticum aestivum* L.) plants inoculated with arbuscular mycorrhizal fungi. Acta Physiol Plant 35:2601–2610

Talaat NB, Shawky BT (2014a) Protective effects of arbuscular mycorrhizal fungi on wheat (*Triticum aestivum* L.) plants exposed to salinity. Environ Exp Bot 98:20–31

Talaat NB, Shawky BT (2014b) Modulation of the ROS-scavenging system in salt-stressed wheat plants inoculated with arbuscular mycorrhizal fungi. J Plant Nutr Soil Sci 177:199–207

Talaat NB, Shawky BT (2015) Plant-microbe interaction and salt stress tolerance in plants. In: Wani SH, Hossain MA (eds) Managing salt tolerance in plants: molecular and genomic perspectives. CRC Press/Taylor & Francis Group, Oxford, pp 267–289

Talaat NB, Ghoniem AE, Abdelhamid MT, Shawky BT (2015a) Effective microorganisms improve growth performance, alter nutrients acquisition and induce compatible solutes accumulation in common bean (*Phaseolus vulgaris* L.) plants subjected to salinity stress. Plant Growth Regul 75:281–295

Talaat NB, Shawky BT, Ibrahim AS (2015b) Alleviation of drought induced oxidative stress in maize (*Zea mays* L.) plants by dual application of 24-epibrassinolide and spermine. Environ Exp Bot 113:47–58

Tani C, Sasakawa H (2003) Salt tolerance of *Casuarina equisetifolia* and *Frankia* Ceql strain isolated from the root nodules of *C. equisetifolia*. Soil Sci Plant Nutr 49:215–222

Tank N, Saraf M (2010) Salinity-resistant plant growth promoting rhizobacteria ameliorates sodium chloride stress on tomato plants. J Plant Interact 5:51–58

Tasseva G, Richard L, Zachowski A (2004) Regulation of phosphatidylcholine biosynthesis under salt stress involves choline kinases in *Arabidopsis thaliana*. FEBS Lett 566:115–120

Tiwari S, Singh P, Tiwari R, Meena KK, Yandigeri M, Singh DP, Arora DK (2011) Salt-tolerant rhizobacteria-mediated induced tolerance in wheat (*Triticum aestivum*) and chemical diversity in rhizosphere enhance plant growth. Biol Fertil Soils 47:907–916

Trivedi R (2017) Ecology of saline soil microorganisms. In: Arora S, Singh AK, Singh YP (eds) Bioremediation of salt affected soils: an Indian perspective. Springer, Cham, pp 157–172

Ullah S, Bano A (2015) Isolation of PGPRs from rhizospheric soil of halophytes and its impact on maize (*Zea mays* L.) under induced soil salinity. Can J Microbiol 11:1–7
Vannier N, Mony C, Bittebière AK, Vandenkoornhuyse P (2015) Epigenetic mechanisms and microbiota as a toolbox for plant phenotypic adjustment to environment. Front Plant Sci 6:1159
Waditee R, Bhuiyan NH, Rai V, Aoki K, Tanaka Y, Hibino T, Suzuki S, Takano J, Jagendorf AT, Takabe T (2005) Proc Natl Acad Sci USA 102:1318–1323
Waditee R, Bhuiyan NH, Hirata E, Hibino T, Tanaka Y, Shikata M, Takabe T (2007) Metabolic engineering for betaine accumulation in microbes and plants. J Biol Chem 282(47):34185–34193
Waditee R, Sittipol D, Tanaka Y, Takabe T (2012) Overexpression of serine hydroxymethyltransferase from halotolerant cyanobacterium in *Escherichia coli* results in increased accumulation of choline precursors and enhanced salinity tolerance. FEMS Microbiol Lett 333:46–53
Weyens N, van der Lelie D, Taghavi S, Newman L, Vangronsveld J (2009) Exploiting plant-microbe partnerships to improve biomass production and remediation. Trends Biotechnol 27:591–598
Yang X, Liang Z, Wen X, Lu C (2008) Genetic engineering of the biosynthesis of glycine betaine leads to increased tolerance of photosynthesis to salt stress in transgenic tobacco plants. Plant Mol Biol 66:73–86
Yokoi S, Quintero FJ, Cubero B, Ruiz MT, Bressan RA, Hasegawa PM, Pardo JM (2002) Differential expression and function of *Arabidopsis thaliana* NHX Na^+/H^+ antiporters in the salt stress response. Plant J 30:529–539
Zhang G, Yi L (2013) Stability of halophilic proteins: from dipeptide attributes to discrimination classifier. Int J Biol Macromol 53:1–6
Zhong NQ, Han LB, Wu XM, Wang LL, Wang F, Ma YH, Xia GX (2012) Ectopic expression of a bacterium NhaD-type Na^+/H^+ antiporter leads to increased tolerance to combined salt/alkali stresses. J Integr Plant Biol 54(6):412–421
Zhou H, Wang H, Huang Y, Fang T (2016) Characterization of pyrene degradation by halophilic *Thalassospira* sp. strain TSL5-1 isolated from the coastal soil of Yellow Sea, China. Int Biodeterior Biodegrad 107:62–69

Chapter 4
Regulation and Modification of the Epigenome for Enhanced Salinity Tolerance in Crop Plants

Minoru Ueda, Kaori Sako, and Motoaki Seki

Abstract Histone modifications (acetylation, methylation, phosphorylation, etc.), histone variants, regulatory RNAs, and DNA methylation represent the functional elements of epigenetics. They serve as a basis for regulating biological processes such as flowering and germination, as well as environmental stress responses in plants. Chromatin modifications can also function to prime plants to respond to adverse environmental conditions and act as short-term or long-term (transgenerational) stress memory, enabling to be preadapted to the prevailing environment. Recognition of the importance of epigenetic regulation in biological processes is increasing, including its role in salinity stress response, although many details are still lacking. To date, only a few studies in crop plants have provided evidence for epigenetic changes that occur in response to salinity and that result in increasing tolerance to salinity stress. In the current review, we discuss insights into the involvement of epigenetic regulatory elements, such as histone modifications, histone variants, regulatory RNAs, and DNA methylation, in salinity stress response based on studies in the model plant, *Arabidopsis*. In particular, knowledge of the involvement of histone acetylation in salt-stress response has increased, and various chemical compounds capable of regulating levels of histone acetylation have been identified.

M. Ueda · K. Sako
Plant Genomic Network Research Team, RIKEN Center for Sustainable Resource Science, Yokohama, Kanagawa, Japan

Core Research for Evolutional Science and Technology (CREST), Japan Science and Technology Agency (JST), Kawaguchi, Saitama, Japan

M. Seki (✉)
Plant Genomic Network Research Team, RIKEN Center for Sustainable Resource Science, Yokohama, Kanagawa, Japan

Plant Epigenome Regulation Laboratory, RIKEN Cluster for Pioneering Research, Wako, Saitama, Japan

Core Research for Evolutional Science and Technology (CREST), Japan Science and Technology Agency (JST), Kawaguchi, Saitama, Japan

Kihara Institute for Biological Research, Yokohama City University, Yokohama, Kanagawa, Japan
e-mail: motoaki.seki@riken.jp

V. Kumar et al. (eds.), *Salinity Responses and Tolerance in Plants, Volume 2*,
https://doi.org/10.1007/978-3-319-90318-7_4

Based on the available evidence, we provide a perspective on the potential use of chemical epigenetic modifiers, which function as histone deacetylase (HDAC) inhibitors, for enhancing stress tolerance in crops such as cassava.

Keywords Histone acetylation · HDAC · HDAC inhibitor · Histone modification · *Arabidopsis* · DNA methylation · Cassava

Abbreviations

HDAC	Histone deacetylase
HDC	Histone deacetylation complex 1 (HDC1)
LEA	Late embryogenesis abundant
lncRNA	Long noncoding RNA
ncRNA	Noncoding RNA
NHX	Sodium hydrogen exchanger
RdDM	RNA-directed DNA methylation
ROS1	Repressor of transcriptional gene silencing 1 (ROS1)
siRNAs	Short interfering RNA

4.1 Introduction

Waddington defined the term *epigenetics* in the early 1940s (referenced in Jablonka and Lamb 2002). Since then, knowledge of the role of epigenetic regulation in abiotic stress response has gradually increased, starting with McClintock (1984) who first recognized the relationship between epigenetics and stress response (McClintock 1984; Kim et al. 2015; Asensi-Fabado et al. 2016; Provart et al. 2016).

Soil salinity is one of the leading factors that hinder crop production globally, and the development of plants that are more tolerant to salinity stress is considered to be a critical goal that needs to be addressed. Soil salinity is known to affect plant cells in two ways. One is inducing a water deficit due to the high concentration of salt in the soil surrounding a plant which results in decreased water uptake by roots, thus inducing an osmotic stress. The other effect is the high accumulation of salt within a plant which alters the Na^+/K^+ ratio, as well as creating excessive levels of Na^+ and Cl^- ions, thus resulting in ion cytotoxicity (Munns and Tester 2008; Julkowska and Testerink 2015). Details on salinity stress response are described in earlier chapters in volumes 1 and 2. Earlier studies revealed that several mechanisms, including maintenance of ion homeostasis, accumulation of compatible solutes, hormonal adjustments, the cellular antioxidant system, and Ca^{2+} signaling, play an essential role in the ability of plants to adapt to and survive salinity stress

(Jia et al. 2015). Key elements that ameliorate the potential damage of high salt levels to plant cells have been reported. These include transporters for ion homeostasis [such as NHX1 (Apse et al. 1999), SOS1/2/3 (Shi et al. 2003; Yang et al. 2009), and HKT1 (Moller et al. 2009)] and compatible solutes that act as osmolytes to maintain osmotic homeostasis [such as proline (Kishor et al. 1995) and glycine-betaine (Sakamoto and Murata 1998)]. Other cellular components that have been reported to play a role in salinity response and adaptation include late embryogenesis abundant (LEA) proteins (Xu et al. 1996) and enzymes that function in the biosynthesis of antioxidants, including GST/GPX (Roxas et al. 1997) and SOD (McKersie et al. 1999). Epigenetic elements that regulate the expression of genes coding for transporters involved in ion homeostasis and osmolyte accumulations have only recently been reported.

In this review, we summarize the studies on how epigenetic elements, including histone modifications, histone variants, regulatory RNAs, and DNA methylation, affect the expression of the key components of salt-stress response described above. In particular, our understanding of the involvement of histone modifications in salinity stress response has significantly progressed. The details of how salt response is mediated through histone variants and regulatory RNAs, however, are still poorly understood. Lastly, we provide a perspective on the potential use of chemical epigenetic modifiers, which function as histone deacetylase (HDAC) inhibitors, for enhancing stress tolerance in crops such as cassava.

4.2 Histone Acetylation

Histones are DNA-packaging proteins that provide stability to the genome by preventing physical genotoxicity (e.g., DNA breaks) (Luger et al. 1997; Downs et al. 2007). Their original function during the course of evolution, even prior to the divergence of the Archaea and Eukarya, was to act as regulators of mRNA expression (Ammar et al. 2012). A variety of chemical modifications (acetylation, methylation, phosphorylation, etc.) can occur to the N-tails of histones, which is one of the properties that enable them to regulate mRNA expression, a function that is generally conserved in eukaryotes (Jenuwein and Allis 2001; Kouzarides 2007). Chromatin possesses a diverse array of chemical moieties, which allows to contain and transmit information independent of the genetic code (i.e., epigenetic) and regulate gene expression. Epigenetic regulation is considered to play a major role in plant development and the ability of plants to adapt to the prevailing environment.

Among histone modifications, the mode of action of histone acetylation is relatively well understood. Positively charged lysine residues within the N-tails of histones are often the targets of histone acetylation. These positive charged amino acid residues can bind to the negatively charged region of the nucleosome (proteins and phosphate groups of DNAs), thus impacting chromatin structure. Acetylation of histones, however, neutralizes the positive charges on lysine residues and thus reduces the ability of histones to bind to the nucleosomes. This results in producing a relaxed, open chromatin structure, which facilitates the recruitment of transcriptional factors to DNA, and the subsequent activation of transcription (Shahbazian and Grunstein 2007). Acetylation levels are regulated by the balanced activity of histone acetyltransferases (HATs), which act as writers of histone acetylation, and histone deacetylases (HDACs), which act as erasers of acetylation. Recent studies have revealed that cross talk between writers and erasers of histone acetylation harmonize the regulation of abiotic stress responses, including salinity stress (Kim et al. 2015; Asensi-Fabado et al. 2016).

In maize (*Zea mays* L.) roots, the upregulation of cell wall-related genes, such as *ZmEXPB2* and *ZmXET1*, has been associated with an increase in H3K9 acetylation in the promoter and coding regions of genes. This acetylation is thought to be necessary for a response to high salinity conditions to occur. The upregulation of *ZmEXPB2* and *ZmXET1* genes has been speculated to be mediated by two *HAT* genes (*ZmHATB* and *ZmGCN5*) because mRNA expression of these *HAT* genes was found to increase under salt-stress conditions (Li et al. 2014). Consistent with this finding, an *Arabidopsis ada2b-1* mutant is hypersensitive to salt stress and is deficient in transcriptional coactivators that complex with GCN5. In the *ada2b-1* mutant, the acetylation levels of histones H3K9/K14 are significantly reduced in areas containing the salt-stress-inducible genes, *RESPONSIVE TO ABA18*, *COLD-RESPONSIVE 6.6*, and *RESPONSIVE TO DESSICATION29B* (Kaldis et al. 2011). There is no evidence, however, that upregulation of these salt-responsive genes results in increased tolerance to salinity stress. These data do at least suggest that HATs may act as positive regulators of salt-stress response in plants.

The role of HDAC in salt-stress response appears to be more complex. Recent evidences indicate that *HDAC* genes have a significant role in regulating salt-stress response. The data indicate that HDAC inhibition contributes to both increasing and decreasing tolerance to salinity stress. Based on their catalytic domain, HDACs are categorized into zinc-dependent and nicotinamide adenine dinucleotide (NAD(+)) types. The reduced potassium deficiency 3 (RPD3)-like and the silent information regulator (SIR) 2-like (sirtuin) gene families are zinc-dependent and NAD(+)-dependent HDACs, respectively. The RPD3-like family is further divided into three classes (I, II, and IV), based on their homology to yeast HDACs (Bolden et al. 2006; Seto and Yoshida 2014; Verdin and Ott 2015). Plants have also evolved a plant-specific HDAC (HD-tuin) family (Brosch et al. 1996; Lusser et al. 1997; Hollender and Liu 2008). The *Arabidopsis* genome encodes 18 HDAC genes, representing 3 HDAC families (Hollender and Liu 2008). This includes 12 RPD3-like family proteins, 2 sirtuin family proteins, and 4 HD-tuin family proteins. Recent studies have reported that several class I and II RPD3-like family genes and HD-tuin family proteins are involved in salt-stress response.

HDA9 negatively regulates salt-stress response (Zheng et al. 2016), whereas HDA6, HD2C, and HD2D positively regulate it (Chen et al. 2010; Chen and Wu 2010; Luo et al. 2012; Perrella et al. 2013; Han et al. 2016; Mehdi et al. 2016). In the case of HDA19, there are controversial data that *athd1*, an *hda19* knockout mutant in the Ws background, shows sensitivity to salt stress (Chen and Wu 2010), whereas an *hda19* repression line in the Col-0 background exhibits the opposite phenotype (Mehdi et al. 2016). Recently, it was reported that two recessive alleles of *hda19* showed tolerance to salinity stress in the Col-0 background (Ueda et al. 2017). At least in Col-0, HDA19 functions as a negative regulator in salt-stress response. Their positive and negative regulation mechanisms underlying salt-stress response are reported as follows.

Under salinity stress conditions, the mRNA expression of the ABA and abiotic stress-responsive genes, *ABI1*, *ABI2*, *KAT1*, *KAT2*, *DREB2A*, *RD29A*, and *RD29B*, was decreased in *axe1-5* and *HDA6* RNA-interfering plants, resulting in a salt-sensitive phenotype (Chen et al. 2010). Complex formation of HD2C with HDA6 was confirmed, and their single and double mutants show increased histone H3K9K14 acetylation: suggesting that HD2C and HDA6 proteins cooperatively regulate salinity stress response (Luo et al. 2012). Consistent with the *hd2c* phenotype, overexpression of *HD2C* resulted in enhanced tolerance to salt stress (Sridha and Wu 2006). Taken together, the functional association of HDA6 with HD2C acts as a positive regulator in salt-stress response.

In *Arabidopsis*, HDA9 and HDA19, which are categorized as class I HDACs, regulate salt-stress response via different pathways. In the case of *hda9*, a suppression of water deprivation-related genes is observed (Zheng et al. 2016). Additionally, HDA9 also forms a complex with a SANT domain-containing protein, POWERDRESS (PWR), and a transcription factor, WRKY53, contributing to the suppression of leaf senescence (Chen et al. 2016). The HDA9/PWR/WRKY53 complex controls leaf senescence through multiple pathways that are coordinated by ABA, JA, and autophagy-related genes, such as NPX1, WRKY57, and APG9 (Chen et al. 2016). ABI3 and ABA receptors (PYL4, PYL5, and PYL6) are positive regulators of ABA signaling and are considered as direct targets of HDA19 (Ryu et al. 2014; Mehdi et al. 2016). Goyal et al. (2005) reported that mRNA levels of LEA proteins, which function in preventing protein aggregation, are enhanced in plants that are deficient in HDA19. Additionally, the level of a rate-limiting enzyme [delta1-pyrroline-5-carboxylate synthetase 1 (P5CS1)], which is involved in the biosynthesis of proline, is also enhanced in HDA19-deficient plants (Ueda et al. 2017). Collectively, these data suggest that HDA9 and HDA19 have different modes of action in the regulation of the ABA signaling pathway. Interactive factors of HDA19, such as MSI1 (multicopy suppressor of ira1), SIN3 (SWI-INDEPENDENT3)-like 2 (SNL2), SNL3, and SNL4, and histone deacetylation complex1 (HDC1) are physically associated with HDA19. Among these interacting factors, MSI1, at least, appears to fine-tune salt-stress response through ABA signaling with HDA19 (Mehdi et al. 2016). There are five MSI family members (MSI1 to MSI5) in *Arabidopsis* (Hennig et al. 2005). The physical interactive partner of HDA6 is not MSI1 but rather MSI4 and MSI5 (Gu et al. 2011). It is plausible that the diversification of interactive partners associated with different HDACs and chromatin components might be responsible for the contrasting phenotypes observed between HDA6 and HDA19 in regard to salt-stress response, even though HDA6 is also a class I HDAC and both HDA6 and HDA19 are able to form a complex with HDC1 (Hollender and Liu 2008; Perrella et al. 2013). Further detailed analysis of complex-forming factors associated with each HDAC protein will provide a greater understanding of how each HDAC protein antagonistically or cooperatively regulates salt-stress response.

The hierarchal regulation of salt-stress response through HDAC isoforms has also been reported. In eukaryotes, class II HDACs are grouped based on their primary structural similarity and the presence of a common motif (LEGGY motif) (Hollender and Liu 2008; Tran et al. 2012). There are four *HDAC* genes in *Arabidopsis* and a quadruple mutant exhibited increased sensitivity to salt stress (Ueda et al. 2017). In the quadruple mutant, the mutagenesis of *HDA19* also resulted in enhanced salt-stress tolerance, indicating that the suppression of HDA19 masks the phenotype produced by the role of class II HDACs in salt-stress response (Ueda et al. 2017). Further analysis is needed to determine whether or not the hierarchal regulation of salt-stress response takes place in a dependent or an independent pathway with HDAC isoforms. Such studies will clarify the actual cross talk that occurs between HDACs in response to salt stress.

4.3 Histone Methylation and Phosphorylation

The mode of action of histone methylation is similar to histone acetylation. The resulting modification of the histone by methylation increases the basicity and hydrophobicity of histone tails and thus alters its affinity to chromatin or transcription factors (Teperino et al. 2010). Histone methylation is regulated by histone methyltransferases (HMTs) and histone demethylases (HDMs). Arginine and lysine residues are methylated by different proteins, namely, arginine methyltransferases (PRMTs) and histone lysine methyltransferases (HKMTs), respectively. The *Arabidopsis* genome encodes 41 genes for SET domain proteins, which are putative candidates for HKMTs, and 9 genes that code for PRMTs (Liu et al. 2010 and references therein). Similar to HMTs, HDMs are divided into two classes, lysine-specific demethylases (LSD) and hydroxylation by Jumonji C (JmjC) domain-containing proteins (JMJ). These proteins facilitate the removal of methyl groups from methylated lysine residues in an independent catalytic reaction. The *Arabidopsis* genome encodes 4 *LSD* and 21 *JMJ* genes (Liu et al. 2010). *HMTs* and *HDMs* comprise a larger gene family than *HAT* and *HDAC* gene families in plants.

In contrast to acetylation, the degree of methylation of a single lysine residue (mono-, di-, or trimethylated) or a single arginine residue (mono- or dimethylated) is linked to distinct biochemical properties and transcriptional response (Teperino et al. 2010; Xiao et al. 2016). For example, H3K4me3, H3K9me3, and H3K36me2 marks tend to be localized on actively transcribed genes, while H3K27me3 marks are generally detected on genes whose transcription is repressed (Roudier et al. 2009; Liu et al. 2010). The variable degrees of methylation and the diversification of writers and erasers of histone methylation are considered to be parameters that allow plants to regulate gene expression in response to stress response more precisely than histone acetylation. To the best of our knowledges, however, there is limited evidence for the involvement of histone methylation in salt-stress response.

The expression of most stress-responsive genes is positively correlated with the addition of H3K4me3 marks, which are often associated with transcriptional activation (Zhang et al. 2009; Kim et al. 2012). An eraser for the H3K4me3 mark, however, is associated with increased tolerance to salinity stress. Gain of function mutants (*jmj15-1* and *jmj15-2*) for the histone demethylase Jumonji C domain-containing protein 15 (JMJ15) exhibited increased tolerance to salinity stress, while a loss of function mutant (*jmj15-3*) exhibited increased sensitivity to salinity stress (Shen et al. 2014). Although JMJ15 preferentially represses gene expression through demethylation of H3K4me2/3, the direct targets responsible for increased tolerance to salinity stress that are JMJ15-dependent are not known at present. An increase in lignin content has been observed in the stems of *jmj15-1* and *jmj15-2* mutants. Shafi et al. (2015) reported that the basal level of H_2O_2 present in the mutants acts as messenger to activate the transcription of genes for lignin biosynthesis in vascular tissue, which is associated with an increase in salt-stress tolerance. This suggests that the gain of function in *JMJ15* may adjust the expression of genes involved in cell response to reactive oxygen species (ROS), thus increasing plant tolerance to oxidative stress, salinity stress, and abiotic stress in general (Choudhury et al. 2017). Significant alterations, such as an increase in H3K4me3 and a decrease in H3K9me2, are observed in salt-sensitive *hda6 Arabidopsis* plants (Luo et al. 2012). Therefore, further identification and analysis of writers or other histone modifications that regulate salt-stress response through modifications of histone H3 on lysine 4 (H3K4) would greatly improve our understanding of the epigenetic regulation of salt-stress response via histone methylation.

The shortening and fractionation of H3K27me3 islands by priming plants with a mild salt stress appear to be the mechanism through which the priming increases drought tolerance (Sani et al. 2013). H3K27me3 is believed to function in repressing gene transcription (Li et al. 2007), and the priming treatment induces a unique epigenetic mark around *HKT1*. Sani et al. (2013) suggest that the shortening and fractionation of the suppressive mark may enhance drought tolerance by inducing of *HKT1* mRNA expression, whose overexpression is associated with an increase in salinity stress tolerance (Moller et al. 2009). The erasers or writers at lysine 27 of histone H3 (H3K27), such as ATXR5/6, SDG1/5/10, and JMJ11/12/30/32, may represent candidates for epigenetic modifications that improve salinity stress tolerance (Xiao et al. 2016).

Phosphorylation of histone H3 is induced by osmotic stress. This mark seems to play a pivotal role in the maintenance of proper heterochromatic organization rather than in the response to salt stress (Wang et al. 2015). Currently, evidence for the involvement of histone modifications in salt-stress response is limited to acetylation and methylation. Further studies, however, will identify other histone modifications that play a pivotal role in salinity stress response.

4.4 DNA Methylation

In plants, cytosine at both CG and non-CG (CHG and CHH where H is A, T or C) sites can be methylated. DNA methylation is a stable and heritable modification and is an important repression mark for transcription and suppression of transposable elements, the latter of which serves to protect the integrity of the genome (Gallusci et al. 2017). Salinity stress has an impact on the pattern of DNA methylation resulting in either hyper- or hypomethylation depending on the plant species (Kovarik et al. 1997; Al-Lawati et al. 2016). In soybean, DNA methylation levels in some salt-responsive transcriptional factors were reduced under salt-stress conditions (Song et al. 2012). Likewise, HPLC analysis of wheat revealed that global cytosine methylation levels also decreased in response to salinity stress conditions (Wang et al. 2014). A comparison of a salt-tolerant wheat cultivar (SR3) and a salt-sensitive wheat cultivar (JN177) revealed that the level of DNA methylation in some salt-responsive genes changed in response to salt stress and that transcription was upregulated in SR3 (Wang et al. 2014). Similarly, Garg et al. performed whole-genome bisulfite sequencing in a salt-sensitive rice cultivar (IR64) and a salt-tolerant rice cultivar (Pokkali) (Garg et al. 2015). DNA methylation patterns were different in the two varieties, and the differences were associated with the elevated transcription of abiotic stress-related genes. These results suggest that changes in the pattern of DNA methylation that occurs in salt-tolerant cultivars may contribute to increased levels of salinity tolerance and represent a feature applicable to crops in general.

DNA demethylation plays an important role in salt tolerance in plants. The Demeter family of 5-methylcytosine DNA glycosylases plays a pivotal role in removing 5-methylcytosine (Zhu 2009). In tobacco, the overexpression of *Arabidopsis* repressor of transcriptional gene silencing 1 (ROS1), a member of the Demeter family, resulted in enhanced salt-stress tolerance. DNA methylation levels in these transgenic tobacco plants decreased in genes encoding enzymes involved in flavonoid biosynthesis and antioxidant pathways, and transcription levels of these genes were upregulated in response to salt stress (Bharti et al. 2015). Decreased DNA methylation levels, however, are not always associated with enhanced tolerance to salt stress. A defective mutant of methyltransferase 1 (MET1) that maintains CG methylation exhibited decreased levels of global DNA methylation but was sensitive to salt stress (Yao et al. 2012). The authors speculated that the reduction in the level of global DNA methylation increased the level of strand breaks that occurred in the high salt-stress condition.

DNA methylation in plants is believed to work as a "transgenerational stress memory" that transmits parental stress memory to nonstressed offspring (Hauser et al. 2011). Wibowo et al. demonstrated that DNA methylation enables "short-term stress memory" in *Arabidopsis*. Repeated exposure to high levels of salt stress induced changes in the pattern of DNA methylation at non-CG sites in sequences located adjacent to stress-responsive genes (Wibowo et al. 2016). Some of those changes were then transmitted to the progeny through the female germ line. These epigenetic changes were associated with a heritable salt-tolerant phenotype. The stress memory, however, was gradually reset in the absence of the salt stress. Epigenetic variation allows plants to adapt to their environment; however, the cost of resetting epigenetic marks also needs to be taken into account. Plants may or may not employ epigenetic plasticity depending on the level of fluctuation in the environment.

4.5 Regulatory RNAs

In addition to messenger RNA (mRNA), RNAs referred to as nonprotein coding RNAs (noncoding RNAs; ncRNA) have wide-ranging regulatory effects on biological processes. ncRNAs are categorized according to their function as follows: transfer RNA functions in translation, ribosomal RNA serves as a component of ribosomes, small nuclear RNAs are involved in splicing, microRNAs function in translational repression, and siRNAs are involved with mRNA cleavage (Eddy 2001). miRNAs and siRNAs play a pivotal role in gene silencing in different ways. Recent evidence revealed that siRNAs recruit DNA and histone methylation in a locus-specific manner (Zilberman et al. 2003; Law and Jacobsen 2010). In the case of DNA methylation, small RNA-directed DNA methylation (RdDM) independently regulates DNA methylation separately from the DDM1-dependent pathway (Law and Jacobsen 2010; Zemach et al. 2013; Matzke and Mosher 2014). Mutant *ddm1* plants exhibit sensitivity to salt stress (Yao et al. 2012), and the salt-stress-responsive transcriptional factor, MYB74, whose overexpression induces increased sensitivity to salt stress during seed germination, is regulated by RdDM-dependent pathway. There are a few examples of epigenetic regulation of salt-stress response that are mediated through siRNAs; however, recent evidence has revealed that ncRNAs also appear to control DNA methylation (Yang et al. 2016; Matsui et al. 2017). Therefore, as more ncRNAs are identified, it is expected that additional examples of the regulation of salt-stress response genes by the RdDM pathway will be uncovered.

Transcriptome analyses have identified numerous long noncoding RNAs (lncRNAs) in eukaryotes. Determining if an RNA sequence is an mRNA or an lncRNA can be problematic because small peptides translated from lncRNA have been occasionally discovered in eukaryotes (Kung et al. 2013 and references therein). It appears that lncRNAs act to control various stages of mRNA processing and stability (Kung et al. 2013); however, they also seem to have a unique role in modulating epigenetic regulators. In animals, lncRNAs participate in recruiting DNA methyltransferases and X-chromosome inactivation (Lee 2012). COLDAIR is an lncRNA in *Arabidopsis* that targets polycomb repressive complex 2 (PRC2) to the FLC locus, resulting in epigenetic silencing (Heo and Sung 2011). Numerous stress-responsive lncRNAs have been identified in *Arabidopsis* (Di et al. 2014), and a drought-induced lncRNA (DRIR) has been reported to be associated with enhancing tolerance to both drought and salinity stress (Qin et al. 2017). Details on the specific function of DRIR are still lacking; however, it is evident that lncRNAs that control the recruitment of epigenetic modulators, such as Xist and COLDAIR, to regions where genes for salinity stress response are encoded will be discovered in future studies.

4.6 Histone Variants

Histones are a core component of chromatin and comprise a large gene family in plants (Sequeira-Mendes and Gutierrez 2015). H2A, H2B, H3, and H4 histones are known as core histones and constitute histone octamers. H1/H5 histones function as linker histones. Although histone proteins are relatively highly conserved in eukaryotes, some differences may exist between variants of histones that influence their affinity to DNA or histone-binding proteins that complex with DNA to form chromatin. The variation in the binding affinity of histones could serve as a basis for developmental stage-specific or stress-specific responses based on histone modifications. For example, among the variants of H2A, H2A.Z has a relatively distinct function in thermosensory and drought stress responses (Kumar and Wigge 2010; Sura et al. 2017). No histone variants have been reported, however, that preferentially mediate salt-stress response.

4.7 Chemical Epigenetic Modifiers: Controlling Epigenetic Modifications to Increase Salinity Stress Tolerance

Inhibiting DNA methylation and the activity of histone-modifying enzymes is a promising therapeutic approach to treat diseases in humans, especially cancer. Thus, research to identify epigenetic inhibitors has rapidly increased. Inhibitors of DNA methylation are classified as nucleoside analogs or non-nucleoside compounds (Xu et al. 2016). The nucleoside analogs are mainly derivatives of cytidine, such as 5-azacytidine (azacitidine) and 5-aza-2′-deoxycytidine (decitabine). Non-nucleoside compounds include inhibitors of DNA methyltransferases in mammals, including RG108 (Fahy et al. 2012). To date, there is no evidence that the use of these inhibitors increases salinity stress tolerance in plants. Development of a selective inhibitor against histone methyltransferases in plants, however, could potentially contribute to enhancing stress tolerance since overexpression of *JMJ15*, an HDM, has been reported to increase stress tolerance (Shen et al. 2014) (see Sect. 4.3).

Chemical inhibition of HDAC proteins is a potential approach that should be evaluated for its ability to increase salinity stress tolerance in plants under field conditions. Genetic mutation analysis of *HDAC* genes revealed that HDAC inhibition in a recessive mutation could both positively and negatively alter salt-stress response. Multiple disruption of class I (*HDA19*) and class II *HDACs* genes resulted in an increase in salinity stress tolerance, suggesting that hierarchal regulation allows the inhibition of even nonclass selective HDACs to increase stress tolerance (see Sect. 4.2). The screening of 13 different selective HDIs [class I selective HDIs, FK228, Ky-2, MC1293, MGCD-0103, and MS-275; class II selective HDIs, MC1568, TMP195, TMP269, and Tubastatin A; and nonclass selective HDIs, JNJ-26481585, LBH-589, sodium butyrate (NaBT), and trichostatin A (TSA)] revealed that 8 of the HDIs (FK228, JNJ-26481585, Ky-2, LBH-589, MC1293, MS-275, sodium butyrate (NaBT), and Trichostatin A (TSA)) clearly increased salinity stress tolerance in *Arabidopsis* (Sako et al. 2016; Ueda et al. 2017). Consistent with the genetic analysis of salt-stress response, a pharmacological approach demonstrated that either selective or nonselective inhibition using class I HDIs enhances salinity stress tolerance. The phenotype obtained through pharmacological or genetic methods strongly suggests that suppression of HDAC activity enhances salinity stress tolerance. Among the pharmacological inhibitors, application of Ky-2 to increase tolerance to salt stress in young seedlings has been analyzed in detail. Treatment of seedlings with Ky-2 resulted in global hyper-acetylation and the upregulation of approximately 2000 genes, including salt-responsive genes. Furthermore, since Ky-2 treatment of *sos1* mutants resulted in a loss of salt-stress tolerance, the upregulation of *AtSOS1* appears to be a critical component of the enhanced salt-stress tolerance exhibited by plants treated with Ky-2. This premise is supported by the observation that histone H4 acetylation levels are elevated in the *AtSOS1* region in response to the Ky-2 treatment. The Ky-2 treatment also decreased the accumulation of intercellular Na^+ ions (Sako et al. 2016). A strategy for using HDIs to increase salt-stress tolerance in cassava has been explored. Treatment of cassava plants with SAHA, a type of HDI, moderated the level of stress induced by high salt conditions. The mechanism by which the SAHA treatment was able to alleviate salt stress in the cassava plants is still unknown (Patanun et al. 2017).

Treatment of plants with HDI compounds is considered to be a potentially useful approach for enhancing salinity stress tolerance in crops, particularly for crops in which the ability to introduce new, economically important traits through genetic transformation or conventional breeding strategies is difficult. Furthermore, the temporal use of HDIs to avoid environmental stresses may be a beneficial aspect of their use since it may limit the growth inhibition resulting from the expression of stress-responsive genes. In order for the use of chemical approaches to increase stress tolerance to become a viable management practice, the development of a plant-specific HDI will be required since most HDIs at present share targets (HDACs) in both plants and animals, which raises significant safety concerns. It is feasible and worthwhile to find a solution that would avoid the inhibitory effect of HDIs on non-plant HDACs because the HDAC that is required to be inhibited to increase salt-stress tolerance has been identified. Increasing salt-stress tolerance in crops is a critical global need that needs to be addressed in order to produce sufficient food for a growing world population.

References

Al-Lawati A, Al-Bahry S, Victor R, Al-Lawati AH, Yaish MW (2016) Salt stress alters DNA methylation levels in alfalfa (*Medicago* spp). Genet Mol Res 15:15018299

Ammar R, Torti D, Tsui K, Gebbia M, Durbic T, Bader GD, Giaever G, Nislow C (2012) Chromatin is an ancient innovation conserved between Archaea and Eukarya. Elife 1:e00078

Apse MP, Aharon GS, Snedden WA, Blumwald E (1999) Salt tolerance conferred by overexpression of a vacuolar Na^+/H^+ antiport in *Arabidopsis*. Science 285:1256–1258

Asensi-Fabado MA, Amtmann A, Perrella G (2016) Plant responses to abiotic stress: the chromatin context of transcriptional regulation. Biochim Biophys Acta 1860:106–122

Bharti P, Mahajan M, Vishwakarma AK, Bhardwaj J, Yadav SK (2015) AtROS1 overexpression provides evidence for epigenetic regulation of genes encoding enzymes of flavonoid biosynthesis and antioxidant pathways during salt stress in transgenic tobacco. J Exp Bot 66:5959–5969

Bolden JE, Peart MJ, Johnstone RW (2006) Anticancer activities of histone deacetylase inhibitors. Nat Rev Drug Discov 5:769–784

Brosch G, Lusser A, Goralik-Schramel M, Loidl P (1996) Purification and characterization of a high molecular weight histone deacetylase complex (HD2) of maize embryos. Biochemistry 35:15907–15914

Chen LT, Wu K (2010) Role of histone deacetylases HDA6 and HDA19 in ABA and abiotic stress response. Plant Signal Behav 5:1318–1320

Chen LT, Luo M, Wang YY, Wu K (2010) Involvement of *Arabidopsis* histone deacetylase HDA6 in ABA and salt stress response. J Exp Bot 61:3345–3353

Chen X, Lu L, Mayer KS, Scalf M, Qian S, Lomax A, Smith LM, Zhong X (2016) POWERDRESS interacts with HISTONE DEACETYLASE 9 to promote aging in *Arabidopsis*. Elife 5:e17214

Choudhury FK, Rivero RM, Blumwald E, Mittler R (2017) Reactive oxygen species, abiotic stress and stress combination. Plant J 90:856–867

Di C, Yuan J, Wu Y, Li J, Lin H, Hu L, Zhang T, Qi Y, Gerstein MB, Guo Y, Lu ZJ (2014) Characterization of stress-responsive lncRNAs in *Arabidopsis thaliana* by integrating expression, epigenetic and structural features. Plant J 80:848–861

Downs JA, Nussenzweig MC, Nussenzweig A (2007) Chromatin dynamics and the preservation of genetic information. Nature 447:951–958

Eddy SR (2001) Non-coding RNA genes and the modern RNA world. Nat Rev Genet 2:919–929

Fahy J, Jeltsch A, Arimondo PB (2012) DNA methyltransferase inhibitors in cancer: a chemical and therapeutic patent overview and selected clinical studies. Expert Opin Ther Pat 22:1427–1442

Gallusci P, Dai Z, Genard M, Gauffretau A, Leblanc-Fournier N, Richard-Molard C, Vile D, Brunel-Muguet S (2017) Epigenetics for plant improvement: current knowledge and modeling avenues. Trends Plant Sci 22:610–623

Garg R, Narayana Chevala V, Shankar R, Jain M (2015) Divergent DNA methylation patterns associated with gene expression in rice cultivars with contrasting drought and salinity stress response. Sci Rep 5:14922

Goyal K, Walton LJ, Tunnacliffe A (2005) LEA proteins prevent protein aggregation due to water stress. Biochem J 388:151–157

Gu X, Jiang D, Yang W, Jacob Y, Michaels SD, He Y (2011) *Arabidopsis* homologs of retinoblastoma-associated protein 46/48 associate with a histone deacetylase to act redundantly in chromatin silencing. PLoS Genet 7:e1002366

Han Z, Yu H, Zhao Z, Hunter D, Luo X, Duan J, Tian L (2016) AtHD2D gene plays a role in plant growth, development, and response to abiotic stresses in *Arabidopsis thaliana*. Front Plant Sci 7:310

Hauser MT, Aufsatz W, Jonak C, Luschnig C (2011) Transgenerational epigenetic inheritance in plants. Biochim Biophys Acta 1809:459–468

Hennig L, Bouveret R, Gruissem W (2005) MSI1-like proteins: an escort service for chromatin assembly and remodeling complexes. Trends Cell Biol 15:295–302

Heo JB, Sung S (2011) Vernalization-mediated epigenetic silencing by a long intronic noncoding RNA. Science 331:76–79
Hollender C, Liu Z (2008) Histone deacetylase genes in *Arabidopsis* development. J Integr Plant Biol 50:875–885
Jablonka E, Lamb MJ (2002) The changing concept of epigenetics. Ann N Y Acad Sci 981:82–96
Jenuwein T, Allis CD (2001) Translating the histone code. Science 293:1074–1080
Jia H, Shao M, He Y, Guan R, Chu P, Jiang H (2015) Proteome dynamics and physiological responses to short-term salt stress in *Brassica napus* leaves. PLoS One 10:e0144808
Julkowska MM, Testerink C (2015) Tuning plant signaling and growth to survive salt. Trends Plant Sci 20:586–594
Kaldis A, Tsementzi D, Tanriverdi O, Vlachonasios KE (2011) *Arabidopsis thaliana* transcriptional co-activators ADA2b and SGF29a are implicated in salt stress responses. Planta 233:749–762
Kim JM, To TK, Ishida J, Matsui A, Kimura H, Seki M (2012) Transition of chromatin status during the process of recovery from drought stress in *Arabidopsis thaliana*. Plant Cell Physiol 53:847–856
Kim JM, Sasaki T, Ueda M, Sako K, Seki M (2015) Chromatin changes in response to drought, salinity, heat, and cold stresses in plants. Front Plant Sci 6:114
Kishor P, Hong Z, Miao GH, Hu C, Verma D (1995) Overexpression of [delta]-pyrroline-5-carboxylate synthetase increases proline production and confers osmotolerance in transgenic plants. Plant Physiol 108:1387–1394
Kouzarides T (2007) Chromatin modifications and their function. Cell 128:693–705
Kovarik A, Koukalova B, Bezdek M, Opatrny Z (1997) Hypermethylation of tobacco heterochromatic loci in response to osmotic stress. Theor Appl Genet 95:301–306
Kumar SV, Wigge PA (2010) H2A.Z-containing nucleosomes mediate the thermosensory response in *Arabidopsis*. Cell 140:136–147
Kung JT, Colognori D, Lee JT (2013) Long noncoding RNAs: past, present, and future. Genetics 193:651–669
Law JA, Jacobsen SE (2010) Establishing, maintaining and modifying DNA methylation patterns in plants and animals. Nat Rev Genet 11:204–220
Lee JT (2012) Epigenetic regulation by long noncoding RNAs. Science 338:1435–1439
Li B, Carey M, Workman JL (2007) The role of chromatin during transcription. Cell 128:707–719
Li H, Yan S, Zhao L, Tan J, Zhang Q, Gao F, Wang P, Hou H, Li L (2014) Histone acetylation associated up-regulation of the cell wall related genes is involved in salt stress induced maize root swelling. BMC Plant Biol 14:105
Liu C, Lu F, Cui X, Cao X (2010) Histone methylation in higher plants. Annu Rev Plant Biol 61:395–420
Luger K, Mader AW, Richmond RK, Sargent DF, Richmond TJ (1997) Crystal structure of the nucleosome core particle at 2.8 A resolution. Nature 389:251–260
Luo M, Wang YY, Liu X, Yang S, Lu Q, Cui Y, Wu K (2012) HD2C interacts with HDA6 and is involved in ABA and salt stress response in *Arabidopsis*. J Exp Bot 63:3297–3306
Lusser A, Brosch G, Loidl A, Haas H, Loidl P (1997) Identification of maize histone deacetylase HD2 as an acidic nucleolar phosphoprotein. Science 277:88–91
Matsui A, Iida K, Tanaka M, Yamaguchi K, Mizuhashi K, Kim JM, Takahashi S, Kobayashi N, Shigenobu S, Shinozaki K, Seki M (2017) Novel stress-inducible antisense RNAs of protein-coding loci are synthesized by RNA-dependent RNA polymerase. Plant Physiol 175:457–472
Matzke MA, Mosher RA (2014) RNA-directed DNA methylation: an epigenetic pathway of increasing complexity. Nat Rev Genet 15:394–408
McClintock B (1984) The significance of responses of the genome to challenge. Science 226:792–801
McKersie BD, Bowley SR, Jones KS (1999) Winter survival of transgenic alfalfa overexpressing superoxide dismutase. Plant Physiol 119:839–848
Mehdi S, Derkacheva M, Ramstrom M, Kralemann L, Bergquist J, Hennig L (2016) The WD40 domain protein MSI1 functions in a histone deacetylase complex to fine-tune abscisic acid signaling. Plant Cell 28:42–54

Moller IS, Gilliham M, Jha D, Mayo GM, Roy SJ, Coates JC, Haseloff J, Tester M (2009) Shoot Na^+ exclusion and increased salinity tolerance engineered by cell type-specific alteration of Na^+ transport in *Arabidopsis*. Plant Cell 21:2163–2178

Munns R, Tester M (2008) Mechanisms of salinity tolerance. Ann Rev Plant Biol 59:651–681

Patanun O, Ueda M, Itouga M, Kato Y, Utsumi Y, Matsui A, Tanaka M, Utsumi C, Sakakibara H, Yoshida M, Narangajavana J, Seki M (2017) The histone deacetylase inhibitor suberoylanilide hydroxamic acid alleviates salinity stress in cassava. Front Plant Sci 7:2039

Perrella G, Lopez-Vernaza MA, Carr C, Sani E, Gossele V, Verduyn C, Kellermeier F, Hannah MA, Amtmann A (2013) Histone deacetylase complex1 expression level titrates plant growth and abscisic acid sensitivity in *Arabidopsis*. Plant Cell 25:3491–3505

Provart NJ, Alonso J, Assmann SM, Bergmann D, Brady SM, Brkljacic J, Browse J, Chapple C, Colot V, Cutler S, Dangl J, Ehrhardt D, Friesner JD, Frommer WB, Grotewold E, Meyerowitz E, Nemhauser J, Nordborg M, Pikaard C, Shanklin J, Somerville C, Stitt M, Torii KU, Waese J, Wagner D, McCourt P (2016) 50 years of Arabidopsis research: highlights and future directions. New Phytol 209:921–944

Qin T, Zhao H, Cui P, Albesher N, Xiong L (2017) A nucleus-localized long non-coding RNA enhances drought and salt stress tolerance. Plant Physiol 175:1321–1336

Roudier F, Teixeira FK, Colot V (2009) Chromatin indexing in *Arabidopsis*: an epigenomic tale of tails and more. Trends Genet 25:511–517

Roxas VP, Smith RK Jr, Allen ER, Allen RD (1997) Overexpression of glutathione S-transferase/glutathione peroxidase enhances the growth of transgenic tobacco seedlings during stress. Nat Biotechnol 15:988–991

Ryu H, Cho H, Bae W, Hwang I (2014) Control of early seedling development by BES1/TPL/HDA19-mediated epigenetic regulation of *ABI3*. Nat Commun 5:4138

Sakamoto A, Murata AN (1998) Metabolic engineering of rice leading to biosynthesis of glycinebetaine and tolerance to salt and cold. Plant Mol Biol 38:1011–1019

Sako K, Kim JM, Matsui A, Nakamura K, Tanaka M, Kobayashi M, Saito K, Nishino N, Kusano M, Taji T, Yoshida M, Seki M (2016) Ky-2, a histone deacetylase inhibitor, enhances high-salinity stress tolerance in *Arabidopsis thaliana*. Plant Cell Physiol 57:776–783

Sani E, Herzyk P, Perrella G, Colot V, Amtmann A (2013) Hyperosmotic priming of Arabidopsis seedlings establishes a long-term somatic memory accompanied by specific changes of the epigenome. Genome Biol 14:R59

Sequeira-Mendes J, Gutierrez C (2015) Links between genome replication and chromatin landscapes. Plant J 83:38–51

Seto E, Yoshida M (2014) Erasers of histone acetylation: the histone deacetylase enzymes. Cold Spring Harb Perspect Biol 6:a018713

Shafi A, Chauhan R, Gill T, Swarnkar MK, Sreenivasulu Y, Kumar S, Kumar N, Shankar R, Ahuja PS, Singh AK (2015) Expression of SOD and APX genes positively regulates secondary cell wall biosynthesis and promotes plant growth and yield in *Arabidopsis* under salt stress. Plant Mol Biol 87:615–631

Shahbazian MD, Grunstein M (2007) Functions of site-specific histone acetylation and deacetylation. Ann Rev Biochem 76:75–100

Shen Y, Conde ESN, Audonnet L, Servet C, Wei W, Zhou DX (2014) Over-expression of histone H3K4 demethylase gene *JMJ15* enhances salt tolerance in *Arabidopsis*. Front Plant Sci 5:290

Shi H, Lee BH, Wu SJ, Zhu JK (2003) Overexpression of a plasma membrane Na^+/H^+ antiporter gene improves salt tolerance in *Arabidopsis thaliana*. Nat Biotechnol 21:81–85

Song Y, Ji D, Li S, Wang P, Li Q, Xiang F (2012) The dynamic changes of DNA methylation and histone modifications of salt responsive transcription factor genes in soybean. PLoS One 7:e41274

Sridha S, Wu K (2006) Identification of *AtHD2C* as a novel regulator of abscisic acid responses in Arabidopsis. Plant J 46:124–133

Sura W, Kabza M, Karlowski WM, Bieluszewski T, Kus-Slowinska M, Paweloszek L, Sadowski J, Ziolkowski PA (2017) Dual role of the histone variant H2A.Z in transcriptional regulation of stress-response genes. Plant Cell 29:791–807

Teperino R, Schoonjans K, Auwerx J (2010) Histone methyl transferases and demethylases; can they link metabolism and transcription? Cell Metab 12:321–327

Tran HT, Nimick M, Uhrig RG, Templeton G, Morrice N, Gourlay R, DeLong A, Moorhead GB (2012) *Arabidopsis thaliana* histone deacetylase 14 (HDA14) is an alpha-tubulin deacetylase that associates with PP2A and enriches in the microtubule fraction with the putative histone acetyltransferase ELP3. Plant J 71:263–272

Ueda M, Matsui A, Tanaka M, Nakamura T, Abe T, Sako K, Sasaki T, Kim JM, Ito A, Nishino N, Shimada H, Yoshida M, Seki M (2017) The distinct roles of class I and II RPD3-like histone deacetylases in salinity stress response. Plant Physiol 175:1760–1773

Verdin E, Ott M (2015) 50 years of protein acetylation: from gene regulation to epigenetics, metabolism and beyond. Nat Rev Mol Cell Biol 16:258–264

Wang M, Qin L, Xie C, Li W, Yuan J, Kong L, Yu W, Xia G, Liu S (2014) Induced and constitutive DNA methylation in a salinity-tolerant wheat introgression line. Plant Cell Physiol 55:1354–1365

Wang Z, Casas-Mollano JA, Xu J, Riethoven JJ, Zhang C, Cerutti H (2015) Osmotic stress induces phosphorylation of histone H3 at threonine 3 in pericentromeric regions of *Arabidopsis thaliana*. Proc Natl Acad Sci USA 112:8487–8492

Wibowo A, Becker C, Marconi G, Durr J, Price J, Hagmann J, Papareddy R, Putra H, Kageyama J, Becker J, Weigel D, Gutierrez-Marcos J (2016) Hyperosmotic stress memory in Arabidopsis is mediated by distinct epigenetically labile sites in the genome and is restricted in the male germline by DNA glycosylase activity. Elife 5:e13546

Xiao J, Lee US, Wagner D (2016) Tug of war: adding and removing histone lysine methylation in *Arabidopsis*. Curr Opin Plant Biol 34:41–53

Xu D, Duan X, Wang B, Hong B, Ho T, Wu R (1996) Expression of a late embryogenesis abundant protein gene, *HVA1*, from barley confers tolerance to water deficit and salt stress in transgenic rice. Plant Physiol 110:249–257

Xu P, Hu G, Luo C, Liang Z (2016) DNA methyltransferase inhibitors: an updated patent review (2012–2015). Expert Opin Ther Pat 26:1017–1030

Yang Q, Chen ZZ, Zhou XF, Yin HB, Li X, Xin XF, Hong XH, Zhu JK, Gong Z (2009) Overexpression of *SOS* (*Salt Overly Sensitive*) genes increases salt tolerance in transgenic *Arabidopsis*. Mol Plant 2:22–31

Yang DL, Zhang G, Tang K, Li J, Yang L, Huang H, Zhang H, Zhu JK (2016) Dicer-independent RNA-directed DNA methylation in *Arabidopsis*. Cell Res 26:1264

Yao Y, Bilichak A, Golubov A, Kovalchuk I (2012) *ddm1* plants are sensitive to methyl methane sulfonate and NaCl stresses and are deficient in DNA repair. Plant Cell Rep 31:1549–1561

Zemach A, Kim MY, Hsieh PH, Coleman-Derr D, Eshed-Williams L, Thao K, Harmer SL, Zilberman D (2013) The *Arabidopsis* nucleosome remodeler DDM1 allows DNA methyltransferases to access H1-containing heterochromatin. Cell 153:193–205

Zhang X, Bernatavichute YV, Cokus S, Pellegrini M, Jacobsen SE (2009) Genome-wide analysis of mono-, di- and trimethylation of histone H3 lysine 4 in *Arabidopsis thaliana*. Genome Biol 10:R62

Zheng Y, Ding Y, Sun X, Xie S, Wang D, Liu X, Su L, Wei W, Pan L, Zhou DX (2016) Histone deacetylase HDA9 negatively regulates salt and drought stress responsiveness in Arabidopsis. J Exp Bot 67:1703–1713

Zhu JK (2009) Active DNA demethylation mediated by DNA glycosylases. Ann Rev Genet 43:143–166

Zilberman D, Cao X, Jacobsen SE (2003) ARGONAUTE4 control of locus-specific siRNA accumulation and DNA and histone methylation. Science 299:716–719

Chapter 5
Manipulating Programmed Cell Death Pathways for Enhancing Salinity Tolerance in Crops

Ahmad Arzani

Abstract One of the key challenges for researchers is to obtain a deeper understanding of the strategies and mechanisms of plant adaptation to environmental stress that help overcome the limitations associated with climate change and loss of biodiversity. In this context, tolerance to salinity stress is one of the main abiotic factors constraining the plant growth, and production is of special importance. Programmed cell death (PCD) plays a protective role against biotic and abiotic stresses. PCD might play an important role in the maintenance of normal tissue homeostasis, regulation of cell metabolism, and remodeling of tissues after injury and infection as well as the elimination of damaged cells. Salinity stress induces an alteration in chloroplasts, mitochondria, cytoplasm, plasma membrane (PM), endoplasmic reticulum (ER), Golgi apparatus, vesicle formation and trafficking, and vacuoles formation which may result in PCD in plants. The overexpression of pro-survival genes including anti-apoptotic genes and those involved in suppression of apoptosis genes in the transgenic plants to enhance abiotic stress tolerance has been the subject of a number of investigations, particularly in the context of salinity tolerance. Therefore, the development of transformed plants for resistance to apoptosis could be an effective approach to improving salinity tolerance, while the use of complementary techniques like RNA-interfering (RNAi)-mediated gene knockdowns has been shown to be an interesting and appealing alternative. The objective of this review is to summarize the current state of knowledge on improving salinity tolerance in crop plants through manipulation of PCD pathways.

Keywords Abiotic stress · Molecular networks · Salt · Transformation · Vacuolar processing enzyme

---

A. Arzani (✉)
Department of Agronomy and Plant Breeding, College of Agriculture, Isfahan University of Technology, Isfahan, Iran
e-mail: a_arzani@cc.iut.ac.ir

V. Kumar et al. (eds.), *Salinity Responses and Tolerance in Plants, Volume 2*,
https://doi.org/10.1007/978-3-319-90318-7_5

Abbreviations

AIF	Apoptosis-inducing factor
AL-PCD	Apoptotic-like PCD
ASPP	Apoptosis-stimulating proteins of p53
ACD	Autophagic cell death
BAG	Bcl-2-associated athanogene
Bak	BCL-2 antagonist/killer-1
Bax	Bcl-2-associated X protein
Bcl-2	B-cell lymphoma2
Bcl-xl	BCL-2-like 1
Ca^{2+}	Calcium ion
ER	Endoplasmic reticulum
FB1	Fumonisin B1
GORK	Guard cell outward-rectifying K^+ channel
H_2O_2	Hydrogen peroxide
HR	Hypersensitivity
IAP	Inhibitor of apoptosis
K^+	Potassium ion
MAPK	Mitogen-activated protein kinase
Mcl-1	Myeloid cell leukemia-1
Na^+	Sodium ion
NADPH	Nicotinamide adenine dinucleotide phosphate hydrogen
PM	Plasma membrane
RNAi	RNA interfering
ROS	Reactive oxygen species
PCD	Programmed cell death
PLC	Phospholipase C
SKOR	Outward-rectifying K^+ channel
VPE	Vacuolar processing enzyme

5.1 Introduction

Abiotic stress threatens staple crop production, coupled with the expanding world population necessitate not only efficient breeding strategies for developing abiotic stress tolerance crop plants but also the extension of plant production into the marginal regions including saline soil and water (Arzani and Ashraf 2016). Soil or water salinity is one of the key abiotic stresses that cause plant growth and yield reduction worldwide (Arzani 2008). Abiotic stress can be sensed and appropriate responses triggered implicating changes in growth, development, and metabolism (Conde et al. 2011). Plants' response to salinity stress represents the sum of numerous parallel-distributed processes that act to alleviate hyperosmolarity and reestablish ionic homeostatic conditions in cells (Arzani and Ashraf 2016). Programmed cell death (PCD) is also among the evolved plant strategies to overcome these adverse conditions. PCD operates during growth and development as well as in response to various hostile environmental conditions. In this way the removal of damaged and superfluous cells can be facilitated; thus, cellular differentiation and homeostasis are supported in plants (De Pinto et al. 2012).

Therefore, PCD plays not only a protective role against abiotic and biotic stresses but also a major role in plant development. PCD is a highly coordinated process with series of steps involving specific nucleases and proteases and results in the selective elimination of the cells. In animals, autophagy, apoptosis, and programmed necrosis are the three major PCD forms, clearly characterized by their morphological features (Bialik et al. 2010; Ouyang et al. 2012). Autophagy is usually defined by the accumulation of autophagic vacuoles. Chromatin condensation, the formation of apoptotic bodies and nuclear fragmentation by the caspases as the executioners of apoptosis are the hallmarks of apoptosis. A more passive form of PCD is necrosis, which is distinguished by the presence of PM rupture and cytoplasmic swelling.

Autophagic cell death (ACD) is one of the characterized types of PCD. Autophagy process is initiated with the generation of double membrane-bound autophagosomes, encompassing cytoplasmic organelles and macromolecules, headed for recycling (Huett et al. 2010). There is increasing recognition that autophagic cells commit suicide to prevent excessive stress by undertaking cell death, which discriminates from programmed necrosis and apoptosis (Bialik et al. 2010). Nonetheless, autophagy regulates an enormous number of physiological and pathological functions such as cell differentiation, infections, starvation, cell survival, and death (Liu et al. 2010; Michaeli et al. 2016). The role of autophagy in cell death has been reviewed by Minina et al. (2014). In addition, recent advances in plant autophagy regarding mechanisms of selective autophagy, regulation of autophagy, and role of autophagy in recycling and availability of nutrients have recently been reviewed by Michaeli et al. (2016).

"Apoptosis," as the second form of PCD, comes from a Greek root word that has been used to refer to "dropping off" the leaves or petals from a tree (Kerr et al. 1972). Given the definition roots, apoptosis is likely the most frequent type of PCD, while the biological impact of other non-apoptotic types may also be a driving force of the PCD especially in plants. Apoptosis is characterized by morphological alterations of nucleus and cytoplasm including cell shrinkage, pyknosis (DNA condensation), and karyorrhexis (nuclear fragmentation) as well as biochemical changes such as internucleosomal cleavage of DNA, a number of intracellular substrate cleavages by specific proteolysis, and phosphatidylserine externalization (Ouyang et al. 2012).

Programmed necrosis as the third type of PCD contributes to cell swelling, cell lysis, and organelle dysfunction (Wu et al. 2012). Therefore, PCD may have a role in the maintenance of tissue homeostasis, regulation of cell metabolism, and remodeling of tissues after injury and elimination of damaged cells (Wynn et al. 2013). In contrast to the wealth of knowledge regarding the molecular mechanisms of PCD, in plants the molecular networks regulating PCD are still in their infancy, and information on this topic is scarce. This is in spite of the abundance and the importance of PCD throughout plant life span occurring as a conspicuous part of development (dPCD) as well as a response to abiotic and biotic stresses (ePCD) (Lam 2004; Huysmans et al. 2017). Although plants react differently to various abiotic stresses, the initial recognizing and induction of reactive oxygen species (ROS) generation are a common set of response to abiotic factors in all plant species (Sewelam et al. 2016). Nonetheless, production of ROS is a crucial factor in plant stress response and is also associated to in signaling of PCD (Chen et al. 2009a; Kumar et al. 2016).

In animals, ICE-/CED-like family proteases, named caspases, play a central role in PCD such as apoptosis and pyroptosis (Green 2011). In spite of the absence of caspases (abbreviation of cysteinyl aspartate-specific proteases) in plants, the metacaspases were postulated as the functional caspase homologs in plants (Bonneau et al. 2008). In recent years, significant knowledge has been gained in these areas including the characterization of two PCD types: vacuolar PCD and necrotic PCD. The apoptotic cells can be eliminated in the animal using macrophages, whereas in plant lytic vacuoles progressively engulf and digest the cytoplasmic content during vacuolar cell death. On the other hand, necrosis is an alternative form of cell death which is triggered by severe stress and characterized by mitochondrial dysfunction, premature rupture of the plasma membrane, and organized cell disassembly. Vacuolar processing enzyme (VPE) is a plant cysteine proteinase that is mediator driving the execution of various PCD and is considered as a counterpart of animal caspase 1 (Hatsugai et al. 2015).

Climate change and biodiversity loss create new challenges for developing dynamic strategies of plant adaptation to the changing environment. Stress-induced PCD markedly influences plant growth and yield, and it is an important threat to agriculture production (Mittler and Blumwald 2010). The applied and basic research on stress-induced PCD and stress responses, with the eventual goal of manipulating them for practical use, are incredibly challenging areas that attract the growing interest. Therefore, research on PCD-induced abiotic stress and stress responses in plants has strengthened significantly during the past years, and thereby understanding of regulatory mechanisms and knowledge of the immunity role will undoubtedly help to reach the eventual goal to lessen yield losses (Petrov et al. 2015; Wang et al. 2015). The objective of this review is to summarize the current state of knowledge on improving salinity tolerance in crop plants through manipulation of PCD pathways.

5.2 PCD in Response to Abiotic Stress

Plants tolerate the adverse environmental conditions by employing various adaptation mechanisms including toxin exclusion and dramatic amelioration of susceptibility (hypersensitivity) where the abiotic stress is extreme. The monitored level of applied heat stress-induced PCD in plant cells, where heat shock could be responsible for the cell death morphology, is reported in Arabidopsis (Hogg et al. 2011), tobacco (Vacca et al. 2004), soybean (Zuppini et al. 2006), maize (Wang et al. 2015), and lace plant (Dauphinee et al. 2014). PCD in plant species has been induced by low or high temperature in tobacco (Koukalova et al. 1997), cucumber (Balk et al. 1999), *Arabidopsis thaliana* (Swidzinski et al. 2002), and maize (Wang et al. 2015).

Plant symptoms illustrating either undesirable or desirable response to salinity stress can be visually rated in the field. Nevertheless, a reduction in growth which is manifested by leaf burn and necrotic lesions on the leaves is a well-known indicator of exposure of plants to salinity (Tanou et al. 2009). It is suggested that leaf necrosis could be caused by the failure of the cells to avert the accumulation of Na^+ ions into the cytoplasm (Greenway and Munns 1980); in other words, leaf necrosis may be a symptom of the breakdown of ionic regulation (Subbarao and Johansen 1994). In barley, Patterson et al. (2009) compared two barley cultivars (Sahara and Clipper) exposed to 100 mM NaCl treatment and observed that "Sahara" cultivar had significantly less leaf necrosis and higher leaf Na^+ concentrations than "Clipper," concluding that "Sahara" has a higher tolerance to accumulated Na^+. However, despite a general consensus attributing leaf necrosis to an undesirable reaction in plant salinity stress, it is probably most disputable, and it could also be considered as the fundamental lack of knowledge about the reaction at the cellular level and entirely limited to macroscopic observations. Nevertheless, pathogen-induced HR cell death is one of the most efficient plant defense strategies, whereas pathogen- secreted toxin-induced cell death is a necrotrophic pathogen tactic for infection. Interestingly, although distinct mechanisms may regulate toxin-induced cell death and pathogen-induced cell death, both were mediated by the same VPE (Kuroyanagi et al. 2005).

Leaf margin, leaf tip burn, and leaf necrosis are among the plant responses to drought stress which can be found at the late vegetative stage. It was suggested that drought-induced leaf necrosis can be illustrated by the lack of anthocyanin pigmentation (Rosenow et al. 1983). Therefore, leaf necrosis is considered distinctly different from that of the disease symptom, where leaf necrosis is known as desirable plant reaction of host resistance named as "hypersensitivity (HR)." HR is a plant-specific PCD which is essential for defense response to restrict the spread of pathogens. Apoptosis is generally regarded as a critical physiological cell death program required for the tissue homeostasis as well as an active suicidal response to various pathological or physiological stimuli in the mammalian organism (Kabbage et al. 2017). Among the several cell death pathways that have been postulated, apoptotic-like PCD (AL-PCD) seems to be an interesting operational mode in plants leading to a corpse morphology that is similar to the apoptotic morphology perceived in animal cells (Reape and McCabe 2008). It is now established that AL-PCD is an essential cellular process in plants that have a crucial role in the developmental, stress-induced, and senescence processes as well as in response to pathogen infection (Lam et al. 1999). Apart from the developmental and biotic stimuli, it has been shown that AL-PCD is induced by abiotic stresses such as high-fluence UV radiation and heat stress (Foyer and Noctor 2005; Doyle et al. 2010).

Caspases are either involved or not involved in PCD. Accordingly, PCDs can be categorized into two groups, caspase-independent and caspase-dependent PCD

(Kroemer and Martin 2005). Apoptosis is entirely contingent upon caspase activation and thus caspase-dependent PCD represents typical apoptosis. Caspase-independent mechanism of cell death comprises paraptosis, autophagy, necrosis-like PCD, apoptosis-like PCD, and mitotic cataclysm. The non-caspase PCD was found to be associated with caspase-independent elimination, including the use of mitochondrial protein apoptosis-inducing factor (AIF) (Cande et al. 2002; Kroemer and Martin 2005; Zanna et al. 2005). Analysis of the *Arabidopsis* genome indicated the incidence of five close homologs of AIF which detected monodehydroascorbate reductases (MDARs) (Lisenbee et al. 2005) while AIF initially characterized in mammalian mitochondrial DNA (Susin et al. 1999). Of AL-PCD regulation especially relevant to plant cells is the affirming dual target sites of MDAR that is to both chloroplasts and mitochondria.

In plants, the role of mitochondrial proteins triggering cell death is still in its infancy and debatable (Reape and McCabe 2010). However, a pivotal role of the mitochondrion in plant PCD has also been implicated in plant responses to salinity stress (Yao et al. 2004; Lin et al. 2006; Chen et al. 2009b; Wang et al. 2010; Monetti et al. 2014; Hamed-laouti et al. 2016). ROS produced from the electron transport chain in mitochondrion causes dysfunction of mitochondrial lipids and proteins (Yao et al. 2004) leading to the opening of a nonspecific pore in the inner mitochondrial membrane, also called the permeability transition pore (PTP) and release of "caspase-like" proteins (Yao et al. 2004; Reape and McCabe 2010; Sirisha et al. 2014). The dysfunction of mitochondria has been proposed as a prerequisite for the establishment of NaCl-induced PCD in several plant species comprising both glycophyte (*A. thaliana*, rice, tobacco) and halophyte (*Cakile maritima*, *Thellungiella halophila*) (Lin et al. 2006; Chen et al. 2009b; Wang et al. 2010; Monetti et al. 2014; Hamed-laouti et al. 2016). There are considerable evidence and speculation that interaction between ROS and antioxidants would supply a boundary for the environmental metabolic signals mediating activation of the acclimation of the cells to stress or alternatively induction of PCD (Foyer and Noctor 2005).

A dual biological role for ROS might be attributed to the leaf senescence including regulation of the expression of senescence-associated genes and elevation of the program of cell death by direct oxidizing target macromolecules. Interestingly, taking in account the chloroplasts is one of the sources of ROS production in plants (Doyle et al. 2010) would help to resolve the question as to what extent the PCD reaction is responsive to the environmental stimuli in the plant kingdom. In addition, photoreduction of oxygen and energy transfer from triplet excited chlorophyll to oxygen, respectively, are responsible for generating superoxide radicals (O^{-2}) and singlet oxygen ($^1O^2$) in chloroplasts (Kim et al. 2012). The PCD is induced with increasing singlet oxygen ($^1O^2$) concentration in chloroplasts, but the output of $1O^2$-mediated chloroplast leakage and liberate of chloroplastic proteins to the cytosol on the $1O^2$-mediated collapse of cells needs to be elucidated.

PCD, a genetically controlled cell response, has evolved under selective pressure and thus should be advantageous to the plant. Despite the recent progress in the understanding expression of the ROS-responsive genes which induced in response to abiotic and biotic stress, many challenges remain, particularly with regard to the beneficial effects of the ROS-dependent genes influencing PCD on plant growth and resistance to both abiotic and biotic stresses. Hence, it appears possible that induction of the ROS-dependent PCD pathway in plants can be part of physiological changes that normally occur during an acclimation response to enhance stress resistance.

PCD has been perceived traditionally as a vital protective mechanism for disease resistance in plants. Today, it appears that PCD plays a fundamental role in the regulation of much more diverse cellular functions, such as in response to biotic and abiotic stress as well as developmental processes (see a recent review by Huysmans et al. 2017). It should be acknowledged that since the dissection of the PCD at the whole-plant level is difficult, most of the attempts have been made at in vitro cell assays. Here a new scenario for the biological roles of PCD at the whole-plant level to facilitate the possible explanation contributing to the induction of PCD in response to abiotic stress is presented. The results of assessment of abiotic stress tolerance in the C_4 model plant, *Setaria viridis* (L.) Beauv. accessions originated from diverse geographical areas of the world, a portion of which has been published elsewhere (Saha et al. 2016), suggested to us that PCD might have occurred in response to salinity stress (Saha et al. unpublished data). Interestingly, only one accession showed leaf necrosis after 4 weeks of treatment at 300 mM NaCl concentration (Fig. 5.1) and astonishingly ranked as one of the most salinity-tolerant genotypes. Further in vitro positron emission tomography (PET) study showed a clear difference in Na^{22} uptake and transport in this accession compared to a sensitive accession (Ariño-Estrada et al. 2017). However, observations at the cellular level are more pertinent for assessing the possible role of PCD in salinity tolerance than

Fig. 5.1 Leaf necrosis resulted from 300 mM NaCl treatment for 4 weeks in one of the *Setaria viridis* (L.) Beauv. accessions

those at the whole-plant level. It is also important to note that leaf necrosis was observed at the reproductive growth stage, while the PET imaging has been conducted at the seedling stage. The inconsistency with the previously characterized leaf necrosis, as being regarded typical for the sensitive plant to salinity stress, can be explained by developmental stage differences in high levels of NaCl accumulation occurred in the leaf cells at either the reproduction (present study) or the vegetative/seedling stage (previous studies). In addition, a complex combination of differential expression of genes encoding the photosynthetic enzymes and anatomical characterization was functionally essential for evolving more effective photosynthetic mechanism in the C_4 plants. C_4 plants exhibit higher adaptation to tropical regions than C_3 plants and assumed to have an evolutionary adaptation in hot areas of the world in response to diminishing ambient CO_2 concentration (Sage 2004). Therefore, this finding inspires us to look for an alternative interpretation. Further work on the macroscopic, microscopic, and molecular aspects of the salt-tolerant leaf necrosis structure is underway to test the hypothesis that leaf necrosis might be a favorable plant response (i.e., HR) to salinity.

As illustrated in Fig. 5.2, plant cells undergoing PCD exhibit the following features: condensation of the cytoplasm and the nucleus, the retraction of the plasma membrane from the cell wall, loss of membrane integrity, DNA laddering, release of cytochrome c from mitochondria, increase in activity of the proteases of caspase-1-like and caspase-3-like, and alterations in the K^+ efflux and ion homeostasis (Wang et al. 2010; Poor et al. 2013; Reape and McCabe 2013; Reape et al. 2015).

5.3 PCD in Response to Salinity Stress

Although to date no report on HR-like response has been documented for plant salinity tolerance, several researchers have investigated PCD at the cellular level. The influences of NaCl stress primarily on chloroplasts, mitochondria, cytoplasm, plasma membrane (PM), endoplasmic reticulum (ER), Golgi apparatus, vesicle formation and trafficking, and vacuoles have been investigated in plants. The degradation of the inner chloroplast membrane due to the NaCl-induced stress on the ultrastructure of plant leaves has been reported by Hernandez et al. (1995). The disintegration of organellar membranes (particularly the degradation of thylakoid membrane of chloroplast) in cells was found to be one of the major effects of salinity stress (Mitsuya et al. 2000). Salinity stress caused swelling of thylakoid as one of the main alterations of chloroplast ultrastructure in barley (Zahra et al. 2014) and rice (Yamane et al. 2012). The wrinkled effects of salinity on chloroplast ultrastructure have been observed at 100 mM NaCl treatment in tomato (Khavari-Nejad and Mostofi 1998) and at 200 mM NaCl treatment potato (Fidalgo et al. 2004) cells under in vitro conditions. The chloroplasts can play a similarly important role as do the mitochondria in triggering PCD, by regulating ROS signaling. The chloroplasts generate more ROS in a less efficient photosynthesis caused by salinity stress, and hence the ROS leads to cell death (Doyle et al. 2010; Kim et al. 2012; Aken and Breusegem 2015; Reape et al. 2015).

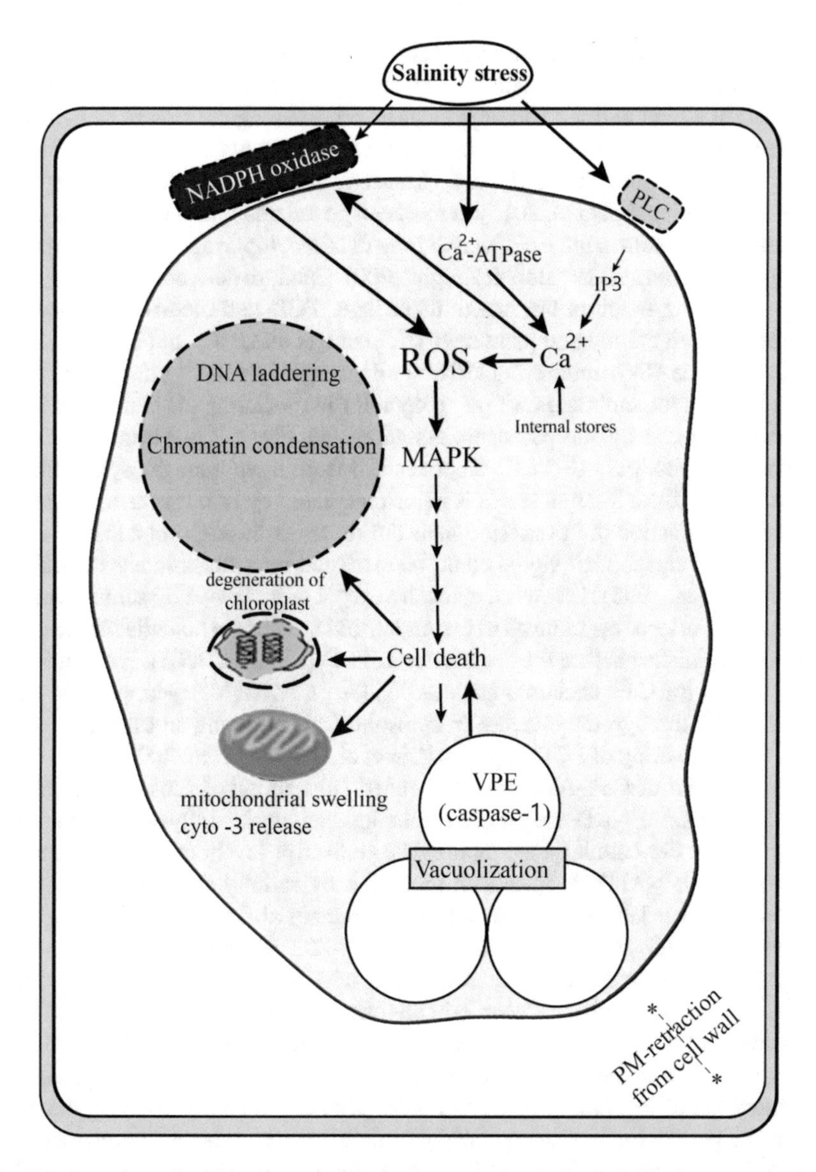

Fig. 5.2 Overview of salinity stress-induced programmed cell death (PCD) in plants. Salinity stress causes the following changes in the plant cell: (1) phospholipase C (PLC), a plasma membrane (PM) enzyme, liberates IP3 from membrane phospholipid, which results in release of Ca^{2+} from internal stores; (2) an increased Ca^{2+}-ATPase gene expression leads to increase in this membrane-bound enzyme, which provides energy to drive the cellular Ca^{2+} pump. An increase Ca^{2+} in cytosol triggers PM-bound NADPH oxidase activity, which produces superoxide in the apoplast. This reactive oxygen species (ROS) transmits death signals through mitogen-activated protein kinases (MAPK) signaling pathway. In addition, death signals can be emitted by vacuolar processing enzyme (VPE) releasing from vacuole. The ROS levels could also be increased by a sense of release of either cytochrome c (Cyt c) or cytochrome f (Cyt f) from mitochondrion. Signals are transmitted to nucleus and ultimately cell execution proteins is synthesized, which results in PCD

A profound downgrading in cytoplasmic streaming was observed at 100 mM NaCl treatment. In plants, cytoplasmic streaming is a marked attribute of cell compartment, in which vesicles and organelles transport along the strands of cytoplasm containing actin filaments. Cytoplasmic streaming indirectly explores some features of the metabolic function in the cell (Mansour and Salama 2004; Shimmen and Yokota 2004; Pieuchot et al. 2015). It has been postulated that salinity stress causes an increase in cytoplasmic Ca^{2+}, which may detain cytoplasmic streaming through the support of internal Ca^{2+} stores (Knight 2000). Calcium does not solely play a key role in signaling function but might also trigger PCD and mediate death-specific enzymes in both animal and plant cells (Boursiac et al. 2010). In plants, Zhu et al. (2010) used the RNA-interfering (RNAi) silencing of the Ca^{2+} pump *NbCA1* and showed that endomembrane Ca^{2+} pump operates in mediating the kinetics of a PCD pathway triggered by the pathogen. As shown in Fig. 5.2, a plasma membrane enzyme, phospholipase C (PLC), liberates IP3 from membrane phospholipid. The release of Ca^{2+} from internal stores is in turn mediated by IP3. It is hence suggested that the amplification of the stress signals during stress through enhancing the level of the stress-induced Ca^{2+} signal could be mediated by a stress-induced PLC gene (Hirayama et al. 1995). Likewise, cytosolic Ca^{2+} was increased by salinity stress in tobacco cells after a few minutes of treatment, and membrane potential of mitochondria was diminished before the occurrence of PCD (Lin et al. 2005). Pretreatment of protoplasts with Ca^{2+} chelators such as EGTA or LaCl3 delayed salinity stress-induced PCD through the increase in cytosolic Ca^{2+} implying an essential role for Ca^{2+} in the triggering of PCD in plants (Lin et al. 2005; Li et al. 2007a).

The function of Ca^{2+}-ATPases is amended substantially in response to abiotic stress in plants. Ca^{2+}-ATPase membrane-bound enzyme hydrolyzes ATP to supply energy to run the cellular Ca^{2+} pump. The transcript levels of genes encoding a putative ER Ca^{2+}-ATPase have been increased by salinity stress in tobacco cells (Perez-Prat et al. 1992) and in tomato (Wimmers et al. 1992). In tobacco, Perez-Prat et al. (1992) observed an increased Ca^{2+}-ATPase gene expression in both adapted and unadapted cells cultured at 428 mM NaCl, while the levels of transcripts were much higher in adapted cells than in unadapted cells. Likewise, an enhanced Ca^{2+}-ATPase transcript has been observed in plants treated with 50 mM NaCl for 24 h in tomato (Wimmers et al. 1992). In addition, it was suggested that the regulation of expression of Ca^{2+}-ATPase gene could be regulated by the RNAi such as 22 nt miR4376 in tomato (Wang et al. 2011). It has been demonstrated that Ca^{2+}-ATPase comprises five functional domains which are named based on their function or position. Hence, they include P-domain (the catalytic core), A-domain (actuator domain), N-domain (nucleotide-binding domain), as well as S- and T-domains (membrane-embedded domains). The ATP hydrolysis is performed by the cytoplasmic domains (A, P, and N), whereas the T- and S-domains play role in the ion transport, together with conformational changes through tertiary contacts and linkers (Palmgren and Nissen 2011).

It is argued that the burden of PCD categorizations should be put on the tonoplast disruption relating to cytoplasmic clearing since the vacuole elaborates on various plant PCD types including HR, differentiation of tracheary elements, senescence of various plant tissues, and so on (van Doorn 2011). Accordingly, the only terminology of two classes of PCD comprising necrosis and vacuolar cell death was suggested by van Doorn (2011). Vacuolar cell death is caused by a progression of an autophagy-associated phenomenon and the release of hydrolases from ruptured vacuoles (Bagniewska-Zadworna and Arasimowicz-Jelonek 2016). In addition, vacuole disintegration and tonoplast disruption are extremely rapid and irreversible processes and represent an unequivocal step in a cell headed for death in plant roots (Bagniewska-Zadworna and Arasimowicz-Jelonek 2016). In *Physcomitrella patens*, it was shown that knockout of vacuolar ACA pump (*PCA1*) gene could lead to higher sensitivity to salinity stress, because of diminished level of NaCl-triggered Ca^{2+} in the cytosol (Qudeimat et al. 2008). Furthermore, the rapid enlargement of vacuolar volume has been observed under in vitro salinity conditions in mangrove [*Bruguiera sexangula* (Lour.) Poir.] cells and barley (*Hordeum vulgare* L. cv. Doriru) root meristematic cells (Mimura et al. 2003). Paradoxically, this phenomenon was not confirmed in pea (*Pisum sativum* L.) (Mimura et al. 2003). The accumulation of Na^{+} ions in the central vacuole causes enhanced vacuolar volumes and is considered as one of the strategies employed by the cell in response to salinity stress.

The detrimental effects of ROS on plant tissues are being increasingly recognized, but the biochemical mechanism linking the ROS production and PCD is poorly known. The main enzymes responsible for superoxide anion generation are cell wall-associated peroxidases and apoplastic plasma membrane-bound NADPH oxidases which are regulated by various environmental and developmental stimuli (Gechev et al. 2006; Sagi and Fluhr 2006). Salinity stress causes an increase in Ca^{2+} in the cytosol and triggers PM-bound NADPH oxidase activity, which produces superoxide in the apoplast (Monetti et al. 2014). Nevertheless, ROS are involved in signaling pathways and mediating PCD activation (Chen et al. 2009a; Mittler 2017) as they influence the activity of mitogen-activated protein kinase (MAPK), which is able to induce several nuclear transcription factors (Fig. 5.2). The overaccumulation of ROS in the cells causes oxidative cellular damage and cell death through reacting with different cellular components. In addition, it is now appreciated that ROS involves in triggering a programmed or physiological pathway for cell death that were not previously thought to be associated straightforwardly with executing cells through oxidation (Mittler 2017). In general, the quantity of ROS accumulation can activate opposing pathways leading to either survival or PCD. A supportive data was obtained by *Chlamydomonas reinhardtii* subjected to hydrogen peroxide (H_2O_2) treatment. The programmed cell suicide event is shown to be triggered by enhanced level of H_2O_2 which resulted in caspase-3-like protein recruitment, DNA laddering, and increased cleavage of PARP (a poly-(ADP)-ribose polymerase-like enzyme) (Vavilala et al. 2015). In general, a variety of genes, transcription factors, and signaling molecules associated with the inducible expression of genes mediating salinity-induced PCD. Certainly, this series includes some of those appointed to control ROS accumulation, release of Cyt c and Ca^{2+}, and mitochondrial permeability transition (Lin et al. 2005, 2006; Li et al. 2007b; Chen et al. 2009a; Monetti et al. 2014; Biswas and Mano 2015; Bahieldin et al. 2016; Pan et al. 2016).

PM and its proteins involved in a wide spectrum of cellular processes including signal perception-transduction and cellular homeostasis, which are regulated by various developmental and environmental stimuli (Mansour 2014; Mansour et al. 2015). Salinity stress-induced PCD has also associated with the retraction of the PM from the cell wall, most likely due to raising the osmotic pressure leading to plasmolysis (Dauphinee et al. 2014; Zhang et al. 2016). Salinity stress is also known to cause membrane disturbance resulting in the loss of membrane integrity, which allows intracellular components to leak out of the cells. NaCl-induced K^+ efflux is believed to be responsible for the effect of salinity on the loss of membrane integrity and non-specific membrane damage in a number of species (Shabala et al. 2006; Cuin et al. 2008; Demidchik et al. 2010, 2014).

The cell membrane is the first living tissue that perceives signals of abiotic stresses including salinity and because of important role and abundance of lipids, which is one of the most sensitive ROS targets (Mansour 2014; Mansour et al. 2015). Salinity stress may cause electrolyte leakage as it is one of the integral parts of the plant's response to stress. Demidchik et al. (2014) suggested that the main consequence of electrolyte leakage is stress-induced K^+ release which outwardly rectifying K^+ channels activated by ROS are responsible for this in plant cells. The K^+ loss results from ion channel-mediated K^+ efflux can induce PCD (Demidchik et al. 2010, 2014). The phenomenon of ROS generation, leading also to PCD, is not an independent process but may largely be influenced by the K^+ loss in conditions of stress-induced electrolyte leakage. In plant cells, highly selective outward-rectifying potassium channel (SKOR), guard cell outward-rectifying K+ channel (GORK), and annexins catalyzing K^+ efflux can be activated by ROS (•OH and H_2O_2). In addition, under salinity and oxidative stress, PCD could be induced by GORK-mediated K^+ efflux (Demidchik et al. 2014).

The mitochondrion has recently acquired renewed attention in toxicology because of its crucial role in signaling and mediating cell death in certain cell types. It was proposed that mitochondria can also be associated in signaling pathways relevant to PCD induction, which is the mitochondrial release of cytochrome c and Ca^{2+} into the cytosol where they trigger cell death caspases (Lin et al. 2005, 2006; Reape et al. 2015). In plants, the permeability of the mitochondrial membrane increases due to mitochondrial generated ROS which releases apoptotic mediators such as cytochrome c (see Fig. 5.2; Tiwari et al. 2002). Salinity stress caused partially or fully inactivation of the photosynthetic reaction centers which results in the downgraded conversion of light energy into chemical energy (Turan and Tripathy 2015), leading to increased ROS formation (Ambastha et al. 2017). In rice, a recent study of the ultrastructure of seedling leaves has proposed the involvement of chloroplasts in PCD induced by salinity stress (Ambastha et al. 2017). Salinity stress also reported inducing cell death in isolated protoplasts of tobacco (Lin et al. 2006) and rice (Ambastha et al. 2017).

Salinity stress may cause substantial amendments in the Golgi bodies and disordered vesicle formation and trafficking in plant cells. In *Arabidopsis thaliana*, high NaCl levels promoted vesicle formation, which may imply elevated levels of macroautophagy, plausibly to recycle degenerated intracellular elements (Liu et al. 2009). The ultrastructural alterations were observed in not only mitochondria but also Golgi bodies, which eventually resulted in autophagy in a halophyte plant, *Thellungiella halophila* (Wang et al. 2010).

5.4 Types of NaCl-Induced Cell Death in Plants

The first approach taken by plant cells for apoptosis-like PCD through DNA fragmentation does not include the NaCl-induced osmotic stress because osmotic stress cannot be accounted for the activation of endonucleases. DNA laddering results in PCD were only found in cells with NaCl and KCl treatment and not in sorbitol-treated cells, indicating that ionic component is to be associated with the PCD (Affenzeller et al. 2009; Vavilala et al. 2016). Hence, the effects of Na^+ ion toxicity have attracted much greater interest as an apt target for dissecting the PCD and the salinity tolerance mechanisms than the osmotic effects (Arzani and Ashraf 2016). Potassium serves as a macronutrient with important roles in a variety of physiological processes in plants, including nucleic acid and protein synthesis. The second approach taken for NaCl-induced PCD is related to ion homeostasis disturbance that results from an excessive amount of Na^+ and a K^+ deficit in the cytosol. It is postulated that reduction of cytosol K^+/Na^+ ratio in the cells would be an essential component in triggering PCD (Joseph and Jini 2010). Under saline conditions, the influx of Na^+ through plasma membrane by the nonselective cation channels (NSCC) causes plasma membrane depolarization which leads to K^+ leakage from the cell through depolarization-activated potassium outward-rectifying channels (KORs) (Shabala 2009; Demidchik 2014; Kim et al. 2014). K^+ deficit results from the release of K^+ from the cytoplasm, which in turn may trigger the effectors of PCD, cysteine proteases (Shabala 2009; Demidchik et al. 2010). The final way taken for the PCD is associated with NaCl-induced oxidative stress, generating ROS, which causes PCD through the deleterious effects to nucleic acids, proteins, lipids, and enzymes, as well as increased peroxidation of membrane lipids and membrane leakage. The enhanced ROS and reduced mitochondrial membrane potential were observed in protoplasts of *Nicotiana tabacum* treated with salinity stress. Similarly, increase in cytosolic Ca^{2+} was found a few minutes after salinity treatment, and decreased membrane potential of mitochondria was also noticed before the occurrence of PCD in tobacco BY2 cells (Monetti et al. 2014). In *Thellungiella halophila*, salinity stress-induced PCD through caspase-like proteases under in vitro conditions was observed. Cells undergoing PCD exhibited attributes such as DNA laddering, retraction of plasma membrane from the cell wall, Cyt c release, and increase in caspase-3-like protease activity (Wang et al. 2010). The cells subjected to in vitro salt stress (500 mM NaCl) showed PCD symptoms such as DNA laddering, nuclear condensation, reduced cell viability, and positive TUNEL in wheat (Rezaei et al. 2013).

5.5 Engineering PCD Pathway to Enhance Salinity Tolerance

Transgenic plants regenerated from the cells transformed with recombinant DNA are becoming increasingly pervasive and will approach ubiquity in research laboratories. The production of transgenic plants has become commonplace and has been employed as a routine tool for the introduction of a foreign or related gene to an agronomically important crop variety and for elucidating mechanisms of gene expression. Transgenic plants expressing novel salinity tolerance genes can be employed to improve crop performance under saline conditions (Arzani and Ashraf 2016). The prosperous development of transgenic plants with the desired trait, such as salinity tolerance, relies on object identification of the genes that are key players in governing that trait. Although overexpression of the majority of salinity tolerance genes being in model plants such as tobacco or *Arabidopsis* plants, the list of candidate genes mainly associated with Na^+ exclusion in the transgenic plants from both *Arabidopsis* and field crops has been compiled by Arzani and Ashraf (2016). The various strategies to engineer PCD pathways that enhance salinity tolerance are as follows:

5.5.1 *Manipulation of Anti-PCD Genes*

The development of transformed plants for resistance to apoptosis could be an effective approach to improving salinity tolerance. It has been revealed that the generation of transgenic plants expressing anti-PCD genes led to enhancing biotic and abiotic tolerance. The family of apoptosis-stimulating proteins of p53 (ASPP) with iASPP, as the most evolutionary conserved member (Sullivan and Lu 2007), is one of the most promising candidates for use as anti-apoptotic factors. The ASSP family members bind to key player proteins regulating cell growth (APCL, PP1) and apoptosis (p53, p63, p73, Bcl-2, and RelA/p65) and most likely regulate the apoptotic function of p53, p63, and p73 (Sullivan and Lu 2007). The iASPP proteins only inhibit the apoptotic function of P53 (including p63 and p73) and do not impact the cell-cycle arrest activity of p53.

The expression of different apoptotic Bcl-2 genes can be activated by p53 as a transcription factor (Levine and Oren 2009).

In mammals, the family of Bcl-2 (B-cell lymphoma2) proteins, localized in the outer mitochondrial membrane, is a key regulator of mitochondrial outer membrane permeabilization (MOMP) and subsequent apoptosis. Bcl-2 proteins comprise both anti-apoptotic member (Bcl-2, Bcl-XL, and Mcl-1) proteins and the pro-apoptotic (Bax, Bak, and Bad) members (Le Pen et al. 2016). They exert influence on balancing the mitochondrial membrane potential. Although the members of Bcl-2 family, caspases, and the members of the inhibitor of apoptosis (IAP) family are important regulators of apoptosis in animals, conservation cycle does not evidently occur in plants. However, plant PCD and animal apoptosis have many common morphological resemblances. The expression of anti-apoptotic (pro-survival) genes has generally been investigated in the model or crop plants using animal and plant target genes. For example, tomato plants were transformed with animal anti-apoptotic *Bcl-xL* and *Ced-9* genes and led to retarded cell death or lack of cucumber mosaic virus symptoms (Xu et al. 2004). Tomato and tobacco plants expressing *SfIAP* gene from an insect (*Spodoptera frugiperda*) preclude cell death caused by the necrotrophic fungus *Alternaria alternata*, salinity, heat, and fungal toxin fumonisin B1 (FB1) treatment (Li et al. 2010). Likewise, expression of *SfIAP* gene has enhanced salinity tolerance in rice (Hoang et al. 2014).

The family of Bcl-2-associated athanogene (BAG) proteins is conserved in the eukaryotic organisms. The anti-cell death activity of BAG has been described through constitutive overexpression of *AtBAG4* in rice (Hoang et al. 2015). All BAG proteins share a common signature motif at the C terminus (BD domain), which directly mediates binding to the Hsp70/Hsc70 heat shock proteins (see the recent review by Kabbage et al. 2017). The ubiquitous 70 kDa Hsp70 family proteins play a crucial role, as molecular chaperones in mediating the refolding of denatured proteins and the folding of newly synthesized proteins. Therefore, Hsp70 proteins can assist anti-apoptotic Bcl-2 proteins through protein-protein interaction at marked essential points to suppress apoptosis pathways (Joly et al. 2010). Overexpression of *Hsp70* derived from *Citrus tristeza* virus in rice conferred tolerance to salinity stress (Hoang et al. 2015). In rice, transgenic plants overexpressing *Bcl-2* gene significantly alleviated PCD symptoms through reduction of NaCl-induced K^+ efflux and inhibition of the expression of VPEs (Kim et al. 2014).

5.5.2 *Overexpression of Inhibitor of Apoptosis (IAP) Genes*

Although plant genomes do not contain IAPs, tolerance to cell death induced by stress has been detected in the ectopic expression of viral and animal IAPs in plants. In tobacco, transgenic plants overexpressing the baculovirus *Orgyia pseudotsugata* nuclear polyhedrosis virus IAP (OpIAP) protein were resistant to tomato-spotted wilt virus and the necrotrophic fungi *Cercospora nicotianae* and *Sclerotinia sclerotiorum* (Dickman et al. 2001).

5.5.3 *Interfering RNA (RNAi)-Induced Apoptosis Gene Silencing*

Alternatively, small interfering RNA (siRNA)-induced transcriptional gene silencing system can be used to knockdown or knockout the expression of apoptotic genes. Long noncoding (lncRNA) and microRNAs miRNA are the two foremost subtypes of regulatory noncoding RNA (ncRNAs). They comprehensively regulate the interrelated steps and mediate the regulated cell death including apoptosis and necrosis through their interaction as well as in association with assorted intracellular components (Su et al. 2016). Cytoplasmic mRNAs can be silenced by miRNAs through either promoting translation repression, expediting mRNA decapping, or triggering an endonuclease cleavage (Bagga et al. 2005; Wu et al. 2006; Pasquinelli 2012; Nam et al. 2014). As such, the alternative cleavage and polyadenylation mechanisms that produce varied 3′-UTR isoforms influence the efficiency of miRNA targeting, while the translation inhibition is dependent on the CCR4-NOT complex and the miRNA-induced silencing complex (miRISC), which causes the recruitment of eIF4A2 and locked on the mRNA region between the start codon and the pre-initiation complex (Nam et al. 2014). In humans, loss of microRNA-mediated repression of *Bcl2* gene expression, in many instances, causes chronic lymphocytic leukemia (CLL) (Anderson et al. 2016). RNAi-mediated silencing of *P69B* a substrate of two matrix metalloproteinases (Sl2/3-MMP) from tomato and located upstream of *Sl2/3-MMP* in tomato transgenic plants led to reduced expression of the cell death marker genes *tpoxC1*, *hsr203j*, and *Hin1* (Zimmermann et al. 2016). VPEs are cysteine proteinases that function as key moderators of stress-induced PCD in plants. Suppression of *OsVPE3* gene in the transgenic lines of rice led to improved salinity tolerance (Lu et al. 2016). Transgenic rice plants overexpressing Bcl-2 resulted in inhibition of salt-induced PCD through a significant reduction of the transient increase in the cytosolic Ca^{2+}, suppression of *OsVPE2* and *OsVPE3*, expression, and inhibited K^+ efflux across the plasma membrane (Kim et al. 2014). In *Arabidopsis*, inhibition of FB1-induced cell death was observed using loss of function mutation in all four VPE (αVPE, βVPE, γVPE, and δVPE) genes (Kuroyanagi et al. 2005). In *Nicotiana benthamiana*, silencing VPE_{1a} and VPE_{1b} diminished sensitivity to cell death caused by the elicitor of bacterial hairpin but did not affect cell death caused by ethylene-inducing peptide1 (Nep1), the fungal necrosis, and the elicitor of oomycete boehmerin (Zhang et al. 2010). Therefore, although VPE_{1a} and VPE_{1b} may involve in elicitor-triggered immunity, they execute cell death in a context-specific manner.

5.5.4 Repression of ROS-Induced PCD

The signaling and biological roles of ROS (e.g., $^{\bullet}O^{-}_{2}$, H_2O_2, $^{\bullet}OH$, $^{1}O_2$) in higher-order eukaryotic cells are still controversial and are unclear. Paradoxically, it is conceivable that both the stimulatory and inhibitory capacities of ROS can be related to its conspicuous biological properties, which comprise half-life, chemical reactivity, and lipid solubility (D'Autreaux and Toledano 2007). ROS, on the one hand, appear to act as signaling molecules that mediate intercellular pathways controlling cell growth, differentiation, inflammation, survival, and immunity when available at a moderate levels (D'Autreaux and Toledano 2007; Foyer and Noctor 2016; Gilroy et al. 2016; Mittler 2017). On the other hand, the excessive generation of ROS results in oxidative damage to essential biological molecules such as DNA, RNA, membranes (lipid peroxidation), and proteins, which causes the demolition of cellular integrity through amending their functionality. During normal homeostasis, endogenous ROS production mainly takes place in the Ero1-PDI oxidative folding system in ER, the electron transport chain in the mitochondrion, and the membrane-bound NADPH oxidase (NOX) complex (Sevier and Kaiser 2008). Considering PCD can be attained by enhanced ROS accumulation and abiotic conditions like salinity stress, genetic programming of cellular metabolism in plants, repressing salinity stress-induced PCD, would lead to an equal relative increase in yield under saline conditions (Xu et al. 2004; Mittler and Blumwald 2010; Hoang et al. 2015). Constitutive overexpression of maize *ABP9* (ABRE-binding protein 9) gene in transgenic *Arabidopsis* plants downregulated cellular ROS content induced by stress and ABA and diminishes cell death (Zhang et al. 2011). Interestingly, aside from the key roles of SOS_1 and SOS_2 in salinity tolerance (see the recent review by Arzani and Ashraf 2016) under salinity stress conditions, they influence the expression of other genes involving in the ROS scavenging activity. Verslues et al. (2007) reported the physical interaction between SOS_2 and $NDPK_2$ (H_2O_2 signaling protein) with CATs. Expression of a baculovirus anti-apoptotic protein, p35, has been observed to suppress PCD induced by H_2O_2 in insect cells through clearly sequestering ROS. It was speculated that the ROS contents can be regulated by either *p35* gene directly or *Hsp70* and *AtBAG4* genes indirectly. In tobacco, transgenic plants expressing p35 (gene from *Autographa californica* multiple nucleopolyhedrovirus (AcMNPV)) enhanced abiotic stress tolerance including salinity, which was associated with the capacity to scavenge ROS by p35 (Wang et al. 2009). As a final overview, Table 5.1 summarizes the reported candidate genes involving in PCD pathway and while overexpressed in the transgenic plants to enhance salinity tolerance.

5.6 Concluding Remark

The molecular mechanisms of salinity-induced PCD via autophagy cell death (ACD) remain to be elucidated by studying the autophagic vacuolization of the cytoplasm and the dynamics of the vacuole in various plant species. Apoptosis and anti-apoptosis phenomena occur as a consequence of the successive development of genetic alterations in multiple genes and epigenetic changes that regulate activities of apoptotic caspases responsible for the execution of various PCD. Therefore, another area of research which illuminates these phenomena is that which explores DNA modifications and dynamic histones related to crucial alterations of genome expression during the PCD. Hence, studies to elucidate the common and innovation features existing between abiotic-induced PCD and pathogen-induced PCD will assist in understanding the physicochemical details of apoptotic-like PCD which needs for selectively manipulating target cell in each of the two conditions.

Nevertheless, improving salinity tolerance through manipulation of PCD pathways in crop plants could be attained by:

1. Upregulation/overexpression of anti-apoptosis genes and downregulation or suppression of pro-apoptosis genes which are functionally indispensable and structurally conserved throughout the plant and animal kingdoms. For instance, the

Table 5.1 Candidate genes involving in programmed cell death (PCD) pathway overexpressed[a] in the transgenic plants to enhance salinity tolerance

Gene	Function	Origin species	Transgenic plant species	Effects on PCD/mechanism	References
Ced-9 homolog of *Bcl-2*	Anti-apoptotic	*Caenorhabditis elegans*	*Nicotiana benthamiana*	Enhanced tolerance to salinity and oxidative stress by altering H^+ and K^+ flux hypothetically by K^+-permeable channels (KOR and NSCC)	Shabala et al. (2007)
SOS1 salt overly sensitive 1	Efflux of Na^+ from cells	*Homo sapiens*	*Arabidopsis thaliana*	Arabidopsis *sos1* mutants displayed higher PCD symptoms, showing salinity-induced PCD is regulated by ion disequilibrium	Huh et al. (2002)
Bcl-2 B-cell lymphoma	Anti-apoptotic	*Homo sapiens*	*Oryza sativa*	Suppressed K^+ efflux across the PM by blocking NSCCs, reduction of cytoplasmic Ca^{2+}, and inhibited the expression of *OsVPE2* and *OsVPE3*, leading to the alleviated salinity-induced PCD	Deng et al. (2011)
Bcl-2	Anti-apoptotic	*Homo sapiens*	*Oryza sativa*	Alleviated PCD symptoms through reduction of NaCl-induced K+ efflux, inhibition of the expression of VPEs	Kim et al. (2014)
Hsp70 heat shock protein	Assist anti-apoptotic *Bcl-2*	*Citrus tristeza* virus	*Oryza sativa*	Enhanced salinity tolerance via alleviation of PCD	Hoang et al. (2015)
IAP	Inhibitor of apoptosis	*Spodoptera frugiperda* (*SfIAP*)	*Oryza sativa*	Enhanced salinity tolerance via precluding PCD	Hoang et al. (2014)
BAG	Bcl-2-associated athanogene	*A. thaliana* (*AtBAG4*)	*Oryza sativa*	Enhanced salinity tolerance via precluding PCD	Hoang et al. (2015)
VPE3 RNAi suppressed	Vacuolar processing enzymes (VPEs)	*Oryza sativa* (*OsVPE3*)	*Oryza sativa*	Enhanced salinity tolerance with downregulated *OsVPE3*	Lu et al. (2016)

Gene	Function	Origin species	Transgenic plant species	Effects on PCD/mechanism	References
p35	Anti-apoptotic	Baculovirus (*Autographa californica* multiple nucleopolyhedrovirus (AcMNPV))	*Nicotiana tabacum*	Enhanced salinity tolerance with ability to scavenge ROS	Wang et al. (2009)
p35	Anti-apoptotic	Same as above	*Oryza sativa*	Enhanced salinity tolerance with ability to scavenge ROS	Hoang et al. (2015)
MKK4	A MAPK kinase	(*GhMKK5*)	*Nicotiana benthamiana*	Reduced salinity tolerance via increase in H_2O_2-induced HR-like PCD	Zhang et al. (2012)
HSPR	Heat shock protein related	*A. thaliana* (*AtHSPR*)	*A. thaliana*	Protect cells from death upon salinity stress	Yang et al. (2015)
OsSRP-LRS RNAi suppressed; *AtSerpin1* homolog	Serine protease inhibitors (serpins)	*Oryza sativa*	*Oryza sativa*	Negatively regulates stress-induced cell death	Bhattacharjee et al. (2015)

[a]Unless otherwise stated

protein members of the Bcl-2 family comprised both pro-apoptosis and anti-apoptosis genes that regulate the release of cytochrome c and other apoptotic alterations in the mitochondrion.

2. The repression of the plant caspase-like enzymes including VPEs, metacaspases, and phytaspases also called subtilisin-like proteases (subtilases) are alternative candidates for "silencing" or "downregulation" by emerging genetic and epigenetic tools.

References

Affenzeller MJ, Darehshouri A, Andosch A, Lu C, Lütz-Meindl U (2009) Salt stress-induced cell death in the unicellular green alga *Micrasterias denticulata*. J Exp Bot 60:939–954

Aken O, Breusegem F (2015) Licensed to kill: mitochondria, chloroplasts, and cell death. Trends Plant Sci 20:754–766

Ambastha V, Sopory SK, Tiwari BS, Tripathy BC (2017) Photo-modulation of programmed cell death in rice leaves triggered by salinity. Apoptosis 22:41–56

Anderson MA, Deng J, Seymour JF, Tam C, Kim SY, Fein J, Yu L, Brown JR, Westerman D, Si EG, Majewski IJ, Segal D, Enschede SLH, Huang DCS, Davids MS, Letai A, Roberts AW (2016) The BCL2 selective inhibitor venetoclax induces rapid onset apoptosis of CLL cells in patients via a TP53-independent mechanism. Blood 127:3215–3224

Ariño-Estrada G, Mitchell GS, Saha P, Arzani A, Cherry SR, Blumwald E, Kyme AZ (2017) Imaging salt transport in plants using PET: a feasibility study. IEEE nuclear science symposium and medical imaging conference 2017 (IEEE NSS/MIC 2017)

Arzani A (2008) Improving salinity tolerance in crop plants: a biotechnological view. In Vitro Cell Dev Biol Plant 44:373–383

Arzani A, Ashraf M (2016) Smart engineering of genetic resources for enhanced salinity tolerance in crop plants. Crit Rev Plant Sci 35:146–189

Bagga S, Bracht J, Hunter S, Massirer K, Holtz J, Eachus R, Pasquinelli AE (2005) Regulation by let-7 and lin-4 miRNAs results in target mRNA degradation. Cell 122:553–563

Bagniewska-Zadworna A, Arasimowicz-Jelonek M (2016) The mystery of underground death: cell death in roots during ontogeny and in response to environmental factors. Plant Biol 18:171–184

Bahieldin A, Atef A, Edris S, Gadalla NO, Ali HM, Hassan SM, Al-Kordy MA, Ramadan AM, Makki RM, Al-Hajar ASM, El-Domyati FM (2016) Ethylene responsive transcription factor ERF109 retards PCD and improves salt tolerance in plant. BMC Plant Biol 16:216. https://doi.org/10.1186/s12870-016-0908-z

Balk J, Leaver CJ, McCabe P (1999) Translocation of cytochrome c from the mitochondria to the cytosol occurs during heat-induced programmed cell death in cucumber plants. FEBS Lett 463:151–154

Bhattacharjee L, Singh PK, Singh S, Nandi AK (2015) Down-regulation of rice serpin gene *OsSRP-LRS* exaggerates stress-induced cell death. J Plant Biol 58:327–332

Bialik S, Zalckvar E, Ber Y, Rubinstein AD, Kimchi A (2010) Systems biology analysis of programmed cell death. Trends Biochem Sci 35:556–564

Biswas MS, Mano J (2015) Lipid peroxide-derived short-chain carbonyls mediate hydrogen peroxide-induced and salt-induced programmed cell death in plants. Plant Physiol 168:885–898

Bonneau L, Ge Y, Drury GE, Gallois P (2008) What happened to plant caspases? J Exp Bot 59:491–499

Boursiac Y, Lee SN, Romanowsky S, Blank R, Sladek C, Chung WS, Harper JF (2010) Disruption of the vacuolar calcium-ATPases in Arabidopsis results in the activation of a salicylic acid-dependent programmed cell death. Plant Physiol 154:1158–1171

Cande C, Cecconi F, Dessen P, Kroemer G (2002) Apoptosis-inducing factor (AIF): key to the conserved caspase-independent pathways of cell death? J Cell Sci 115:4727–4734
Chen R, Sun S, Wang C, Li Y, Liang Y, An F, Li C, Dong H, Yang X, Zhang J, Zuo J (2009a) The Arabidopsis PARAQUAT RESISTANT2 gene encodes an S-nitrosoglutathione reductase that is a key regulator of cell death. Cell Res 19:1377–1387
Chen X, Wang Y, Li J, Jiang A, Cheng Y, Zhang W (2009b) Mitochondrial proteome during salt stress-induced programmed cell death in rice. Plant Physiol Biochem 47:407–415
Conde A, Chaves MM, Geros H (2011) Membrane transport, sensing and signaling in plant adaptation to environmental stress. Plant Cell Physiol 52:1583–1602
Cuin TA, Betts SA, Chalmandrier R, Shabala S (2008) A root's ability to retain K^+ correlates with salt tolerance in wheat. J Exp Bot 59:2697–2706
D'Autreaux B, Toledano MB (2007) ROS as signaling molecules: mechanisms that generate specificity in ROS homeostasis. Nat Rev Mol Cell Biol 8:813–824
Dauphinee AN, Warner S, Gunawardena AH (2014) A comparison of induced and developmental cell death morphologies in lace plant (*Aponogeton madagascariensis*) leaves. BMC Plant Biol 14:389
De Pinto MC, Locato V, De Gara L (2012) Redox regulation in plant programmed cell death. Plant Cell Environ 35:234–244
Demidchik V (2014) Mechanisms and physiological roles of K^+ efflux from root cells. J Plant Physiol 171:696–707
Demidchik V, Cuin TA, Svistunenko D, Smith SJ, Miller AJ, Shabala S, Sokolik A, Yurin V (2010) Arabidopsis root K^+-efflux conductance activated by hydroxyl radicals: single-channel properties, genetic basis and involvement in stress-induced cell death. J Cell Sci 123:1468–1479
Demidchik V, Straltsova D, Medvedev SS, Pozhvanov GA, Sokolik A, Yurin V (2014) Stress-induced electrolyte leakage: the role of K+-permeable channels and involvement in programmed cell death and metabolic adjustment. J Exp Bot 65:1259–1270
Deng M, Bian H, Xie Y, Kim Y, Wang W, Lin E, Zeng Z, Guo F, Pan J, Han N, Wang J, Qian Q, Zhu M (2011) Bcl-2 suppresses hydrogen peroxide-induced programmed cell death via OsVPE2 and OsVPE3, but not via OsVPE1 and OsVPE4, in rice. FEBS J 278:4797–4810
Dickman MB, Park YK, Oltersdorf T, Li W, Clemente T, French R (2001) Abrogation of disease development in plants expressing animal antiapoptotic genes. Proc Natl Acad Sci U S A 98:6957–6962
Doyle SM, Diamond M, McCabe PF (2010) Chloroplast and reactive oxygen species involvement in apoptotic-like programmed cell death in Arabidopsis suspension cultures. J Exp Bot 61:473–482
Fidalgo F, Santos A, Santos I, Salema R (2004) Effects of long-term salt stress on antioxidant defence systems, leaf water relations and chloroplast ultrastructure of potato plants. Ann Appl Biol 145:185–192
Foyer CH, Noctor G (2005) Redox homeostasis and antioxidant signaling: a metabolic interface between stress perception and physiological responses. Plant Cell 17:1866–1875
Foyer CH, Noctor G (2016) Stress-triggered redox signaling: what's in pROSpect? Plant Cell Environ 39:951–964
Gechev TS, Van Breusegem F, Stone JM, Denev I, Laloi C (2006) Reactive oxygen species as signals that modulate plant stress responses and programmed cell death. BioEssays 28:1091–1101
Gilroy S, Białasek M, Suzuki N, Górecka M, Devireddy A, Karpinski S, Mittler R (2016) ROS, calcium and electric signals: key mediators of rapid systemic signaling in plants. Plant Physiol 171:1606–1615
Green DR (2011) Means to an end. Apoptosis and other cell death mechanisms. Cold Spring Harbor Laboratory Press, Cold Spring Harbor
Greenway H, Munns R (1980) Mechanisms of salt tolerance in nonhalophytes. Annu Rev Plant Physiol 31:149–190
Hamed-laouti IB, Arbelet-bonnin D, De Bont L, Biligui B, Gakière B, Abdelly C, Ben Hamed K (2016) Comparison of NaCl-induced programmed cell death in the obligate halophyte *Cakile maritima* and the glycophyte *Arabidopsis thaliana*. Plant Sci 247:49–59

Hatsugai N, Yamada K, Goto-Yamada S, Hara-Nishimura I (2015) Vacuolar processing enzyme in plant programmed cell death. Front Plant Sci 6:234

Hernandez JA, Olmos E, Corpas FJ, Sevilla F, del Rio LA (1995) Salt-induced oxidative stress in chloroplasts of pea plants. Plant Sci 105:151–167

Hirayama T, Ohto C, Mizoguchi T, Shinozaki K (1995) A gene encoding a phosphatidylinositol-specific phospholipase C is induced by dehydration and salt stress in *Arabidopsis thaliana*. Proc Nat Acad Sci USA 92:3903–3907

Hoang TML, Williams B, Khanna H, Dale J, Mundree SG (2014) Physiological basis of salt stress tolerance in rice expressing the anti-apoptotic gene *SfIAP*. Funct Plant Biol 41:1168–1177

Hoang TML, Moghaddam L, Williams B, Khanna H, Dale J, Mundree SG (2015) Development of salinity tolerance in rice by constitutive-overexpression of genes involved in the regulation of programmed cell death. Front Plant Sci 6:175

Hogg B, Kacprzyk J, Molony EM, O'Reilly C, Gallagher TF, Gallois P (2011) An in vivo root hair assay for determining rates of apoptotic-like programmed cell death in plants. Plant Methods 7:45

Huett A, Goel G, Xavier RJ (2010) A systems biology viewpoint on autophagy in health and disease. Curr Opin Gastroenterol 26:302–309

Huh G, Damsz B, Matsumoto TK, Reddy MP, Rus AM, Ibeas JI, Narasimhan ML, Bressan RA, Hasegawa PM (2002) Salt causes ion disequilibrium-induced programmed cell death in yeast and plants. Plant J 29:649–659

Huysmans M, Lema AS, Coll NS, Nowack MK (2017) Dying two deaths—programmed cell death regulation in development and disease. Curr Opin Plant Biol 35:37–44

Joly A, Wettstein G, Mignot G, Ghiringhelli F, Garrido C (2010) Dual role of heat shock proteins as regulators of apoptosis and innate immunity. J Innate Immun 2:238–247

Joseph B, Jini D (2010) Salinity induced programmed cell death in plants: challenges and opportunities for salt-tolerant plants. J Plant Sci 5:376–390

Kabbage M, Kessens R, Bartholomay LC, Williams B (2017) The life and death of a plant cell. Annu Rev Plant Biol 68:1–7. https://doi.org/10.1146/annurev-arplant-043015-111655

Kerr JFR, Wylie AH, Currie AR (1972) Apoptosis: a basic biological phenomenon with wide-ranging implication in tissue kinetics. Br J Cancer 26:239–257

Khavari-Nejad RA, Mostofi Y (1998) Effects of NaCl on photosynthetic pigments, saccharides, and chloroplast ultrastructure in leaves of tomato cultivars. Photosynthetica 35:151–154

Kim C, Meskauskiene R, Zhang S, Lee K, Ashok M, Blajecka K, Herrfurth C, Feussner I, Apela K (2012) Chloroplasts of Arabidopsis are the source and a primary target of a plant-specific programmed cell death signaling pathway. Plant Cell 24:3026–3039

Kim Y, Wang M, Bai Y, Zeng Z, Guo F, Han N, Bian H, Wang J, Pan J, Zhu M (2014) Bcl-2 suppresses activation of VPEs by inhibiting cytosolic Ca^{2+} level with elevated K^+ efflux in NaCl-induced PCD in rice. Plant Physiol Biochem 80:168–175

Knight H (2000) Calcium signaling during abiotic stress in plants. Int Rev Cytol 195:269–325

Koukalova B, Kovarik A, Fajkus J, Siroky J (1997) Chromatin fragmentation associated with apoptotic changes in tobacco cells exposed to cold stress. FEBS Lett 414:289–292

Kroemer G, Martin SJ (2005) Caspase-independent cell death. Nat Med 11:725–730

Kumar SR, Mohanapriya G, Sathishkumar R (2016) Abiotic stress-induced redox changes and programmed cell death in plants—a path to survival or death? In: Gupta DK, Palma JM, Corpas FJ (eds) Redox state as a central regulator of plant-cell stress responses. Springer, Germany, pp 233–252

Kuroyanagi M, Yamada K, Hatsugai N, Kondo M, Nishimura M, Hara-Nishimura I (2005) Vacuolar processing enzyme is essential for mycotoxin-induced cell death in *Arabidopsis thaliana*. J Biol Chem 280:32914–32920

Lam E (2004) Controlled cell death, plant survival and development. Nat Rev Mol Cell Biol 5:305–315

Lam E, Pontier D, del Pozo O (1999) Die and let live: programmed cell death in plants. Curr Opin Plant Biol 2:502–507

Le Pen J, Laurent M, Sarosiek K, Vuillier C, Gautier F, Montessuit S, Martinou JC, Letaï A, Braun F, Juin PP (2016) Constitutive p53 heightens mitochondrial apoptotic priming and favors cell death induction by BH3 mimetic inhibitors of BCL-xL. Cell Death Dis 7:e2083

Levine AJ, Oren M (2009) The first 30 years of p53: growing ever more complex. Nat Rev Cancer 9:749–758

Li J, Jiang A, Chen H, Wang Y, Zhang WPB (2007a) Lanthanum prevents salt stress-induced programmed cell death in rice root tip cells by controlling early induction events. J Integr Biol 49:1024–1031

Li J, Jiang A, Zhang W (2007b) Salt stress-induced programmed cell death in rice root tip cells. J Integr Plant Biol 49:481–486

Li W, Kabbage M, Dickman MB (2010) Transgenic expression of an insect inhibitor of apoptosis gene, *SfIAP*, confers abiotic and biotic stress tolerance and delays tomato fruit ripening. Physiol Mol Plant Pathol 74:363–375

Lin J, Wang Y, Wang G (2005) Salt stress-induced programmed cell death via Ca^{2+}-mediated mitochondrial permeability transition in tobacco protoplasts. Plant Growth Regul 45:243–250

Lin J, Wang Y, Wang G (2006) Salt stress-induced programmed cell death in tobacco protoplasts is mediated by reactive oxygen species and mitochondrial permeability transition pore status. J Plant Physiol 163:731–739

Lisenbee CS, Lingard MJ, Trelease RN (2005) Arabidopsis peroxisomes possess functionally redundant membrane and matrix isoforms of monodehydroascorbate reductase. Plant J 43:900–914

Liu Y, Xiong Y, Bassham DC (2009) Autophagy is required for tolerance of drought and salt stress in plants. Autophagy 5:954–963

Liu B, Cheng Y, Liu Q, Bao JK, Yang JM (2010) Autophagic pathways as new targets for cancer drug development. Acta Pharmacol Sin 31:1154–1164

Lu W, Deng M, Fu G, Wang M, Zeng Z, Han N, Yang Y, Zhu M, Bian H (2016) Suppression of *OsVPE3* enhances salt tolerance by attenuating vacuole rupture during programmed cell death and affects stomata development in rice. Rice 9:65. https://doi.org/10.1186/s12284-016-0138-x

Mansour MMF (2014) Plasma membrane transport systems and adaptation to salinity. J Plant Physiol 171:1787–1800

Mansour MMF, Salama KHA (2004) Cellular basis of salinity tolerance in plants. Environ Exp Bot 52:113–122

Mansour MMF, Salama KHA, Allam HYH (2015) Role of the plasma membrane in saline conditions: lipids and proteins. Bot Rev 81:416–451

Michaeli S, Galili G, Genschik P, Fernie AR, Avin-Wittenberg T (2016) Autophagy in plants—what's new on the menu? Trends Plant Sci 21:134–144

Mimura T, Kura-Hotta M, Tsujimura T, Ohnishi M, Miura M, Okazaki Y, Mimura M, Maeshima M, Washitani-Nemoto S (2003) Rapid increase of vacuolar volume in response to salt stress. Planta 216:397–402

Minina EA et al (2014) Autophagy as initiator or executioner of cell death. Trends Plant Sci 19:692–697

Mitsuya S, Takeoka Y, Miyake H (2000) Effects of sodium chloride on foliar ultrastructure of sweet potato (*Ipomoea batatas* Lam.) plantlets grown under light and dark conditions in vitro. J Plant Physiol 157:661–667

Mittler R (2017) ROS are good. Trends Plant Sci 22:11–19

Mittler R, Blumwald E (2010) Genetic engineering for modern agriculture: challenges and perspectives. Annl Rev Plant Biol 61:443–462

Monetti E, Kadono T, Tran D, Azzarello E, Arbelet-Bonnin D, Biligui B, Briand J, Kawano T, Mancuso S, Bouteau F (2014) Deciphering in early events involved in hyperosmotic stress-induced programmed cell death in tobacco BY-2 cells. J Exp Bot 65:1361–1375

Nam JW, Rissland OS, Koppstein D, Abreu-Goodger C, Jan CH, Agarwal V, Yildirim MA, Rodriguez A, Bartel DP (2014) Global analyses of the effect of different cellular contexts on micro RNA targeting. Mol Cell 53:1031–1043

Ouyang L, Shi Z, Zhao S, Wang FT, Zhou TT, Liu B, Bao JK (2012) Programmed cell death pathways in cancer: a review of apoptosis, autophagy and programmed necrosis. Cell Prolif 45:487–498

Palmgren MG, Nissen P (2011) P-type ATPases. Annu Rev Biophys 40:243–266

Pan YJ, Liu L, Lin YC, Zu YG, Li LP, Tang ZH (2016) Ethylene antagonizes salt-induced growth retardation and cell death process via transcriptional controlling of ethylene-, BAG-and senescence-associated genes in Arabidopsis. Front Plant Sci 7:696. https://doi.org/10.3389/fpls.2016.00696

Pasquinelli AE (2012) MicroRNAs and their targets: recognition, regulation and an emerging reciprocal relationship. Nat Rev Genet 13:271–282

Patterson JH, Newbigin E, Tester M, Bacic A, Roessner U (2009) Metabolic responses to salt stress are described for two barley (*Hordeum vulgare* L.) cultivars, Sahara and Clipper, which differed in salinity tolerance. J Exp Bot 60:4089–4103

Perez-Prat E, Narashimhan ML, Binzel ML, Botella MA, Chen Z, Valpuesta V, Bressan RA, Hasegawa PM (1992) Induction of a putative Ca^{2+}-ATPase mRNA in NaCl adapted cells. Plant Physiol 100:1471–1478

Petrov V, Hille J, Mueller-Roeber B, Gechev TS (2015) ROS-mediated abiotic stress-induced programmed cell death in plants. Front Plant Sci 6:69

Pieuchot L, Lai J, Loh RA, Leong FY, Chiam K, Stajich J, Jedd G (2015) Cellular subcompartments through cytoplasmic streaming. Dev Cell 34:410–420

Poor P, Kovacs J, Szopko D, Tari I (2013) Ethylene signaling in salt stress- and salicylic acid-induced programmed cell death in tomato suspension cells. Protoplasma 250:273–284

Qudeimat E, Faltusz AM, Wheeler G, Lang D, Holtorf H, Brownlee C, Reski R, Frank W (2008) A PIIB-type Ca^{2+}-ATPase is essential for stress adaptation in *Physcomitrella patens*. Proc Nat Acad Sci USA 105:19555–19560

Reape TJ, McCabe PF (2008) Apoptotic-like programmed cell death in plants. New Phytol 180:13–26

Reape TJ, McCabe PF (2010) Apoptotic-like regulation of programmed cell death in plants. Apoptosis 15:249–256

Reape TJ, McCabe PF (2013) Commentary: the cellular condensation of dying plant cells: programmed retraction or necrotic collapse? Plant Sci 207:135–139

Reape T, Brogan N, McCabe P (2015) Mitochondrion and chloroplast regulation of plant programmed cell death. In: Gunawardena A, McCabe P (eds) Plant programmed cell death. Springer, New York, pp 33–53

Rezaei A, Amirjani M, Mahdiyeh M (2013) Programmed cell death induced by salt stress in wheat cell suspension. Int J Forest Soil Eros 3:35–39

Rosenow DT, Quisenberry JE, Wendt CW, Clark LE (1983) Drought tolerant sorghum and cotton germplasm. Agric Water Manag 7:207–222

Sage RF (2004) The evolution of C_4 photosynthesis. New Phytol 161:341–370

Sagi M, Fluhr R (2006) Production of reactive oxygen species by plant NADPH oxidases. Plant Physiol 141:336–340

Saha P, Sade N, Arzani A, Wilhelmi MMR, Coe KM, Li B, Blumwald E (2016) Effects of abiotic stress on physiological plasticity and water use of *Setaria viridis* (L.). Plant Sci 251:128–138

Sevier C, Kaiser C (2008) Ero1 and redox homeostasis in the endoplasmic reticulum. Biochim Biophys Acta 1783:549–556

Sewelam N, Kazan K, Schenk PM (2016) Global plant stress signaling: reactive oxygen species at the cross-road. Front Plant Sci 7:187

Shabala S (2009) Salinity and programmed cell death: unravelling mechanisms for ion specific signaling. J Exp Bot 60:709–712

Shabala S, Demidchik V, Shabala L, Cuin TA, Smith SJ, Miller AJ, Davies JM, Newman IA (2006) Extracellular Ca2+ ameliorates NaCl-induced K^+ loss from Arabidopsis root and leaf cells by controlling plasma membrane K^+-permeable channels. Plant Physiol 141:1653–1665

Shabala S, Cuin TA, Prismall L, Nemchinov LG (2007) Expression of animal *CED-9* anti-apoptotic gene in tobacco modifies plasma membrane ion fluxes in response to salinity and oxidative stress. Planta 227:189–197

Shimmen T, Yokota E (2004) Cytoplasmic streaming in plants. Curr Opin Cell Biol 16:68–72
Sirisha VL, Sinha M, D'Souza JS (2014) Menadione-induced caspase-dependent programmed cell death in the green chlorophyte *Chlamydomonas reinhardtii*. J Phycol 50:587–601
Su Y, Wu H, Pavlosky A, Zou LL, Deng X, Zhang ZX, Jevnikar AM (2016) Regulatory non-coding RNA: new instruments in the orchestration of cell death. Cell Death Dis 7:e2333. https://doi.org/10.1038/cddis.2016.210
Subbarao GV, Johansen C (1994) Strategies and scope for improving salinity tolerance in crop plants. In: Pessarakli M (ed) Handbook of plant crop stress. Marcel Dekker, New York, pp 559–579
Sullivan A, Lu X (2007) ASPP: a new family of oncogenes and tumour suppressor genes. Br J Cancer 96:196–200
Susin SA, Lorenzo HK, Samzami N, Marzo I, Snow BE, Brothers GM, Mangion J, Jacotot E, Costantini P, Loeffler M, Larochette N, Goodlett DR, Aebersold R, Siderovski DP, Penninger JM, Kroemer G (1999) Molecular characterization of mitochondrial apoptosis-inducing factor. Nature 397:441–446
Swidzinski JA, Sweetlove LJ, Leaver CJ (2002) A custom microarray analysis of gene expression during programmed cell death in *Arabidopsis thaliana*. Plant J 30:431–446
Tanou G, Molassiotis A, Diamantidis G (2009) Induction of reactive oxygen species and necrotic death-like destruction in strawberry leaves by salinity. Environ Exp Bot 65:270–281
Tiwari BS, Belenghi B, Levine A (2002) Oxidative stress increased respiration and generation of reactive oxygen species, resulting in ATP depletion, opening of mitochondrial permeability transition, and programmed cell death. Plant Physiol 128:1271–1281
Turan S, Tripathy BC (2015) Salt-stress induced modulation of chlorophyll biosynthesis during de-etiolation of rice seedlings. Physiol Plant 153:477–491
Vacca RA, de Pinto MC, Valenti D, Passarella S, Marra E, De Garra L (2004) Production of reactive oxygen species, alteration of cytoplasmic ascorbate peroxidase, and impairment of mitochondrial metabolism are early events in heat-shock induced cell death in tobacco bright yellow 2 cells. Plant Physiol 134:1100–1112
Van Doorn WG (2011) Classes of programmed cell death in plants, compared to those in animals. J Exp Bot 62:4749–4761
Vavilala SL, Gawde KK, Sinha M, D'Souza S (2015) Programmed cell death is induced by hydrogen peroxide but not by excessive ionic stress of sodium chloride in the unicellular green alga *Chlamydomonas reinhardtii*. Eur J Phycol 50:422–438
Vavilala SL, Sinha M, Gawde KK, Hirolikar SM, D'Souza S (2016) KCl induces a caspase-independent programmed cell death in the unicellular green chlorophyte *Chlamydomonas reinhardtii* (Chlorophyceae). Phycologia 55:378–392
Verslues PE, Batelli G, Grillo S, Agius F, Kim YS, Zhu J, Agarwal M, Katiyar-Agarwal S, Zhu JK (2007) Interaction of SOS_2 with nucleoside diphosphate kinase 2 and catalases reveals a point of connection between salt stress and H_2O_2 signaling in *Arabidopsis thaliana*. Mol Cell Biol 27:7771–7780
Wang Z, Song J, Zhang Y, Yang B, Chen S (2009) Expression of baculovirus anti-apoptotic p35 gene in tobacco enhances tolerance to abiotic stress. Biotechnol Lett 31:585–589
Wang J, Li X, Liu Y, Zhao X (2010) Salt stress induces programmed cell death in *Thellungiella halophila* suspension-cultured cells. J Plant Physiol 167:1145–1151
Wang Y, Itaya A, Zhong X, Wu Y, Zhang J, Knaap EV, Olmstead R, Qi Y, Ding B (2011) Function and evolution of a microRNA that regulates a Ca^{2+}-ATPase and triggers the formation of phased small interfering RNAs in tomato reproductive growth. Plant Cell 23:3185–3203
Wang P, Zhao L, Hou H, Zhang H, Huang Y, Wang Y, Li H, Gao F, Yan S, Li L (2015) Epigenetic changes are associated with programmed cell death induced by heat stress in seedling leaves of *Zea mays*. Plant Cell Physiol 56:965–976
Wimmers LE, Ewing NN, Bennett AB (1992) Higher plant Ca2C-ATPase: primary structure and regulation of mRNA abundance by salt. Proc Nat Acad Sci USA 89:9205–9209
Wu L, Fan J, Belasco JG (2006) MicroRNAs direct rapid deadenylation of mRNA. Proc Nat Acad Sci USA 103:4034–4039

Wu W, Liu P, Li J (2012) Necroptosis: an emerging form of programmed cell death. Crit Rev Oncol Hematol 82:249–258

Wynn TA, Chawla A, Pollard JW (2013) Macrophage biology in development, homeostasis and disease. Nature 496:445–455

Xu P, Rogers SJ, Roossinck MJ (2004) Expression of antiapoptotic genes *bcl-xL* and *ced-9* in tomato enhances tolerance to viral-induced necrosis and abiotic stress. Proc Nat Acad Sci USA 101:15805–15810

Yamane K, Mitsuya S, Taniguchi W, Miyake H (2012) Salt-induced chloroplast protrusion is the process of exclusion of ribulose-1,5-bisphosphate carboxylase/oxygenase from chloroplasts into cytoplasm in leaves of rice. Plant Cell Environ 35:1663–1671

Yang T, Zhang L, Hao H, Zhang P, Zhu H, Cheng W, Wang Y, Wang X, Wang C (2015) Nuclear-localized AtHSPR links abscisic acid-dependent salt tolerance and antioxidant defense in Arabidopsis. Plant J 84:1274–1294

Yao N, Bartholomew JE, James M, Greenberg JT (2004) The mitochondrion: an organelle commonly involved in programmed cell death in *Arabidopsis thaliana*. Plant J 40:596–610

Zahra J, Nazim H, Cai S, Han Y, Wu D, Zhang B, Haider SI, Zhang G (2014) The influence of salinity on cell ultrastructures and photosynthetic apparatus of barley genotypes differing in salt stress tolerance. Acta Physiol Plant 36:1261–1269

Zanna C, Ghelli A, Porcelli AM, Martinuzzi A, Carelli V, Rugolo M (2005) Caspase-independent death of Leber's hereditary optic neuropathy cybrids is driven by energetic failure and mediated by AIF and Endonuclease G. Apoptosis 10:997–1007

Zhang H, Dong S, Wang M, Wang W, Song W, Dou X, Zheng X, Zhang Z (2010) The role of vacuolar processing enzyme (VPE) from *Nicotiana benthamiana* in the elicitor-triggered hypersensitive response and stomatal closure. J Exp Bot 61:3799–3812

Zhang X, Wang L, Meng H, Wen H, Fan Y, Zhao J (2011) Maize ABP9 enhances tolerance to multiple stresses in transgenic Arabidopsis by modulating ABA signaling and cellular levels of reactive oxygen species. Plant Mol Biol 75:365–378

Zhang L, Li Y, Lu W, Meng F, Wu C, Guo X (2012) Cotton GhMKK5 affects disease resistance, induces HR-like cell death, and reduces the tolerance to salt and drought stress in transgenic *Nicotiana benthamiana*. J Exp Bot 63:3935–3951

Zhang H, Fan X, Wang B, Song L (2016) Calcium ion on membrane fouling reduction and bioflocculation promotion in membrane bioreactor at high salt shock. Bioresour Technol 200:535–540

Zhu X, Caplan J, Mamillapalli P, Czymmek K, Dinesh-Kuma SP (2010) Function of endoplasmic reticulum calcium ATPase in innate immunity-mediated programmed cell death. EMBO J 29:1007–1018

Zimmermann D, Gomez-Barrera JA, Pasule C, Brack-Frick UB, Sieferer E, Nicholson TM, Pfannstiel J, Stintzi A, Schaller A (2016) Cell death control by matrix metalloproteinases. Plant Physiol 171:1456–1469

Zuppini A, Bugno V, Baldan B (2006) Monitoring programmed cell death triggered by mild heat shock in soybean-cultured cells. Funct Plant Biol 33:617–627

Chapter 6
Helicases and Their Importance in Abiotic Stresses

Zeba I. Seraj, Sabrina M. Elias, Sudip Biswas, and Narendra Tuteja

Abstract Helicases are a ubiquitous class of ATP-dependent nucleic acid unwinding enzymes crucial for life processes in all living organisms. There are six classes of helicases based on their conserved amino acid sequences. All eukaryotic RNA helicases belong to the SF1 and SF2 groups. Groups SF3–SF5 are mainly viral and bacterial DNA helicases, while SF6 includes the ubiquitous mini-chromosome maintenance or MCM group of helicases. SF3–SF6 are also characterized as hexameric ring-forming, whereas SF1 and SF2 groups are usually monomeric. The SF2 class is the largest group of helicases, including both DNA and RNA helicases with the widest range of function in replication, transcription, translation, repair, as well as chromatin remodeling. There is no clear sequence-based separation between DNA and RNA helicases. SF2 also includes both RNA and DNA helicases that are involved in biotic and abiotic stresses. While both DNA and RNA helicases play important roles in normal cellular function, the latter are more markedly involved in stress alleviation. This functional divergence was also evident in promoter sequence comparisons of the 113 *A. thaliana* helicases. Some DNA helicases like those from SF6 (MCM) and SF2 (CHR) are also active under stressed conditions. However, the most prominent stress-activated helicases are those with the conserved amino acid motifs, DEAD/H. Overexpression of DEAD/H helicases in many crops confers a growth advantage in the transgenic plants and has resulted in their protection against major abiotic stresses, such as salinity, drought, and oxidative stresses with minimal loss in yield potential.

Keywords Ubiquitous · ATP-dependent nucleic acid · DNA and RNA helicases · Hexameric · Monomeric · Chromatin remodeling · Conserved amino acid motifs · Transgenic plants · Abiotic stresses · Yield potential

---

Z. I. Seraj (✉) · S. M. Elias · S. Biswas
Department of Biochemistry and Molecular Biology, Dhaka University, Dhaka, Bangladesh
e-mail: zebai@du.ac.bd

N. Tuteja
Amity Institute of Microbial Technology, Amity University, Noida, Uttar Pradesh, India

V. Kumar et al. (eds.), *Salinity Responses and Tolerance in Plants, Volume 2*,
https://doi.org/10.1007/978-3-319-90318-7_6

Abbreviations

BAT 1	HLA-B-associated transcript 1
BRM	BRAHMA
BZR1	Brassinazole-resistant 1
CBL	Calcineurin B-like
CHR31	Chromatin remodeling31
CHX	Cation/H^+ antiporter
CIPK	CBL-interacting protein kinase
DEAD	D-E-A-D (Asp-Glu-Ala-Asp)
DRH	Dicer-related RNA helicase
EDA	Embryo sac development arrest
eIF	Eukaryotic initiation factor
EMBL	European Molecular Biology Laboratory
KEA	K^+/H^+ antiporter
KEGG	Kyoto encyclopedia of genes and genomes
LOS4	Low osmotic response gene 4
MCM	Mini-chromosome
MEE29	Mutant effect embryo 29
MEGA7	Molecular evolutionary genetics analysis version 7
MER	Oligonucleotide-mer replacement
MINU	Minuscule
NDPK	Nucleoside diphosphate kinase
PIF	Petite integration fequency
PTGS	Posttranscriptional gene silencing
RdDM	RNA-directed DNA methylase
REQ	Recombinant specific
RH	RNA helicase
ROS	Reactive oxygen species
SDE	Silencing defective
SF1-SF6	Superfamily1-superfamily6
SOS	Salt overly sensitive
Swi/snf	Switch/sucrose non-fermented DNA helicase
SYD	SPAYED (stem cell maintainence in SAM and enhances expression of the pluripotency gene WUSCHEL)
ToGR1	Thermotolerant growth required 1
Upf1-like	Up-frameshift-1, triggers nonsense-mediated decay

6.1 Introduction

Helicases, comprising both DNA and RNA helicases are universal enzymes performing basal functions from prokaryotes to eukaryotes. Moreover there is a strong conservation of the core enzyme structure in all organisms in both the animal and plant kingdoms. Since plants are sessile, some of these helicases have also evolved to perform important functions in both biotic and abiotic stresses along with their basic cellular functions (Kim et al. 2008; Xu et al. 2011; Tuteja et al. 2012; Barak et al. 2014; Liang and Deutscher 2016; Nawaz and Kang 2017; Asensi-Fabado et al. 2017).

In general helicases are motor proteins which catalyze the unwinding of stable duplex DNA or RNA molecules by using the energy of ATP hydrolysis. DNA helicases therefore play important roles in replication, recombination, repair, and transcription as well as in maintenance of chromosome stability (Tuteja 2003). RNA helicases unwind local secondary RNA structures, act as RNA chaperones during RNA cellular movement through the nucleus, and help in transcription, ribosome assembly, RNA processing, and translation. RNA helicases may also function in oligomerization or as RNPases to displace protein from structured or unstructured RNA (Rajkowitsch et al. 2007). This is because RNA molecules are more prone to forming stable nonfunctional secondary structures, particularly during stress (Umate and Tuteja 2010).

6.2 Sequence Conservation, Classification and Broad Functions

Sequence homologies based on conserved amino acid motifs classify helicases into six superfamilies from SF1 to SF6 (Fig. 6.1a). SF1 and SF2 are generally monomeric, while SF3–SF6 form hexameric ring-like structures used in DNA replication

a

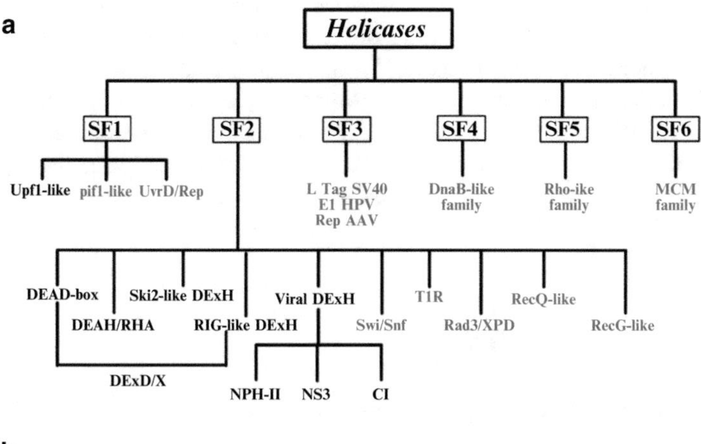

b

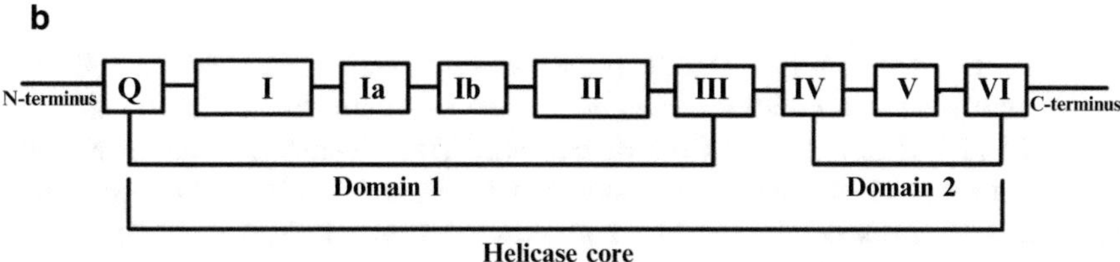

Fig. 6.1 (**a**) Classification of helicases. RNA helicase families are shown in black, and DNA helicase families are shown in *gray* and (**b**) conserved sequence motif organization of SF1 and SF2 RNA helicases

(Singleton et al. 2007; Georgescu et al. 2017). SF3–SF5 mainly comprise small groups of viral and bacterial DNA helicases, while SF6 consists of a larger group of DNA helicases from the archaeal mini-chromosome maintenance (MCM) to the eukaryotic Mcm2–7 (Abdelhaleem 2004; Sun et al. 2014). The helicase core is conserved with six motifs of amino acid residues comprising domain 1 and three grouped into domain 2 (Fig. 6.1b). The nucleic acid is held together by the folding of these two domains which form a cleft that juxtaposes the ATP-binding and RNA-binding sites together (Jankowsky 2011). The motifs are named starting at the N-terminal end as Q, I, Ia, Ib, II, and III in domain 1 and IV, V, and VI in domain 2 near the C-terminal end. Motifs, I, Ia, and Ib are also referred to as Walker A and motif II as Walker B. The Walker A- and B-domains are present in all SF1 and SF2 helicases and function in NTP binding and hydrolysis. Motif II or Walker B characteristically consists of the DEAD motif (Asp-Glu-Ala-Asp) in DEAD-box helicases, which is a large subclass of both DNA and RNA helicases implicated to be differentially expressed in stress in addition to their normal cellular functions. The DEAD motif is responsible for coupling ATPase to nucleic acid unwinding. The amino acid Q (responsible for the name of nine amino acid sequence Q motif) or Gln is invariant in case of all DEAD-box helicases. Important roles played by the other motifs include ATP coupling with hydrolysis by motif III, RNA binding by motifs IV and V, and nucleic acid-dependent NTP hydrolysis by motif VI (Tuteja et al. 2012; Chen et al. 2013).

The N-terminal and C-terminal flanking regions are poorly conserved and may be much larger than the helicase core. The flanking domains serve diverse functions by having additional DNA-/RNA-binding domains, protein-binding domains, or oligomerization modules (Umate et al. 2010). SF1 and SF2 superfamilies are further subdivided into three and ten members, respectively. Among the SF1 subdivisions are the two DNA helicases, UvrD and Pif1-like (petite integration frequency named from mitochondrial mutant discovery). The third subclass forms the Upf1 RNA helicases. UvrD enzymes are involved in replication, Pif1-like in telomere regulation, and Upf1 in nonsense-mediated decay (Chen et al. 2013).

6.3 Stress-Responsive Helicases

6.3.1 Differential Regulation of the Helicases

The only reported systematic study of the differential regulation of plant helicases under multiple stress conditions has been done by Umate et al. (2010) using the total 113 identified helicases from *A. thaliana*. They documented the expression pattern of these 113 helicase DNA and RNA helicase genes in *Arabidopsis* under ten different stress conditions like anoxia, cold (three independent replicates), drought, genotoxic, heat (two independent replicates), hypoxia, osmotic, oxidative, salt, and wounding followed by differential expression analysis of the log transformed data from an oligonucleotide array on Affymetrix 22K ATH1. A few of the reported genes were previously shown to have similar expression patterns by Kim et al. (2008). We could also verify the observed patterns for some of these genes from the gene expression atlas database from EMBL. The expression pattern was generally found to be random across the different families of DNA and RNA helicases, although many more RNA helicases responded to stress. A gene ontology enrichment analysis of the upregulated expression pattern revealed RNA helicase, zinc ion, and transition metal ion binding in addition to general functions like the ATP-dependent helicase activities shared with all other helicases. This may reflect a more prominent role of RNA helicase activity in stress adaptation compared to DNA helicases, even though some of the latter were also upregulated under stress.

Two major clusters of gene expression under multiple stresses were observed in the heat map of Umate et al. (2010), where one cluster showed high upregulation and the other an equal downregulation. In the cluster where the helicase genes were overexpressed, unique GO enrichment was found in DNA helicase activity, DNA replication, DNA repair, response to DNA damage, and cellular response to stress and stimulus compared to other clusters. Therefore it appears that the DNA helicases are activated to repair any DNA damage caused by multiple stresses.

We tried to correlate the gene expression pattern of the *Arabidopsis* helicase genes with their promoter motifs in order to group them in any functional pattern. For this purpose the available upstream 1500 bp flanking regions of 112 helicase genes were aligned and clustered using MEGA7. No specific pattern in the promoter sequence-based phylogeny was observed that related with the stress responsiveness. But the tree was separated into six groups (Fig. 6.2), where group 1 and 2 formed a major cluster A and the rest (group 3–6) a major cluster B. Cluster A was found to be enriched in DNA metabolic processes, DNA recombination, DNA helicase activity, cellular response to stimulus and cluster B in RNA helicase activity, nucleus and organelle lumen, and GO terms (Fig. 6.3). Both groups however showed common helicase, ATPase, hydrolase, and pyrophosphatase activities. For ease of understanding, the functional classification of the helicases has been organized in Table 6.1 along with their accession numbers.

Upstream sequence motif search revealed several stress-responsive cis elements in the helicase genes corresponding to their function under multiple stresses. A major portion of the drought-responsive genes are in groups 5 and 6. Hypoxia, wounding, and salt-responsive genes cluster mainly into group 1. Multiple stress response was observed in some genes (Table 6.2). These multifunctional genes were enriched with motifs like ARE (anaerobic response element), ERE (ethylene response element), MBS (drought responsive), and MeJA responsive with motifs like TGACG, CGTCA, and TGACG in their upstream regions. Other motifs included HSE (heat stress element); light-responsive elements like box 4, GAG, G box, ATCT, TCT, and fungal elicitor; and wound-responsive motifs like W box, Box

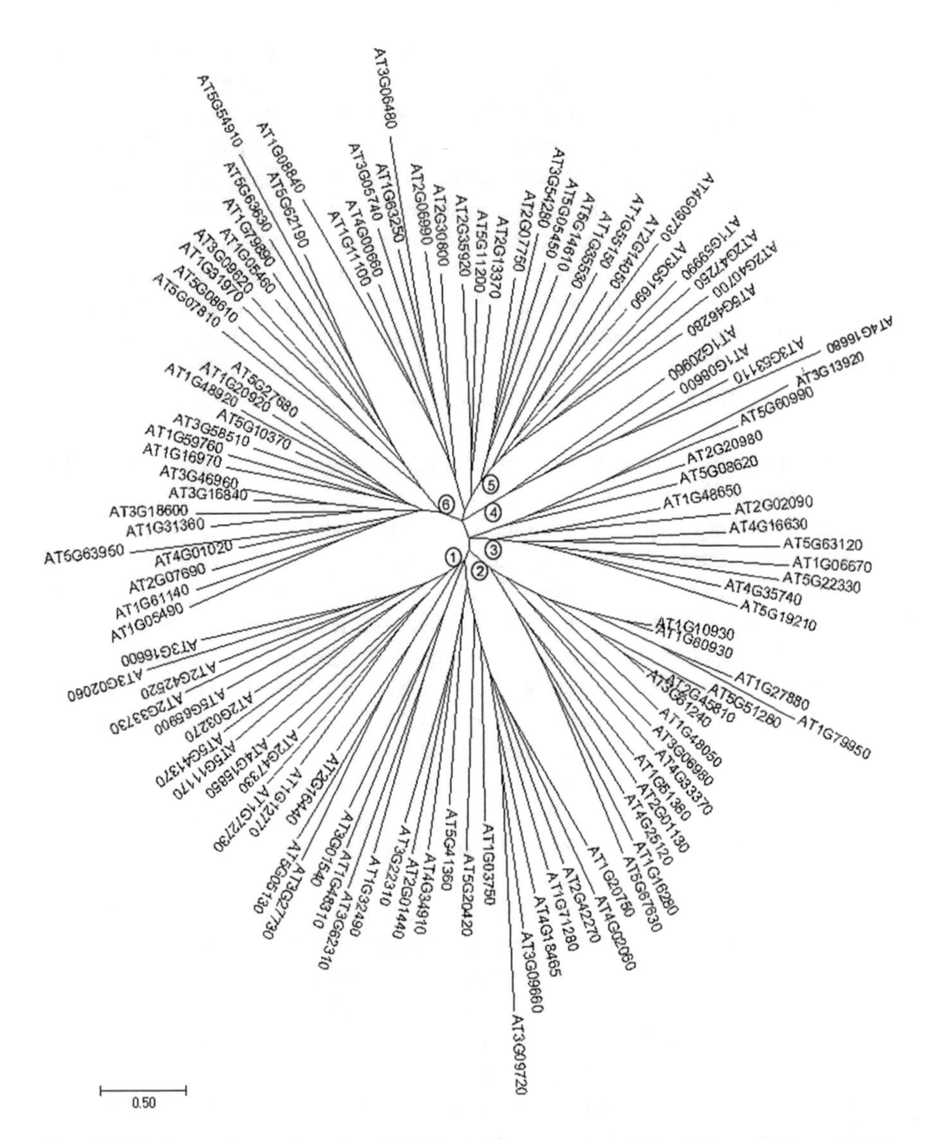

Fig. 6.2 Phylogenetic tree based on 1500 bp upstream region of *A. thaliana* DNA and RNA helicases

W1, and WUN motif. They also contained TC-rich repeat involved in defense and stress responsiveness, P box (gibberellin responsive element), TCA element (salicylic acid responsive), and 5′ UTR pyrimidine-rich stretch which is responsible for high-transcription level. ABRE element was also present in the upstream region of some of the multi-stress-responsive genes.

No	GO type	Description	A	B
1	P	DNA metabolic process		
2	P	DNA recombination		
3	P	nucleobase, nucleoside, nucleotide and nucleic acid metabolic process		
4	P	nitrogen compound metabolic process		
5	P	cellular response to stimulus		
6	F	helicase activity		
7	F	nucleoside-triphosphatase activity		
8	F	hydrolase activity, acting on acid anhydrides, in phosphorus-containing anhydrides		
9	F	hydrolase activity, acting on acid anhydrides		
10	F	pyrophosphatase activity		
11	F	ATP-dependent helicase activity		
12	F	purine NTP-dependent helicase activity		
13	F	ATPase activity, coupled		
14	F	ATPase activity		
15	F	hydrolase activity		
16	F	ATP binding		
17	F	adenyl ribonucleotide binding		
18	F	purine nucleoside binding		
19	F	nucleoside binding		
20	F	adenyl nucleotide binding		
21	F	purine ribonucleotide binding		
22	F	ribonucleotide binding		
23	F	purine nucleotide binding		
24	F	nucleotide binding		
25	F	catalytic activity		
26	F	nucleic acid binding		
27	F	DNA helicase activity		
28	F	DNA-dependent ATPase activity		
29	F	binding		
30	F	DNA binding		
31	P	DNA replication		
32	F	RNA helicase activity		
33	C	nucleolus		
34	C	nuclear part		
35	C	nuclear lumen		
36	C	intracellular organelle lumen		
37	C	membrane-enclosed lumen		
38	C	nucleus		
39	C	organelle lumen		

Fig. 6.3 Heatmap showing differences of enrichment in gene ontology terms in major clusters of upstream sequence phylogenetic tree (*A = cluster A, B = cluster B, P = biological pathway, F = molecular function, and C = cellular component, gradient from red to yellow means high to low level of enrichment*)

6.3.2 SF1 and SF6 Helicases

Many RNA/DNA helicases of the SF2 group and specific DNA helicases in the SF1 and SF6 groups have been reported to be differentially expressed mainly under abiotic stresses but also under biotic stress. The large group of SF2 helicases implicated in stress are discussed separately below. The Pif1-like DNA helicase from the SF1 group has been shown to be involved in suppression of telomere formation at

Table 6.1 List of *Arabidopsis* helicases based on the phylogeny of their upstream sequences clustered according to groups in Fig. 6.2

Group	Accession	Description	DNA/ RNA[a]
1	AT1G20750	RAD3-like DNA-binding helicase protein	DNA
1	AT4G02060	Mini-chromosome maintenance (MCM2/3/5) family protein	DNA
1	AT2G42270	U5 small nuclear ribonucleoprotein helicase	DNA
1	AT1G71280	DEA(D/H)-box RNA helicase family protein	
1	AT4G18465	RNA helicase family protein	
1	AT3G09660	Mini-chromosome maintenance 8	DNA
1	AT3G09720	P-loop containing nucleoside triphosphate hydrolase	
1	AT1G03750	Switch 2, chromatin remodeling 9	DNA
1	AT5G20420	Chromatin remodeling 42	DNA
1	AT5G41360	Homolog of xeroderma pigmentosum complementation protein	DNA
1	AT4G34910	P-loop containing nucleoside triphosphate hydrolases	
1	AT2G01440	DEAD-/DEAH-box RNA helicase family protein	
1	AT3G22310	Putative mitochondrial RNA helicase 1	
1	AT1G32490	RNA helicase family protein	
1	AT3G62310	RNA helicase family protein	
1	AT1G48310	Chromatin remodeling factor18	DNA
1	AT3G01540	DEAD-box RNA helicase 1	
1	AT3G27730	ATP binding; ATP-dependent helicases; DNA helicases	DNA
1	AT5G05130	DNA/RNA helicase protein	
1	AT2G16440	Mini-chromosome maintenance (MCM2/3/5) family protein	DNA
1	AT1G12770	P-loop containing nucleoside triphosphate hydrolases	
1	AT1G72730	DEA(D/H)-box RNA helicase family protein	
1	AT2G47330	P-loop containing nucleoside triphosphate hydrolases superfamily protein	
1	AT4G15850	RNA helicase 1	
1	AT5G11170	DEAD-/DEAH-box RNA helicase family protein	
1	AT5G41370	Homolog of xeroderma pigmentosum complementation gro	DNA
1	AT2G03270	DNA-binding protein, putative	DNA
1	AT5G65900	DEA(D/H)-box RNA helicase family protein	
1	AT2G33730	P-loop containing nucleoside triphosphate hydrolases	
1	AT2G42520	P-loop containing nucleoside triphosphate hydrolases	
1	AT3G02060	DEAD-/DEAH-box helicase, putative	
1	AT3G16600	SNF2 domain-containing protein/helicase domain con	
2	AT1G10930	DNA helicase (RECQl4A)	DNA
2	AT1G60930	RECQ helicase L4B	DNA
2	AT1G27880	DEAD-/DEAH-box RNA helicase family protein	
2	AT1G79950	RAD3-like DNA-binding helicase protein	DNA
2	AT5G51280	DEAD-box protein abstract, putative	
2	AT2G45810	DEA(D/H)-box RNA helicase family protein	

(continued)

Table 6.1 (continued)

Group	Accession	Description	DNA/ RNA[a]
2	AT3G61240	DEA(D/H)-box RNA helicase family protein	
2	AT1G48050	Ku80 family protein	DNA
2	AT3G06980	DEA(D/H)-box RNA helicase family protein	
2	AT4G33370	DEA(D/H)-box RNA helicase family protein	
2	AT1G51380	DEA(D/H)-box RNA helicase family protein	
2	AT2G01130	DEA(D/H)-box RNA helicase family protein	
2	AT4G25120	P-loop containing nucleoside triphosphate	DNA
2	AT1G16280	RNA helicase 36	
2	AT5G67630	P-loop containing nucleoside triphosphate	DNA
3	AT3G13920	Eukaryotic translation initiation factor	
3	AT5G60990	DEA(D/H)-box RNA helicase family protein	
3	AT2G20980	Mini-chromosome maintenance protein 10	DNA
3	AT5G08620	DEA(D/H)-box RNA helicase family protein	
3	AT1G48650	DEA(D/H)-box RNA helicase family protein	
3	AT2G02090	SNF2 domain-containing protein/helicase	
3	AT4G16630	DEA(D/H)-box RNA helicase family protein	
3	AT5G63120	P-loop containing nucleoside triphosphate	
3	AT1G06670	Nuclear DEIH-box helicase	
3	AT5G22330	P-loop containing nucleoside triphosphate	
3	AT4G35740	DEAD-/DEAH-box RNA helicase family protein	
3	AT5G19210	P-loop containing nucleoside triphosphate	
4	AT4G16680	P-loop containing nucleoside triphosphate	
4	AT3G53110	P-loop containing nucleoside triphosphate	
4	AT1G08600	P-loop containing nucleoside triphosphate	
4	AT1G20960	U5 small nuclear ribonucleoprotein helicase	DNA
5	AT1G08840	DNA replication helicase, putative	DNA
5	AT1G11100	SNF2 domain-containing protein/helicase	
5	AT4G00660	RNA helicase-like 8	
5	AT3G05740	RECQ helicase l1	DNA
5	AT1G63250	DEA(D/H)-box RNA helicase family protein	
5	AT3G06480	DEAD-box RNA helicase family protein	
5	AT2G06990	RNA helicase, ATP-dependent, SK12/DOB1 pr	
5	AT2G30800	Helicase in vascular tissue and tapetum	
5	AT2G35920	RNA helicase family protein	
5	AT5G11200	DEAD-/DEAH-box RNA helicase family protein	
5	AT2G13370	Chromatin remodeling 5	DNA
5	AT2G07750	DEA(D/H)-box RNA helicase family protein	
5	AT3G54280	DNA binding; ATP binding; nucleic acid bind	DNA
5	AT5G05450	P-loop containing nucleoside triphosphate	
5	AT5G14610	DEAD-box RNA helicase family protein	
5	AT1G35530	DEAD-/DEAH-box RNA helicase family protein	

(continued)

Table 6.1 (continued)

Group	Accession	Description	DNA/ RNA[a]
5	AT1G55150	DEA(D/H)-box RNA helicase family protein	
5	AT2G14050	Mini-chromosome maintenance 9	DNA
5	AT4G09730	RH39	
5	AT3G51690	PIF1 helicase	DNA
5	AT1G59990	DEA(D/H)-box RNA helicase family protein	
5	AT2G47250	RNA helicase family protein	
5	AT2G40700	P-loop containing nucleoside triphosphate	
5	AT5G46280	Mini-chromosome maintenance (MCM2/3/5) fam	DNA
6	AT1G05490	Chromatin remodeling 31	DNA
6	AT1G61140	SNF2 domain-containing protein/helicase domain-containing protein/zin	
6	AT2G07690	Mini-chromosome maintenance (MCM2/3/5) family protein	DNA
6	AT4G01020	Helicase domain-containing protein/IBR domain-containing protein/zinc	
6	AT5G63950	Chromatin remodeling 24	DNA
6	AT1G31360	RECQ helicase L2	DNA
6	AT3G18600	P-loop containing nucleoside triphosphate hydrolases superfamily protein	
6	AT3G16840	P-loop containing nucleoside triphosphate hydrolases superfamily protein	
6	AT3G46960	RNA helicase, ATP-dependent, SK12/DOB1 protein	
6	AT1G16970	KU70 homolog	DNA
6	AT1G59760	DExH-box ATP-dependent RNA helicase DExH9[b]	
6	AT3G58510	DEA(D/H)-box RNA helicase family protein	
6	AT5G10370	Helicase domain-containing protein/IBR domain-containing protein/zinc	
6	AT1G48920	Nucleolin like 1	DNA
6	AT1G20920	P-loop containing nucleoside triphosphate hydrolases superfamily protein	
6	AT5G27680	RECQ helicase SIM	DNA
6	AT5G07810	SNF2 domain-containing protein/helicase	
6	AT5G08610	P-loop containing nucleoside triphosphate hydrolases superfamily protein	
6	AT1G31970	DEA(D/H)-box RNA helicase family protein	
6	AT3G09620	P-loop containing nucleoside triphosphate	
6	AT1G05460	P-loop containing nucleoside triphosphate	
6	AT1G79890	RAD3-like DNA-binding helicase protein	DNA
6	AT5G63630	P-loop containing nucleoside triphosphate	
6	AT5G54910	DEA(D/H)-box RNA helicase family protein	
6	AT5G62190	DEAD-box RNA helicase (PRH75)	

[a]DNA helicases clearly identified in the literature have been labeled DNA in the last column. The naming of most enzymes showing RNA helicase activity reflects that their major property is probably RNA unwinding. Enzymes not grouped in either category above may be both RNA/DNA helicases

[b]Source: UniProtKB/Swiss-Prot;Acc:Q9XIF2

Table 6.2 List of multi-stress-responsive helicases in *Arabidopsis*

Accession	Gene description	Stress	Group[a]
AT4G16680	P-loop containing nucleoside triphosphate; RNA/DNA helicase	Salt, osmotic, hypoxia, anoxia	Group 4
AT1G05490	Chromatin remodeling 31; DNA helicase	Salt, osmotic, genotoxic, drought, heat, anoxia	Group 5
AT1G71280	DEA(D/H)-box RNA helicase family protein	Salt, osmotic, hypoxia, drought, heat, cold	Group 1
AT3G01540	DEAD-box RNA helicase 1	Salt, oxidative, wounding, cold	Group 1
AT3G16600	SNF2 domain-containing protein/ helicase domain; DNA helicase	Salt, wounding, hypoxia, heat, anoxia	Group 1
AT3G27730	ATP binding; ATP-dependent helicases; DNA helicase	Wounding, hypoxia, anoxia	Group 1
AT1G05460	P-loop containing nucleoside triphosphate; RNA/DNA helicase	Oxidative, drought, heat	Group 6
AT2G35920	RNA helicase family protein	Oxidative, drought, cold, anoxia	Group 5
AT1G60930	RECQ helicase L4B; DNA helicase	Salt, hypoxia, heat, anoxia	Group 2
AT1G11100	SNF2 domain-containing protein; DNA helicase	Salt, hypoxia, heat, anoxia	Group 5

[a]Group refers to the cluster of the coalesced helicases in Fig. 6.2

DNA double-strand breaks (Boulé and Zakian 2006). The *Arabidopsis* homolog of Pif1-like DNA helicase, At3g51690, has been shown to be upregulated under wounding stress by Umate and Tuteja (2010).

The Upf1 RNA helicases of SF1 are important for nonsense-mediated decay (NMD) of transcripts under both control and stress conditions. NMD is a eukaryotic quality control mechanism that controls the stability of both normal and aberrant transcripts. Plant NMD is downregulated by stress, thereby enhancing the expression of defense response genes (Shaul 2015). Loss of function mutations in Upf1 had enhanced pathogen tolerance due to the constitutive upregulation of defense responses, including high levels of the plant hormone, salicylic acid (SA) (Jeong et al. 2011; Riehs-Kearnan et al. 2012). Coupling of splicing regulators such as the RNA-binding SR proteins (Ser-Arg) with NMD has been shown. SR proteins are important in alternative splicing which is in turn important for adaptation to stress conditions (Palusa and Reddy 2010). The eukaryotic prereplicative complex, including hetero-hexameric mini-chromosome maintenance proteins (MCM2–7), ensures that DNA is replicated only once per cell division cycle. MCM8–10 helicases are needed during DNA replication fork assembly (de la Paz Sanchez et al. 2012). Binding of these proteins and subsequent activation by kinases provide the license to chromosomal DNA to undergo replication (Tuteja et al. 2011). The actual mechanism of the role of the MCM proteins when plant encounters any stress remains unknown. It was shown by Ni et al. (2009) that overexpression of *A. thaliana* MCM2 increases lateral root formation in *Arabidopsis*. Moreover overexpression of MCM6 from pea conferred high salt tolerance to tobacco with better growth, photosynthetic activity, and minimal reduction in yield, most likely by activating the expression of stress-related genes (Dang et al. 2011; Brasil et al. 2017). Contrary to the function of MCMs as multimeric proteins, which perform their function by encircling DNA, a single subunit of AtMCM3 on the other hand was shown to be able to unwind DNA by Rizvi et al. (2016). The authors suggest that this could be a mechanism for DNA unwinding and replication under specific conditions, such as biotic or abiotic stress. Both stresses are known to activate reactive oxygen species (ROS) which has been shown to act as a sensitive signal for many downstream targets (Sewelam et al. 2016).

6.3.3 The SF2 Group of Helicases

6.3.3.1 SF2 DNA Helicases

The SF2 group of helicases is by far the largest and diverse group of helicases consisting of ten subfamilies. Out of these five are predominantly RNA helicases and five are DNA helicases. The DNA helicases are called the Swi/Snf, T1R, Rad3/XPD, RecQ-like, and RecG-like. These DNA helicases are involved in chromosome remodeling (Tang et al. 2010) and DNA recombination and damage response (Tuteja 2003; Tang et al. 2010). Among these DNA helicases, the Swi/Snf (switch/sucrose non-fermenting helicases) have been reported to be differentially expressed during abiotic stress and are likely downregulated as reported by a number of research groups. Recruitment of Swi/Snf to target loci is done by long noncoding RNAs (lncRNAs), where the latter are regulated by environmental stress (Tang et al. 2010). Enzymes in this group like ISWI and CHD show nucleosome-specific activity and are responsible for stable alterations in chromosome assembly after replication. A specific DNA helicase, CHD1, interacts with ABI3 (abscisic acid insensitive3 transcription factor) and another protein FUSCA3 during plant embryo development. Another protein, BRM, stabilizes the nucleosome at the start of ABI5 transcription in the absence of stress. It is evident that mutants in this group of helicases are hypersensitive to abscisic acid (ABA). Another protein in this group MINU1 promotes growth arrest in drought and heat stress. Yet another protein of this group SYD which maintains stem cells in shoot apices also wards off the fungus *Botrytis cinerea*, since *syd* mutants are hypersensitive to this fungus (Han et al. 2015).

The RecQ-like (Kwon et al. 2013) and Rad3/XPD (Manova and Gruszka 2015) DNA helicases have been shown to be upregulated during DNA replication stress and hence indirectly in abiotic and biotic stresses. For instance, plants in their sedentary lifestyles are threatened by excessive salinity, drought, extreme low or high temperatures, as well as fungal, viral, or bacterial infections. These stresses in turn mobilize defenses like activation of cell cycle checks and DNA repair factors, inhibition of cell growth, and activation of the apoptosis pathway (Deckert et al. 2009).

The subgroups among the RNA helicases of the SF2 family are DEAD-box, DEAH/RHA, Ski2-likeDExH, RIG-1-like DExH, and viral DExH (Fairman-Williams et al. 2010; Jankowsky 2011; Chen et al. 2013). The DEAD-box family is by far the largest subgroup of helicases with more than hundred members already identified (Umate and Tuteja 2010; Umate et al. 2010). This subgroup has mainly been associated mostly with abiotic stress but also biotic and will be discussed separately below.

A DEAH RNA helicase was identified by a defective root initiation mutant called *rid* 1–1 (root initiation defective). This mutant was previously identified by severe defects in hypocotyl dedifferentiation and de novo meristem formation in tissue culture under high-temperature conditions. *RID1* was identified as encoding a DEAH-box RNA helicase implicated in pre-mRNA splicing (Ohtani et al. 2013). No other DEAH RNA helicase has been directly related to stress, although the mutant *prp16/cuv* coding for a DEAH RNA helicase differentially facilitates expression of genes involved in auxin biosynthesis, transport, perception, and signaling, thereby collectively influencing auxin-mediated development (Tsugeki et al. 2015).

All the RNA helicases of the SF2 group, except the viral DExH, are called the DExD/H as a group. An RNA helicase (*AtHELPS*) with a conserved DEVH motif has been reported to be involved in signaling in K^+ starvation due to salt stress. *AtHELPS* regulates the AKT-1-mediated and CBL-/CIPK-regulated K^+ uptake under limiting concentrations. Under potassium stress, there was an increased influx of K^+ and enhanced AKT-1, CBL1/9, and CIPK23 activities in the *helps* mutants. Interestingly this RNA helicase was also induced by cold and application of zeatin (Xu et al. 2011). The presence of several DExD/H helicases has been identified by Umate et al. (2010). Some of these have been found to be linked to P-bodies and stress granules (Maldonado-Bonilla 2014) when searched for under KEGG pathways in RiceNetDB (http://bis.zju.edu.cn/ricenetdb/genedetails).

Ski2-like RNA helicases are large multi-subunit proteins involved in RNA processing. Some evidences also indicate that these proteins can integrate into larger protein assemblies (Johnson and Jackson 2013). Overexpression of a rice *suv3* (suppressor allele var1 of the subgroup ski2-like) in rice was shown to improve photosynthesis and production of antioxidants for amelioration of stress (Tuteja et al. 2013). Moreover the transgenic rice overexpressing *suv3* was shown to regulate the expression of various stress-induced genes (Sahoo et al. 2014).

RIG-1-like (retinoic acid inducible-1-like) RNA helicases, on the other hand act as sensors for virus infection for conferring immunity in Eukaryotes (Guo et al. 2013).

6.3.3.2 DEAD-Box Helicases

The DEAD-box helicases form a very large subgroup of mostly RNA helicases in the SF2 family. Moreover members of the DEAD-box subfamily of helicases have been implicated as playing significant roles in both biotic and abiotic stresses in plants with much greater frequency than helicases of other subgroups. Stress-related activities of the other helicases have been outlined above. The effect of DEAD-box helicases in plant stress has been validated by endogenous study of their activities in mutants (Kant et al. 2007; Pascuan et al. 2016), by ectopic over- and underexpression studies (Table 6.3) (Li et al. 2008; Macovei et al. 2012; Tuteja et al. 2014; Zhu

Table 6.3 Capacity to confer salt tolerance by SF2 helicases tested by their endogenous and ectopic expression

Sl. No	Gene	Accession No.	Helicase type	Family	Isolated from	Expressed in	Observation/remarks	Reference	PMID/DOI
01	Pea p68	AF271892.1	RNA	DEAD	Pea	Tobacco	Improved growth, photosynthesis, and antioxidant machinery and higher K^+/Na^+	Tuteja et al. (2014)	PMC4039504
02	PgeIF4A	–	DNA/ RNA	DEAD	*P. Glaucum*	Ground nut	Higher shoot, tap root, and lateral root formation	Rao et al. (2017)	PMC5383670
03	OsSUV3	GQ982584	DNA/ RNA	Ski2p	Rice	Rice	Improves photosynthesis and antioxidant machinery	Tuteja et al. (2013)	PMC4091554
04	OsSUV3	GQ982584	DNA/ RNA	Ski2p	Rice	Rice	Higher plant hormone levels that regulate the expression of several stress-induced genes	Sahoo et al. (2014)	PMC4151020
05	AvDH1	–	DNA/ RNA	DEAD	*A. Venetum*	–	Unwinds DNA and RNA and salinity tolerance	Liu et al. (2008)	Doi:https://doi. org/10.1093/jxb/erm355
06	eIF4AIII	AT3G19760	RNA	–	*A. Thaliana*	–	Ortholog of the stress-related helicases PDH45 from *P. sativum* and MH1 from *M. sativa*	Pascuan et al. (2016)	Doi:https://doi. org/10.1007/s00299-016 1947-5
07	OsABP	Os06g33520	RNA	DEAD	Rice	Rice	Upregulated in response to multiple abiotic stresses	Macovei et al. (2012)	PMC3489646
08	SlDEAD31	KJ713393	RNA	DEAD	Tomato	Tomato	Enhanced salt and drought tolerance and expression of multiple stress-related genes	Zhu et al. (2015)	PMC4524616
09	OsBIRH1	Os03g01830	RNA	DEAD	Rice	*A. Thaliana*	Resistance against *A. brassicicola* and *P. syringae* and tolerance to oxidative stress	Li et al. (2008)	PMC2413282
10	MH1	–	–	–	Alfalfa	*A. Thaliana*	Drought and salt-stress tolerance by ROS scavenging and osmotic adjustment	Luo et al. (2009)	https://doi.org/10.1016/j. jplph.2008.06.018.

Sl. No	Gene	Accession No.	Helicase type	Family	Isolated from	Expressed in	Observation/remarks	Reference	PMID/DOI
11	RCF1	At1g20920	RNA		*A. Thaliana*	–	Pre-mRNA splicing and cold tolerance	Guan et al. (2013)	PMC3584546
12	AtHELPS	–	RNA	DExH/D	*A. Thaliana*	–	AtHELPS, negative regulator, plays a role in K+ deprivation stress	Xu et al. (2011)	https://doi.org/10.1111/j.17424658.2011.08147.x.
13	AtRH9 and AtRH25	At1g31970 At5g08620	RNA	–	*A. Thaliana*	–	Both genes were upregulated under drought stress and downregulated under salt stress	Kant et al. (2007)	PMC2048787
14	BAT1/UAP56	Os01g36890.2	RNA/DNA	DExH/D	Rice	–	Upregulated by salt and ABA. Bipolar ATPase	Tuteja et al. (2015)	https://doi.org/10.1007/s00709-015-0791-8

Table 6.4 Effects of the ectopic overexpression of the pea DNA/RNA helicase (PDH45) gene

SN	Expressed	Observation	Reference	PMID/DOI
01	Tobacco	Growth, flowering, and seed set 200 at 200 mMNaCl (~20 dS/m) stress compared to death of WT	Sanan-Mishra et al. (2005)	PMC544286
02	Rice	Better fertility and higher-grain yield by 16% compared to WT plants continuous stress of 6 dS/m from 30 days till maturity	Amin et al. (2012)	https://doi.org/10.1007/s11032-011-9625-3
03	Rice	Decrease in the miRNAs expression concomitant with lower expression of target genes	Macovei and Tuteja (2012)	PMC3502329
04	Rice	No effect of transgene on enzymatic activities and functional diversity of the rhizosphere soil microbial community	Sahoo and Tuteja (2013)	https://doi.org/10.4161/psb.24950
05	Peanut	Improved root volume, higher chlorophyll, and reduced membrane damage in PEG dehydration stress	Manjulatha et al. (2014)	https://doi.org/10.1007/s12033-013-9687-z
06	Sugarcane	Higher cell membrane thermostability, relative water and chlorophyll content, and photosynthetic efficiency under water and salt stress	Augustine et al. (2015a)	https://doi.org/10.1007/s12033-015-9841-x
07	Sugarcane	Co-transformed EaDREB2 and PDH45 show higher-salinity tolerance but lower drought tolerance than EaDREB2 alone	Augustine et al. (2015b)	Doi:https://doi.org/10.1007/s00299-014-1704-6
08	Rice	Better regulation of Na^+ level, ROS production, Ca^{2+} homeostasis, cell viability, and cation transporters (SOS1, NHX1) in roots	Nath et al. (2016)	https://doi.org/10.1016/j.jplph.2015.11.008
09	Chili	Alleviation of senescence and the multiple stresses, drought, salinity, and oxidative	Shivakumara et al. (2017)	PMC5459802
10	Tobacco	Pyramiding of PDH45 with EPSPS renders tolerance to salinity and herbicide by enhancing the antioxidant machinery	Garg et al. (2017)	PMC5364135
11	Rice	PDH45 and SUV transgenes showed delayed leaf senescence associated with changes in telomere length	Macovei et al. (2017)	https://doi.org/10.1007/s00709-016-1017-4

et al. 2015; Rao et al. 2017) as well as microarray and other transcriptional assays (Umate et al. 2010). These studies have been done in multiple crops (Tables 6.3 and 6.4) and provide an overall idea about the importance of DEAD-box helicases in adaptation of plants to stressed conditions.

The work with the pea DNA helicase (PDH45) has been most extensive with overexpression in tobacco, rice, peanut, sugarcane, and chili where the transgene confers salt tolerance to the host plant (Table 6.4). Interestingly, PDH45 is a unique member in that it contains DESD and SRT instead of DEAD/Hand SAT domains in motifs V and VI, respectively. PDH45 is localized in both nucleus and cytosol and exhibits 3–5′ directional unwinding activity in an ATP-dependent manner. Moreover it shows both DNA and RNA helicase activities and is highly homologous to the eukaryotic initiation factor eIF4A as are most other DEAD-box helicases (Pham et al. 2000; Gill et al. 2013). Na^+-, Ca^{2+}-, and ROS-specific dyes showed the gradual reduction of all three of these molecules with time only in overexpressing IR64 rice lines (Nath et al. 2016). They also hypothesized events after salt stress, using networking of downstream enzymes upregulated by PDH45, where they propose the following mechanism. In root of overexpressing PDH45 transgenic rice, upregulation of OsSOS1, i.e., salt overly sensitive (SOS) pathway, may lead to less Na^+ accumulation via exclusion. Such, low Na^+ accumulation results in low ROS (H_2O_2) generation and ultimately cell viability followed by balanced Ca^{2+}homeostasis under salinity stress. Further, upregulation of cation transporters such as OsCHX11, OsCAX, OsTPC1, and OsCNGC1 collectively contributed to provide tolerance to the overexpressing PDH45 transgenic lines. These are all cationic transporters with rice accession numbers as CHX11 (Os05g31730), OsCAX (Os02g04630), OsTPC1 (Os01g48680), and OsCNGC1 (Os06g33600) (Nath et al. 2016).

A more detailed hypothetical network could be formed when we used the GeneMANIA web interface (Warde-Farley et al. 2010) after plugging in PDH45 (AtRH45) and its downstream gene homologs (from *Arabidopsis thaliana*) from the work of Augustine et al. (2015a, b) as well as Nath et al. (2016) (Fig. 6.4). In concert with the work of Nath et al. (2016), we could also show (Fig. 6.4) that PDH45 (AtRH45) mitigates salinity stress by directly or indirectly influencing four different biological pathways: (a) Na^+/K^+ homeostasis, (b) salt overly sensitive (SOS) pathway, (c) Ca^+ homeostasis, and (d) ROS-scavenging pathways. A physical interaction between RH45 and ELF9 was observed. Moreover ELF9 was co-expressed with NHX1. ELF9 is an RNA-binding protein, and *elf9* mutants show premature transition to flowering under stress (Meyer et al. 2015).

Interestingly, the Na^+/K^+ homeostasis network showed interaction between nine cation/antiporter proteins CHX13–23 with the exception of CHX22, 3 K^+/H^+ antiporters, KEA4–6 and 4 Na^+/H^+ antiporters, and NHX1, 2, 5, and 6. AtCHX17 has been shown to be important for K^+ acquisition (Cellier et al. 2004), while KEA5 was reported to be activated by K^+ deficiency (Shin and Schachtman 2004). The NHX1 antiporter is observed to physically interact with the calmodulin protein CML18 as previously reported by Yamaguchi et al. (2005).

The SOS (salt overly sensitive) pathway proteins were a major part of the network, and SOS1 was shown to be physically interacting with RCD1 (radical-induced cell death1). RCD1 has been shown to interact with SOS1, and a large number of other transcription factors predicted to be involved in both development- and stress-related processes. RCD1 is possibly involved in signaling networks that regulate quantitative changes in gene expression in response to ROS (You and Chan 2015). A major genetic interaction between SOS2 and NDPK2 (nucleoside diphosphate kinase 2) was observed in the constructed network. Expression of *Arabidopsis*

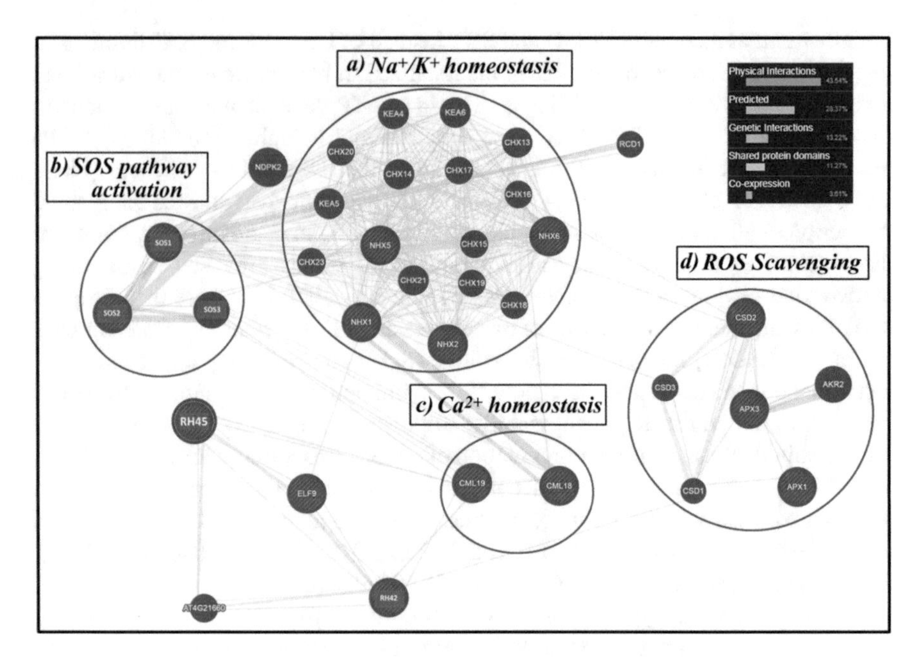

Fig. 6.4 Hypothetical networking of *PDH45* (*AtRH45*) during salinity stress in *Arabidopsis* generated by GeneMANIA web interface (Warde-Farley et al. 2010). *PDH45 (AtRH45)* mitigates salinity stress by controlling directly or indirectly through four different biological pathways. (**a**) Na^+/K^+ homeostasis, (**b**) salt overly sensitive (SOS) pathway 1, (**c**) Ca^{2+} homeostatis, and (**d**) ROS-scavenging pathways

NDPK2 has been shown to increase antioxidant enzyme activities and enhance tolerance to multiple environmental stresses in transgenic sweet potato plants (Kim et al. 2009).

Among the ROS scavengers, APX3 was observed to physically interact with AKR2. Ankyrin repeat-containing protein 2A has been reported to be an essential molecular chaperone for peroxisomal membrane-bound ascorbate peroxidase3 in *Arabidopsis* (Shen et al. 2010).

6.3.3.3 Organelle DEAD-Box Helicases

Search for appropriate transit peptides helped Nawaz and Kang (2017) identify 7–12 chloroplast RNA helicases (RHs) and 4–7 mitochondrial helicases. Among those that responded to multiple abiotic stresses, Umate et al. (2010) documented several chloroplast-localized DEAD-box RHs (such as RH58, RH47, RH26, RH50, RH41, and RH52) and mitochondria-targeted DEAD-box RHs (such as RH33 and RH31). Mitochondrial PLT1 (PLETHORA1) and ABO6 (ABA overly sensitive) have been found to play important roles in the regulation of primary root growth and root meristem activity by modulating ABA and auxin signaling. A chloroplast-targeted RH3 is reported to be essential for carbon fixation and the maintenance of ABA level in *Arabidopsis* under environmental stresses (Nawaz and Kang 2017; Lee et al. 2013). Interestingly, the expression of majority of chloroplast- or mitochondria-targeted DEAD-box RHs in rice, maize, and wheat are downregulated under diverse abiotic stresses. It would be interesting to observe the targets of RHs which are actually upregulated under stressed conditions. Cold treatment has been shown to cause upregulation of DEAD-box RHs like the chloroplast OsRH7, AtRH26, and AtRH50 as well as mitochondrial OsRH17, ZmRH30, and AtRH33. Heat on the other hand causes upregulation of AtRH50 from the chloroplast and AtRH31 in the mitochondria (Nawaz and Kang 2017; Umate et al. 2010). Thus it is reasonable to predict that RNA helicases targeted to either chloroplasts or mitochondria are crucial for the regulation of gene expression and RNA metabolism in these cellular organelles in order to minimize the effects of stress on photosynthesis and respiration, respectively.

6.4 Conclusions

DNA and RNA helicases mainly function as ATP-dependent nucleic acid unwinding enzymes crucial for cellular function and survival. DNA helicases mainly function in maintaining genomic stability, including telomere length and function. This is in addition to their essential role in the replication and transcription machinery. The DNA helicases which are activated under environmental stresses mainly function to repair DNA and maintain chromosomal configuration. RNA helicases are important for transcription, RNA processing, formation of RNP bodies, and translation. However the RNA helicases also act as chaperones which help during RNA transport. They unwind RNA which is more prone to form secondary structures, particularly under stressed conditions. Fewer DNA helicases are activated under stress, even though their expression levels are high. On the other hand, many more RNA helicases, many of which are DEAD-box helicases, are activated even though their expression levels are comparatively lower. Organelle-specific helicases are mostly DEAD-box RNA helicases and show more pronounced differential regulation under cold and heat stress compared to salt, drought, and osmotic stresses. A handful of both DNA and RNA helicases respond to multiple stresses. Ectopic expression of RNA helicases, particularly DEAD-box ones, gives a growth advantage to the transgenic plants. This vigor in growth gives them an edge over wild type for survival and tolerance. This vigorous growth could be due to maintenance of RNA in a physically unwinded state for ease of transcription, particularly under stressed conditions.

References

Abdelhaleem M (2004) Do human RNA helicases have a role in cancer? Biochim Biophys Acta (BBA)-Rev Cancer 1704(1):37–46

Amin M, Elias SM, Hossain A, Ferdousi A, Rahman MS, Tuteja N, Seraj ZI (2012) Over-expression of a DEAD-box helicase, PDH45, confers both seedling and reproductive stage salinity tolerance to rice (Oryza sativa L.). Mol Breed 30:345–354

Asensi-Fabado M-A, Amtmann A, Perrella G (2017) Plant responses to abiotic stress: the chromatin context of transcriptional regulation. Biochim Biophys Acta (BBA)-Gene Regul Mech 1860(1):106–122

Augustine SM, Narayan JA, Syamaladevi DP, Appunu C, Chakravarthi M, Ravichandran V, Tuteja N, Subramonian N (2015a) Introduction of Pea DNA Helicase 45 into Sugarcane (Saccharum spp. Hybrid) enhances cell membrane thermostability and upregulation of stress-responsive genes leads to abiotic stress tolerance. Mol Biotechnol 57(5):475–488

Augustine SM, Narayan JA, Syamaladevi DP, Appunu C, Chakravarthi M, Ravichandran V, Tuteja N, Subramonian N (2015b) Overexpression of EaDREB2 and pyramiding of EaDREB2 with the pea DNA helicase gene (PDH45) enhance drought and salinity tolerance in sugarcane (Saccharum spp. hybrid). Plant Cell Rep 34(2):247–263

Barak S, Singh Yadav N, Khan A (2014) DEAD-box RNA helicases and epigenetic control of abiotic stress-responsive gene expression. Plant Signal Behav 9(12):e977729

Boulé J-B, Zakian VA (2006) Roles of Pif1-like helicases in the maintenance of genomic stability. Nucleic Acids Res 34(15):4147–4153

Brasil JN, Costa CNM, Cabral LM, Ferreira PC, Hemerly AS (2017) The plant cell cycle: pre-replication complex formation and controls. Genet Mol Biol 40(1):276–291

Cellier F, Conéjéro G, Ricaud L, Luu DT, Lepetit M, Gosti F, Casse F (2004) Characterization of AtCHX17, a member of the cation/H+ exchangers, CHX family, from *Arabidopsis thaliana* suggests a role in K+ homeostasis. Plant J 39(6):834–846

Chen C-Y, Liu X, Boris-Lawrie K, Sharma A, Jeang K-T (2013) Cellular RNA helicases and HIV-1: insights from genome-wide, proteomic, and molecular studies. Virus Res 171(2):357–365

Dang HQ, Tran NQ, Gill SS, Tuteja R, Tuteja N (2011) A single subunit MCM6 from pea promotes salinity stress tolerance without affecting yield. Plant Mol Biol 76(1–2):19–34

de la Paz SM, Costas C, Sequeira-Mendes J, Gutierrez C (2012) Regulating DNA replication in plants. Cold Spring Harb Perspect Biol 4(12):a010140

Deckert J, Pawlak S, Rybaczek D (2009) The nucleus as a 'headquarters' and target in plant cell stress reactions. In: Maksymiec W (ed) Compartmentation of responses to stresses in higher plants, true or false, Research Signpost. Transworld Research Network, Trivandrum, pp 61–90

Fairman-Williams ME, Guenther U-P, Jankowsky E (2010) SF1 and SF2 helicases: family matters. Curr Opin Struct Biol 20(3):313–324

Garg B, Gill SS, Biswas DK, Sahoo RK, Kunchge NS, Tuteja R, Tuteja N (2017) Simultaneous Expression of PDH45 with EPSPS Gene Improves Salinity and Herbicide Tolerance in Transgenic Tobacco Plants. Front Plant Sci 8:364

Georgescu R, Yuan Z, Bai L, Santos RLA, Sun J, Zhang D, Yurieva O, Li H, O'Donnell ME (2017) Structure of eukaryotic CMG helicase at a replication fork and implications to replisome architecture and origin initiation. Proc Natl Acad Sci 114(5):E697–E706

Gill SS, Tajrishi M, Madan M, Tuteja N (2013) A DESD-box helicase functions in salinity stress tolerance by improving photosynthesis and antioxidant machinery in rice (*Oryza sativa* L. cv. PB1). Plant Mol Biol 82(1–2):1–22

Guan Q, Wu J, Zhang Y, Jiang C, Liu R, Chai C, Zhu J (2013) A DEAD Box RNA Helicase Is Critical for PremRNA Splicing, Cold-Responsive Gene Regulation, and Cold Tolerance in Arabidopsis. Plant Cell 25:342–356

Guo X, Zhang R, Wang J, Ding S-W, Lu R (2013) Homologous RIG-I–like helicase proteins direct RNAi-mediated antiviral immunity in *C. elegans* by distinct mechanisms. Proc Natl Acad Sci 110(40):16085–16090

Han SK, Wu MF, Cui S, Wagner D (2015) Roles and activities of chromatin remodeling ATPases in plants. Plant J 83(1):62–77

Jankowsky E (2011) RNA helicases at work: binding and rearranging. Trends Biochem Sci 36(1):19–29

Jeong H-J, Kim YJ, Kim SH, Kim Y-H, Lee I-J, Kim YK, Shin JS (2011) Nonsense-mediated mRNA decay factors, UPF1 and UPF3, contribute to plant defense. Plant Cell Physiol 52(12):2147–2156

Johnson SJ, Jackson RN (2013) Ski2-like RNA helicase structures: common themes and complex assemblies. RNA Biol 10(1):33–43

Kant P, Kant S, Gordon M, Shaked R, Barak S (2007) STRESS RESPONSE SUPPRESSOR1 and STRESS RESPONSE SUPPRESSOR2, two DEAD-box RNA helicases that attenuate Arabidopsis responses to multiple abiotic stresses. Plant Physiol 145(3):814–830

Kim JS, Kim KA, Oh TR, Park CM, Kang H (2008) Functional characterization of DEAD-box RNA helicases in *Arabidopsis thaliana* under abiotic stress conditions. Plant Cell Physiol 49(10):1563–1571

Kim Y-H, Lim S, Yang K-S, Kim CY, Kwon S-Y, Lee H-S, Wang X, Zhou Z, Ma D, Yun D-J (2009) Expression of ArabidopsisNDPK2 increases antioxidant enzyme activities and enhances tolerance to multiple environmental stresses in transgenic sweet potato plants. Mol Breed 24(3):233–244

Kwon Y-I, Abe K, Endo M, Osakabe K, Ohtsuki N, Nishizawa-Yokoi A, Tagiri A, Saika H, Toki S (2013) DNA replication arrest leads to enhanced homologous recombination and cell death in meristems of rice OsRecQl4 mutants. BMC Plant Biol 13(1):62

Lee KH, Park J, Williams DS, Xiong Y, Hwang I, Kang BH (2013) Defective chloroplast development inhibits maintenance of normal levels of abscisic acid in a mutant of the Arabidopsis RH3 DEAD-box protein during early post-germination growth. Plant J 73(5):720–732

Li D, Liu H, Zhang H, Wang X, Song F (2008) OsBIRH1, a DEAD-box RNA helicase with functions in modulating defence responses against pathogen infection and oxidative stress. J Exp Bot 59(8):2133–2146

Liang W, Deutscher MP (2016) REP sequences: mediators of the environmental stress response? RNA Biol 13(2):152–156

Liu HH, Liu J, Fan SL, Song MZ, Han XL, Liu F, Shen FF (2008) Molecular cloning and characterization of a salinity stress-induced gene encoding DEAD-box helicase from the halophyte Apocynum venetum. J Exp Bot 59:633–644

Luo Y, Liu YB, Dong YX, Gao XQ, Zhang XS (2009) Expression of a putative alfalfa helicase increases tolerance to abiotic stress in Arabidopsis by enhancing the capacities for ROS scavenging and osmotic adjustment. J Plant Physiol 166:385–394

Macovei A, Vaid N, Tula S, Tuteja N (2012) A new DEAD-box helicase ATP-binding protein (OsABP) from rice is responsive to abiotic stress. Plant Signal Behav 7(9):1138–1143

Macovei A, Tuteja N (2012) MicroRNAs targeting DEAD-box helicases are involved in salinity stress response in rice (Oryza sativa L.). BMC Plant Biol 12:183. https://doi.org/10.1186/1471-2229-12-183

Macovei A, Sahoo RK, Fae M, Balestrazzi A, Carbonera D, Tuteja N (2017) Overexpression of PDH45 or SUV3 helicases in rice leads to delayed leaf senescence-associated events. Protoplasma 254:1103–1113

Maldonado-Bonilla LD (2014) Composition and function of P bodies in *Arabidopsis thaliana*. Front Plant Sci 5:201

Manjulatha M, Sreevathsa R, Kumar AM, Sudhakar C, Prasad TG, Tuteja N, Udayakumar M (2014) Overexpression of a pea DNA helicase (PDH45) in peanut (Arachis hypogaea L.) confers improvement of cellular level tolerance and productivity under drought stress. Mol Biotechnol 56:111–125

Manova V, Gruszka D (2015) DNA damage and repair in plants–from models to crops. Front Plant Sci 6:885

Meyer K, Koester T, Staiger D (2015) Pre-mRNA splicing in plants: in vivo functions of RNA-binding proteins implicated in the splicing process. Biomol Ther 5(3):1717–1740

Nath M, Yadav S, Sahoo RK, Passricha N, Tuteja R, Tuteja N (2016) PDH45 transgenic rice maintain cell viability through lower accumulation of Na+, ROS and calcium homeostasis in roots under salinity stress. J Plant Physiol 191:1–11

Nawaz G, Kang H (2017) Chloroplast-or mitochondria-targeted DEAD-Box RNA helicases play essential roles in organellar RNA metabolism and abiotic stress responses. Front Plant Sci 8:871

Ni DA, Sozzani R, Blanchet S, Domenichini S, Reuzeau C, Cella R, Bergounioux C, Raynaud C (2009) The Arabidopsis MCM2 gene is essential to embryo development and its overexpression alters root meristem function. New Phytol 184(2):311–322

Ohtani M, Demura T, Sugiyama M (2013) Arabidopsis root initiation defective1, a DEAH-box RNA helicase involved in pre-mRNA splicing, is essential for plant development. Plant Cell 25(6):2056–2069

Palusa SG, Reddy AS (2010) Extensive coupling of alternative splicing of pre-mRNAs of serine/arginine (SR) genes with nonsense-mediated decay. New Phytol 185(1):83–89

Pascuan C, Frare R, Alleva K, Ayub ND, Soto G (2016) mRNA biogenesis-related helicase eIF4AIII from *Arabidopsis thaliana* is an important factor for abiotic stress adaptation. Plant Cell Rep 35(5):1205–1208

Pham XH, Reddy MK, Ehtesham NZ, Matta B, Tuteja N (2000) A DNA helicase from *Pisum sativum* is homologous to translation initiation factor and stimulates topoisomerase I activity. Plant J 24(2):219–229

Rajkowitsch L, Chen D, Stampfl S, Semrad K, Waldsich C, Mayer O, Jantsch MF, Konrat R, Bläsi U, Schroeder R (2007) RNA chaperones, RNA annealers and RNA helicases. RNA Biol 4(3):118–130

Rao TSRB, Naresh JV, Reddy PS, Reddy MK, Mallikarjuna G (2017) Expression of *Pennisetum glaucum* Eukaryotic Translational Initiation Factor 4A (PgeIF4A) confers improved drought, salinity, and oxidative stress tolerance in groundnut. Front Plant Sci 8:453

Riehs-Kearnan N, Gloggnitzer J, Dekrout B, Jonak C, Riha K (2012) Aberrant growth and lethality of Arabidopsis deficient in nonsense-mediated RNA decay factors is caused by autoimmune-like response. Nucleic Acids Res 40(12):5615–5624

Rizvi I, Choudhury NR, Tuteja N (2016) *Arabidopsis thaliana* MCM3 single subunit of MCM2–7 complex functions as 3′ to 5′ DNA helicase. Protoplasma 253(2):467–475

Sahoo RK, Tuteja N (2013) Effect of salinity tolerant PDH45 transgenic rice on physicochemical properties, enzymatic activities and microbial communities of rhizosphere soils. Plant Signal Behav 8:e24950

Sahoo RK, Ansari MW, Tuteja R, Tuteja N (2014) OsSUV3 transgenic rice maintains higher endogenous levels of plant hormones that mitigates adverse effects of salinity and sustains crop productivity. Rice 7(1):17

Sanan-Mishra N, Pham XH, Sopory SK, Tuteja N (2005) Pea DNA helicase 45 overexpression in tobacco confers high salinity tolerance without affecting yield. Proc Natl Acad Sci U S A 102:509–514

Sewelam N, Kazan K, Schenk PM (2016) Global plant stress signaling: reactive oxygen species at the cross-road. Front Plant Sci 7:187

Shaul O (2015) Unique aspects of plant nonsense-mediated mRNA decay. Trends Plant Sci 20(11):767–779

Shen G, Kuppu S, Venkataramani S, Wang J, Yan J, Qiu X, Zhang H (2010) ANKYRIN REPEAT-CONTAINING PROTEIN 2A is an essential molecular chaperone for peroxisomal membrane-bound ASCORBATE PEROXIDASE3 in Arabidopsis. Plant Cell 22(3):811–831

Shin R, Schachtman DP (2004) Hydrogen peroxide mediates plant root cell response to nutrient deprivation. Proc Natl Acad Sci U S A 101(23):8827–8832

Singleton MR, Dillingham MS, Wigley DB (2007) Structure and mechanism of helicases and nucleic acid translocases. Annu Rev Biochem 76:23–50

Shivakumara TN, Sreevathsa R, Dash PK, Sheshshayee MS, Papolu PK, Rao U, Tuteja N, Udayakumar M (2017) Overexpression of Pea DNA Helicase 45 (PDH45) imparts tolerance to multiple abiotic stresses in chili (Capsicum annuum L.). Sci Rep 7:2760

Sun J, Fernandez-Cid A, Riera A, Tognetti S, Yuan Z, Stillman B, Speck C, Li H (2014) Structural and mechanistic insights into Mcm2–7 double-hexamer assembly and function. Genes Dev 28(20):2291–2303

Tang L, Nogales E, Ciferri C (2010) Structure and function of SWI/SNF chromatin remodeling complexes and mechanistic implications for transcription. Prog Biophys Mol Biol 102(2):122–128

Tsugeki R, Tanaka-Sato N, Maruyama N, Terada S, Kojima M, Sakakibara H, Okada K (2015) CLUMSY VEIN, the Arabidopsis DEAH-box Prp16 ortholog, is required for auxin-mediated development. Plant J 81(2):183–197

Tuteja N (2003) Plant DNA helicases: the long unwinding road. J Exp Bot 54(391):2201–2214

Tuteja N, Tran NQ, Dang HQ, Tuteja R (2011) Plant MCM proteins: role in DNA replication and beyond. Plant Mol Biol 77(6):537–545

Tuteja N, Singh S, Tuteja R (2012) Helicases in improving abiotic stress tolerance in crop plants. Improv Crop Resist Abiotic Stress 1 & 2:435–449

Tuteja N, Sahoo RK, Garg B, Tuteja R (2013) OsSUV3 dual helicase functions in salinity stress tolerance by maintaining photosynthesis and antioxidant machinery in rice (*Oryza sativa* L. cv. IR64). Plant J 76(1):115–127

Tuteja N, Banu MSA, Huda KMK, Gill SS, Jain P, Pham XH, Tuteja R (2014) Pea p68, a DEAD-box helicase, provides salinity stress tolerance in transgenic tobacco by reducing oxidative stress and improving photosynthesis machinery. PLoS One 9(5):e98287

Tuteja N, Tarique M, Trivedi DK, Sahoo RK, Tuteja R (2015) Stress-induced Oryza sativa BAT1 dual helicase exhibits unique bipolar translocation. Protoplasma 252:1563–1574

Umate P, Tuteja R (2010) Architectures of the unique domains associated with the DEAD-box helicase motif. Cell Cycle 9(20):4228–4235

Umate P, Tuteja R, Tuteja N (2010) Genome-wide analysis of helicase gene family from rice and Arabidopsis: a comparison with yeast and human. Plant Mol Biol 73(4):449–465

Warde-Farley D, Donaldson SL, Comes O, Zuberi K, Badrawi R, Chao P, Franz M, Grouios C, Kazi F, Lopes CT (2010) The GeneMANIA prediction server: biological network integration for gene prioritization and predicting gene function. Nucleic Acids Res 38(suppl_2):W214–W220

Xu RR, Qi SD, Lu LT, Chen CT, Wu CA, Zheng CC (2011) A DExD/H box RNA helicase is important for K+ deprivation responses and tolerance in *Arabidopsis thaliana*. FEBS J 278(13):2296–2306

Yamaguchi T, Aharon GS, Sottosanto JB, Blumwald E (2005) Vacuolar Na+/H+ antiporter cation selectivity is regulated by calmodulin from within the vacuole in a Ca2+-and pH-dependent manner. Proc Natl Acad Sci U S A 102(44):16107–16112

You J, Chan Z (2015) ROS regulation during abiotic stress responses in crop plants. Front Plant Sci 6:1092

Zhu M, Chen G, Dong T, Wang L, Zhang J, Zhao Z, Hu Z (2015) SlDEAD31, a putative DEAD-Box RNA helicase gene, regulates salt and drought tolerance and stress-related genes in tomato. PLoS One 10(8):e0133849

Chapter 7
miRNAs: The Game Changer in Producing Salinity Stress-Tolerant Crops

Ratanesh Kumar, Sudhir Kumar, and Neeti Sanan-Mishra

Abstract The problem of soil salinity is emerging as the most damaging factor for crop growth and productivity. An increased salt concentration causes both osmotic and ionic stress to the plants, which inhibits their growth rate and yields. The genetic processes that coordinate the plant's response to salinity have been explained by mechanisms that regulate the signaling cascades, cellular homeostasis, osmoticum maintenance and metabolic processes. This involves modulation of the gene expression. The microRNAs (miRNAs) represent an important class of endogenous post-transcriptional regulators. They play an important role in plant biology by regulating every aspect of development, metabolism and environmental responses. This review discusses the available information on the salt stress-responsive genes and their role in the adaptive response to salt stress. It also discusses the role of miRNAs as potential regulators in salt stress response. The application of miRNA-based strategies for improving plants is also described.

Keywords MicroRNA · Origin · Function · *Oryza sativa* · Salt stress · Target gene · Posttranscriptional regulation · Signaling molecules · Transporters · miRNA application · Osmolytes · Hormones

Abbreviations

ABA	Abscisic acid
ABREs	ABA-responsive cis-elements
AGO	Argonaute
AP2	Apetala 2

R. Kumar · S. Kumar · N. Sanan-Mishra (✉)
Plant RNAi Biology Group, International Centre for Genetic Engineering and Biotechnology, New Delhi, India
e-mail: neeti@icgeb.res.in

V. Kumar et al. (eds.), *Salinity Responses and Tolerance in Plants, Volume 2*,
https://doi.org/10.1007/978-3-319-90318-7_7

AREB/ABF	ABA-responsive element-binding protein/ABRE-binding factor
AtHK1	*Arabidopsis thaliana* histidine kinase1
Ca^{2+}	Calcium ion
CAD	C-terminal CDPK activation domain
CBC	Cap-binding complex
CBL	Calcineurin B-like
CDPKs.	Ca^{2+}-dependent protein kinases
CIPKs	CBL-interacting protein kinases
CKs	Cytokinins
CSDs	Cu/Zn superoxide dismutases
D bodies	Dicing bodies
DAG	Diacylglycerol
DCL	Dicer-like
GAs	Gibberellins
GST	Glutathione S-transferase
H_2O_2	Hydrogen peroxide
HD-ZIP	Homeodomain-leucine zipper protein
HEN1	Hua enhancer1
HST	Hasty
HYL1	Hyponastic leaves1
IAA	Indole acetic acid
IP_3	Inositol 1,4,5-trisphosphate
JA	Jasmonic acid
MAPK	Mitogen-activated protein kinase
miRNAs	MicroRNAs
NAC	NAM, ATAF1 2, CUC2
NF-YA5	Nuclear factor YA5
NO	Nitric oxide
O_2^-	Superoxide
OsSPL14	*Oryza sativa* squamosa promoter binding protein-like 14
PI	Phosphatidylinositol
PI-4,5-P_2 or PIP_2	PI-4,5-biphosphate
PI-4-P or PIP	PI-4-monophosphate
Pi-PLC	Phosphoinositide phospholipase C
Pre-miRNA	Precursor miRNAs
Pri-miRNA	Primary miRNAs
RISC	RNA-induced silencing complex
RLKs	Receptor-like kinases
RNA Pol II	RNA polymerase II
RNAi	RNA interference
ROS	Reactive oxygen species
SA	Salicylic acid
SE	Serrate
SIMK	Salt stress-inducible MAPK
siRNAs	Short interfering RNAs

SmD3-bodies	Small nuclear RNA-binding protein D3 bodies
SnRK2	Sucrose non-fermenting 1-related protein kinase2
SOD	Superoxide dismutase
SOS	Salt overly sensitive
TGH	Tough
UBC	Ubiquitin-conjugating enzyme

7.1 Introduction

Agriculture is an indispensable aspect of our lives and most of the developing countries are dependent on agriculture for their progress. Crop productivity needs to be considerably enhanced to meet the growing demands of an increasing population. However decrease in the availability of arable land and global climatic changes are posing an inviolable stress on the existing agriculture. To deal with such adverse climatic stresses, plants have to adjust their physiology accordingly. This can be achieved by induction and/or repression of a large array of genes that governs the adaptability/susceptibility response of each plant.

Increasing irrigation networks have accentuated the problem of soil salinization and about 20% of irrigated agricultural land is considered to be saline (Flowers and Yeo 1995). Imbalance in water cycle accompanied by poor drainage facilities of the irrigated land results in salt accumulation when water dries up (Serrano et al. 1999). It is predicted that increased salinization of arable land is expected to have devastating global effects, resulting in 30% land loss within next 25 years (Wang et al. 2003). India has 8.6 million ha salt-affected area including 3.4 million ha sodic soils (Blumwald and Grover 2006). Salt stress is one of the most serious abiotic stresses in crop plants that limits plant's ability to extract water and also leads to high cellular concentrations of salts within the cells (Hasanuzzaman et al. 2013). This results in reduced osmotic potential, followed by ion toxicity, which in turn disrupts the normal functioning of enzymatic and metabolic activities including cell wall damage, accumulation of electron-dense proteinaceous particles, plasmolysis, cytoplasmic lysis, damage to endoplasmic reticulum and reduction in photosynthesis (Garcia et al. 1997; Khan et al. 1997; Pareek et al. 1997; Sivakumar et al. 1998; Munns and Tester 2008). These in turn inhibit crop growth and development.

Several blueprints have been applied to cope up with salinity problem in crops. The most prominent strategies involve irrigation and soil management, but these are restricted to sustainable farm operations. However there is ample scope for crop management through conventional crop breeding and molecular biotechnology techniques. It is therefore a serious priority to identify the molecular regulators of salt stress response and understand the mechanism behind modifications in gene expression to cope up with the adverse effects of high soil salt. Recent advances in plant molecular biology have led to the adoption of strategic and systematic research toward the identification, functional analysis and analytical studies of genes and downstream molecules that control salt tolerance processes (Hasegawa et al. 2000;

Yu et al. 2012). With the emergence of high-throughput genomics, plant biologists were able to identify a number of genes that are differentially expressed under salinity stress. It was envisaged that the complex gene regulatory networks are coordinated at the transcriptional, posttranscriptional and translation levels in a spatiotemporal fashion (Seki et al. 2002; Barrera-Figueroa et al. 2012). With the unearthing of the small RNA molecules as critical regulatory components, a new depth was added to the regulatory modules. The major stakeholders in this mode of regulation are the microRNAs (miRNAs) and short interfering RNAs (siRNAs). These molecules bind to their cognate sites on the transcripts to repress their function; hence, their mode of action is called RNA interference (RNAi) or gene silencing. This discovery entailed the requirement to revisit the mechanisms underlying the response to salt stress. In this review we discuss about the different cellular networks that are activated in response to salt stress and the recent insights in the role of miRNAs as crucial regulators of this response.

7.2 Genetic Networks in Salt Stress

High salinity results in both hyper-ionic and hyper-osmotic stress that causes potential injury to the plant species. The injury can be either a reversible inhibition of metabolism and growth or an irreversible injury involving death of the cells. Response to salt stress is a complex mechanism involving numerous adaptations such as osmoregulation and osmotic adjustment, hormonal regulation, activation of the antioxidant defense system and ion homeostasis. Several excellent reviews are available on the different aspects of salt stress signaling and response mechanisms (Flowers et al. 2010; Deinlein et al. 2014; Farooq et al. 2015; Parihar et al. 2015). We briefly describe the various genetic components involved in this process.

7.2.1 Signaling Molecules

Salt stress, like any other environmental factor, involves the activation of stress signaling pathways as a first step toward activation of plant's response machinery. The plasma membrane forms the physical barrier between the living cell and its environment. It also helps to perceive and transmit external information to the genetic module. NtC7 is a novel transmembrane protein that functions in response to osmotic stress. It resembles the receptor domain of receptor-like kinases (RLKs). Experimental evidence corroborates that overexpression of NtC7 in transgenic tobacco activates the production of cellular metabolites which confer tolerance to osmotic stress (Tamura et al. 2003). Arabidopsis histidine kinase1 (AtHK1) is also involved in the process. It senses the salt signal and activates the mitogen activated protein kinase (MAPK) pathway that leads to increased osmolyte synthesis and accumulation (Maeda et al. 1994).

MAPKs are known to control the stress responses in plants by activating the antioxidative genes and the transcription factors to control expression of a large number of genes. SIMK (salt stress-inducible MAPK), a 42-kDa protein, has been shown to be activated in response to hyperosmotic stress, but its role in the salt stress signaling is not clear (Jonak et al. 1993). The MAPK pathway is also activated in response to stress-induced reactive oxygen species (ROS) generation and accumulation. This is best exemplified by the ROS-responsive MAPKKK, MEKK1, MPK4 and MPK6 (Xing et al. 2008; Jammes et al. 2009).

The ROS such as superoxide (O_2^-), hydrogen peroxide (H_2O_2) and nitric oxide (NO) cause extensive damage to the cellular membranes and extensively impact the ion homeostasis (Baier et al. 2005). Moreover, increased ROS levels can cause salicylic acid accumulation contributing to cell death and stomatal closure (Khokon et al. 2011). The regulation of the ROS levels is one of the key mechanisms for increasing the adaptation to adverse environmental conditions in plants (Gill and Tuteja 2010). The ROS are scavenged by antioxidant metabolites such as ascorbate, glutathione, and tocopherols and by ROS detoxifying enzymes such as superoxide dismutase (SOD), ascorbate peroxidase, and catalase. The expression of *Triticum turgidum* MnSOD in *Arabidopsis thaliana* enhanced tolerance to multiple abiotic stresses by promoting proline accumulation and lowering H_2O_2 content (Kaouthar et al. 2016). Recent studies have demonstrated the role for a SUMO conjugating enzyme in ROS- and ABA (abscisic acid)-dependent signaling during salt and drought tolerance (Karan and Subudhi 2012).

The major constituent of cell membrane are lipids which can be phosphorylated at several different positions generating multiple phosphorylated species like phosphatidylinositol (PI), PI-4-monophosphate (PI-4-P or PIP) and PI-4, 5-biphosphate (PI-4,5-P_2 or PIP_2) (Gaude et al. 2008; Munnik and Testerink 2009; Peters et al. 2010; Pokotylo et al. 2014). These act as precursor for generation of intracellular secondary messenger molecules like inositol 1,4,5-trisphosphate (IP_3) and diacylglycerol (DAG). The perception of drought and salt stress activates specific phosphoinositide phospholipase C (Pi-PLC) isoforms (Drøbak and Watkins 2000), which cleave PIP_2 to produce IP_3 and DAG (Hirayama et al. 1995; Kopka et al. 1998; Mikami et al. 1998; DeWald et al. 2001; Hunt et al. 2003; Tuteja and Sopory 2008; Peters et al. 2010). They act as second messengers to activate protein kinase C and trigger calcium ion (Ca^{2+}) release, respectively, thereby activating a downstream cascade that results in manipulation of gene expression (Sanders et al. 2002).

The Ca^{2+} plays a central role in regulating and specifying the cellular responses to various environmental stresses (Sanders et al. 2002; White and Broadley 2003; Dodd et al. 2010). It represents a convergence point of many disparate signaling pathways (Xiong et al. 2006). Each stress stimulus is linked with specific Ca^{2+} transients or Ca^{2+} signatures that differ in amplitude, frequency, oscillation duration and spatiotemporal patterns (McAinsh and Hetherington 1998). The concentrations of Ca^{2+} are delicately balanced by loading it in cellular "Ca^{2+} stores" like vacuoles, endoplasmic reticulum, mitochondria and cell wall from where it can be easily released whenever required by the cell (Mahajan et al. 2006). High salinity results in increased influx of cytosolic Ca^{2+} in the cytoplasm from the apoplast as well as the intracellular compartments (Knight et al. 1997). This transient increase in cytosolic Ca^{2+} initiates the stress signal transduction leading to activation of salt adaptation responses.

The specificity is further conferred by the presence of Ca^{2+}-binding proteins that may differ in cell types and their subcellular localization (Sanders et al. 1999; Knight and Knight 2001; Kudla et al. 2010). There are three main families of Ca^{2+} sensor proteins, viz., calmodulin, the Ca^{2+}-sensor protein calcineurin B-like (CBL) and the Ca^{2+}-dependent protein kinases (CDPKs). CBL is an important protein that participates in salt stress signal transduction pathway via the CBL-interacting protein kinases (CIPKs) to control the efflux of Na^+. Expression of *cbl* gene from maize and soybean in *Arabidopsis* enhanced salt tolerance (Li et al. 2012). The CDPKs are serine/threonine protein kinases composed of an N-terminal variable domain and a C-terminal CDPK activation domain (CAD). The CAD consists of a pseudo-substrate segment with up to four Ca^{2+}-binding EF-hand motifs (Rutschmann et al. 2002; Christodoulou et al. 2004; Chandran et al. 2006; Wernimont et al. 2011). These can directly bind Ca^{2+} and get autophosphorylated (Chaudhuri et al. 1999; Rutschmann et al. 2002; Franz et al. 2011) to activate the downstream stress signal transduction. It has been reported that CDPKs are involved in salt, cold and drought stress signaling (Urao et al. 1994; Tähtiharju et al. 1997). Overexpression of oscdpk7 resulted in enhanced osmotic stress tolerance in rice (Saijo et al. 2000).

Among the most well worked out downstream events is the activation of the SOS (salt overly sensitive) pathway (Knight et al. 1999; Halfter et al. 2000; Zhu 2002; Guo et al. 2004; Chinnusamy et al. 2005). Salt stress-induced increase in intercellular Ca^{2+} is perceived by myristoylated calcium-binding protein, SOS3 (Liu and Zhu 1998; Halfter et al. 2000; Liu et al. 2000; Kim et al. 2007; Quan et al. 2007). It functions as a primary sensor to sense the Na^+-induced increase in Ca^{2+} levels in the cytoplasm. Upon binding with Ca^{2+}, it physically interacts with a calcium-dependent serine/threonine protein kinase, SOS2. This enzyme has a carboxyl terminal regulatory domain and an amino terminal catalytic domain (Liu and Zhu 1998; Halfter et al. 2000; Liu et al. 2000; Hrabak et al. 2003). It activates a transmembrane Na^+/H^+ antiporter, SOS1, which helps to remove Na^+ from cytoplasm (Shi et al. 2000; Qiu et al. 2002, 2003; Quintero et al. 2002; Lin et al. 2009).

7.2.2 Transporters

The ionic homeostasis inside the cell involves interplay of the Na^+, K^+, H^+ and Ca^{2+} ions through the coordinated regulation of their transporters. High soil salinity disturbs this equilibrium resulting in hyper-ionic stress, which causes osmotic imbalance, membrane disorganization, reduction in growth, inhibition of cell division and expansion. It also has deleterious effect on the functioning of some enzymes (Niu et al. 1995). High Na^+ levels also lead to reduction in photosynthesis and production of reactive oxygen species (Flowers et al. 1997; Greenway and Munns 1980; Yeo 1998).

Salt stress increases the Na^+ influx through non-selective cation channels. This results in membrane depolarization that activates the K^+ efflux through outward rectifying K^+ channels. Most cells normally maintain relatively high K^+ and low Na^+ concentrations in the cytosol. This is facilitated by the Na^+/K^+ symporters that have high affinity for Na^+ and favor intracellular influx of K^+ (Niu et al. 1995; Blumwald 2000). K^+ is an essential co-factor of important enzymes like pyruvate kinase (Mahajan and Tuteja 2005). Increase in salt around roots dissipates the membrane potential by affecting the Na^+/K^+ symporters and Na^+ competes with K^+ for entry (Niu et al. 1995; Blumwald 2000). It has been shown that externally supplied Ca^{2+} reduces the toxic effects of NaCl, by facilitating higher K^+/Na^+ selectivity (Cramer et al. 1987; Lauchli and Schubert 1989; Liu and Zhu 1998).

Many other transporters are also activated to counter the salt stress by reducing K^+ efflux. The plasma membrane Na^+/H^+ antipoter (SOS1) and the vacuolar Na^+/H^+ exchanger are upregulated in response to osmotic stress-induced ABA signaling pathway and work to oppose NaCl-induced membrane depolarization. At the same time, the K^+–Na^+ co-transporter (HKT1) is negatively regulated by SOS pathway so that it reduces Na^+ influx in cytoplasm. Sucrose non-fermenting 1-related protein kinase 2 (SnRK2) is another enzyme that regulates expression of genes involved in ion homeostasis and oxidative stress response. Overexpression of the *sapk4* in rice conferred increased tolerance to salt (Diédhiou et al. 2008).

7.2.3 Osmolytes

A major consequence of salt stress is the loss of intracellular water. To prevent this water loss from the cell and to protect the cellular proteins, plants accumulate many metabolites that are known as “compatible solutes” or “osmolytes.” These solutes do not inhibit the normal metabolic reactions but facilitate osmotic adjustments (Yancey et al. 1982; Ford 1984; McCue and Hanson 1990; Delauney and Verma 1993; Louis and Galinski 1997; Bressan et al. 1998). Frequently observed metabolites with an osmolytic function are sugars, mainly fructose and sucrose; sugar alcohols like mannitol, sorbitol, ononitol and pinnitol; and complex sugars like trehalose and fructans. In addition, charged metabolites like glycine, glycine betaine, betaine, proline and ectoine are also accumulated. Water moves from high water potential to low water potential and accumulation of these osmolytes makes the water potential low inside the cell and prevents the intracellular water loss.

Mannitol, a sugar alcohol, is the most widely distributed osmolyte and is widely used to control osmotic potential in culture media or nutrient media (Lewis and Smith 1967). It is a primary photosynthetic product, found in 70 higher plant families and many marine algae (Bieleski 1982). Its accumulation, storage, and utilization are intricately balanced in response to salt and drought stress. The study of this adaptive mechanism under high salt concentration in Celery (*Apium graveolens* L.) revealed that mannitol biosynthesis increases under salt stress (Everard et al. 1994; Pharr et al. 1995) with a concomitant downregulation of the catabolic enzyme, mannitol dehydrogenase (MTD), that oxidizes mannitol to mannose (Everard et al. 1994; Stoop and Pharr 1994; Landouar-Arsivaud et al. 2011). Mannitol also acts as an antioxidant by quenching ROS that are generated as a consequence of salt and drought stress (Smirnoff 1998). Mannitol-1P dehydrogenase is an NADP-dependent enzyme that converts fructose to mannitol and mannitol-1-phosphatase that act as substrates for the production of mannitol. The *E. coli mt1D* (mannitol-1-phosphatase) when introduced into *Arabidopsis*, tobacco and wheat plants lead to enhanced seed germination and less biomass reduction under salt stress (Tarczynski et al. 1993; Karakas et al. 1997; Abebe et al. 2003). *Arabidopsis* plants, which generally do not contain mannitol, when transformed with celery mannose-6P reductase (M6PR) gene, accumulated substantial amounts of mannitol (Zhifang and Loescher 2003). These transgenic plants were protected against salt-related damage to the chloroplasts and as result could maintain photosystem II and carboxylation efficiencies under salt stress (Sickler et al. 2007).

Trehalose is a nonreducing disaccharide and has high water retention capabilities. It acts as an osmoprotectant in resurrection plants such as *Selaginella lepidophylla*, which undergo desiccation during drought, salt, heat, or freezing stress conditions (Adams et al. 1990; Crowe et al. 1992). It protects biomolecules whose conformation is changed due to osmotic stress by water replacement. Introduction of *E. coli otsA* and *ScTPS1* (trehalose-6-phosphate synthase from yeast) in *Arabidopsis*, alfalfa, rice and tomato enhanced tolerance to salt stress in plants (Holmstrom et al. 1996; Goddijn and Smeekens 1998; Pilon-Smits et al. 1998; Garg et al. 2002; Nelson et al. 2004). The co-transformation of *otsA* (encoding trehalose phosphate synthase) and *otsB* (encoding trehalose phosphate phosphatase) genes in rice also increased tolerance to salt (Garg et al. 2002).

Glycine betaine (N, N, N trimethylglycine-betaine) is a nontoxic, electrical neutral amphoteric compound that forms a hydration shell around the protein complexes and enzymes to protect their native conformation under osmotic stress (Carillo et al. 2011). It is a major osmolyte synthesized by many plants in response to abiotic stresses (Rhodes and Hanson 1993). Accrual of glycine betaine under salinity stress improved stomatal conductance and relative leaf water content in common beans (Lopez et al. 2002). It is synthesized from choline by sequential action of two enzymes choline dehydrogenase (CDH) and betaine aldehyde dehydrogenase (BADH). The *E. coli betA* and *betB* genes coding for CDH and BADH enzymes, respectively, were introduced in tobacco resulting in increased biomass and faster recovery from photoinhibition under salt stress (Holmström et al. 2000). It has been found that the co-expression of N-methyl transferase gene in cyanobacteria caused accumulation of betaine in significant amounts and conferred salt tolerance to a freshwater cyanobacterium sufficient for it to become capable of growth in seawater (Waditee et al. 2005). *Arabidopsis* plants expressing N-methyltransferase gene also accumulated betaine to high levels and exhibited improved seed yield under stress conditions (Waditee et al. 2005).

Proline is one of the amino acids, which appears most commonly in response to stress to maintain cell turgor pressure or osmatic balance. It acts as a metal chelator, an antioxidative defense molecule and signaling agent (Kohl et al. 1988; Saradhi and Mohanty 1993; Smirnoff 1993; Kishor et al. 1995; Peng et al. 1996;

Hua et al. 1997; Zhang et al. 1997). Plant leaves synthesize proline from glutamine (Rejeb et al. 2014). The overexpression of P5CS (pyrroline-5-carboxylate synthase) gene from *Vigna aconitifolia* in tobacco leads to increased levels of proline and consequently improved growth under drought stress (Kishor et al. 1995). In some crop plants, for instance, wheat, the corresponding accumulation and mobilization of proline was found to increase tolerance toward water deficit stress (Nayyar and Walia 2003).

Polyamines (PA) like spermidine, spermine and its diamine precursor putrescine are low molecular weight natural organic compounds with aliphatic nitrogen structure which includes at least two primary amino groups and one or more internal amino groups (Edreva 1996; Groppa and Benavides 2008; Gill and Tuteja 2010). PAs are positively charged at physiological pH and interact with various other negatively charged organic molecules. This interaction contributes to different important growth and developmental processes in plants. Endogenous PAs possess free radical scavenging properties and antioxidant activity that confer tolerance to plants against different biotic and abiotic stresses (Groppa and Benavides 2008; Gill and Tuteja 2010; Fariduddin et al. 2013). Exogenous salt application induces the expression of polyamine biosynthetic genes such as *adc2* (arginine decarboxylase) and *spms* (Spm synthases) and results in increased putrescine and spermine levels (Urano et al. 2003). Therefore, genetic manipulation of crop plants with genes encoding enzymes of polyamine biosynthetic pathways may provide better stress tolerance to crop plants (Bagni and Tassoni 2001; Capell et al. 2004; Liu et al. 2007). The heterogonous overexpression of ornithine decarboxylase, arginine decarboxylase, S-adenosyl methionine decarboxylase and spermidine synthase (in rice, tobacco and tomato) has shown tolerance against stress conditions (Roy and Wu 2002; Waie and Rajam 2003; Liu et al. 2007; Cuevas 2008; Wen et al. 2008; Cheng et al. 2009).

7.2.4 Hormones

Research on the mechanism of salt resilience in plants has indicated that the severe impact of stress on seed germination and plant development may be due to decline in endogenous phytohormone levels (Zholkevich and Pustovoytova 1993; Jackson 1997; Debez et al. 2001). Thus, salt stress alters the levels of stress hormones, which induce changes in photosynthesis, osmotic adjustment and plant growth. It was demonstrated that under stress the levels of abscisic acid (ABA) and jasmonic acid (JA) increase, while those of auxin, indole acetic acid (IAA) and salicylic acid (SA) decrease (Wang et al. 2001). The exogenous application of auxins (Khan et al. 2010), gibberellins (Afzal et al. 2005), or cytokinins (Gul et al. 2000) can mitigate the impact of salinity stress to enhance germination and increase seed yield and quality (Egamberdieva 2009), whereas the exogenous use of ABA reduces the accumulation of ethylene and regulates leaf abscission (Gómez-Cadenas et al. 2002). Application of 0.5 mM SA in mung bean plants alleviated salinity by increasing N, P, K and Ca contents and enhancing antioxidant enzyme activity (Khan et al. 2010).

Auxins and Cytokinins (CKs) play an important role in regulating various aspects of plant development including cell division, apical dominance, leaf senescence, nutrient mobilization, photomorphogenic development, shoot differentiation, vascular tissue development, etc. (Mok and Mok 2001; Wang et al. 2001; Davies 2004). Sakhabutdinova et al. (2003) reported that salt causes a dynamic decrease in the level of IAA in the plant roots. The variation in IAA content under stress conditions could be related with development retardation (Ribaut and Pilet 1994; Nilsen and Orcutt 1996). CKs assume the role of antagonists of ABA and antagonists/synergists of auxins in stomatal movement (Blackman and Davies 1984), cotyledon development and seed germination (Pospíšilová 2003). Investigations of CK receptor mutants in stress-response assays demonstrated that they act as negative controllers in ABA signaling and osmotic imbalance responses (Tran et al. 2007; Merchan et al. 2007). Kinetin is involved in breaking the stress-induced dormancy during germination of tomato, barley and cotton seeds (Bozcuk 1981). In addition, it was observed that CK levels are reduced as an early response to salt stress; however, the effects of NaCl on salt-sensitive varieties are not interceded by CKs (Walker and Dumbroff 1981). Kinetin also acts as an immediate free radical scavenger (Chakrabarti and Mukherji 2003). According to Chakrabarti and Mukherji (2003), a foliar spray of IAA and kinetin enhanced activity of various antioxidant enzymes to restore metabolic alterations imposed by salinity stress.

Gibberellins (GAs) regulate seed germination, leaf development, stem elongation and flowering (Yamaguchi and Kamiya 2000; Olszewski et al. 2002; Magome et al. 2004). The GA homeostasis and its cross talk with other hormones are crucial to regulate the plant growth and development. It has been shown that destabilization of DELLA proteins is promoted by the GA and modulated by environmental signals (e.g., salt and light) and other plant hormones (e.g., auxin and ethylene). When plants are exposed to biotic (McConn et al. 1997) or abiotic stresses (Lehmann et al. 1995), GA accumulates rapidly. For example, GA3 accumulation enhances growth and development in wheat and rice under saline conditions (Parashar and Varma 1988; Prakash and Prathapasenan 1990; Kumar and Singh 1996). However the mechanism by which GA3-priming can prompt salt resistance in plants is still unclear (Iqbal and Ashraf 2010). GA application also improves the catabolism of ABA (Gonai et al. 2004).

ABA is the major hormone that arbitrates plants to survive unfavorable ecological conditions, like salinity (Keskin et al. 2010). Endogenous levels of ABA increase proportionally in response to exposure of plants to salt stress, but this increase in ABA concentration was attributed to water deficiency rather than osmotic imbalance resulting from salt stress (Zhang et al. 2006). It was observed that acute increment in endogenous concentration of ABA in rice occurs on exposure to NaCl (Kang et al. 2005). The stress responses seem to be coordinated by increase in ABA movement in the xylem sap indicating "root-to-shoot" signal transmission (Davies et al. 1994; Jia et al. 2002). ABA also resulted in stomatal closure in the leaf (Hwang and Lee 2001) and limited leaf expansion (Cabot et al. 2009). The stress hormone

ABA also induces genes required to maintain the salt and osmotic balance within the cells (Wang et al. 2001) by upregulating expression of NHX1 (Shi and Zhu 2002) and vacuolar H^+-inorganic pyrophosphatase (Fukuda and Tanaka 2006). Keskin et al. (2010) showed that the MAPK4-like, TIP1 and GLP1 genes were induced much faster by ABA application in wheat. These reports indicate that ABA assumes crucial role in plant response to salinity.

Ethylene is a gaseous hormone, which serves as a key modulator of plant response to ecological factors (Cao et al. 2007; Abeles et al. 2012). Under salt stress and some other environmental stresses, ethylene synthesis is stimulated (Morgan and Drew 1997). It was observed that *Arabidopsis thaliana* amassed less ACC in its leaves and roots under high salt as compared to the halophyte plants *Cakile maritima* (European searocket) and *Thellungiella salsuginea* (Ellouzi et al. 2014). In soybean, a study utilizing 2D gel electrophoresis revealed that some components of ethylene biosynthesis in the salt-tolerant genotype Lee 68 were more abundant than that in the salt-sensitive genotype Jackson (Tyczewska et al. 2017). The application of ethylene on rice seedlings increased salt sensitivity, while 1-MCP (an ethylene perception blocker) treatment enhanced salinity tolerance. Rice *mhz7/osein2, mhz6/oseil1* and *oseil2* exhibited reduced tolerance to salt (Yang et al. 2015). MHZ6/OsEIL1 and OsEIL2 could bind to the promoter of OsHKT2 (a Na^+ transporter gene) and increase its expression resulting in enhanced Na^+ uptake (Yang et al. 2015). It can therefore be generalized that inherent ethylene synthesis is vital for salt acclimation.

7.2.5 Transcription Factors

The abiotic stress-responsive transcription factors play a key role in coordinated regulation of gene expression for gearing the plant machinery to respond to stress conditions (Golldack et al. 2011). The control of these factors by ABA, methyl jasmonate, GA and ethylene indicates that they may serve as crucial regulatory nodes at the end of distinct signal transduction pathways (Ma et al. 2013) to regulate primary metabolism of cell, energy supply and allocation and growth and development. Various families of transcription factors, viz., AP2/ERF, bHLH, WRKY, NAC, etc., have been linked with salinity tolerance (Tripathi et al. 2014). A list of various stress-regulated transcription factors is provided in Table 7.1.

The genes involved in the ABA-dependent pathway contain the conserved ABA-responsive cis-elements (ABREs) that are recognized by the AREB/ABF (ABA-responsive element-binding protein/ABRE-binding factor) family proteins. It was shown that AREB1, AREB2 and AREB3 are bZIP-type of transcription factors, and they synergistically control the sucrose non-fermenting 1-related protein kinase 2 (SnRK2) (Yoshida et al. 2010).

The DREB/CBF subfamily of the AP2/ERF transcription factors also plays a vital role in managing the adaptations to adverse conditions through ABA-dependent

Table 7.1 List of stress-associated transcription factors and their regulatory miRNAs

Transcription factors						miRNA		
Family name	Abiotic stresses							
	Salt	Drought	Heat	Cold	Ions	Family	Role in stress	References
SBP-like	√	√	√	√	√	miR156/157	Upregulated in response to salinity, drought response, tolerance to heat stress, heat stress memory, reduced cold tolerance, Cd, Al, Mn, As stress	Wu and Poethig (2006), Huang et al. (2010), Valdés-López et al. (2010), Xin et al. (2010), Zhou et al. (2010), Ding et al. (2011), Kantar et al. (2011), Lima et al. (2011), Eldem et al. (2012), Ren et al. (2012), Srivastava et al. (2012), Yu et al. (2012), Zeng et al. (2012), Zhou et al. (2012a, b), Stief et al. (2014), Cui et al. (2015) and Sun et al. (2015)
MYB, TCP	√	√	√			miR159	Salt stress responses, heat stress tolerance, ABA hypersensitivity, osmotic stress tolerance	Abe et al. (2003), Reyes and Chua (2007), Alonso-Peral et al. (2010), Xin et al. (2010), Frazier et al. (2011), Barrera-Figueroa et al. (2012), Chen et al. (2012), Sun et al. (2012), Wang et al. (2012), Hivrale et al. (2016) and Roy (2016)

(continued)

Table 7.1 (continued)

Transcription factors						miRNA		
Family name	Abiotic stresses							
	Salt	Drought	Heat	Cold	Ions	Family	Role in stress	References
ARF	√	√	√		√	miR160	Salinity stress, heat tolerance, drought resistance, Cd, Al, Mn stresses	Liu et al. (2007), Huang et al. (2010), Valdés-López et al. (2010), Xin et al. (2010), Lima et al. (2011), Lu et al. (2011), Barrera-Figueroa et al. (2012), Chen et al. (2012), May et al. (2013), Khan et al. (2014), Kruszka et al. (2014), Kumar (2014), Li et al. (2014), Khaksefidi et al. (2015), Sun et al. (2015), Xie et al. (2015) and Hivrale et al. (2016)
NAC	√	√	√		√	miR164	Salt stress response, drought resistance, heat tolerance, mechanical stress, Cd stress	Lu et al. (2005), Amor et al. (2009), May et al. (2013), Fang et al. (2014), Li et al. (2014), Sun et al. (2015), Hivrale et al. (2016) and Qiu et al. (2016)

(continued)

Table 7.1 (continued)

Transcription factors						miRNA		References
Family name	Abiotic stresses					Family	Role in stress	
	Salt	Drought	Heat	Cold	Ions			
HD-zip	√	√	√	√	√	miR166	Salt stress response, drought resistance, heat tolerance, upregulated under cold, Cd, Al, As stress response	Sunkar and Zhu (2004), Boualem et al. (2008), Liu et al. (2008), Zhou et al. (2008a, b), Trindade et al. (2010), Xin et al. (2010), Zhou et al. (2010), Ding et al. (2011), Lima et al. (2011), Kantar et al. (2011), Yu et al. (2012), May et al. (2013), Kruszka et al. (2014) and Hivrale et al. (2016)
ARF	√	√	√			miR167	ABA response, hypoxia response, heat stress response, cold stresses response	Liu et al. (2008, 2009), Xin et al. (2010), Chen et al. (2012), Tang et al. (2012), Gupta et al. (2014), Kruszka et al. (2014), Sailaja et al. (2014), Wang et al. (2014), Khaksefidi et al. (2015) and Hivrale et al. (2016)

(continued)

Table 7.1 (continued)

Transcription factors						miRNA		
Family name	Abiotic stresses							
	Salt	Drought	Heat	Cold	Ions	Family	Role in stress	References
HAP12–CCAAT-box TF complex, NF-YA or HAP2	√	√	√	√	√	miR169	Upregulated under salt stress, drought tolerance, abscisic acid response, sensitivity to nitrogen deficiency, heat tolerance, response to cold stress, Cd, Al, nanoparticles, As	Sunkar and Zhu (2004), Zhao et al. (2007, 2009, 2011), Li et al. (2008), Liu et al. (2008), Lu et al. (2008), Huang et al. (2009), Zhang et al. (2009, 2011), Xin et al. (2010), Lima et al. (2011), Wang et al. (2011), Burklew et al. (2012), Chen et al. (2012), Contreras-Cubas et al. (2012), Srivastava et al. (2012), Zeng et al. (2012), Yin et al. (2012), Ding et al. (2013), Guan et al. (2013), Kong et al. (2014), Cheng et al. (2016)
Scarecrow-like/ GRAS		√				miR170	Drought tolerance	Zhou et al. (2010), Chauhan and Kumar (2016)

(continued)

Table 7.1 (continued)

Transcription factors						miRNA		
Family name	Abiotic stresses							
	Salt	Drought	Heat	Cold	Ions	Family	Role in stress	References
	√	√	√		√	miR171	Salinity tolerance, heat tolerance, drought tolerance, response in Cd, Hg, Al stress	Xie et al. (2007), Lu et al. (2008), Chen et al. (2012), Zhou et al. 2012a, b), Mahale et al. (2014), Wang et al. (2014), Deng et al. (2015), Hivrale et al. (2016) and Esmaeili et al. (2017)
AP2, bZIP	√	√	√	√	√	miR172	Enhanced water deficit and salt tolerance, responsive to heat, resistance to cold stress, upregulated in Hg, response to Mn	Liu et al. (2008), Valdés-López et al. (2010), Xin et al. (2010), Zhou et al. (2010, 2012a, b), May et al. (2013), Khaksefidi et al. (2015) and Li et al. (2016)
TCP family TF21	√	√	√	√	√	miR319	Tolerance to salinity and drought, tolerance to chilling temperature, heat resistance, As, Cd, Al and Hg stress responsive	Schommer et al. (2008), Tuli et al. (2010), Chen et al. (2012), Liu and Zhang (2012), Thiebaut et al. (2012), Yang et al. (2013), Zhou et al. (2013), Kumar (2014) and Hivrale et al. (2016)

(continued)

Table 7.1 (continued)

Transcription factors						miRNA		
Family name	Abiotic stresses							
	Salt	Drought	Heat	Cold	Ions	Family	Role in stress	References
TIR1, AFBs, MYB family	√	√	√	√	√	miR393	Enhanced salt tolerance, drought resistance, heat response regulation, regulated by cold stress, regulated by Cd, Hg and Al toxicities	Sunkar and Zhu (2004), Xie et al. (2007), Zhao et al. (2007), Liu et al. (2008), Zhou et al. (2008a, b, 2010) Xin et al. (2010), Gao et al. (2011a, b), Barrera-Figueroa et al. (2012), Guan et al. (2013) and Hivrale et al. (2016)
GRF, WRKY	√	√	√	√	√	miR396	More sensitive to salinity and alkalinity, response to arsenic treatment, downregulated in Cd exposure, response to high temperature, drought tolerance, response to cold exposure	Liu et al. (2008), Lu et al. (2008), Yang and Yu (2009), Gao et al. (2010), Zhou et al. (2010), Ding et al. (2011), Giacomelli et al. (2012), Liu and Zhang (2012), Tang et al. (2012), Hivrale et al. (2016) and Song et al. (2017a, b)
BCP	√	√			√	miR408	Enhanced heavy metal, high-salinity stress tolerance, aluminum resistance, enhanced drought tolerance,	Ezaki et al. (2000), Zhang et al. (2013) and Hajyzadeh et al. (2015)

(continued)

Table 7.1 (continued)

Transcription factors						miRNA		
Family name	Abiotic stresses							
	Salt	Drought	Heat	Cold	Ions	Family	Role in stress	References
MADS-box	√	√		√	√	miR444	Adapt to nitrogen-limiting conditions, enhanced P_i accumulation, downregulated under salinity, dehydration, drought and cold stress responses, downregulated by Cd stress	Ding et al. (2011), Kantar et al. (2011), Yan et al. (2014), Deng et al. (2015), Ma et al. (2015), Gao et al. (2016) and Song et al. (2017a, b)
SBP		√		√	√	miR529	Downregulated in response to drought, upregulated in cold, upregulated in response to Hg, Cd and Al	Zhou et al. (2008a, b, 2010), Khraiwesh et al. (2012) and Wang et al. (2016)
MYB			√			miR828 and miR858	Oxidative stress, response to high temperature	Lin et al. (2012) and Wang et al. (2016)

and ABA-independent pathways (Yamaguchi-Shinozaki and Shinozaki 2005, 2006). Arabidopsis plants overexpressing DREB/CBF or At-bZIP24 exhibited enhanced tolerance to drought, salinity and cold stresses (Kasuga et al. 1999; Yang et al. 2009). Moreover CBF1 directed GA biosynthesis and accumulation of the DELLA protein, RGA, thus indicating a role for AP2/ERF transcription factors and GA-regulated plant growth and development in abiotic stress signaling pathways (Achard et al. 2008). At-*DREB1A* overexpression also conferred salinity tolerance in transgenic seedlings by upregulation of other stress-inducible downstream genes (Sarkar et al. 2014; Khatib et al. 2011). *At-DREB2A*, *AtB7* and *AtBF3* genes also modulate salt tolerance by accumulating salt-responsive solutes such as proline and glycine betaine (Pruthvi et al. 2014). Furthermore, an AP2-type transcription factor, CAP2, alleviated salt-stress tolerance in transgenic chickpea plants (Frugier et al. 2000).

The NAC (NAM, ATAF1 2, CUC2) transcription factor family also plays an important role in regulating the developmental networks operating in the plants in response to abiotic and biotic stresses (Ma et al. 2013; Hernandez and Sanan-Mishra 2017). The drought and ABA responses are regulated by the NAC10- and WRKY-type transcription factors (Jeong et al. 2010). The ABA-responsive NAC transcription factor, VND-INTERACTING1 (VNI2), is a repressor of xylem vessel development and has a role in leaf maturing and aging (Yang et al. 2011). Similarly H_2O_2-responsive JUNGBRUNNEN1 (JUB1) regulates leaf senescence in response to hyperosmotic and salt stress (Wu et al. 2012). A putative NAC-type chickpea transcription factor (CarNAC4) induced salt tolerance by reducing MDA content (Yu et al. 2016).

Intriguingly, drought, salt and ABA induce the transcriptional expression of the MYB and WRKY transcription factors (Cominelli et al. 2008; Lippold et al. 2009). Altered drought sensitivity of At-MYB41 overexpressing line of *Arabidopsis* was related to lipid metabolism, cell wall expansion and cuticle synthesis indicating its role in drought tolerance and survival (Cominelli et al. 2008). At-MYB41 was likewise related to primary carbon metabolism demonstrating an accord between cuticle deposition, plant resistance against desiccation and in addition cellular carbon and lipid metabolism (Cominelli et al. 2008; Lippold et al. 2009). The salt-responsive R2R3 type of MYB, Os-MPS (MULTIPASS), targets genes associated with biosynthesis of plant hormones and cell wall development (Schmidt et al. 2013). Similarly At-WRKY63 loss of function resulted in drought hypersensitivity and reduction in guard cell closure (Ren et al. 2010). ABA- and salt-responsive At-WRKY33 and Th-WRKY4 induce genes involved in the detoxification of ROS such as glutathione S-transferase (GST) U11, peroxidases and lipoxygenase LOX1 (Jiang and Deyholos 2009; Zheng et al. 2013).

7.3 miRNA Origin and Function

The silencing of transcription factors and other cellular transcripts by the small RNAs added an interesting angle to the cellular regulatory networks. Within this class, the miRNAs have emerged as important regulatory molecules. They comprise a class of endogenously expressed 20–24 nucleotide long, noncoding RNAs, which play crucial roles as regulators of gene expression in plants, animals and some viruses (Bartel 2004; Pfeffer et al. 2004; Filipowicz et al. 2008; He et al. 2008; Siomi and Siomi 2010). They can silence gene expression at the posttranscriptional level through degradation of mRNA transcripts or translational arrest or at the transcriptional level by chromatin methylation (Vaucheret 2006; Sanan-Mishra et al. 2009; Sharma et al. 2017; Djami-Tchatchou et al. 2017). The miRNAs are involved in regulating almost all biological and metabolic processes such as growth of cells, meristem maintenance, cell differentiation, organ development, signaling pathways, disease resistance and response to environmental stress (Jones-Rhoades et al. 2006; Bushati and Cohen 2007; Willmann and Poethig 2007; Zhang et al. 2007; Leung and Sharp 2010; Sharma et al. 2017).

The first miRNA, *lin-4,* was discovered in the soil nematode *Caenorhabditis elegans* as a regulator of a heterochronic gene (Lee et al. 1993; Pasquinelli et al. 2000; Reinhart et al. 2002). In plants, the first report of miRNA was concurrent with the finding of the second miRNA, *let-7,* in *C. elegans* (Pasquinelli et al. 2000;

Lagos-Quintana et al. 2001; Lau et al. 2001). In the consecutive years, the pathway of miRNA biogenesis and their mode of action were elucidated in *Arabidopsis thaliana* (Llave et al. 2002; Park et al. 2002; Reinhart et al. 2002). Since then there has been an exponential increase in the identification of miRNAs as well as in understanding the importance of their involvement in biological processes (Aukerman and Sakai 2003; Carrington and Ambros 2003; Brodersen et al. 2008; Meyers et al. 2008). Research focus has now shifted to get newer wisdom in the role of miRNAs in plant gene regulatory functions (Chuck et al. 2009; Zhang et al. 2013).

The miRNAs are produced endogenously from *miRNA* genes, in a highly coordinated multistep process that is restricted within the nucleus (Bartel 2004; Zhang et al. 2007). Most of the plant *miRNA* genes are primarily situated at the intergenic regions. These are transcribed by the DNA-dependent RNA polymerase II (RNA Pol II) into large transcripts called primary (pri) miRNAs (Faller and Guo 2008). The processing of pri-miRNAs to hair-pin-structured precursor (pre) miRNAs and further to functionally mature miRNAs is performed by a protein complex containing RNase III endonuclease, Dicer-like (DCL) (Bartel 2004; Jung et al. 2009; Voinnet 2009). This sequential reaction is carried out in the dicing bodies (D bodies) or small nuclear RNA-binding protein D3 bodies (SmD3-bodies) (Han et al. 2004; Kurihara and Watanabe 2004; Fang and Spector 2007; Fujioka et al. 2007). DCL1 interacts with different proteins like SE, HYL1, TGH and CPL1 that were detected in subnuclear foci by bimolecular fluorescence complementation (Fang and Spector 2007; Song et al. 2007; Manavella et al. 2012; Ren et al. 2012). The function of the proteins forming the core complex is described below:

DICER-LIKE 1 (DCL1) is part of a large gene family of RNase III-like endoribonucleases which are involved in processing of both the pri- and pre-miRNA in *Arabidopsis* (Margis et al. 2006). DCL protein contains PAZ (PIWI/AGO1/ZWILLE) domain, a DExHbox RNA helicase, two RNAse III domains and at least one dsRNA-binding domain.

SERRATE (SE) is a C_2H_2-zinc finger protein which controls leaf development, meristem activity and inflorescence architecture. It physically interacts with HYL1 and DCL1, to form the protein complex essential for pri-miRNA processing in plants (Yang et al. 2006). It was shown that SE is involved in increasing the accuracy of pri-miRNA processing by DCL1 (Kurihara et al. 2006; Dong et al. 2008; Manavella et al. 2012; Ren et al. 2012).

HYPONASTIC LEAVES1 (HYL1) is the plant homolog of R2D2 (Han et al. 2004). It is a nuclear dsRNA-binding protein that interacts with DCL1 to mediate the first step of pri-miRNA processing (Kurihara et al. 2006). The *hyl1* knockout mutants show pleiotropic developmental defects with increased amounts of unprocessed pri-miRNA molecules indicating that HYL1 is required for miRNA accumulation (Han et al. 2004; Vazquez et al. 2004). It was shown that HYL1 displays hypersensitivity to abscisic acid and hyposensitivity to cytokinin and auxin. These studies served to establish a link between phytohormones and miRNAs in *Arabidopsis* (Lu and Fedoroff 2000; Vazquez et al. 2004).

TOUGH (TGH) is an RNA-binding protein, having a G-patch and SWAP domain. It is an evolutionarily conserved protein among eukaryotes that directly interacts with DCL1. It increases DCL1 activity in pri-miRNA processing (Ren et al. 2012).

CAP-BINDING COMPLEX (CBC) is a nuclear cap-binding, heterodimeric riboprotein complex that comprises of two subunits, CBP20 and CBP80. Null mutants of *cbp80/abh1* and *cbp20* displayed lesser accumulation of mature miRNAs and exhibited hypersensitivity to abscisic acid (Hugouvieux et al. 2001; Kim et al. 2008). This shows that CBC is probably involved in miRNA maturation pathway, but the exact mechanism of its action is not known. It is postulated to stabilize the pri-miRNA molecules by binding to their m7G cap (Ren and Yu 2012).

HUA ENHANCER1 (HEN1) is a methyl transferase that methylates the 3′ terminal sugar residue of nucleotides on each strand of the miRNA:miRNA* duplex to prevent their uridylation and subsequent degradation (Yu et al. 2005). It has two dsRNA-binding domains and a nuclear localization signal (Boutet et al. 2003). It was observed that miRNAs fail to accumulate or only accumulate at a lower level in *hen1* mutants, thereby suggesting its role in stabilization and protection of miRNAs from degradation (Li et al. 2005).

HASTY (HST) is a plant homolog of exportin-5 that transports miRNA:miRNA* duplex from the nucleus into the cytoplasm through the nuclear pore in an ATP-dependent manner (Park et al. 2005). The *hst* mutants show reduction in accumulation of partial miRNAs in cytoplasm, which suggests that miRNAs may be also transported by other mechanisms (Voinnet 2009).

The mature miRNAs are loaded in a large ribonucleoprotein complex called RISC (RNA Induced Silencing complex) to target mRNAs for silencing (Chen et al. 2005). The slicer component of the RISC is the evolutionary conserved protein, ARGONAUTE (AGO) (Wu et al. 2009). It is required for amassing of miRNAs and for slicing the target mRNAs. AGO family members are defined by the presence of a PAZ domain and a PIWI domain. The PAZ domain may interface with DCL and its interacting proteins to align miRNA on its targets, while the PIWI domain contains the RNase H endonuclease activity required for target cleavage (Cerutti et al. 2000; Song et al. 2004). The miRNA function is affected by mutations in AGO family proteins. This causes phenotypic abnormalities like disorder of axillary shoot meristem and leaf development (Bohmert et al. 1998; Vaucheret et al. 2004; Kidner and Martienssen 2005).

Plant mRNAs bind to their targets with near-perfect complementarity (Bartel 2004; Pillai et al. 2004; Aleman et al. 2007). The miRNA-mediated regulation predominantly operates through transcript cleavage though translational repression has also been observed (Aukerman and Sakai 2003; Chen 2004; Bari et al. 2006; Gandikota et al. 2007). It was shown that SQUINT (SQN), the orthologue of immunophilin cyclophilin 40 (Cyp40) in *Arabidopsis*, is required for miRNA-mediated repression by promoting AGO1 activity (Smith et al. 2009). Some proteins like AGO1, AGO10, the microtubule-severing enzyme KATANIN, the decapping component VARICOSE (VCS)/Ge-1, 3-hydroxy-3-methylglutaryl CoA reductase (HMG1), sterol C-8 isomerase HYDRA1 (Brodersen et al. 2012) and SUO (Yang et al. 2012) are shown to be required for miRNA-mediated translational repression. In addition, miRNA and AGO1 are associated with polysomes (Lanet et al. 2009). These observations suggested that translational repression is distinct from slicing and is more widespread in plants than previously thought. However, the mechanism underlying miRNA-mediated translational repression still remains largely unknown in plants.

It was hypothesized that regulation of target mRNA by cleavage was important in regulating developmental processes, which require permanent determination of cell fates. In contrast to on-off switching of cleavage, the mode of translational repression enables fine-tuning of targets and might be important in reversible modulation of the negative regulators of stress responses. By repressing translation of negative regulators, it is guaranteed that expression of the regulators will reappear when the stress disappears and ensures reducing the fitness loss due to prolonged stress response activation (Voinnet 2009). This idea was supported by miRNAs controlling phosphate starvation (Sunkar and Zhu 2004) and basal defense against bacterial infection (Navarro et al. 2006).

7.4 Salt Stress-Associated miRNAs

To endure the presence of salinity in their sessile way of life, plants have evolved a significant level of formative versatility, including adaptations, i.e., exclusion of ions through molecular networks. A number of genes and their products in plants are influenced because of salinity (Zhu 2002) and the miRNAs have been shown to play an important role in this process (Sunkar and Zhu 2004; Fujii et al. 2005; Phillips et al. 2007; Ruiz-Ferrer and Voinnet 2009; Pérez-Quintero et al. 2010; Zhou et al. 2010; Meng et al. 2011; Yu et al. 2012; Zheng et al. 2012; Mittal et al. 2016; Sharma et al. 2017). Since most of the miRNAs are conserved across plant species, it is likely that they may regulate similar targets in all the plants. The targets for the conserved miRNAs include several transcription factors (TFs) like MYB, NAC1 and homeodomain-leucine zipper protein (HD-ZIP) involved in plant development and organ formation (Jones-Rhoades and Bartel 2004). Many of these proteins have been reported as stress-responsive factors in plants (Fang and Grzymala-Busse 2006; Xu et al. 2008). In addition, the miRNAs target transcripts of proteins involved in diverse metabolic pathways or physiological processes of plants such as NADP-dependent malic enzyme and cytochrome oxidase, which are known to be involved in the salt stress responses (Cheng and Long 2007; Yan et al. 2005).

The first direct evidence for the involvement of miRNAs in plant stress responses came from the work of Jones-Rhoades and Bartel (2004) in *Arabidopsis*, where they identified several novel miRNAs, including miR395 to be upregulated upon sulfate starvation. miR395 targets a low-affinity sulfate transporter, AST68 and three ATP sulfurylases (APS1, APS3, and APS4) involved in sulfate assimilation. This was followed by the identification of phosphate deficiency-induced expression of miR399 with a corresponding decrease in the target ubiquitin-conjugating enzyme (UBC) in *Arabidopsis* (Fujii et al. 2005; Aung et al. 2006; Bari et al. 2006). The identification of miR398, which targets two Cu/Zn superoxide dismutases (CSDs), linked miRNAs to the ROS pathway (Sunkar et al. 2006). Till date 217 miRNAs have been reported in salinity stress in different plant species including *Arabidopsis*, *Glycine max*, *Glycine soja*, *Gossypium hirsutum*, *Medicago truncatula*, *Nicotiana tabacum*, *Oryza sativa*, *Panicum virgatum*, *Phaseolus vulgaris*, *Populus euphratica*, *Saccharum officinarum*, *Triticum aestivum* and *Zea mays*.

Various miRNAs are differentially regulated under salt stress conditions in different tissues. The miR156, miR158, miR159, miR165, miR167, miR168, miR169, miR171, miR319, miR393, miR394, miR396 and miR397 were upregulated, while the miR398 was downregulated under salt stress (Liu et al. 2008). In common bean miRS1 and miR159.2 were upregulated under salinity stress (Arenas-Huertero et al. 2009). Similarly salt stress induced miR530a, miR1445, miR1446a-e, miR1447 and miR1711-n expression and downregulated miR482.2 and miR1450 in *P. trichocarpa* (Lu et al. 2008). In our lab about 23 new miRNA sequences were cloned from salt stressed basmati rice variety (Sanan-Mishra et al. 2009). In artichoke, expression level of miR159 and miR319 increased after salt treatment (De Paola et al. 2012). High-throughput sequencing and bioinformatics analysis identified ten miRNAs in rice inflorescences that were involved in the response to salt stress (Barrera-Figueroa et al. 2012).

The comparative analysis of salt-tolerant and salt-sensitive maize roots showed that miR156, miR164, miR167 and miR396 families were downregulated, although miR162, miR168, miR395, and miR474 families were upregulated in salinity. The analysis suggested gene networks that manage and cope up with abiotic stresses (Ding and Zhu 2009). The comparative and integrated analysis of miRNAs and their targets were used to obtain a global picture of the regulatory networks operative in salt-susceptible and salt-tolerant rice varieties (Goswami et al. 2017). It was seen that salt stress induced the expression of a large number of miRNAs in the salt-tolerant rice while repressed the expression of several miRNAs in salt-sensitive rice. The upregulation of the miRNAs seemed to play an important role in enhancing the tolerance of the plants to salt stress. By comparing the data within the genetically similar backgrounds using the Gly-transgenics in which salt tolerance was artificially engineered, the changes in the regulatory networks are apparent (Tripathi et al. 2017). It also indicates that manipulating one pathway alone may not be sufficient to completely alter the physiology of the plants (Goswami et al. 2017). The detailed expression profile of Osa-miR820 was compared across various tissues of two indica rice cultivars exhibiting a contrasting response to salt stress (Sharma et al. 2015a). Recently, it was shown to be downregulated under arsenic stress in two contrasting arsenic-responsive rice cultivars (Sharma et al. 2015b).

Moreover contrasting behavior of conserved miRNAs was observed in different plant species. For example, the salt-induced deregulation of miR167 is known in *Arabidopsis* but not in rice (Sunkar and Zhu 2004; Lv et al. 2010). Using NGS technology and bioinformatics tools, around 130 conserved miRNAs belonging to 95 miRNA families were found to be differentially expressed under various stress (Li et al. 2011).

Analysis of individual family members showed that Tas-miR169g and Tas-miR169n demonstrated improved expression under salt response. This downregulated the expression of its target transcript nuclear factor Y subunit A (NF-YA), a transcription factor that was previously shown to be downregulated in drought-affected wheat leaves (Stephenson et al. 2007). The authors reported a cis-acting ABA-responsive element (ABRE) in the upstream region of Tas-miR169n, which suggested that its expression may be ABA regulated (Zhou et al. 2010). Ath-miR398 was downregulated under salt stress leading to upregulation of CSD1 and CSD2, which probably helps in maintaining cellular oxygen metabolism (Jagadeeswaran et al. 2009). Likewise the putative targets for salt stress deregulated artichoke miRNAs, Cca-miR397 and Cca-miR399, are homologous to members of laccase gene family (De Paola et al. 2012). It has been reported that the expression level of laccase genes is enhanced by high concentrations of NaCl in tomato, maize and *Arabidopsis* roots (Cai et al. 2006; Liang et al. 2006; Wei et al. 2000).

Several other studies unraveled the overlapping regulation of miRNAs by stress and hormonal signaling as exemplified by the induction of miR159 by the stress hormone ABA under drought condition (Reyes and Chua 2007). Recently, it was shown that expression of Osa-miR393 was downregulated under salt stress and its overexpression in rice negatively regulated the salt-alkali stress tolerance of the plant (Gao et al. 2011a, b). Moreover levels of this miRNA are altered in response to many heavy metals like aluminum (Lima et al. 2011), arsenite (Liu et al. 2012; Yu et al. 2012), cadmium (Huang et al. 2009; Ding et al. 2011), etc.

In rice, expression profiling analysis employing stem-loop reverse transcription quantitative PCR was used to identify the expression patterns of 41 miRNAs in response to drought, salt, cold, or ABA treatments (Shen et al. 2010). In another study comparison of the expression profiles between drought-tolerant N22 and drought-sensitive PB1 identified miRNAs with variety-specific expression patterns during phase transition (miR164, miR396, miR812 and miR1881) as well as drought stress (Kansal et al. 2015). Many of these miRNAs are also deregulated by salt stress, indicating an evolution of a regulatory mechanism in regulating rice inflorescence development under stress.

Crucial role for miRNAs in maintaining the target gene expression under stress is also indicated by overexpression studies. The drought and salinity stress-induced miR393 targets an auxin transporter gene (OsaUX1) and a rice tiller inhibitor gene (OsTIR1). Transgenic rice plants overexpressing miR393 showed an increase in tillers and early flowering, together with decreased tolerance to salt and hypersensitiveness to auxin (Xia et al. 2012). Under salt stressed condition, expression of Osa-miR396c also decreased in ABA-dependent manner and overexpression of the miRNA resulted in reduced salt stress tolerance (Gao et al. 2011a, b).

7.5 Application of miRNAs in Crop Improvement

miRNAs are indispensable for the sustenance of proper growth and development. They act in a precise manner to rapidly coordinate the gene expression by elevation or reduction in the expression of specific genes. They also regulate key signaling components by directly or indirectly influencing the cellular machinery to respond to the incoming stress. miRNA-based gene manipulation strategies are being extensively applied in functional genomics to achieve various qualitative or quantitative effects. If the candidate miRNA is a positive regulator for a negative or undesirable trait, then the strategy includes miRNA overexpression for knocking out the targeted pathway. This may be achieved by overexpressing the natural precursor sequence or by engineering an existing pre-miRNA sequence to generate artificial miRNAs (amiRs). On the contrary, if miRNA controls a favorable trait, then the approach involves generating cleavage-resistant miRNA target transcripts or employing artificial target mimics to bypass the miRNA regulation.

Studies on miRNAs have identified specific molecules that can be easily manipulated to regulate the plant phenotypes. It is well known that sulfur is indispensable for the growth and development of plants. Its limitation or starvation induces the expression of miR395 to bring down the levels of ATP sulfurylases and sulfate transporter. Temporal and spatial expression of miR395 can be modulated to regulate the sulfate metabolism pathway so that plants can survive in low sulfur-containing soils. Another excellent example is that of miR398, which is downregulated under stress such as drought, salt and bacterial infection, resulting in upregulation of its target genes CSD1 and CSD2 (Jagadeeswaran et al. 2009). Superoxide dismutases sequester reactive oxygen species and protect plant from damage.

There are some excellent studies on the manipulation of miRNAs to meet the agriculture demands. The expression of miR399f is induced by salt and ABA treatment (Jia et al. 2009). Transgenic *Arabidopsis* overexpressing *miR399f* displayed tolerance to salt stress and ABA treatment, but were hypersensitive to drought (Baek et al. 2016). Similarly drought stress downregulates miR169 to upregulate its target nuclear factor YA5 (NF-YA5). Transgenic plants overexpressing NF-YA5 showed enhanced drought tolerance, while plants overexpressing miR169a were more sensitive to drought stress (Li et al. 2008). Similar observation was found in the case of soybean, where overproduction of GmNFYA3 gene, a newly identified target gene for miR169a, leads to improved drought tolerance (Ni et al. 2013), though the salinity tolerance was reduced (Ni et al. 2013).

The role of miRNA-mediated regulation was also observed in the case of biotic stress. miR393 negatively regulates a F-box auxin receptor, TIR1, which was reported to promote resistance against *Pseudomonas syringae*. Overexpression of miR393 in *Arabidopsis* showed increased resistance but with some developmental abnormalities (Navarro et al. 2006). Fungal resistance against *Magnaporthe oryzae* was demonstrated due to overexpression of Osa-miR7696 (Campo et al. 2013).

The functional knowledge of the miRNAs under manipulation is a prerequisite to these strategies as it may affect plant morphology. This was observed in the case of cold stress-downregulated miR319. Its targets OsPCF5, OsPCF6, OsPCF7, OsPCF8 and OsTCP21 were upregulated under stress (Sunkar and Zhu 2004; Liu et al. 2008; Lv et al. 2010). The overexpression of Osa-miR319 led to increased cold stress tolerance, but the plants had developmental defects. To avoid this effect, two RNAi lines for OsPCF5 and OsTCP21 were generated which showed better cold tolerance than wild-type controls. Another study identified miR396 and its targets GRF (growth-regulating factor) to be involved in the cell division and differentiation during leaf development in *Arabidopsis* (Wang et al. 2001; Jones-Rhoades and Bartel 2004). Due to repression of the expression of GRF genes, transgenic plant shows narrow-leaf phenotypes because of reduction in cell number and it became the cause for lower stomatal density. However this made the transgenic plants more tolerant to drought (Liu et al. 2009). Overexpression of miR396 in tobacco also confirmed that miR396 plays important roles not only in leaf development but also in drought tolerance of plants (Yang and Yu 2009).

The basic molecular mechanism of miRNAs can be employed to enhance the agriculture traits. Improving grain yield is a major concern for agriculture scientists and this area needs to exploit the miRNA technology. Several miRNAs have been reported to be involved in regulating grain yield in crop plants. miR397 is one such molecule which increases grain yield by downregulating the OsLAC gene. It was recently shown that plants overexpressing Os-miR393 showed high tillering and early flowering (Jiao et al. 2010), but these plants were more susceptible to drought as well as salt stress (Xia et al. 2012). Later it was demonstrated that overexpression of Os-miR397 enhanced brassinosteroid signaling which in turn increased grain yield (Zhang et al. 2013). OsSPL14 (squamosa promoter binding protein-like 14), a master regulator of plant architecture, is targeted by Os-miR156. Thus, miR156 overexpression increases the leaf and tiller initiation rates (Xie et al. 2012). In another case miR444a was shown to interact with OsMADS57, OsTB1(TEOSINTEBRANCHED1) and D14 (Dwarf14), which regulated tillering in rice thereby directly affecting the grain yields (Guo et al. 2013).

The miR156 is an excellent candidate for increasing plant biomass and altering their lignin content (Schwab et al. 2005; Fu et al. 2012; Rubinelli et al. 2013). Overexpression of the maize Corngrass1 (Cg1) miRNA that belongs to the miR156 family caused prolonged vegetative phase and delayed flowering (Chuck et al. 2011) resulting in increased biomass. The transgenic plants also showed up to 250% more starch and improved digestibility indicating their use in biofuel production. It was shown that miR156 regulates SPL transcription factor which promotes transcription of miR172 (Wu et al. 2009), to target Apetala 2 (AP2) gene, a crucial regulator of flower development (Mehrpooyan et al. 2012; Liu et al. 2013). Thus relative expression levels of miR156 and miR172 can be regulated to control juvenile-to-adult phase transitions to obtain plants of a specific developmental stage as per the requirement of the biofuel industry (Galli et al. 2014; Jeong et al. 2013; Wei et al. 2009; Zeng et al. 2012).

Likewise miR164 which targets NAC and MYB transcription factors (Johnson et al. 2014) to regulate plant development and metabolic processes and also response to drought and salt stress (Wei et al. 2009) can be manipulated for obtaining plants with enhanced tolerance to stress and high yield feedstock for biofuel purposes. Parthenocarpy or seedlessness is a highly desirable agronomic trait that can be achieved by manipulating phytohormone (specially auxins and gibberellins) by controlling activities of miRNAs or their targets.

7.6 Concluding Remarks

Salinity stress severely affects plant growth development and research during the last few decades has unraveled several aspects of the molecular mechanisms controlling the salt stress tolerance. Many traits related to salt stress response have been identified and have been effectively employed for engineering stress tolerance in model plants. The discovery of miRNAs as potent regulators of gene expression added a new dimension to the regulatory modules. This also provided novel gene manipulation tools for successfully improving crop yield under salt stress conditions.

It is thus envisaged that the basic knowledge on molecular genetic resources including mutation analysis, gene discovery and genome-wide association studies along with the understanding of the miRNA-mediated regulation of genetic networks will provide acceptable alternatives to classical plant breeding and transgenic methods for improving plant responses to salt stress. Understanding the regulation of genetic reprogramming will enable us to understand the link between changes in morphology and salt stress tolerance. It is imminent that research on miRNAs will showcase them as the game changer in producing salinity stress-tolerant crops.

Acknowledgments We apologize to colleagues whose work could not be included owing to space constraints. The study on miRNAs in our lab was supported by financial grants from the Department of Biotechnology.

References

Abe H, Urao T, Ito T, Seki M, Shinozaki K, Yamaguchi-Shinozaki K (2003) *Arabidopsis* AtMYC2 (bHLH) and AtMYB2 (MYB) function as transcriptional activators in abscisic acid signaling. Plant Cell 15(1):63–78

Abebe T, Guenzi AC, Martin B, Cushman JC (2003) Tolerance of mannitol-accumulating transgenic wheat to water stress and salinity. Plant Physiol 131(4):1748–1755

Abeles FB, Morgan PW, Saltveit ME Jr (2012) Ethylene in plant biology. Academic, San Diego

Achard P, Gong F, Cheminant S, Alioua M, Hedden P, Genschik P (2008) The cold-inducible CBF1 factor–dependent signaling pathway modulates the accumulation of the growth-repressing DELLA proteins via its effect on gibberellin metabolism. Plant Cell 20(8):2117–2129

Adams RP, Kendall E, Kartha K (1990) Comparison of free sugars in growing and desiccated plants of *Selaginella lepidophylla*. Biochem Syst Ecol 18(2–3):107–110
Afzal I, Basra S, Iqbal A (2005) The effect of seed soaking with plant growth regulators on seedling vigor of wheat under salinity stress. Stress Physiol Biochem 1:6–14
Aleman L, Sun Y, Fokar M, Allen R (2007) Role of phytohormone signaling pathways in cotton fiber development. In: World cotton research conference-4, Lubbock, Texas, USA, 10–14 September 2007. International Cotton Advisory Committee (ICAC)
Alonso-Peral MM, Li J, Li Y, Allen RS, Schnippenkoetter W, Ohms S, White RG, Millar AA (2010) The microRNA159-regulated GAMYB-like genes inhibit growth and promote programmed cell death in *Arabidopsis*. Plant Physiol 154(2):757–771
Amor BB, Wirth S, Merchan F, Laporte P, d'Aubenton-Carafa Y, Hirsch J, Maizel A, Mallory A, Lucas A, Deragon JM (2009) Novel long non-protein coding RNAs involved in *Arabidopsis* differentiation and stress responses. Genome Res 19(1):57–69
Arenas-Huertero C, Pérez B, Rabanal F, Blanco-Melo D, De la Rosa C, Estrada-Navarrete G, Sanchez F, Covarrubias AA, Reyes JL (2009) Conserved and novel miRNAs in the legume *Phaseolus vulgaris* in response to stress. Plant Mol Biol 70(4):385–401
Aukerman MJ, Sakai H (2003) Regulation of flowering time and floral organ identity by a microRNA and its APETALA2-like target genes. Plant Cell 15(11):2730–2741
Aung K, Lin S-I, Wu C-C, Huang Y-T, C-l S, Chiou T-J (2006) pho2, a phosphate over accumulator, is caused by a nonsense mutation in a microRNA399 target gene. Plant Physiol 141(3):1000–1011
Baek D, Chun HJ, Kang S, Shin G, Park SJ, Hong H, Kim C, Kim DH, Lee SY, Kim MC (2016) A role for *Arabidopsis* miR399f in salt, drought, and ABA signaling. Mol Cell 39(2):111
Bagni N, Tassoni A (2001) Biosynthesis, oxidation and conjugation of aliphatic polyamines in higher plants. Amino Acids 20(3):301–317
Baier M, Kandlbinder A, Golldack D, Dietz KJ (2005) Oxidative stress and ozone: perception, signalling and response. Plant Cell Environ 28(8):1012–1020
Bari R, Pant BD, Stitt M, Scheible W-R (2006) PHO2, microRNA399, and PHR1 define a phosphate-signaling pathway in plants. Plant Physiol 141(3):988–999
Barrera-Figueroa BE, Gao L, Wu Z, Zhou X, Zhu J, Jin H, Liu R, Zhu J-K (2012) High throughput sequencing reveals novel and abiotic stress-regulated microRNAs in the inflorescences of rice. BMC Plant Biol 12(1):132
Bartel DP (2004) MicroRNAs: genomics, biogenesis, mechanism, and function. Cell 116(2):281–297
Bieleski R (1982) Sugar alcohols. In: Plant carbohydrates. Springer, New York, pp 158–192
Blackman P, Davies W (1984) Age-related changes in stomatal response to cytokinins and abscisic acid. Ann Bot 54(1):121–126
Blumwald E (2000) Sodium transport and salt tolerance in plants. Curr Opin Cell Biol 12(4):431–434
Blumwald E, Grover A (2006) Salt tolerance. Plant biotechnology: current and future uses of genetically modified crops. Wiley, UK, pp 206–224
Bohmert K, Camus I, Bellini C, Bouchez D, Caboche M, Benning C (1998) AGO1 defines a novel locus of *Arabidopsis* controlling leaf development. EMBO J 17(1):170–180
Boualem A, Laporte P, Jovanovic M, Laffont C, Plet J, Combier JP, Niebel A, Crespi M, Frugier F (2008) MicroRNA166 controls root and nodule development in *Medicago truncatula*. Plant J 54(5):876–887
Boutet S, Vazquez F, Liu J, Béclin C, Fagard M, Gratias A, Morel J-B, Crété P, Chen X, Vaucheret H (2003) *Arabidopsis* HEN1: a genetic link between endogenous miRNA controlling development and siRNA controlling transgene silencing and virus resistance. Curr Biol 13(10):843–848
Bozcuk S (1981) Effects of kinetin and salinity on germination of tomato, barley and cotton seeds. Ann Bot 48(1):81–84
Bressan RA, Hasegawa PM, Pardo JM (1998) Plants use calcium to resolve salt stress. Trends Plant Sci 3(11):411–412

Brodersen P, Sakvarelidze-Achard L, Bruun-Rasmussen M, Dunoyer P, Yamamoto YY, Sieburth L, Voinnet O (2008) Widespread translational inhibition by plant miRNAs and siRNAs. Science 320(5880):1185–1190

Brodersen P, Sakvarelidze-Achard L, Schaller H, Khafif M, Schott G, Bendahmane A, Voinnet O (2012) Isoprenoid biosynthesis is required for miRNA function and affects membrane association of Argonaute 1 in *Arabidopsis*. Proc Natl Acad Sci 109(5):1778–1783

Burklew CE, Ashlock J, Winfrey WB, Zhang B (2012) Effects of aluminum oxide nanoparticles on the growth, development, and microRNA expression of tobacco (*Nicotiana tabacum*). PLoS One 7(5):34783

Bushati N, Cohen SM (2007) microRNA functions. Annu Rev Cell Dev Biol 23:175–205

Cabot C, Sibole JV, Barceló J, Poschenrieder C (2009) Abscisic acid decreases leaf Na+ exclusion in salt-treated *Phaseolus vulgaris* L. J Plant Growth Regul 28(2):187–192

Cai X, Davis EJ, Ballif J, Liang M, Bushman E, Haroldsen V, Torabinejad J, Wu Y (2006) Mutant identification and characterization of the laccase gene family in *Arabidopsis*. J Exp Bot 57(11):2563–2569

Campo S, Peris-Peris C, Siré C, Moreno AB, Donaire L, Zytnicki M, Notredame C, Llave C, San Segundo B (2013) Identification of a novel microRNA (miRNA) from rice that targets an alternatively spliced transcript of the Nramp6 (Natural resistance-associated macrophage protein 6) gene involved in pathogen resistance. New Phytol 199(1):212–227

Cao W-H, Liu J, He X-J, Mu R-L, Zhou H-L, Chen S-Y, Zhang J-S (2007) Modulation of ethylene responses affects plant salt-stress responses. Plant Physiol 143(2):707–719

Capell T, Bassie L, Christou P (2004) Modulation of the polyamine biosynthetic pathway in transgenic rice confers tolerance to drought stress. Proc Natl Acad Sci U S A 101(26):9909–9914

Carillo P, Annunziata MG, Pontecorvo G, Fuggi A, Woodrow P (2011) Salinity stress and salt tolerance. In: Abiotic stress in plants-mechanisms and adaptations. InTech

Carrington JC, Ambros V (2003) Role of microRNAs in plant and animal development. Science 301(5631):336–338

Cerutti L, Mian N, Bateman A (2000) Domains in gene silencing and cell differentiation proteins: the novel PAZ domain and redefinition of the Piwi domain. Trends Biochem Sci 25(10):481–482

Chakrabarti N, Mukherji S (2003) Alleviation of NaCl stress by pretreatment with phytohormones in *Vigna radiata*. Biol Plant 46(4):589–594

Chandran V, Stollar EJ, Lindorff-Larsen K, Harper JF, Chazin WJ, Dobson CM, Luisi BF, Christodoulou J (2006) Structure of the regulatory apparatus of a calcium-dependent protein kinase (CDPK): a novel mode of calmodulin-target recognition. J Mol Biol 357(2):400–410

Chaudhuri S, Seal A, Gupta MD (1999) Autophosphorylation-dependent activation of a calcium-dependent protein kinase from groundnut. Plant Physiol 120(3):859–866

Chauhan N, Kumar VH (2016) Gender responsive climate change strategies for sustainable development. Productivity 57(2):182

Chen X (2004) A microRNA as a translational repressor of APETALA2 in *Arabidopsis* flower development. Science 303(5666):2022–2025

Chen C, Ridzon DA, Broomer AJ, Zhou Z, Lee DH, Nguyen JT, Barbisin M, Xu NL, Mahuvakar VR, Andersen MR (2005) Real-time quantification of microRNAs by stem–loop RT–PCR. Nucleic Acids Res 33(20):e179–e179

Chen L, Zhang Y, Ren Y, Xu J, Zhang Z, Wang Y (2012) Genome-wide identification of cold-responsive and new microRNAs in *Populus tomentosa* by high-throughput sequencing. Biochem Biophys Res Commun 417(2):892–896

Cheng S, Long JS (2007) Testing for IIA in the multinomial logit model. Sociol Methods Res 35(4):583–600

Cheng L, Zou Y, Ding S, Zhang J, Yu X, Cao J, Lu G (2009) Polyamine accumulation in transgenic tomato enhances the tolerance to high temperature stress. J Integr Plant Biol 51(5):489–499

Cheng HY, Wang Y, Tao X, Fan YF, Dai Y, Yang H, Ma XR (2016) Genomic profiling of exogenous abscisic acid-responsive microRNAs in tomato (*Solanum lycopersicum*). BMC Genomics 17(1):423

Chinnusamy V, Jagendorf A, Zhu J-K (2005) Understanding and improving salt tolerance in plants. Crop Sci 45(2):437–448

Christodoulou J, Malmendal A, Harper JF, Chazin WJ (2004) Evidence for differing roles for each lobe of the calmodulin-like domain in a calcium-dependent protein kinase. J Biol Chem 279(28):29092–29100

Chuck G, Candela H, Hake S (2009) Big impacts by small RNAs in plant development. Curr Opin Plant Biol 12(1):81–86

Chuck GS, Tobias C, Sun L, Kraemer F, Li C, Dibble D, Arora R, Bragg JN, Vogel JP, Singh S (2011) Overexpression of the maize Corngrass1 microRNA prevents flowering, improves digestibility, and increases starch content of switchgrass. Proc Natl Acad Sci 108(42):17550–17555

Cominelli E, Sala T, Calvi D, Gusmaroli G, Tonelli C (2008) Over-expression of the *Arabidopsis* AtMYB41 gene alters cell expansion and leaf surface permeability. Plant J 53(1):53–64

Contreras-Cubas C, Palomar M, Arteaga-Vázquez M, Reyes JL, Covarrubias AA (2012) Non-coding RNAs in the plant response to abiotic stress. Planta 236(4):943–958

Cramer GR, Lynch J, Läuchli A, Epstein E (1987) Influx of Na+, K+, and Ca2+ into roots of salt-stressed cotton seedlings effects of supplemental Ca2+. Plant Physiol 83(3):510–516

Crowe JH, Hoekstra FA, Crowe LM (1992) Anhydrobiosis. Annu Rev Physiol 54(1):579–599

Cuevas J, Lopez-Cobollo R, Alcazar R et al (2008) Putrescine is involved in *Arabidopsis* freezing tolerance and cold acclimation by regulating abscisic acid levels in response to low temperature. Plant Physiol 148(2):1094–1105

Cui N, Sun X, Sun M, Jia B, Duanmu H, Lv D, Duan X, Zhu Y (2015) Overexpression of OsmiR156k leads to reduced tolerance to cold stress in rice (*Oryza Sativa*). Mol Breed 35(11):214

Davies P (2004) Plant hormones: biosynthesis, signal transduction. Action Kluwer Academic Publishers, London

Davies WJ, Tardieu F, Trejo CL (1994) How do chemical signals work in plants that grow in drying soil? Plant Physiol 104(2):309

De Paola D, Cattonaro F, Pignone D, Sonnante G (2012) The miRNAome of globe artichoke: conserved and novel micro RNAs and target analysis. BMC Genomics 13(1):41

Debez A, Chaibi W, Bouzid S (2001) Effet du NaCl et de régulateurs de croissance sur la germination *d'Atriplex halimus* L. Cah Agric 10(2):135–138

Deinlein U, Stephan AB, Horie T, Luo W, Xu G, Schroeder JI (2014) Plant salt-tolerance mechanisms. Trends Plant Sci 19(6):371–379

Delauney AJ, Verma DPS (1993) Proline biosynthesis and osmoregulation in plants. Plant J 4(2):215–223

Deng P, Wang L, Cui L, Feng K, Liu F, Du X, Tong W, Nie X, Ji W, Weining S (2015) Global identification of microRNAs and their targets in barley under salinity stress. PLoS One 10(9):e0137990

DeWald DB, Torabinejad J, Jones CA, Shope JC, Cangelosi AR, Thompson JE, Prestwich GD, Hama H (2001) Rapid accumulation of phosphatidylinositol 4, 5-bisphosphate and inositol 1, 4, 5-trisphosphate correlates with calcium mobilization in salt-stressed *Arabidopsis*. Plant Physiol 126(2):759–769

Diédhiou CJ, Golldack D, Dietz K-J, Popova OV (2008) The SNF1-type serine-threonine protein kinase SAPK4 regulates stress-responsive gene expression in rice. BMC Plant Biol 8(1):49

Ding Y-F, Zhu C (2009) The role of microRNAs in copper and cadmium homeostasis. Biochem Biophys Res Commun 386(1):6–10

Ding Y, Chen Z, Zhu C (2011) Microarray-based analysis of cadmium-responsive microRNAs in rice (*Oryza sativa*). J Exp Bot 62(10):3563–3573

Ding Y, Tao Y, Zhu C (2013) Emerging roles of microRNAs in the mediation of drought stress response in plants. J Exp Bot 64(11):3077–3086

Djami-Tchatchou AT, Sanan-Mishra N, Ntushelo K, Dubery IA (2017) Functional roles of microRNAs in agronomically important plants—potential as targets for crop improvement and protection. Front Plant Sci 8:378

Dodd AN, Kudla J, Sanders D (2010) The language of calcium signaling. Annu Rev Plant Biol 61:593–62074
Dong Z, Han M-H, Fedoroff N (2008) The RNA-binding proteins HYL1 and SE promote accurate in vitro processing of pri-miRNA by DCL1. Proc Natl Acad Sci 105(29):9970–9975
Drøbak BK, Watkins PA (2000) Inositol (1, 4, 5) trisphosphate production in plant cells: an early response to salinity and hyperosmotic stress. FEBS Lett 481(3):240–244
Edreva A (1996) Polyamines in plants. Bulg J Plant Physiol 22(1–2):73–101
Egamberdieva D (2009) Alleviation of salt stress by plant growth regulators and IAA producing bacteria in wheat. Acta Physiol Plant 31(4):861–864
Eldem V, Akçay UÇ, Ozhuner E, Bakır Y, Uranbey S, Unver T (2012) Genome-wide identification of miRNAs responsive to drought in peach (*Prunus persica*) by high-throughput deep sequencing. PLoS One 7(12):e50298
Ellouzi H, Hamed KB, Hernández I, Cela J, Müller M, Magné C, Abdelly C, Munné-Bosch S (2014) A comparative study of the early osmotic, ionic, redox and hormonal signaling response in leaves and roots of two halophytes and a glycophyte to salinity. Planta 240(6):1299–1317
Esmaeili F, Shiran B, Fallahi H, Mirakhorli N, Budak H, Martínez-Gómez P (2017) In silico search and biological validation of microRNAs related to drought response in peach and almond. Funct Integr Genom 17(2–3):189–201
Everard JD, Gucci R, Kann SC, Flore JA, Loescher WH (1994) Gas exchange and carbon partitioning in the leaves of celery (*Apium graveolens* L.) at various levels of root zone salinity. Plant Physiol 106(1):281–292
Ezaki B, Gardner RC, Ezaki Y, Matsumoto H (2000) Expression of aluminum-induced genes in transgenic *Arabidopsis* plants can ameliorate aluminum stress and/or oxidative stress. Plant Physiol 122(3):657–666
Faller M, Guo F (2008) MicroRNA biogenesis: there's more than one way to skin a cat. BBA-Gene Regul Mech 1779(11):663–667
Fang J, Grzymala-Busse JW (2006) Mining of microRNA expression data-a rough set approach. In: Rough sets and knowledge technology, pp 758–765
Fang Y, Spector DL (2007) Identification of nuclear dicing bodies containing proteins for microRNA biogenesis in living *Arabidopsis* plants. Curr Biol 17(9):818–823
Fang Y, Xie K, Xiong L (2014) Conserved miR164-targeted NAC genes negatively regulate drought resistance in rice. J Exp Bot 65(8):2119–2135
Fariduddin Q, Varshney P, Yusuf M, Ahmad A (2013) Polyamines: potent modulators of plant responses to stress. J Plant Interact 8(1):1–16
Farooq M, Hussain M, Wakeel A, Siddique KH (2015) Salt stress in maize: effects, resistance mechanisms, and management. A review. Agron Sustain Dev 35(2):461–481
Filipowicz W, Bhattacharyya SN, Sonenberg N (2008) Mechanisms of post-transcriptional regulation by microRNAs: are the answers in sight? Nat Rev Genet 9(2):102–114
Flowers T, Yeo A (1995) Breeding for salinity resistance in crop plants: where next? Funct Plant Biol 22(6):875–884
Flowers TJ, Garcia A, Koyama M, Yeo AR (1997) Breeding for salt tolerance in crop plants—the role of molecular biology. Acta Physiol Plant 19(4):427–433
Flowers TJ, Galal HK, Bromham L (2010) Evolution of halophytes: multiple origins of salt tolerance in land plants. Funct Plant Biol 37(7):604–612
Ford CW (1984) Accumulation of low molecular weight solutes in water-stressed tropical legumes. Phytochemistry 23(5):1007–1015
Franz S, Ehlert B, Liese A, Kurth J, Cazalé A-C, Romeis T (2011) Calcium-dependent protein kinase CPK21 functions in abiotic stress response in *Arabidopsis thaliana*. Mol Plant 4(1):83–96
Frazier TP, Sun G, Burklew CE, Zhang B (2011) Salt and drought stresses induce the aberrant expression of microRNA genes in tobacco. Mol Biotechnol 49(2):159–165
Frugier F, Poirier S, Satiat-Jeunemaître B, Kondorosi A, Crespi M (2000) A Krüppel-like zinc finger protein is involved in nitrogen-fixing root nodule organogenesis. Genes Dev 14(4):475–482

Fu C, Sunkar R, Zhou C, Shen H, Zhang JY, Matts J, Wolf J, Mann DG, Stewart CN, Tang Y (2012) Overexpression of miR156 in switchgrass (*Panicum virgatum* L.) results in various morphological alterations and leads to improved biomass production. Plant Biotechnol J 10(4):443–452

Fujii H, Chiou T-J, Lin S-I, Aung K, Zhu J-K (2005) A miRNA involved in phosphate-starvation response in *Arabidopsis*. Curr Biol 15(22):2038–2043

Fujioka Y, Utsumi M, Ohba Y, Watanabe Y (2007) Location of a possible miRNA processing site in SmD3/SmB nuclear bodies in *Arabidopsis*. Plant Cell Physiol 48(9):1243–1253

Fukuda A, Tanaka Y (2006) Effects of ABA, auxin, and gibberellin on the expression of genes for vacuolar H+-inorganic pyrophosphatase, H+-ATPase subunit A, and Na+/H+ antiporter in barley. Plant Physiol Biochem 44(5):351–358

Galli V, Guzman F, de Oliveira LF, Loss-Morais G, Körbes AP, Silva SD, Margis-Pinheiro MM, Margis R (2014) Identifying microRNAs and transcript targets in Jatropha seeds. PLoS One 9(2):e83727

Gandikota M, Birkenbihl RP, Höhmann S, Cardon GH, Saedler H, Huijser P (2007) The miRNA156/157 recognition element in the 3′ UTR of the *Arabidopsis* SBP box gene SPL3 prevents early flowering by translational inhibition in seedlings. Plant J 49(4):683–693

Gao P, Bai X, Yang L, Lv D, Li Y, Cai H, Ji W, Guo D, Zhu Y (2010) Over-expression of osa-MIR396c decreases salt and alkali stress tolerance. Planta 231(5):991–1001

Gao C, Wang Y, Jiang B, Liu G, Yu L, Wei Z, Yang C (2011a) A novel vacuolar membrane H+-ATPase c subunit gene (ThVHAc1) from *Tamarix hispida* confers tolerance to several abiotic stresses in *Saccharomyces cerevisiae*. Mol Biol Rep 38(2):957–963

Gao P, Bai X, Yang L, Lv D, Pan X, Li Y, Cai H, Ji W, Chen Q, Zhu Y (2011b) osa-MIR393: a salinity-and alkaline stress-related microRNA gene. Mol Biol Rep 38(1):237–242

Gao S, Guo C, Zhang Y, Zhang F, Du X, Gu J, Xiao K (2016) Wheat microRNA member tamir444a is nitrogen deprivation-responsive and involves plant adaptation to the nitrogen-starvation stress. Plant Mol Biol Report 34(5):931–946

Garcia AB, Engler JDA, Iyer S, Gerats T, Van Montagu M, Caplan AB (1997) Effects of osmoprotectants upon NaCl stress in rice. Plant Physiol 115(1):159–169

Garg AK, Kim J-K, Owens TG, Ranwala AP, Do Choi Y, Kochian LV, Wu RJ (2002) Trehalose accumulation in rice plants confers high tolerance levels to different abiotic stresses. Proc Natl Acad Sci 99(25):15898–15903

Gaude N, Nakamura Y, Scheible WR, Ohta H, Dörmann P (2008) Phospholipase C5 (NPC5) is involved in galactolipid accumulation during phosphate limitation in leaves of *Arabidopsis*. Plant J 56(1):28–39

Giacomelli JI, Weigel D, Chan RL, Manavella PA (2012) Role of recently evolved miRNA regulation of sunflower HaWRKY6 in response to temperature damage. New Phytol 195(4):766–773

Gill SS, Tuteja N (2010) Polyamines and abiotic stress tolerance in plants. Plant Signal Behav 5(1):26–33

Goddijn O, Smeekens S (1998) Sensing trehalose biosynthesis in plants. Plant J 14(2):143–146

Golldack D, Lüking I, Yang O (2011) Plant tolerance to drought and salinity: stress regulating transcription factors and their functional significance in the cellular transcriptional network. Plant Cell Rep 30(8):1383–1391

Gómez-Cadenas A, Arbona V, Jacas J, Primo-Millo E, Talon M (2002) Abscisic acid reduces leaf abscission and increases salt tolerance in citrus plants. J Plant Growth Regul 21(3):234–240

Gonai T, Kawahara S, Tougou M, Satoh S, Hashiba T, Hirai N, Kawaide H, Kamiya Y, Yoshioka T (2004) Abscisic acid in the thermoinhibition of lettuce seed germination and enhancement of its catabolism by gibberellin. J Exp Bot 55(394):111–118

Goswami K, Tripathi A, Sanan-Mishra N (2017) Comparative miRomics of salt-tolerant and salt-sensitive rice. J Integr Bioinform 14(1)

Greenway H, Munns R (1980) Mechanisms of salt tolerance in nonhalophytes. Annu Rev Plant Physiol 31(1):149–190

Groppa M, Benavides M (2008) Polyamines and abiotic stress: recent advances. Amino Acids 34(1):35–45

Guan Q, Lu X, Zeng H, Zhang Y, Zhu J (2013) Heat stress induction of miR398 triggers a regulatory loop that is critical for thermotolerance in *Arabidopsis*. Plant J 74(5):840–851

Gul B, Khan MA, Weber DJ (2000) Alleviation of salinity and dark-enforced dormancy in *Allenrolfea occidentalis* seeds under various thermoperiods. Aust J Bot 48(6):745–752

Guo Y, Qiu Q-S, Quintero FJ, Pardo JM, Ohta M, Zhang C, Schumaker KS, Zhu J-K (2004) Transgenic evaluation of activated mutant alleles of SOS2 reveals a critical requirement for its kinase activity and C-terminal regulatory domain for salt tolerance in *Arabidopsis thaliana*. Plant Cell 16(2):435–449

Guo S, Xu Y, Liu H, Mao Z, Zhang C, Ma Y, Zhang Q, Meng Z, Chong K (2013) The interaction between OsMADS57 and OsTB1 modulates rice tillering via DWARF14. Nat Commun 4:1566

Gupta O, Sharma P, Gupta R, Sharma I (2014) MicroRNA mediated regulation of metal toxicity in plants: present status and future perspectives. Plant Mol Biol 84(1–2):1–18

Hajyzadeh M, Turktas M, Khawar KM, Unver T (2015) miR408 overexpression causes increased drought tolerance in chickpea. Gene 555(2):186–193

Halfter U, Ishitani M, Zhu J-K (2000) The *Arabidopsis* SOS2 protein kinase physically interacts with and is activated by the calcium-binding protein SOS3. Proc Natl Acad Sci 97(7):3735–3740

Han M-H, Goud S, Song L, Fedoroff N (2004) The *Arabidopsis* double-stranded RNA-binding protein HYL1 plays a role in microRNA-mediated gene regulation. Proc Natl Acad Sci U S A 101(4):1093–1098

Hasanuzzaman M, Nahar K, Fujita M, Ahmad P, Chandna R, Prasad M, Ozturk M (2013) Enhancing plant productivity under salt stress: relevance of poly-omics. In: Salt stress in plants. Springer, New York, pp 113–156

Hasegawa PM, Bressan RA, Zhu J-K, Bohnert HJ (2000) Plant cellular and molecular responses to high salinity. Annu Rev Plant Biol 51(1):463–499

He L, Ban Y, Inoue H, Matsuda N, Liu J, Moriguchi T (2008) Enhancement of spermidine content and antioxidant capacity in transgenic pear shoots overexpressing apple spermidine synthase in response to salinity and hyperosmosis. Phytochemistry 69(11):2133–2141

Hernandez Y, Sanan-Mishra N (2017) miRNA mediated regulation of NAC transcription factors in plant development and environment stress response. Plant Gene 11:190–198

Hirayama T, Ohto C, Mizoguchi T, Shinozaki K (1995) A gene encoding a phosphatidylinositol-specific phospholipase C is induced by dehydration and salt stress in *Arabidopsis thaliana*. Proc Natl Acad Sci 92(9):3903–3907

Hivrale V, Zheng Y, Puli COR, Jagadeeswaran G, Gowdu K, Kakani VG, Barakat A, Sunkar R (2016) Characterization of drought-and heat-responsive microRNAs in switchgrass. Plant Sci 242:214–223

Holmstrom K-o, Mantyla E, Wellin B, Mandal A (1996) Drought tolerance in tobacco. Nature 379(6567):683

Holmström KO, Somersalo S, Mandal A, Palva TE, Welin B (2000) Improved tolerance to salinity and low temperature in transgenic tobacco producing glycine betaine. J Exp Bot 51(343):177–185

Hrabak EM, Chan CW, Gribskov M, Harper JF, Choi JH, Halford N, Kudla J, Luan S, Nimmo HG, Sussman MR (2003) The Arabidopsis CDPK-SnRK superfamily of protein kinases. Plant Physiol 132(2):666–680

Hua X-J, Van de Cotte B, Van Montagu M, Verbruggen N (1997) Developmental regulation of pyrroline-5-carboxylate reductase gene expression in Arabidopsis. Plant Physiol 114(4):1215–1224

Huang DW, Sherman BT, Zheng X, Yang J, Imamichi T, Stephens R, Lempicki RA (2009) Extracting biological meaning from large gene lists with DAVID. Curr Protoc Bioinformatics 27:13.11:13.11.1–13.1113.11.13

Huang SQ, Xiang AL, Che LL, Chen S, Li H, Song JB, Yang ZM (2010) A set of miRNAs from *Brassica napus* in response to sulphate deficiency and cadmium stress. Plant Biotechnol J 8(8):887–899

Hugouvieux V, Kwak JM, Schroeder JI (2001) An mRNA cap binding protein, ABH1, modulates early abscisic acid signal transduction in *Arabidopsis*. Cell 106(4):477–487
Hunt L, Mills LN, Pical C, Leckie CP, Aitken FL, Kopka J, Mueller-Roeber B, McAinsh MR, Hetherington AM, Gray JE (2003) Phospholipase C is required for the control of stomatal aperture by ABA. Plant J 34(1):47–55
Hwang J-U, Lee Y (2001) Abscisic acid-induced actin reorganization in guard cells of dayflower is mediated by cytosolic calcium levels and by protein kinase and protein phosphatase activities. Plant Physiol 125(4):2120–2128
Iqbal M, Ashraf M (2010) Changes in hormonal balance: a possible mechanism of pre-sowing chilling-induced salt tolerance in spring wheat. J Agron Crop Sci 196(6):440–454
Jackson M (1997) Hormones from roots as signals for the shoots of stressed plants. Trends Plant Sci 2(1):22–28
Jagadeeswaran G, Saini A, Sunkar R (2009) Biotic and abiotic stress down-regulate miR398 expression in *Arabidopsis*. Planta 229(4):1009–1014
Jammes F, Song C, Shin D, Munemasa S, Takeda K, Gu D, Cho D, Lee S, Giordo R, Sritubtim S (2009) MAP kinases MPK9 and MPK12 are preferentially expressed in guard cells and positively regulate ROS-mediated ABA signaling. Proc Natl Acad Sci 106(48):20520–20525
Jeong JS, Kim YS, Baek KH, Jung H, Ha S-H, Do Choi Y, Kim M, Reuzeau C, Kim J-K (2010) Root-specific expression of OsNAC10 improves drought tolerance and grain yield in rice under field drought conditions. Plant Physiol 153(1):185–197
Jeong JS, Kim YS, Redillas MC, Jang G, Jung H, Bang SW, Choi YD, Ha SH, Reuzeau C, Kim JK (2013) OsNAC5 overexpression enlarges root diameter in rice plants leading to enhanced drought tolerance and increased grain yield in the field. Plant Biotechnol J 11(1):101–114
Jia W, Wang Y, Zhang S, Zhang J (2002) Salt-stress-induced ABA accumulation is more sensitively triggered in roots than in shoots. J Exp Bot 53(378):2201–2206
Jia X, Wang W-X, Ren L, Chen Q-J, Mendu V, Willcut B, Dinkins R, Tang X, Tang G (2009) Differential and dynamic regulation of miR398 in response to ABA and salt stress in *Populus tremula* and *Arabidopsis thaliana*. Plant Mol Biol 71(1–2):51–59
Jiang Y, Deyholos MK (2009) Functional characterization of *Arabidopsis* NaCl-inducible WRKY25 and WRKY33 transcription factors in abiotic stresses. Plant Mol Biol 69(1–2):91–105
Jiao Y, Wang Y, Xue D, Wang J, Yan M, Liu G, Dong G, Zeng D, Lu Z, Zhu X (2010) Regulation of OsSPL14 by OsmiR156 defines ideal plant architecture in rice. Nat Genet 42(6):541–544
Johnson JM, Reichelt M, Vadassery J, Gershenzon J, Oelmüller R (2014) An *Arabidopsis* mutant impaired in intracellular calcium elevation is sensitive to biotic and abiotic stress. BMC Plant Biol 14(1):162
Jonak C, Páy A, Börge L, Hirt H, Heberle-Bors E (1993) The plant homologue of MAP kinase is expressed in a cell cycle-dependent and organ-specific manner. Plant J 3(4):611–617
Jones-Rhoades MW, Bartel DP (2004) Computational identification of plant microRNAs and their targets, including a stress-induced miRNA. Mol Cell 14(6):787–799
Jones-Rhoades MW, Bartel DP, Bartel B (2006) MicroRNAs and their regulatory roles in plants. Annu Rev Plant Biol 57:19–53
Jung J-H, Seo PJ, Park C-M (2009) MicroRNA biogenesis and function in higher plants. Mol Biol Rep 3(2):111–126
Kang DJ, Seo YJ, Lee JD, Ishii R, Kim K, Shin D, Park S, Jang S, Lee IJ (2005) Jasmonic acid differentially affects growth, ion uptake and abscisic acid concentration in salt-tolerant and salt-sensitive rice cultivars. J Agron Crop Sci 191(4):273–282
Kansal S, Devi RM, Balyan SC, Arora MK, Singh AK, Mathur S, Raghuvanshi S (2015) Unique miRNome during anthesis in drought-tolerant indica rice var. Nagina 22. Planta 241(6):1543–1559
Kantar M, Lucas SJ, Budak H (2011) miRNA expression patterns of *Triticum dicoccoides* in response to shock drought stress. Planta 233(3):471–484
Kaouthar F, Ameny F-K, Yosra K, Walid S, Ali G, Faical B (2016) Responses of transgenic *Arabidopsis* plants and recombinant yeast cells expressing a novel durum wheat manganese superoxide dismutase TdMnSOD to various abiotic stresses. J Plant Physiol 198:56–68

Karakas B, Ozias-Akins P, Stushnoff C, Suefferheld M, Rieger M (1997) Salinity and drought tolerance of mannitol-accumulating transgenic tobacco. Plant Cell Environ 20(5):609–616

Karan R, Subudhi PK (2012) A stress inducible SUMO conjugating enzyme gene (SaSce9) from a grass halophyte *Spartina alterniflora* enhances salinity and drought stress tolerance in *Arabidopsis*. BMC Plant Biol 12(1):187

Kasuga M, Liu Q, Miura S, Yamaguchi-Shinozaki K, Shinozaki K (1999) Improving plant drought, salt, and freezing tolerance by gene transfer of a single stress-inducible transcription factor. Nat Biotechnol 17(3)

Keskin BC, Yuksel B, Memon AR, Topal-Sarıkaya A (2010) Abscisic acid regulated gene expression in bread wheat (*Triticum aestivum* L.). Aust J Crop Sci 4(8):617

Khaksefidi RE, Mirlohi S, Khalaji F, Fakhari Z, Shiran B, Fallahi H, Rafiei F, Budak H, Ebrahimie E (2015) Differential expression of seven conserved microRNAs in response to abiotic stress and their regulatory network in *Helianthus annuus*. Front Plant Sci 6:741

Khan M, Hamid A, Salahuddin A, Quasem A, Karim M (1997) Effect of sodium chloride on growth, photosynthesis and mineral ions accumulation of different types of rice (*Oryza sativa* L.). J Agron Crop Sci 179(3):149–161

Khan NA, Syeed S, Masood A, Nazar R, Iqbal N (2010) Application of salicylic acid increases contents of nutrients and antioxidative metabolism in mungbean and alleviates adverse effects of salinity stress. Int J Plant Biol 1(1):e1

Khan Y, Yadav A, Bonthala VS, Muthamilarasan M, Yadav CB, Prasad M (2014) Comprehensive genome-wide identification and expression profiling of foxtail millet (*Setaria italica* L.) miRNAs in response to abiotic stress and development of miRNA database. Plant Cell Tissue Organ Cult 118(2):279–292

Khatib F, Makris A, Yamaguchi-Shinozaki K, Kumar S, Sarker A, Erskine W, Baum M (2011) Expression of the DREB1A gene in lentil (*Lens culinaris Medik.* subsp. *culinaris*) transformed with the *Agrobacterium* system. Crop Pasture Sci 62(6):488–495

Khokon M, Okuma E, Hossain MA, Munemasa S, Uraji M, Nakamura Y, Mori IC, Murata Y (2011) Involvement of extracellular oxidative burst in salicylic acid-induced stomatal closure in *Arabidopsis*. Plant Cell Environ 34(3):434–443

Khraiwesh B, Zhu J-K, Zhu J (2012) Role of miRNAs and siRNAs in biotic and abiotic stress responses of plants. BBA-Gene Regul Mech 1819(2):137–148

Kidner CA, Martienssen RA (2005) The developmental role of microRNA in plants. Curr Opin Plant Biol 8(1):38–44

Kim BG, Waadt R, Cheong YH, Pandey GK, Dominguez-Solis JR, Schültke S, Lee SC, Kudla J, Luan S (2007) The calcium sensor CBL10 mediates salt tolerance by regulating ion homeostasis in *Arabidopsis*. Plant J 52(3):473–484

Kim S, Yang J-Y, Xu J, Jang I-C, Prigge MJ, Chua N-H (2008) Two cap-binding proteins CBP20 and CBP80 are involved in processing primary MicroRNAs. Plant Cell Physiol 49(11):1634–1644

Kishor PK, Hong Z, Miao G-H, Hu C-AA, Verma DPS (1995) Overexpression of [delta]-pyrroline-5-carboxylate synthetase increases proline production and confers osmotolerance in transgenic plants. Plant Physiol 108(4):1387–1394

Knight H, Knight MR (2001) Abiotic stress signalling pathways: specificity and cross-talk. Trends Plant Sci 6(6):262–267

Knight H, Trewavas AJ, Knight MR (1997) Calcium signalling in *Arabidopsis thaliana* responding to drought and salinity. Plant J 12(5):1067–1078

Knight H, Veale EL, Warren GJ, Knight MR (1999) The sfr6 mutation in *Arabidopsis* suppresses low-temperature induction of genes dependent on the CRT/DRE sequence motif. Plant Cell 11(5):875–886

Kohl DH, Schubert KR, Carter MB, Hagedorn CH, Shearer G (1988) Proline metabolism in N2-fixing root nodules: energy transfer and regulation of purine synthesis. Proc Natl Acad Sci 85(7):2036–2040

Kong X, Zhang M, Xu X, Li X, Li C, Ding Z (2014) System analysis of microRNAs in the development and aluminium stress responses of the maize root system. Plant Biotechnol J 12(8):1108–1121

Kopka J, Pical C, Hetherington AM, Müller-Röber B (1998) Ca 2+/phospholipid-binding (C 2) domain in multiple plant proteins: novel components of the calcium-sensing apparatus. Plant Mol Biol 36(5):627–637

Kruszka K, Pacak A, Swida-Barteczka A, Nuc P, Alaba S, Wroblewska Z, Karlowski W, Jarmolowski A, Szweykowska-Kulinska Z (2014) Transcriptionally and post-transcriptionally regulated microRNAs in heat stress response in barley. J Exp Bot 65(20):6123–6135

Kudla J, Batistič O, Hashimoto K (2010) Calcium signals: the lead currency of plant information processing. Plant Cell 22(3):541–563

Kumar R (2014) Role of microRNAs in biotic and abiotic stress responses in crop plants. Appl Biochem Biotechnol 174(1):93–115

Kumar B, Singh B (1996) Effect of plant hormones on growth and yield of wheat irrigated with saline water. Ann Agric Res 17:209–212

Kurihara Y, Watanabe Y (2004) *Arabidopsis* micro-RNA biogenesis through Dicer-like 1 protein functions. Proc Natl Acad Sci U S A 101(34):12753–12758

Kurihara Y, Takashi Y, Watanabe Y (2006) The interaction between DCL1 and HYL1 is important for efficient and precise processing of pri-miRNA in plant microRNA biogenesis. RNA 12(2):206–212

Lagos-Quintana M, Rauhut R, Lendeckel W, Tuschl T (2001) Identification of novel genes coding for small expressed RNAs. Science 294(5543):853–858

Landouar-Arsivaud L, Juchaux-Cachau M, Jeauffre J, Biolley J-P, Maurousset L, Lemoine R (2011) The promoters of 3 celery salt-induced phloem-specific genes as new tools for monitoring salt stress responses. Plant Physiol Biochem 49(1):2–8

Lanet E, Delannoy E, Sormani R, Floris M, Brodersen P, Crété P, Voinnet O, Robaglia C (2009) Biochemical evidence for translational repression by *Arabidopsis* microRNAs. Plant Cell 21(6):1762–1768

Lau NC, Lim LP, Weinstein EG, Bartel DP (2001) An abundant class of tiny RNAs with probable regulatory roles in *Caenorhabditis elegans*. Science 294(5543):858–862

Lauchli A, Schubert S (1989) The role of calcium in the regulation of membrane and cellular growth processes under salt stress. In: Environmental stress in plants. Springer, New York, pp 131–138

Lee RC, Feinbaum RL, Ambros V (1993) The *C. elegans* heterochronic gene lin-4 encodes small RNAs with antisense complementarity to lin-14. Cell 75(5):843–854

Lehmann J, Atzorn R, Brückner C, Reinbothe S, Leopold J, Wasternack C, Parthier B (1995) Accumulation of jasmonate, abscisic acid, specific transcripts and proteins in osmotically stressed barley leaf segments. Planta 197(1):156–162

Leung AK, Sharp PA (2010) MicroRNA functions in stress responses. Mol Cell 40(2):205–215

Lewis D, Smith D (1967) Sugar alcohols (polyols) in fungi and green plants. New Phytol 66(2):143–184

Li J, Yang H, Peer WA, Richter G, Blakeslee J, Bandyopadhyay A, Titapiwantakun B, Undurraga S, Khodakovskaya M, Richards EL (2005) *Arabidopsis* H+-PPase AVP1 regulates auxin-mediated organ development. Science 310(5745):121–125

Li W-X, Oono Y, Zhu J, He X-J, Wu J-M, Iida K, Lu X-Y, Cui X, Jin H, Zhu J-K (2008) The *Arabidopsis* NFYA5 transcription factor is regulated transcriptionally and post transcriptionally to promote drought resistance. Plant Cell 20(8):2238–2251

Li H, Jiang H, Bu Q, Zhao Q, Sun J, Xie Q, Li C (2011) The *Arabidopsis* RING finger E3 ligase RHA2b acts additively with RHA2a in regulating ABA signaling and drought response. Plant Physiol 156:550. 111.176214

Li Z-Y, Xu Z-S, He G-Y, Yang G-X, Chen M, Li L-C, Ma Y-Z (2012) Overexpression of soybean GmCBL1 enhances abiotic stress tolerance and promotes hypocotyl elongation in *Arabidopsis*. Biochem Biophys Res Commun 427(4):731–736

Li M-Y, Wang F, Xu Z-S, Jiang Q, Ma J, Tan G-F, Xiong A-S (2014) High throughput sequencing of two celery varieties small RNAs identifies microRNAs involved in temperature stress response. BMC Genomics 15(1):242

Li SB, Xie ZZ, Hu CG, Zhang JZ (2016) A review of auxin response factors (ARFs) in plants. Front Plant Sci 7:46
Liang M, Haroldsen V, Cai X, Wu Y (2006) Expression of a putative laccase gene, ZmLAC1, in maize primary roots under stress. Plant Cell Environ 29(5):746–753
Lima J, Arenhart R, Margis-Pinheiro M, Margis R (2011) Aluminum triggers broad changes in microRNA expression in rice roots. Genet Mol Res 10(4):2817–2832
Lin H, Yang Y, Quan R, Mendoza I, Wu Y, Du W, Zhao S, Schumaker KS, Pardo JM, Guo Y (2009) Phosphorylation of SOS3-like calcium binding Protein8 by SOS2 protein kinase stabilizes their protein complex and regulates salt tolerance in *Arabidopsis*. Plant Cell 21(5):1607–1619
Lin JS, Lin CC, Lin HH, Chen YC, Jeng ST (2012) MicroR828 regulates lignin and H2O2 accumulation in sweet potato on wounding. New Phytol 196(2):427–440
Lippold F, Sanchez DH, Musialak M, Schlereth A, Scheible W-R, Hincha DK, Udvardi MK (2009) AtMyb41 regulates transcriptional and metabolic responses to osmotic stress in *Arabidopsis*. Plant Physiol 149(4):1761–1772
Liu Q, Zhang H (2012) Molecular identification and analysis of arsenite stress-responsive miRNAs in rice. J Agric Food Chem 60(26):6524–6536
Liu J, Zhu J-K (1998) A calcium sensor homolog required for plant salt tolerance. Science 280(5371):1943–1945
Liu S, Calderwood DA, Ginsberg MH (2000) Integrin cytoplasmic domain-binding proteins. J Cell Sci 113(20):3563–3571
Liu J-H, Kitashiba H, Wang J, Ban Y, Moriguchi T (2007) Polyamines and their ability to provide environmental stress tolerance to plants. Plant Biotechnol 24(1):117–126
Liu H-H, Tian X, Li Y-J, Wu C-A, Zheng C-C (2008) Microarray-based analysis of stress-regulated microRNAs in *Arabidopsis thaliana*. RNA 14(5):836–843
Liu D, Song Y, Chen Z, Yu D (2009) Ectopic expression of miR396 suppresses GRF target gene expression and alters leaf growth in *Arabidopsis*. Physiol Plant 136(2):223–236
Liu Y-Q, Zhang M, Yin B-C, Ye B-C (2012) Attomolar ultrasensitive microRNA detection by DNA-scaffolded silver-nanocluster probe based on isothermal amplification. Anal Chem 84(12):5165–5169
Liu N, Yang J, Guo S, Xu Y, Zhang M (2013) Genome-wide identification and comparative analysis of conserved and novel microRNAs in grafted watermelon by high-throughput sequencing. PLoS One 8(2):e57359
Llave C, Xie Z, Kasschau KD, Carrington JC (2002) Cleavage of scarecrow-like mRNA targets directed by a class of *Arabidopsis* miRNA. Science 297(5589):2053–2056
Lopez C, Takahashi H, Yamazaki S (2002) Plant–water relations of kidney bean plants treated with NaCl and foliarly applied glycinebetaine. J Agron Crop Sci 188(2):73–80
Louis P, Galinski EA (1997) Characterization of genes for the biosynthesis of the compatible solute ectoine from *Marinococcus halophilus* and osmoregulated expression in *Escherichia coli*. Microbiology 143(4):1141–1149
Lu C, Fedoroff N (2000) A mutation in the *Arabidopsis* HYL1 gene encoding a dsRNA binding protein affects responses to abscisic acid, auxin, and cytokinin. Plant Cell 12(12):2351–2365
Lu S, Sun Y-H, Shi R, Clark C, Li L, Chiang VL (2005) Novel and mechanical stress–responsive microRNAs in *Populus trichocarpa* that are absent from *Arabidopsis*. Plant Cell 17(8):2186–2203
Lu S, Sun YH, Chiang VL (2008) Stress-responsive microRNAs in *Populus*. Plant J 55(1):131–151
Lu Y, Feng Z, Bian L, Xie H, Liang J (2011) miR398 regulation in rice of the responses to abiotic and biotic stresses depends on CSD1 and CSD2 expression. Funct Plant Biol 38(1):44–53
Lv D-K, Bai X, Li Y, Ding X-D, Ge Y, Cai H, Ji W, Wu N, Zhu Y-M (2010) Profiling of cold-stress-responsive miRNAs in rice by microarrays. Gene 459(1):39–47
Ma NN, Zuo YQ, Liang XQ, Yin B, Wang GD, Meng QW (2013) The multiple stress-responsive transcription factor SlNAC1 improves the chilling tolerance of tomato. Physiol Plant 149(4):474–486

Ma X, Xin Z, Wang Z, Yang Q, Guo S, Guo X, Cao L, Lin T (2015) Identification and comparative analysis of differentially expressed miRNAs in leaves of two wheat (*Triticum aestivum* L.) genotypes during dehydration stress. BMC Plant Biol 15(1):21

Maeda T, Wurgler-Murphy SM, Saito H (1994) A two-component system that regulates an osmosensing MAP kinase cascade in yeast. Nature 369(6477):242–245

Magome H, Yamaguchi S, Hanada A, Kamiya Y, Oda K (2004) Dwarf and delayed-flowering 1, a novel *Arabidopsis* mutant deficient in gibberellin biosynthesis because of overexpression of a putative AP2 transcription factor. Plant J 37(5):720–729

Mahajan S, Tuteja N (2005) Cold, salinity and drought stresses: an overview. Arch Biochem Biophys 444(2):139–158

Mahajan S, Sopory SK, Tuteja N (2006) CBL-CIPK paradigm: role in calcium and stress signaling in plants. Proc Indian Natl Sci Acad 72(2):63

Mahale BM, Fakrudin B, Ghosh S, Krishnaraj P (2014) LNA mediated in situ hybridization of miR171 and miR397a in leaf and ambient root tissues revealed expressional homogeneity in response to shoot heat shock in *Arabidopsis thaliana*. J Plant Biochem Biotechnol 23(1):93–103

Manavella PA, Hagmann J, Ott F, Laubinger S, Franz M, Macek B, Weigel D (2012) Fast-forward genetics identifies plant CPL phosphatases as regulators of miRNA processing factor HYL1. Cell 151(4):859–870

Margis R, Fusaro AF, Smith NA, Curtin SJ, Watson JM, Finnegan EJ, Waterhouse PM (2006) The evolution and diversification of Dicers in plants. FEBS Lett 580(10):2442–2450

May P, Liao W, Wu Y, Shuai B, McCombie WR, Zhang MQ, Liu QA (2013) The effects of carbon dioxide and temperature on microRNA expression in *Arabidopsis* development. Nat Commun 4:2145

McAinsh MR, Hetherington AM (1998) Encoding specificity in Ca2+ signalling systems. Trends Plant Sci 3(1):32–36

McConn M, Creelman RA, Bell E, Mullet JE (1997) Jasmonate is essential for insect defense in *Arabidopsis*. Proc Natl Acad Sci 94(10):5473–5477

McCue KF, Hanson AD (1990) Drought and salt tolerance: towards understanding and application. Trends Biotechnol 8:358–362

Mehrpooyan F, Othman R, Harikrishna J (2012) Tissue and temporal expression of miR172 paralogs and the AP2-like target in oil palm (*Elaeis guineensis Jacq*.). Tree Genet Genomes 8(6):1331–1343

Meng Y, Shao C, Wang H, Chen M (2011) The regulatory activities of plant microRNAs: a more dynamic perspective. Plant Physiol 157(4):1583–1595

Merchan F, Lorenzo LD, Rizzo SG, Niebel A, Manyani H, Frugier F, Sousa C, Crespi M (2007) Identification of regulatory pathways involved in the reacquisition of root growth after salt stress in *Medicago truncatula*. Plant J 51(1):1–17

Meyers BC, Axtell MJ, Bartel B, Bartel DP, Baulcombe D, Bowman JL, Cao X, Carrington JC, Chen X, Green PJ (2008) Criteria for annotation of plant MicroRNAs. Plant Cell 20(12):3186–3190

Mikami K, Katagiri T, Iuchi S, Yamaguchi-Shinozaki K, Shinozaki K (1998) A gene encoding phosphatidylinositol-4-phosphate 5-kinase is induced by water stress and abscisic acid in *Arabidopsis thaliana*. Plant J 15(4):563–568

Mittal D, Sharma N, Sharma V, Sopory S, Sanan-Mishra N (2016) Role of microRNAs in rice plant under salt stress. Ann Appl Biol 168(1):2–18

Mok DW, Mok MC (2001) Cytokinin metabolism and action. Annu Rev Plant Biol 52(1):89–118

Morgan PW, Drew MC (1997) Ethylene and plant responses to stress. Physiol Plant 100(3):620–630

Munnik T, Testerink C (2009) Plant phospholipid signaling: "in a nutshell". J Lipid Res 50(Supplement):S260–S265

Munns R, Tester M (2008) Mechanisms of salinity tolerance. Annu Rev Plant Biol 59:651–681

Navarro L, Dunoyer P, Jay F, Arnold B, Dharmasiri N, Estelle M, Voinnet O, Jones JD (2006) A plant miRNA contributes to antibacterial resistance by repressing auxin signaling. Science 312(5772):436–439

Nayyar H, Walia D (2003) Water stress induced proline accumulation in contrasting wheat genotypes as affected by calcium and abscisic acid. Biol Plant 46(2):275–279

Nelson PT, Baldwin DA, Scearce LM, Oberholtzer JC, Tobias JW, Mourelatos Z (2004) Microarray-based, high-throughput gene expression profiling of microRNAs. Nat Methods 1(2)
Ni Z, Hu Z, Jiang Q, Zhang H (2013) GmNFYA3, a target gene of miR169, is a positive regulator of plant tolerance to drought stress. Plant Mol Biol 82(1–2):113–129
Nilsen E, Orcutt D (1996) The physiology of plants under deficit, Abiotic Factors. Willey, New York, p 689
Niu X, Bressan RA, Hasegawa PM, Pardo JM (1995) Ion homeostasis in NaCl stress environments. Plant Physiol 109(3):735
Olszewski N, T-p S, Gubler F (2002) Gibberellin signaling biosynthesis, catabolism, and response pathways. Plant Cell 14(suppl 1):S61–S80
Parashar A, Varma S (1988) Effect of presowing seed soaking in gibberellic acid, duration of soaking, different temperatures and their interaction on seed germination and early seedling growth of wheat under saline conditions. Plant Physiol Biochem 15:189
Pareek A, Singla S, Grover A (1997) Salt responsive proteins/genes in crop plants
Parihar P, Singh S, Singh R, Singh VP, Prasad SM (2015) Effect of salinity stress on plants and its tolerance strategies: a review. Environ Sci Pollut Res Int 22(6):4056–4075
Park W, Li J, Song R, Messing J, Chen X (2002) Carpel factory, a Dicer homolog, and HEN1, a novel protein, act in microRNA metabolism in *Arabidopsis thaliana*. Curr Biol 12(17):1484–1495
Park MY, Wu G, Gonzalez-Sulser A, Vaucheret H, Poethig RS (2005) Nuclear processing and export of microRNAs in *Arabidopsis*. Proc Natl Acad Sci U S A 102(10):3691–3696
Pasquinelli AE, Reinhart BJ, Slack F, Martindale MQ, Kuroda MI, Maller B, Hayward DC, Ball EE, Degnan B, Müller P (2000) Conservation of the sequence and temporal expression of let-7 heterochronic regulatory RNA. Nature 408(6808):86–89
Peng Z, Lu Q, Verma D (1996) Reciprocal regulation of Δ 1-pyrroline-5-carboxylate synthetase and proline dehydrogenase genes controls proline levels during and after osmotic stress in plants. Mol Gen Genet 253(3):334–341
Pérez-Quintero ÁL, Zapata A, López C, Neme R (2010) Plant microRNAs and their role in defense against viruses: a bioinformatics approach. BMC Plant Biol 10(1):138
Peters C, Li M, Narasimhan R, Roth M, Welti R, Wang X (2010) Nonspecific phospholipase C NPC4 promotes responses to abscisic acid and tolerance to hyperosmotic stress in *Arabidopsis*. Plant Cell 22(8):2642–2659
Pfeffer S, Zavolan M, Grässer FA, Chien M, Russo JJ, Ju J, John B, Enright AJ, Marks D, Sander C (2004) Identification of virus-encoded microRNAs. Science 304(5671):734–736
Pharr D, Stoop J, Williamson J, Feusi MS, Massel M, Conkling M (1995) The dual role of mannitol as osmoprotectant and photoassimilate in celery. Hort Sci 30:1182
Phillips JR, Dalmay T, Bartels D (2007) The role of small RNAs in abiotic stress. FEBS Lett 581(19):3592–3597
Pillai RS, Artus CG, Filipowicz W (2004) Tethering of human ago proteins to mRNA mimics the miRNA-mediated repression of protein synthesis. RNA 10(10):1518–1525
Pilon-Smits EA, Terry N, Sears T, Kim H, Zayed A, Hwang S, van Dun K, Voogd E, Verwoerd TC, Krutwagen RW (1998) Trehalose-producing transgenic tobacco plants show improved growth performance under drought stress. J Plant Physiol 152(4–5):525–532
Pokotylo I, Kolesnikov Y, Kravets V, Zachowski A, Ruelland E (2014) Plant phosphoinositide-dependent phospholipases C: variations around a canonical theme. Biochimie 96:144–157
Pospíšilová J (2003) Interaction of cytokinins and abscisic acid during regulation of stomatal opening in bean leaves. Photosynthetica 41(1):49–56
Prakash L, Prathapasenan G (1990) NaCl-and gibberellic acid-induced changes in the content of auxin and the activities of cellulase and pectin lyase during leaf growth in rice (*Oryza sativa*). Ann Bot 65(3):251–257
Pruthvi V, Narasimhan R, Nataraja KN (2014) Simultaneous expression of abiotic stress responsive transcription factors, AtDREB2A, AtHB7 and AtABF3 improves salinity and drought tolerance in peanut (*Arachis hypogaea* L.). PLoS One 9(12):e111152
Qiu Q-S, Guo Y, Dietrich MA, Schumaker KS, Zhu J-K (2002) Regulation of SOS1, a plasma membrane Na+/H+ exchanger in *Arabidopsis thaliana*, by SOS2 and SOS3. Proc Natl Acad Sci 99(12):8436–8441

Qiu Q-S, Barkla BJ, Vera-Estrella R, Zhu J-K, Schumaker KS (2003) Na+/H+ exchange activity in the plasma membrane of *Arabidopsis*. Plant Physiol 132(2):1041–1052
Qiu Z, Hai B, Guo J, Li Y, Zhang L (2016) Characterization of wheat miRNAs and their target genes responsive to cadmium stress. Plant Physiol Biochem 101:60–67
Quan R, Lin H, Mendoza I, Zhang Y, Cao W, Yang Y, Shang M, Chen S, Pardo JM, Guo Y (2007) SCABP8/CBL10, a putative calcium sensor, interacts with the protein kinase SOS2 to protect *Arabidopsis* shoots from salt stress. Plant Cell 19(4):1415–1431
Quintero FJ, Ohta M, Shi H, Zhu J-K, Pardo JM (2002) Reconstitution in yeast of the *Arabidopsis* SOS signaling pathway for Na+ homeostasis. Proc Natl Acad Sci 99(13):9061–9066
Reinhart BJ, Weinstein EG, Rhoades MW, Bartel B, Bartel DP (2002) MicroRNAs in plants. Genes Dev 16(13):1616–1626
Rejeb KB, Abdelly C, Savouré A (2014) How reactive oxygen species and proline face stress together. Plant Physiol Biochem 80:278–284
Ren G, Yu B (2012) Critical roles of RNA-binding proteins in miRNA biogenesis in *Arabidopsis*. RNA Biol 9(12):1424–1428
Ren X, Chen Z, Liu Y, Zhang H, Zhang M, Liu Q, Hong X, Zhu JK, Gong Z (2010) ABO3, a WRKY transcription factor, mediates plant responses to abscisic acid and drought tolerance in *Arabidopsis*. Plant J 63(3):417–429
Ren G, Xie M, Dou Y, Zhang S, Zhang C, Yu B (2012) Regulation of miRNA abundance by RNA binding protein TOUGH in *Arabidopsis*. Proc Natl Acad Sci 109(31):12817–12821
Reyes JL, Chua NH (2007) ABA induction of miR159 controls transcript levels of two MYB factors during *Arabidopsis* seed germination. Plant J 49(4):592–606
Rhodes D, Hanson A (1993) Quaternary ammonium and tertiary sulfonium compounds in higher plants. Annu Rev Plant Biol 44(1):357–384
Ribaut J, Pilet P (1994) Water stress and indol-3yl-acetic acid content of maize roots. Planta 193(4):502–507
Roy S (2016) Function of MYB domain transcription factors in abiotic stress and epigenetic control of stress response in plant genome. Plant Signal Behav 11(1):e1117723
Roy M, Wu R (2002) Overexpression of S-adenosylmethionine decarboxylase gene in rice increases polyamine level and enhances sodium chloride-stress tolerance. Plant Sci 163(5):987–992
Rubinelli PM, Chuck G, Li X, Meilan R (2013) Constitutive expression of the Corngrass1 microRNA in poplar affects plant architecture and stem lignin content and composition. Biomass Bioenergy 54:312–321
Ruiz-Ferrer V, Voinnet O (2009) Roles of plant small RNAs in biotic stress responses. Annu Rev Plant Biol 60:485–510
Rutschmann F, Stalder U, Piotrowski M, Oecking C, Schaller A (2002) LeCPK1, a calcium-dependent protein kinase from tomato. Plasma membrane targeting and biochemical characterization. Plant Physiol 129(1):156–168
Saijo Y, Hata S, Kyozuka J, Shimamoto K, Izui K (2000) Over-expression of a single Ca2+-dependent protein kinase confers both cold and salt/drought tolerance on rice plants. Plant J 23(3):319–327
Sailaja B, Anjum N, Prasanth VV, Sarla N, Subrahmanyam D, Voleti S, Viraktamath B, Mangrauthia SK (2014) Comparative study of susceptible and tolerant genotype reveals efficient recovery and root system contributes to heat stress tolerance in rice. Plant Mol Biol Report 32(6):1228–1240
Sakhabutdinova A, Fatkhutdinova D, Bezrukova M, Shakirova F (2003) Salicylic acid prevents the damaging action of stress factors on wheat plants. Bulg J Plant Physiol 21:314–319
Sanan-Mishra N, Kumar V, Sopory SK, Mukherjee SK (2009) Cloning and validation of novel miRNA from basmati rice indicates cross talk between abiotic and biotic stresses. Mol Gen Genomics 282(5):463
Sanders D, Brownlee C, Harper JF (1999) Communicating with calcium. Plant Cell 11(4):691–706
Sanders D, Pelloux J, Brownlee C, Harper JF (2002) Calcium at the crossroads of signaling. Plant Cell 14(suppl 1):S401–S417

Saradhi PP, Mohanty P (1993) Proline in relation to free radical production in seedlings of *Brassica juncea* raised under sodium chloride stress. Plant Soil 155(1):497–500
Sarkar T, Thankappan R, Kumar A, Mishra GP, Dobaria JR (2014) Heterologous expression of the AtDREB1A gene in transgenic peanut-conferred tolerance to drought and salinity stresses. PLoS One 9(12):e110507
Schmidt R, Schippers JH, Mieulet D, Obata T, Fernie AR, Guiderdoni E, Mueller-Roeber B (2013) Multipass, a rice R2R3-type MYB transcription factor, regulates adaptive growth by integrating multiple hormonal pathways. Plant J 76(2):258–273
Schommer C, Palatnik JF, Aggarwal P, Chételat A, Cubas P, Farmer EE, Nath U, Weigel D (2008) Control of jasmonate biosynthesis and senescence by miR319 targets. PLoS Biol 6(9):230
Schwab R, Palatnik JF, Riester M, Schommer C, Schmid M, Weigel D (2005) Specific effects of microRNAs on the plant transcriptome. Dev Cell 8(4):517–527
Seki M, Narusaka M, Ishida J, Nanjo T, Fujita M, Oono Y, Kamiya A, Nakajima M, Enju A, Sakurai T (2002) Monitoring the expression profiles of 7000 *Arabidopsis* genes under drought, cold and high-salinity stresses using a full-length cDNA microarray. Plant J 31(3):279–292
Serrano R, Mulet JM, Rios G, Marquez JA, De Larrinoa IF, Leube MP, Mendizabal I, Pascual-Ahuir A, Proft M, Ros R (1999) A glimpse of the mechanisms of ion homeostasis during salt stress. J Exp Bot 50:1023–1036
Sharma D, Tiwari M, Lakhwani D, Tripathi RD, Trivedi PK (2015a) Differential expression of microRNAs by arsenate and arsenite stress in natural accessions of rice. Metallomics 7(1):174–187
Sharma N, Tripathi A, Sanan-Mishra N (2015b) Profiling the expression domains of a rice-specific microRNA under stress. Front Plant Sci 6:333
Sharma N, Mittal D, Mishra NS (2017) Micro-regulators of hormones and Stress. In: Mechanism of plant hormone signaling under stress. Wiley, New York, pp 319–351
Shen J, Xie K, Xiong L (2010) Global expression profiling of rice microRNAs by one-tube stem-loop reverse transcription quantitative PCR revealed important roles of microRNAs in abiotic stress responses. Mol Gen Genomics 284(6):477–488
Shi H, Zhu J-K (2002) Regulation of expression of the vacuolar Na+/H+ antiporter gene AtNHX1 by salt stress and abscisic acid. Plant Mol Biol 50(3):543–550
Shi H, Ishitani M, Kim C, Zhu J-K (2000) The *Arabidopsis thaliana* salt tolerance gene SOS1 encodes a putative Na+/H+ antiporter. Proc Natl Acad Sci 97(12):6896–6901
Sickler CM, Edwards GE, Kiirats O, Gao Z, Loescher W (2007) Response of mannitol-producing *Arabidopsis thaliana* to abiotic stress. Funct Plant Biol 34(4):382–391
Siomi H, Siomi MC (2010) Posttranscriptional regulation of microRNA biogenesis in animals. Mol Cell 38(3):323–332
Sivakumar P, Sharmila P, Saradhi PP (1998) Proline suppresses Rubisco activity in higher plants. Biochem Biophys Res Commun 252(2):428–432
Smirnoff N (1993) The role of active oxygen in the response of plants to water deficit and desiccation. New Phytol 125(1):27–58
Smirnoff N (1998) Plant resistance to environmental stress. Curr Opin Biotechnol 9(2):214–219
Smith JL, De Moraes CM, Mescher MC (2009) Jasmonate-and salicylate-mediated plant defense responses to insect herbivores, pathogens and parasitic plants. Pest Manag Sci 65(5):497–503
Song J-J, Smith SK, Hannon GJ, Joshua-Tor L (2004) Crystal structure of Argonaute and its implications for RISC slicer activity. Science 305(5689):1434–1437
Song L, Han M-H, Lesicka J, Fedoroff N (2007) *Arabidopsis* primary microRNA processing proteins HYL1 and DCL1 define a nuclear body distinct from the Cajal body. Proc Natl Acad Sci 104(13):5437–5442
Song G, Zhang R, Zhang S, Li Y, Gao J, Han X, Chen M, Wang J, Li W, Li G (2017a) Response of microRNAs to cold treatment in the young spikes of common wheat. BMC Genomics 18(1):212
Song Z, Xu Q, Lin C, Tao C, Zhu C, Xing S, Fan Y, Liu W, Yan J, Li J, Sang T (2017b) Transcriptomic characterization of candidate genes responsive to salt tolerance of Miscanthus energy crops. GCB Bioenergy 9(7):1222–1237

Srivastava S, Srivastava AK, Suprasanna P, D'souza S (2012) Identification and profiling of arsenic stress-induced microRNAs in *Brassica juncea*. J Exp Bot 64(1):303–315
Stephenson TJ, McIntyre CL, Collet C, Xue G-P (2007) Genome-wide identification and expression analysis of the NF-Y family of transcription factors in *Triticum aestivum*. Plant Mol Biol 65(1–2):77–92
Stief A, Altmann S, Hoffmann K, Pant BD, Scheible W-R, Bäurle I (2014) *Arabidopsis* miR156 regulates tolerance to recurring environmental stress through SPL transcription factors. Plant Cell 26(4):1792–1807
Stoop JM, Pharr DM (1994) Mannitol metabolism in celery stressed by excess macronutrients. Plant Physiol 106(2):503–511
Sun G, Stewart CN Jr, Xiao P, Zhang B (2012) MicroRNA expression analysis in the cellulosic biofuel crop switchgrass (*Panicum virgatum*) under abiotic stress. PLoS One 7(3):e32017
Sun X, Xu L, Wang Y, Yu R, Zhu X, Luo X, Gong Y, Wang R, Limera C, Zhang K (2015) Identification of novel and salt-responsive miRNAs to explore miRNA-mediated regulatory network of salt stress response in radish (*Raphanus sativus* L.). BMC Genomics 16(1):197
Sunkar R, Zhu J-K (2004) Novel and stress-regulated microRNAs and other small RNAs from *Arabidopsis*. Plant Cell 16(8):2001–2019
Sunkar R, Kapoor A, Zhu J-K (2006) Posttranscriptional induction of two Cu/Zn superoxide dismutase genes in *Arabidopsis* is mediated by downregulation of miR398 and important for oxidative stress tolerance. Plant Cell 18(8):2051–2065
Tähtiharju S, Sangwan V, Monroy AF, Dhindsa RS, Borg M (1997) The induction of kin genes in cold-acclimating *Arabidopsis thaliana*. Evidence of a role for calcium. Planta 203(4):442–447
Tamura T, Hara K, Yamaguchi Y, Koizumi N, Sano H (2003) Osmotic stress tolerance of transgenic tobacco expressing a gene encoding a membrane-located receptor-like protein from tobacco plants. Plant Physiol 131(2):454–462
Tang Z, Zhang L, Xu C, Yuan S, Zhang F, Zheng Y, Zhao C (2012) Uncovering small RNA-mediated responses to cold stress in a wheat thermosensitive genic male-sterile line by deep sequencing. Plant Physiol 159(2):721–738
Tarczynski MC, Jensen RG, Bohnert HJ (1993) Stress protection of transgenic tobacco by production of the osmolyte mannitol. Sci N Y Then Wash 259:508–508
Thiebaut F, Grativol C, Carnavale-Bottino M, Rojas CA, Tanurdzic M, Farinelli L, Martienssen RA, Hemerly AS, Ferreira PCG (2012) Computational identification and analysis of novel sugarcane microRNAs. BMC Genomics 13(1):290
Tran L-SP, Urao T, Qin F, Maruyama K, Kakimoto T, Shinozaki K, Yamaguchi-Shinozaki K (2007) Functional analysis of AHK1/ATHK1 and cytokinin receptor histidine kinases in response to abscisic acid, drought, and salt stress in *Arabidopsis*. Proc Natl Acad Sci 104(51):20623–20628
Trindade I, Capitão C, Dalmay T, Fevereiro MP, Dos Santos DM (2010) miR398 and miR408 are up-regulated in response to water deficit in *Medicago truncatula*. Planta 231(3):705–716
Tripathi DK, Singh VP, Gangwar S, Prasad SM, Maurya JN, Chauhan DK (2014) Role of silicon in enrichment of plant nutrients and protection from biotic and abiotic stresses. In: Improvement of crops in the era of climatic changes. Springer, p 39–56
Tripathi, A., Chacon, O., Singla-Pareek, S.L., Sopory, S.K., Sanan-Mishra N. (2017) Mapping the microRNA expression profiles in glyoxalase over-expressing salinity tolerant rice. Curr Genomics 18(999):1
Tuli R, Chakrabarty D, Trivedi PK, Tripathi RD (2010) Recent advances in arsenic accumulation and metabolism in rice. Mol Breed 26(2):307–323
Tuteja N, Sopory SK (2008) Chemical signaling under abiotic stress environment in plants. Plant Signal Behav 3(8):525–536
Tyczewska A, Gracz J, Kuczyński J, Twardowski T (2017) Deciphering soybean molecular stress response via high-throughput approach. Acta Biochim Pol 63 (4):631–643
Urano K, Yoshiba Y, Nanjo T, Igarashi Y, Seki M, Sekiguchi F, Yamaguchi-Shinozaki K, Shinozaki K (2003) Characterization of *Arabidopsis* genes involved in biosynthesis of polyamines in abiotic stress responses and developmental stages. Plant Cell Environ 26(11):1917–1926

Urao T, Katagiri T, Mizoguchi T, Yamaguchi-Shinozaki K, Hayashida N, Shinozaki K (1994) Two genes that encode Ca 2+−dependent protein kinases are induced by drought and high-salt stresses in *Arabidopsis thaliana*. Mol Gen Genet 244(4):331–340

Valdés-López O, Yang SS, Aparicio-Fabre R, Graham PH, Reyes JL, Vance CP, Hernández G (2010) MicroRNA expression profile in common bean (*Phaseolus vulgaris*) under nutrient deficiency stresses and manganese toxicity. New Phytol 187(3):805–818

Vaucheret H (2006) Post-transcriptional small RNA pathways in plants: mechanisms and regulations. Genes Dev 20(7):759–771

Vaucheret H, Vazquez F, Crété P, Bartel DP (2004) The action of Argonaute1 in the miRNA pathway and its regulation by the miRNA pathway are crucial for plant development. Genes Dev 18(10):1187–1197

Vazquez F, Vaucheret H, Rajagopalan R, Lepers C, Gasciolli V, Mallory AC, Hilbert J-L, Bartel DP, Crété P (2004) Endogenous trans-acting siRNAs regulate the accumulation of *Arabidopsis* mRNAs. Mol Cell 16(1):69–79

Voinnet O (2009) Origin, biogenesis, and activity of plant microRNAs. Cell 136(4):669–687

Waditee R, Bhuiyan MNH, Rai V, Aoki K, Tanaka Y, Hibino T, Suzuki S, Takano J, Jagendorf AT, Takabe T (2005) Genes for direct methylation of glycine provide high levels of glycinebetaine and abiotic-stress tolerance in *Synechococcus* and *Arabidopsis*. Proc Natl Acad Sci U S A 102(5):1318–1323

Waie B, Rajam MV (2003) Effect of increased polyamine biosynthesis on stress responses in transgenic tobacco by introduction of human S-adenosylmethionine gene. Plant Sci 164(5):727–734

Walker M, Dumbroff E (1981) Effects of salt stress on abscisic acid and cytokinin levels in tomato. Z Pflanzenphysiol 101(5):461–470

Wang Y, Mopper S, Hasenstein KH (2001) Effects of salinity on endogenous ABA, IAA, JA, and SA in *Iris hexagona*. J Chem Ecol 27(2):327–342

Wang W, Vinocur B, Altman A (2003) Plant responses to drought, salinity and extreme temperatures: towards genetic engineering for stress tolerance. Planta 218(1):1–14

Wang T, Chen L, Zhao M, Tian Q, Zhang W-H (2011) Identification of drought-responsive microRNAs in *Medicago truncatula* by genome-wide high-throughput sequencing. BMC Genomics 12(1):367

Wang Y, Sun F, Cao H, Peng H, Ni Z, Sun Q, Yao Y (2012) TamiR159 directed wheat TaGAMYB cleavage and its involvement in anther development and heat response. PLoS One 7(11):e48445. genomics 9 (4):499

Wang B, Sun YF, Song N, Wei JP, Wang XJ, Feng H, Yin ZY, Kang ZS (2014) MicroRNAs involving in cold, wounding and salt stresses in *Triticum aestivum* L. Plant Physiol Biochem 80:90–96

Wang Q, Liu N, Yang X, Tu L, Zhang X (2016) Small RNA-mediated responses to low-and high-temperature stresses in cotton. Sci Rep 6:35558

Wei J-Z, Tirajoh A, Effendy J, Plant AL (2000) Characterization of salt-induced changes in gene expression in tomato (*Lycopersicon esculentum*) roots and the role played by abscisic acid. Plant Sci 159(1):135–148

Wei L, Zhang D, Xiang F, Zhang Z (2009) Differentially expressed miRNAs potentially involved in the regulation of defense mechanism to drought stress in maize seedlings. Int J Plant Sci 170(8):979–989

Wen X-P, Pang X-M, Matsuda N, Kita M, Inoue H, Hao Y-J, Honda C, Moriguchi T (2008) Over-expression of the apple spermidine synthase gene in pear confers multiple abiotic stress tolerance by altering polyamine titers. Transgenic Res 17(2):251–263

Wernimont AK, Amani M, Qiu W, Pizarro JC, Artz JD, Lin YH, Lew J, Hutchinson A, Hui R (2011) Structures of parasitic CDPK domains point to a common mechanism of activation. Proteins 79(3):803–820

White PJ, Broadley MR (2003) Calcium in plants. Ann Bot 92(4):487–511

Willmann MR, Poethig RS (2007) Conservation and evolution of miRNA regulatory programs in plant development. Curr Opin Plant Biol 10(5):503–511

Wu G, Poethig RS (2006) Temporal regulation of shoot development in *Arabidopsis thaliana* by miR156 and its target SPL3. Development 133(18):3539–3547

Wu G, Park MY, Conway SR, Wang J-W, Weigel D, Poethig RS (2009) The sequential action of miR156 and miR172 regulates developmental timing in *Arabidopsis*. Cell 138(4):750–759

Wu Y, Vulić M, Keren I, Lewis K (2012) Role of oxidative stress in persister tolerance. Antimicrob Agents Chemother 56(9):4922–4926

Xia Q, Shi T, Liu S, Wang MY (2012) A level set solution to the stress-based structural shape and topology optimization. Comput Struct 90:55–64

Xie FL, Huang SQ, Guo K, Xiang AL, Zhu YY, Nie L, Yang ZM (2007) Computational identification of novel microRNAs and targets in *Brassica napus*. FEBS Lett 581(7):1464–1474

Xie K, Shen J, Hou X, Yao J, Li X, Xiao J, Xiong L (2012) Gradual increase of miR156 regulates temporal expression changes of numerous genes during leaf development in rice. Plant Physiol 158(3):1382–1394

Xie F, Jones DC, Wang Q, Sun R, Zhang B (2015) Small RNA sequencing identifies miRNA roles in ovule and fibre development. Plant Biotechnol J 13(3):355–369

Xin M, Wang Y, Yao Y, Xie C, Peng H, Ni Z, Sun Q (2010) Diverse set of microRNAs are responsive to powdery mildew infection and heat stress in wheat (*Triticum aestivum* L.). BMC Plant Biol 10(1):123

Xing D-H, Lai Z-B, Zheng Z-Y, Vinod K, Fan B-F, Chen Z-X (2008) Stress-and pathogen-induced *Arabidopsis* WRKY48 is a transcriptional activator that represses plant basal defense. Mol Plant 1(3):459–470

Xiong L, Wang R-G, Mao G, Koczan JM (2006) Identification of drought tolerance determinants by genetic analysis of root response to drought stress and abscisic acid. Plant Physiol 142(3):1065–1074

Xu J, Li Y, Wang Y, Liu H, Lei L, Yang H, Liu G, Ren D (2008) Activation of MAPK kinase 9 induces ethylene and camalexin biosynthesis and enhances sensitivity to salt stress in *Arabidopsis*. J Biol Chem 283(40):26996–27006

Yamaguchi S, Kamiya Y (2000) Gibberellin biosynthesis: its regulation by endogenous and environmental signals. Plant Cell Physiol 41:251

Yamaguchi-Shinozaki K, Shinozaki K (2005) Organization of cis-acting regulatory elements in osmotic-and cold-stress-responsive promoters. Trends Plant Sci 10(2):88–94

Yamaguchi-Shinozaki K, Shinozaki K (2006) Transcriptional regulatory networks in cellular responses and tolerance to dehydration and cold stresses. Annu Rev Plant Biol 57:781–803

Yan S, Tang Z, Su W, Sun W (2005) Proteomic analysis of salt stress-responsive proteins in rice root. Proteomics 5(1):235–244

Yan Y, Wang H, Hamera S, Chen X, Fang R (2014) miR444a has multiple functions in the rice nitrate-signaling pathway. Plant J 78(1):44–55

Yancey PH, Clark ME, Hand SC, Bowlus RD, Somero GN (1982) Living with water stress: evolution of osmolyte systems. Science 217(4566):1214–1222

Yang F, Yu D (2009) Overexpression of *Arabidopsis* MiR396 enhances drought tolerance in transgenic tobacco plants. Acta Bot Yunnanica 31(5):421–426

Yang C-W, González-Lamothe R, Ewan RA, Rowland O, Yoshioka H, Shenton M, Ye H, O'Donnell E, Jones JD, Sadanandom A (2006) The E3 ubiquitin ligase activity of *Arabidopsis* PLANT U-BOX 17 and its functional tobacco homolog ACRE276 are required for cell death and defense. Plant Cell 18(4):1084–1098

Yang J, Kloepper JW, Ryu C-M (2009) Rhizosphere bacteria help plants tolerate abiotic stress. Trends Plant Sci 14(1):1–4

Yang H, Jin X, Lam CWK, Yan S-K (2011) Oxidative stress and diabetes mellitus. Clin Chem Lab Med 49(11):1773–1782

Yang Y, He M, Zhu Z, Li S, Xu Y, Zhang C, Singer SD, Wang Y (2012) Identification of the dehydrin gene family from grapevine species and analysis of their responsiveness to various forms of abiotic and biotic stress. BMC Plant Biol 12(1):140

Yang C, Li D, Mao D, Liu X, Ji C, Li X, Zhao X, Cheng Z, Chen C, Zhu L (2013) Overexpression of microRNA319 impacts leaf morphogenesis and leads to enhanced cold tolerance in rice (*Oryza sativa* L.). Plant Cell Environ 36(12):2207–2218

Yang C, Ma B, He S-J, Xiong Q, Duan K-X, Yin C-C, Chen H, Lu X, Chen S-Y, Zhang J-S (2015) MHZ6/OsEIL1 and OsEIL2 regulate ethylene response of roots and coleoptiles and negatively affect salt tolerance in rice. Plant Physiol 169:148. pp. 00353.02015

Yeo A (1998) Molecular biology of salt tolerance in the context of whole-plant physiology. J Exp Bot 49(323):915–929

Yin Z, Li Y, Yu J, Liu Y, Li C, Han X, Shen F (2012) Difference in miRNA expression profiles between two cotton cultivars with distinct salt sensitivity. Mol Biol Rep 39(4):4961–4970

Yoshida S, Maruyama S, Nozaki H, Shirasu K (2010) Horizontal gene transfer by the parasitic plant *Striga hermonthica*. Science 328(5982):1128–1128

Yu B, Yang Z, Li J, Minakhina S, Yang M, Padgett RW, Steward R, Chen X (2005) Methylation as a crucial step in plant microRNA biogenesis. Science 307(5711):932–935

Yu S, Wang W, Wang B (2012) Recent progress of salinity tolerance research in plants. Russ J Genet 48(5):497–505

Yu F, Wu Y, Xie Q (2016) Ubiquitin–proteasome system in ABA signaling: from perception to action. Mol Plant 9(1):21–33

Zeng Q-Y, Yang C-Y, Ma Q-B, Li X-P, Dong W-W, Nian H (2012) Identification of wild soybean miRNAs and their target genes responsive to aluminum stress. BMC Plant Biol 12(1):182

Zhang C-s, Lu Q, Verma DPS (1997) Characterization of Δ 1-pyrroline-5-carboxylate synthetase gene promoter in transgenic *Arabidopsis thaliana* subjected to water stress. Plant Sci 129(1):81–89. (1):98-109

Zhang J, Jia W, Yang J, Ismail AM (2006) Role of ABA in integrating plant responses to drought and salt stresses. Field Crop Res 97(1):111–119

Zhang A, Jiang M, Zhang J, Ding H, Xu S, Hu X, Tan M (2007) Nitric oxide induced by hydrogen peroxide mediates abscisic acid-induced activation of the mitogen-activated protein kinase cascade involved in antioxidant defense in maize leaves. New Phytol 175(1):36–50

Zhang J, Xu Y, Huan Q, Chong K (2009) Deep sequencing of *Brachypodium* small RNAs at the global genome level identifies microRNAs involved in cold stress response. BMC Genomics 10(1):449

Zhang X, Wang L, Meng H, Wen H, Fan Y, Zhao J (2011) Maize ABP9 enhances tolerance to multiple stresses in transgenic *Arabidopsis* by modulating ABA signaling and cellular levels of reactive oxygen species. Plant Mol Biol 75(4–5):365–378

Zhang Y-C, Yu Y, Wang C-Y, Li Z-Y, Liu Q, Xu J, Liao J-Y, Wang X-J, Qu L-H, Chen F (2013) Overexpression of microRNA OsmiR397 improves rice yield by increasing grain size and promoting panicle branching. Nat Biotechnol 31(9):848–852

Zhao B, Liang R, Ge L, Li W, Xiao H, Lin H, Ruan K, Jin Y (2007) Identification of drought-induced microRNAs in rice. Biochem Biophys Res Commun 354(2):585–590

Zhao B, Ge L, Liang R, Li W, Ruan K, Lin H, Jin Y (2009) Members of miR-169 family are induced by high salinity and transiently inhibit the NF-YA transcription factor. BMC Mol Biol 10(1):29

Zhao M, Ding H, Zhu JK, Zhang F, Li WX (2011) Involvement of miR169 in the nitrogen-starvation responses in *Arabidopsis*. New Phytol 190(4):906–915

Zheng X-y, Spivey NW, Zeng W, Liu P-P, Fu ZQ, Klessig DF, He SY, Dong X (2012) Coronatine promotes *Pseudomonas syringae* virulence in plants by activating a signaling cascade that inhibits salicylic acid accumulation. Cell Host Microbe 11(6):587–596

Zheng L, Liu G, Meng X, Liu Y, Ji X, Li Y, Nie X, Wang Y (2013) A WRKY gene from *Tamarix hispida*, ThWRKY4, mediates abiotic stress responses by modulating reactive oxygen species and expression of stress-responsive genes. Plant Mol Biol 82(4–5):303–320

Zhifang G, Loescher W (2003) Expression of a celery mannose 6-phosphate reductase in *Arabidopsis thaliana* enhances salt tolerance and induces biosynthesis of both mannitol and a glucosyl-mannitol dimer. Plant Cell Environ 26(2):275–283

Zholkevich V, Pustovoytova T (1993) The role of *Cucumis sativum* L leaves and content of phytohormones under soil drought. Russ J Plant Physiol 40:676–680

Zhou X, Wang G, Sutoh K, Zhu J-K, Zhang W (2008a) Identification of cold-inducible microRNAs in plants by transcriptome analysis. BBA-Gene Regul Mech 1779(11):780–788

Zhou ZS, Huang SQ, Yang ZM (2008b) Bioinformatic identification and expression analysis of new microRNAs from *Medicago truncatula*. Biochem Biophys Res Commun 374(3):538–542

Zhou J, Wang X, He K, Charron J-BF, Elling AA, Deng XW (2010) Genome-wide profiling of histone H3 lysine 9 acetylation and dimethylation in *Arabidopsis* reveals correlation between multiple histone marks and gene expression. Plant Mol Biol 72(6):585–595

Zhou J, Liu M, Jiang J, Qiao G, Lin S, Li H, Xie L, Zhuo R (2012a) Expression profile of miRNAs in *Populus cathayana* L. and *Salix matsudana* Koidz under salt stress. Mol Biol Rep 39(9):8645–8654

Zhou ZS, Song JB, Yang ZM (2012b) Genome-wide identification of *Brassica napus* microRNAs and their targets in response to cadmium. J Exp Bot 63(12):4597–4613

Zhou M, Li D, Li Z, Hu Q, Yang C, Zhu L, Luo H (2013) Constitutive expression of a miR319 gene alters plant development and enhances salt and drought tolerance in transgenic creeping bentgrass. Plant Physiol 161(3):1375–1391

Zhu J-K (2002) Salt and drought stress signal transduction in plants. Annu Rev Plant Biol 53(1):247–273

Chapter 8
Genomic Roadmaps for Augmenting Salinity Stress Tolerance in Crop Plants

P. Suprasanna, S. A. Ghuge, V. Y. Patade, S. J. Mirajkar, and G. C. Nikalje

Abstract Serious antagonistic impacts of saline environment on plant growth, development, and yield are well established. In this regard, researchers and breeders have been utilizing many conventional as well as modern approaches to aid the process of developing salt-tolerant crops. Biotechnological tools have made the task of engineering salinity tolerance in plants easier. Currently, two major annexes are effectively employed to develop salt-tolerant crops, first, investigation of genetic variation via marker-assisted selection (MAS) and second the transgenic technology. Sustenance of plants under dynamically growth-limiting saline environment depends on alterations and/or switching between multiple biochemical pathways involved in response. A number of key regulatory genes have been successfully identified and characterized in this context which can be explored to serve the purpose of alleviation in salt-tolerant nature of plants. Several genomics-abetted approaches have been reported aiming toward improvement in growth and yield of crops under saline environment. Present chapter focuses on genomic roadmaps for augmentation of crop salt tolerance by various methods including MAS, transgenic breeding, manipulations in small non-coding RNAs, and genome editing. These approaches utilize key players involved in salinity-mediated plant defense mechanisms, such as ion transporters,

P. Suprasanna (✉)
Nuclear Agriculture and Biotechnology Division, Bhabha Atomic Research Centre (BARC), Mumbai, India

S. A. Ghuge
Division of Biochemical Sciences, National Chemical Laboratory (NCL), Pune, India

V. Y. Patade
Defence Research & Development Organisation (DRDO), Defence Institute of Bio-Energy Research (DIBER) Field Station, Pithoragarh, Uttarakhand, India

S. J. Mirajkar
Division of Vegetable Science, ICAR-Indian Agricultural Research Institute (IARI), Pusa Campus, New Delhi, India

G. C. Nikalje
Department of Botany, R.K. Talreja College of Arts, Science and Commerce, Ulhasnagar, Thane, India

V. Kumar et al. (eds.), *Salinity Responses and Tolerance in Plants, Volume 2*,
https://doi.org/10.1007/978-3-319-90318-7_8

osmolytes, antioxidants, transcription factors, signaling proteins, and microRNA. The chapter attempts to summarize the effective targets and exploration of these key entities to raise salt-tolerant plants through various genomics-related tools.

Keywords Marker assisted selection · Ion transporters · Osmolytes · Antioxidants · Transcription factors · microRNA · Transgenics

Abbreviations

AFLP	Amplified fragment length polymorphisms
AOX	Alternate oxidase
APX	Ascorbate peroxidase
AtNHX1	Na+/H+ antiporter
CaM	calmodulin
CAT	Catalase
CBL	Calcineurin B-like proteins
CDPKs	Calcium-dependent protein kinases
CML	CaM-related proteins
GPX	Glutathione peroxidase
ILs	Introgression lines
MAS	Marker-assisted selection
MQTL	Meta-QTL
mt1D	Mannitol-1-phosphate dehydrogenase
P5CS	delta1-pyrroline-5-carboxylate synthetase
QTL	Quantitative trait loci
RAPD	Random amplified polymorphic DNA
RFLP	Restriction fragment length polymorphisms
RNAi	RNA interference
SNPs	Single nucleotide polymorphisms
SOD	Superoxide dismutase
SOS	Salt overly sensitive
SSR	Simple sequence repeats
STMS	Sequence-tagged microsatellite site
TFs	Transcription factors
TPSP	Trehalose-6-phosphate synthase/phosphatase

8.1 Introduction

Among the abiotic stresses, salinity stress is one of the most important environmental factors which considerably affect plant growth and productivity. Salinity affects about one third of the world's irrigated land (Munns and Tester 2008), and it negatively influences water and nutrient homeostasis within living tissues. The deleterious effects on agricultural crops primarily include growth reduction and yield loss. In this context, both the conventional and modern crop improvement approaches are employed to facilitate development of novel genetic resources for use in direct or indirect breeding for improving salinity tolerance in crop plants. Currently, a wide range of mutational, biotechnological, and genomics-assisted tools are available which are more or less focused on gene discovery and boosting up the process of novel gene introduction or modification (Nongpiur et al. 2016).

Apparently, two main approaches are used to improve and impart salinity tolerance in crop plants. The first is through exploring natural genetic variation, either through selection under stress conditions or through quantitative trait loci (QTL) followed by marker-assisted selection (MAS). The other one is through transgenic technology by modifying the expression of endogenous genes or introducing novel genes (of plant or non-plant origin) to impart stress tolerance. Crop improvement via conventional breeding approaches has yielded limited success due to complexity of the trait since the process is time and labor intensive and requires well-characterized germplasm. In this regard, genetic engineering methods have become useful to develop transgenic crops tolerant to abiotic stresses (Yamaguchi and Blumwald 2005). The primary step before proceeding to make transgenics is the identification of functional and regulator genes serving to control different metabolic pathways, including ion homeostasis, antioxidant defense system, osmolyte synthesis, and other signaling pathways.

Salt stress increases ion toxicity and also affects uptake and movement of other essential nutrients such as potassium in the cell. This may occur either in a monophasic or biphasic manner depending on the duration and extent of exposure to saline conditions. A short exposure usually leads to osmotic or oxidative stress which would be followed by ionic stress upon long-term exposure (Munns and Tester 2008). To sustain under such dynamic growth-limiting situations, plants need to incur switching between multiple biochemical pathways that are much more complex when combined with other biotic and abiotic stresses. A general view of plant responses to salinity stress is presented in Fig. 8.1.

Significant progress has been made in the identification of genes involved in plant salt-stress responses (Hanin et al. 2016). Till date, a number of key genes involved in salinity tolerance have been isolated, characterized, and validated by using different transgenic methods. The candidate genes for salt tolerance are categorized into genes with functional and regulatory role (Shinozaki et al. 2003). The first group includes those involved in osmolyte biosynthesis, ion transporters, water channels, antioxidant systems, sugars, polyamines, heat shock proteins, and late embryogenesis abundant proteins. The second group are involved in the regulation of transcriptional and posttranscriptional machinery besides genes of signaling pathways. Some of these are transcription factors (TFs), protein kinases and phosphatases. In addition, there are several other strategies for attaining abiotic stress tolerance which are being tested for salt tolerance such as, using the stress-inducible promoters to avoid the pleiotropic effects (Checker et al. 2012), employing the pro-

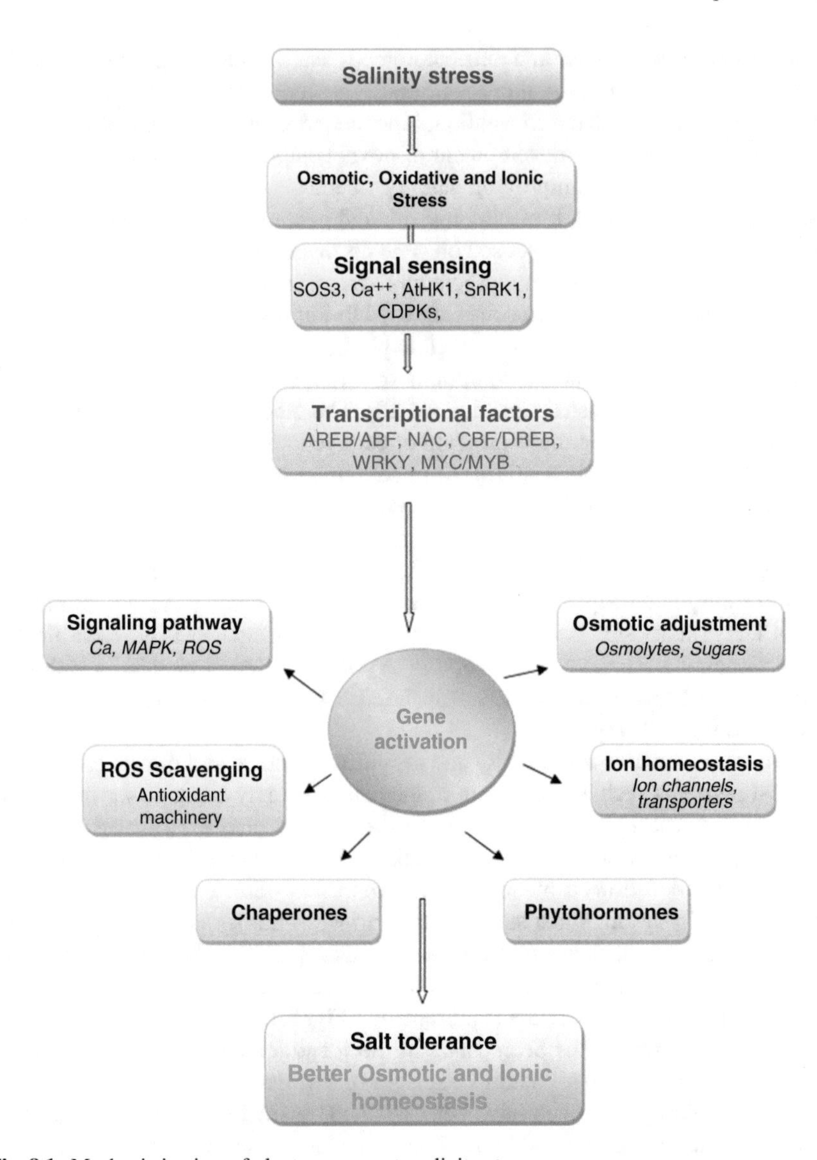

Fig. 8.1 Mechanistic view of plant responses to salinity stress

tein post-translational modifications such as ubiquitination (Lyzenga and Stone 2012; Guo et al. 2008) and the use of halophyte gene resources.

Several state-of-the-art genomics-assisted approaches (Fig. 8.2), such as transgenic overexpression, RNAi, microRNA, genome editing, and genome-wide association studies, are being used for improving salt tolerance in crop plants (Mickelbart et al. 2015; Nongpiur et al. 2016). Overexpression of these genes has been shown as a successful strategy to improve plant tolerance to different abiotic stresses including salinity (Türkan and Demiral 2009; Cominelli et al. 2013; Hanin et al. 2016). In

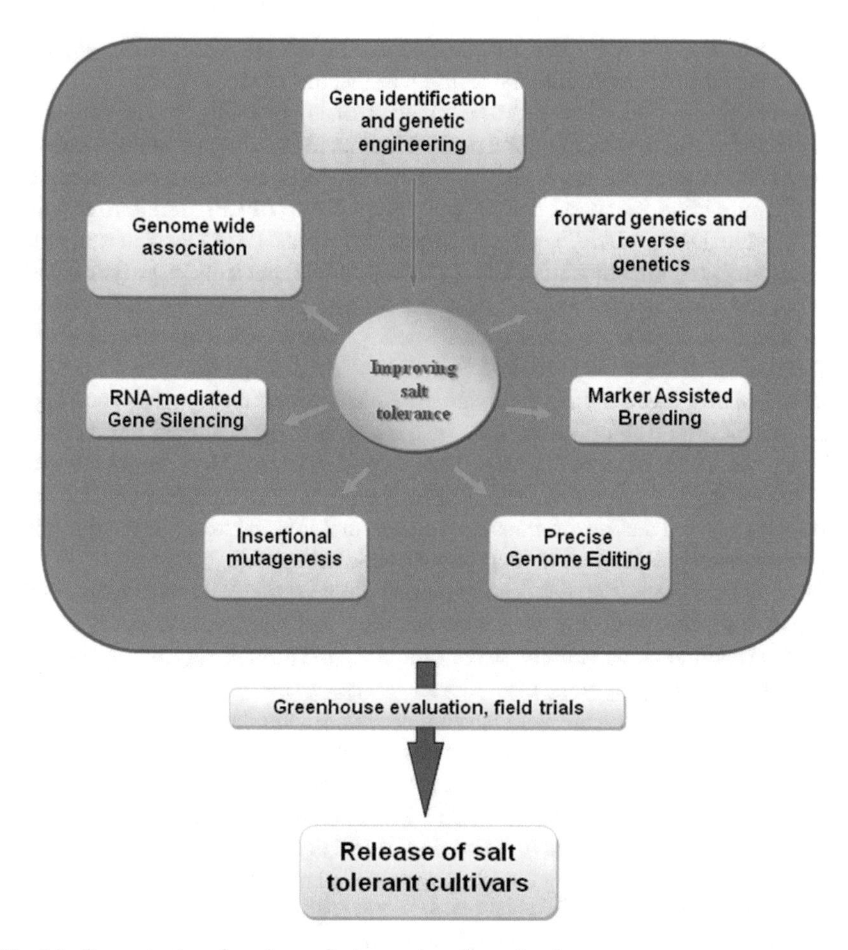

Fig. 8.2 Genomics-based roadmaps for improving plant salt tolerance

this article, we present an overview of the different genomics-based molecular genetic approaches (marker-assisted selection and transgenic breeding) that have contributed to improve the salt tolerance in crop plants.

8.2 Marker-Assisted Selection for Enhancement of Crop Salinity Stress Tolerance

Despite of availability of broad genetic resources, the slow progress in the genetic improvement for salt tolerance through conventional breeding is attributed to the complex nature of the trait accompanied with the high environmental influence and requirement of huge experimental fields (Flowers and Yeo 1997; Munns 2002; Thomson et al. 2010; Munns and Tester 2008). Further, phenotypic screening for salinity tolerance in the large sample population has remained a big challenge through the laborious conventional techniques (Mantri et al. 2014).

Marker-assisted selection (MAS) is a new precision breeding tool that allows the indirect and accurate selection for a desired trait from breeding population based on tightly linked molecular markers, viz., restriction fragment length polymorphisms (RFLP), amplified fragment length polymorphisms (AFLP), random amplified polymorphic DNA (RAPD), simple sequence repeats (SSR) or microsatellites, sequence-tagged microsatellite site (STMS), single nucleotide polymorphisms (SNPs), etc. It enables rapid and accurate screening for complex and polygenic traits which are difficult to score phenotypically through mapping or tagging of the trait linked quantitative trait loci (QTL). The successful applications of molecular marker-assisted breeding have proved its enhanced efficiency and accuracy in improved biotic and abiotic stress tolerance in rice and several other important crops (Singh et al. 2012; Ellur et al. 2016; Babu et al. 2017a, b). MAS offers advantage over the other genetic improvement tools as having relaxed biosafety regulations at development, field testing, commercial release, and import/export of the developed improved genotypes as well as their wider public acceptance.

Among field crops, rice being an important global staple food crop, considerable progress has been made for molecular breeding for improvement in tolerance to abiotic stresses such as salinity stress (Table 8.1). Through rigorous research on molecular breeding programs, the molecular marker maps for important agricultural crops have been constructed with varying density among the species. For development of molecular breeding tools, sources for abiotic stress tolerance have been identified through rigorous screening of genotypes in various crops. Ravikiran et al. (2017) identified two rice genotypes, CST 7–1 and Arvattelu as source for seedling-stage salinity tolerance, based on screening of 192 diverse genotypes under salinity stress (EC ~ 12 dS m^{-1}) using morphophysiological markers. Screening of the genotypes with SSR markers associated with Saltol region on chromosome 1 revealed RM 493 and RM 10793 as good candidates for marker-assisted selection of seedling-stage salinity tolerance. Linh et al. (2012) reported improved salt tolerance in high-yielding Bac Thom 7 rice cultivar through introgression of the Saltol QTL from donor parent FL478. The microsatellite markers, viz., RM493 and RM3412b tightly linked to the Saltol QTL, were used for foreground selection. The selected back-cross lines displayed salt tolerance with agronomic performance similar to that of the original Bac Thom 7. The marker-assisted selection enabled rapid and efficient background (for the recurrent parent genome) and foreground (for target locus Saltol) selections in early generations with minimum linkage drag.

Another study by Babu et al. (2017a) reported use of marker-assisted backcrossing to transfer seedling-stage salt-stress tolerance by transferring a QTL, Saltol, into an elite salinity-sensitive rice cultivar Pusa Basmati 1121. RM 3412 STMS marker linked tightly to the QTL was used for indirect foreground selection. Back cross (BC) lines homozygous for the QTL were advanced to develop four improved near isogenic lines (NILs) of PB1121 with the salt tolerance. The field evaluation confirmed effect of QTL integration into the improved NILs in terms of greater salt

Table 8.1 Successful examples of breeding for salinity stress tolerance through marker-assisted selection

Sr. No.	Crop species	Type of DNA marker	Donor source	QTL/Allele	Mapping population	Tolerance achieved	References
1	Rice	SSR	FL478	qSaltol	Back cross	Improved salt tolerance in high-yielding cultivar bac Thom 7	Linh et al. (2012)
2	Rice	SSR	FL478	qSaltol	Back cross	Improved salt tolerance in BRRI dhan49	Hoque et al. (2015)
3	Soybean	SSR	Tolerant wild accession JWS156–1	*Ncl*	Back cross	Improved salt tolerance in salt-sensitive soybean cultivar Jackson (PI 548657)	Do et al. (2016)
4	Rice	Sequence-tagged microsatellite site (STMS)	FL478	qSaltol	NIL	Salt tolerance in sensitive Pusa Basmati 1121	Babu et al. (2017a)
5	Rice	SNP	Hasawi	qSESI12.1 and qSESF12.1	RIL	Seedling-stage salt tolerance in RIL with IR29	Bizimana et al. (2017)
6	Rice	SNP	DJ15	qST1.2 and qST6	RIL/NIL	Seedling salt tolerance enhanced in sensitive japonica rice variety Koshihikari	Quan et al. (2018)

tolerance at seedling stage and similar or better performance for other agronomic traits than the recurrent parent.

Recently, De Leon et al. (2017) utilized SSR and SNP markers to characterize introgression lines (ILs) of a high salinity-tolerant donor line Pokkali in an elite highly salt-sensitive rice cultivar Bengal and to further identify QTLs for traits contributing to salinity stress tolerance. As expected, because of abundance, more number of QTLs were detected using SNP markers than the SSR. The study emphasized marker-assisted breeding through introgression of salt injury score (SIS) QTLs, in addition to other major QTLs Saltol or qSKC1, for improved salinity tolerance. The identified tolerant ILs can be used as donor breeding lines for selective transfer of salinity tolerance without any linkage drag of undesirable traits from Pokkali to other recipient-sensitive varieties as well as for mapping and further positional cloning of the genes responsible for the trait.

Babu et al. (2017b) identified seedling-stage salt-tolerant *indica* landraces (Badami, Shah Pasand and Pechi Badam), *Oryza rufipogon* accessions (NKSWR2 and NKSWR17) and one each of Basmati rice (second Basmati) and *japonica* cultivars (Tompha Khau) based on phenotypic screening under hydroponics. The salt tolerance level was similar to that of high salt-tolerant genotypes Pokkali and FL478. Molecular diversity study of the diverse rice genotypes using polymorphic SSR markers linked with *Saltol* QTL revealed weak linkage disequilibrium-LD, suggesting its low usefulness in MAS, if the target foreground markers chosen are wide apart. LD mapping identified markers (RM10927, RM10871) linked with QTLs associated with salt tolerance traits. The study also identified *Saltol* marker, RM27, positioned on chromosome 10, associated with root Na/K ratio.

Efforts are also being made to identify novels QTLs for salinity tolerance from different sources in rice and other crops. The enhanced salt tolerance can be achieved through pyramiding of different novel QTLs in one genetic background through MAS. Bizimana et al. (2017) used Hasawi rice genotype, which conferred seedling-stage salinity tolerance due to novel QTLs other than Saltol, as a source to develop 300 recombinant inbred lines with high-yielding salt-sensitive cultivar-IR29. Further for identification of QTLs linked to salinity tolerance, a genetic linkage map was constructed using 194 polymorphic SNP markers. The study reported identification of 20 new QTLs on different chromosomes for salt tolerance through composite interval mapping.

In addition to rice, efforts are also being made for analysis of QTLs for breeding salt tolerance in other crops. In cotton, Zhao et al. (2016) identified salt-tolerant and salt-sensitive upland cotton cultivars through screening based on seedling emergence rates in response to 0.3% salt-NaCl. Seventy-four SSR markers were used to scan the genomes of these diverse cultivars, and eight markers associated with salt tolerance were identified through association analysis for further application in marker-assisted breeding. Similarly, Kere et al. (2017) screened salt-sensitive-11439S and salt-tolerant-11411S inbred parental lines with SSR markers to identify the QTLs for application in MAS for breeding salinity tolerance in cucumber. The analysis confirmed significant association of SSR markers with salt tolerance traits such as survival rate, relative leaf numbers, and percent green leaves, and salinity tolerance was evaluated by visual scoring. Recently, Luo et al. (2017) made efforts to map the critical QTLs contributing to salt tolerance in field-grown mature maize plants using a permanent doubled-haploid (DH) population and high-density SNP markers. Major QTLs responsible for salt tolerance and two candidate genes involving in ion homeostasis were mapped on chromosome 1. The mapped QTLs can be used in breeding salt-tolerant maize varieties through MAS.

Physiological and molecular studies on tolerance to various abiotic (ionic and/or osmotic stresses), viz., salinity, drought, etc., have revealed stress-specific as well as shared stress adaptation mechanisms, highlighting the complexity of stress response and adaptation in plants. In view of this, meta-QTL (MQTL) for tolerance to abiotic stresses including drought, salinity, and water logging through meta-analysis in barley has been recently projected (Zhang et al. 2017). The study conducted meta-analysis to detect and map the major QTL for drought, salinity, and water logging tolerance from different mapping populations on the barley physical map. Fine-mapped QTL for the stress tolerance were validated on MQTLs for further successful MAS in barley breeding.

8.3 Transgenic Breeding: Functional Genes Conferring Salinity Tolerance

8.3.1 Ion Transporters

In general, plants cannot withstand high salt concentration although the plant species differ in their mode of responses to the external salt exposure. Several distinct responses are generated in the plants to avoid high-salinity-induced harmful effects. One of the most distinguishing responses is avoiding salinity stress by compartmentation and the exclusion of detrimental ions like Na^+ and Cl^- from tissues that are very sensitive like the mesophyll and their relocation into the apoplast or vacuole (Sperling et al. 2014). Confinement of harmful ions within a root or apoplastic zone and maintainance of high K^+/Na^+ ratio are the major tolerance strategies for salt tolerance (Shabala and Cuin 2008).

Ion transporters are key players in maintaining ion homeostasis and in salt detoxification processes (Serrano et al. 1999; Hasegawa 2013). Various salts are present in soil out of which sodium chloride (NaCl) is the most significant. The Na^+/H^+ antiporter predominantly transports Na^+ ion from cytoplasm to the vacuole. Therefore, overexpression of genes that are involved in Na^+ transport was studied to a great extent with considerable success. Vacuolar Na^+/H^+ antiporter (*AtNHX1*) from the *Arabidopsis* was among the first and most studied gene. In *Arabidopsis* salt tolerance was conferred by overexpressing vacuolar Na^+/H^+ antiporter (Apse et al. 1999). Following with this initial success, many events were reported where transgenic plants exhibited higher potential for vacuolar sequestration of Na^+ that subsequently avoid its harmful buildup into the cytoplasm. For example, overexpression of *AtNHX1* and other NHX proteins from various hosts in many other plant species like tomato, *B. napus*, wheat, and cotton has been shown to increase salt tolerance (Zhang and Blumwald 2001; Zhang et al. 2001; Xue et al. 2004; He et al. 2005; Munns and Tester 2008). Vacuolar-type H^+-ATPase and the vacuolar pyrophosphatase are the two types of H^+ pumps that are present in vacuolar membrane (Dietz et al. 2001; Otoch et al. 2001; Wang et al. 2001). Overexpression of genes from wheat (*Triticum aestivum*) *TaNHX1* and H^+-pyrophosphatase (*TVP1*) resulted in improved salinity stress tolerance in *Arabidopsis* (Brini et al. 2007a). Similarly improved tolerance to salt stress was found in tobacco (Gouiaa et al. 2012) and tomato (Gouiaa and Khoudi 2015) by overexpression of Na(+)/H(+) antiporter H(+)- pyrophosphatase. On the other hand, the *HKT* gene family has a major role in preventing Na^+ ion toxicity in shoots by root-to-shoot partitioning of Na^+. The significant role of *HKT*s in Na^+ transport in plants makes them promising candidates to enhance salinity tolerance. The identification of the wheat *HKT1* (*TaHKT2;1*) gene

(Schachtman and Schroeder 1994; Rubio et al. 1995) has steered the study of many HKT genes from several other crops (Horie et al. 2009). Moller et al. (2009) shown that targeted overexpression of *AtHKT1;1* in the stele enhances salt tolerance in *A. thaliana*. Lately, Do et al. (2016) showed a close syndicate between the higher expression of the *Ncl* gene (homologous to the Na^+/H^+ antiporter gene family) in the root, the lesser buildup of Na^+, K^+, and Cl^- in the shoot under salt stress. Overexpression of *Ncl* into a Japanese soybean cultivar Kariyutaka resulted in enhanced salt tolerance (Do et al. 2016).

Growing evidence was found on the role of salt overly sensitive (*SOS*) stress signaling pathway in ion homeostasis and salinity tolerance (Sanders 2000; Hasegawa et al. 2000). Most of the *SOS* signaling pathway was reported to be involved in exportation of Na^+ out of the cell. The *SOS* signaling pathway includes three main proteins, namely, SOS1, SOS2, and SOS3. Plasma membrane-localized Na^+/H^+ antiporter (SOS1) is also known as NHX7 (Qiu et al. 2002). SOS2 (serine/threonine kinase) activated by salt stress elicited Ca^+ signals, while SOS3 protein is a myristoylated Ca^+ binding protein (Liu et al. 2000; Ishitani et al. 2000). *Arabidopsis* plants overexpressing genes for *SOS1*, *SOS3*, *AtNHX1* + *SOS3*, *SOS2* + *SOS3*, or *SOS1* + *SOS2* + *SOS3* resulted in improved tolerance to salt stress (Yang et al. 2009). Similarly, Kumar et al. (2009) have shown that salinity stress tolerance in *Brassica* is correlated with transcript abundance of the genes related in *SOS* pathway.

8.3.2 Osmolytes

Osmolytes are small organic compounds having a protective role. Osmolytes are important for two functional roles: osmotic adjustment at high concentrations; and it plays unknown protective role at lower concentrations. Under salt-stress conditions, plant cell accumulates various osmolytes along with Na^+ exclusion from the cytoplasm, to counter the osmotic pressure of harmful ions in vacuoles. Osmolytes like proline, glycine betaine, and sucrose accumulating upon salt stress in many plant species including halophytes are well studied and characterized (Flowers et al. 1977). Table 8.2 presents some of the successful examples of transgenic plants developed using different osmolyte genes.

Hu et al. (2015) found experimental evidences for the accumulation of sugars and amino acids. Particularly, sucrose and trehalose sugar and amino acids like proline, valine, glutamate, asparagine, glutamine, phenylalanine, and lysine accumulated under salt-stress conditions. Also, sugars like sucrose and pinitol are accumulated more in leaves, while starch accumulated in roots under salinity stress conditions. It has been found that these sugars (pinitol and sucrose) and starch can also increase in nodules under salt stress (Bertrand et al. 2015). Similarly, Boriboonkaset et al. (2013) found enrichment of soluble starch and soluble sugar in flag leaf of Pokkali genotype (salt tolerant) of rice which may have alternative role in osmotic adjustment in salt defense mechanism. In tomato plants, jasmonic acid and nitric oxide when applied exogenously, either individually or in combination,

Table 8.2 Example of functional genes used in the improvement of salt-stress tolerance of crop plants

Possible role	Gene(s)	Donor	Transgenic plant	References
Proline biosynthesis	*P5CS*	*Arabidopsis*	Tobacco	Kishor et al. (1995)
Proline biosynthesis	*P5CS*	*Vigna aconitifolia*	Tobacco	Hong et al. (2000)
Proline biosynthesis	*P5CS*	Moth bean	Rice	Su and Wu (2004)
Proline biosynthesis	*P5CS*	*Phaseolus vulgaris*	*Arabidopsis*	Chen et al. (2013)
Proline biosynthesis	*P5CSF129A*	*Sorghum bicolor*	Sorghum	Reddy et al. (2015)
Glycine betaine Biosynthesis	*codA*	*E. coli*	Rice	Sakamoto et al. (1998)
Mannitol biosynthesis	*mt1D*	*E. coli*	Tobacco	Tarczynski et al. (1992)
Mannitol biosynthesis	*mt1*	*E. coli*	Wheat	Abebe et al. (2003)
Vacuolar sequestration of Na + and K+	*TNHX1 and H(+)-*	*Triticum*	*Arabidopsis*	Brini et al. (2007a)
	PPase TVP1	*Triticum aestivum*	Tobacco	Gouiaa et al. (2012)
Vacuolar sequestration of Na + and K +?	*Na+/H+ antiporter* *AtNHX1*	*Arabidopsis*	Tomato	Zhang and Blumwald (2001)
			Brassica napus	Zhang et al. (2001)
			Wheat	Xue et al. (2004)
			Cotton	He et al. (2005)
Vacuolar	*H + −pyrophosphatase*		Cotton	Pasapula et al. (2011)
Membrane-bound proton pump	*(AVP1) AtNHX + AVP1*	*Arabidopsis*	Barley	Schilling et al. (2013)
			Cotton	Shen et al. (2014)
			Tomato	Gouiaa and Khoudi (2015)
Homologous to NHX gene family	*Ncl*	*Glycine max*	Soybean	Do et al. (2016)
Enhanced salinity tolerance	*AtNHX1 and SOS*	*Arabidopsis*	*Arabidopsis*	Yang et al. (2009)
Tolerance against salt and chilling stress	*Glutathione S- transferase(GST)*	Tobacco	Tobacco	Roxas et al. (1997)
	Glutathione peroxidase (GPX)			
ROS-scavenging	*Ascorbate peroxidase (AtAPX)*	*Arabidopsis*	Tobacco	Badawi et al. (2004)
Higher activity of SOD	*Cytosolic copper/zinc superoxide dismutase (CuZnSOD)*	*Avicennia marina*	Rice	Prashanth et al. (2008)
SOD and APX	*Cu/Zn sod (cytsod)*	*Spinacia oleracea*	Toabcco	Faize et al. (2011)
Activity	*Cytosolic apx1 (cytapx)*	*Pisum sativum*		
Osmoprotection	*Late embryogenesis abundant protein (HVA7)*	*Hordeum vulgare*	Rice	Xu et al. (1996)

(continued)

Table 8.2 (continued)

Possible role	Gene(s)	Donor	Transgenic plant	References
Ononitol production	*imt1*	*M. crystallinum*	Tobacco	Sheveleva et al. (1997)
Spermine and spermidine decarboxylase	*S-adenosyl methionine*	Tritordeum	Rice	Roy and Wu (2002)
(SAMDC) production				
Salt and osmotic	*Dehydrin (DHN-5)*	*Triticum*	*Arabidopsis*	Brini et al. (2007b)
Stress tolerance				
In ABA biosynthesis and xanthophyll cycle; enhanced salt tolerance	*Zeaxanthin epoxidase (AtZEP)*	*Arabidopsis*	*Arabidopsis*	Park et al. (2008)
Increased proline	*Osmotin*	Tobacco	Tomato	Goel et al. (2010)
C-tocopherol production; Enhanced salt tolerance	*γ-Tocopherol methyl transferase (γ -TMT)*	*Arabidopsis*	*Brassica juncea*	Yusuf et al. (2010)
Maintaining chlorophyll in salt stress	*Xyloglucan endotrans-glucosylase/hydrolase (CaXTH3)*	Hot pepper	Tomato	Choi et al. (2011)
Increase in germination, chlorophyll and osmotic constituents like sugars	*Dehydration-responsive RD22*	*Vitis vinifera*	Tobacco	Jamoussi et al. (2014)

helped to boost the proline, flavonoid, and glycine betaine synthesis under NaCl salt treatments (Ahmad et al. 2018).

Engineering plants by overexpressing the osmolytes was considered as one of the ways to enhance salt tolerance in plants. In *Arabidopsis*, knockout of the *P5CS1*, a key gene in proline biosynthesis, which encodes a *delta1-pyrroline-5-carboxylate synthetase* (*P5CS*), impairs proline synthesis resulting in salt hypersensitivity (Székely et al. 2008). *P5CS* transformed in tobacco and rice has shown increased proline production, linked with increased salt-stress tolerance (Kishor et al. 1995; Su and Wu 2004). Also, transgenic rice expressing the moth bean *P5CS* gene showed enhanced tolerance to higher dose of NaCl (Su and Wu 2004). Recently, mutated *P5CS* (*P5CSF129A*) gene was overexpressed in *Sorghum* and found that transgenic plants accumulated more proline and showed salt-stress tolerance. Moreover over-production of proline through transfer of a *P5CSF129A* gene conferred protection of photosynthetic and antioxidant enzyme activities (Reddy et al. 2015). Glycine betaine is yet another important osmolyte that helps to balance the osmotic potential of intracellular ions under salinity. Under high salinity, glycine betaine accumulation increased in lamina leaves and bladder hairs of *Atriplex gmelini* (Tsutsumi et al. 2015). Overexpressing c*holine oxidase* in rice plant showed increased levels of glycine betaine and improved tolerance to salt and cold stress (Sakamoto et al. 1998). It was found that transgenic *Arabidopsis* and tobacco plants transformed with bacterial *mtlD* gene which encodes for *mannitol-1-phosphate dehydrogenase* conferred salt tolerance and thereby maintained normal growth and development under high salt-stress growth conditions (Binzel et al. 1998; Thomas et al. 1995). Ectopic expression of bacterial gene *mannitol-1-phosphate dehydrogenase* (*mt1D*), an enzyme involved in mannitol biosynthesis, in tobacco successfully enhanced salt tolerance (Tarczynski et al. 1992). Genes for trehalose biosynthesis have also been employed in improving salt tolerance by developing transgenic plants for overproduction of trehalose (Penna 2003; Turan et al. 2012). Garg et al. (2002) demonstrated tolerance to salt and drought stress in rice by using tissue-specific or stress-inducible expression of a bifunctional trehalose-6-phosphate synthase/phosphatase (*TPSP*) fusion gene (comprising the *E. coli* trehalose biosynthetic genes). Li et al. (2011) reported that transgenic plants overexpressing rice trehalose-6-phosphate synthase (*OsTPS1*) showed improved salinity tolerance without much alteration in plant phenotype. It is also suggested that stress-inducible solute accumulation by using stress-specific or stress-inducible promoters may be better to achieve salt-specific expression of genes for osmotic adjustment.

8.3.3 Antioxidants and Protective Proteins

Abiotic stress causes the accumulation of reactive oxygen species (ROS) that can damage sensitive plant tissues during high salt stress by disturbing cell wall, enzymes, and membrane functions. Antioxidant enzymes and nonenzymatic compounds play a crucial role in detoxifying salinity stress-induced ROS. Salt-stress tolerance is positively correlated with antioxidant enzyme activity and with the accumulation of nonenzymatic antioxidant compounds (Gupta et al. 2005). Antioxidants include superoxide dismutase (SOD), catalase (CAT), glutathione peroxidase (GPX), ascorbate peroxidase (APX), glutathione reductase (GR), etc. Hence, overexpressing ROS-scavenging enzymes is shown to induce salinity tolerance in plants (Roxas et al. 1997; Badawi et al. 2004; Miller et al. 2008). Overexpression of ascorbate peroxidase (APX) in tobacco chloroplasts enhances the tolerance to salinity and drought stress (Badawi et al. 2004). The alternate oxidase (AOX) pathway plays a role under stress conditions. Smith et al. (2009) constitutively overexpressed an AOX1a gene in *Arabidopsis* plants and demonstrated superior salt tolerance than wild-type plants suggesting that genes of the AOX pathway can be useful to improve tolerance to stressful environmental conditions including salinity.

Other proteins like polyamines, osmotin, and LEA proteins mitigate salt stress by protecting macromolecules like nucleic acids, proteins, and carbohydrates from damages caused by ion toxicity and by drought conditions. Polyamines play a critical role in salinity and other abiotic stress tolerance by increasing level of polyamines which shows positive correlation of increased level of polyamines with stress tolerance in plants (Yang et al. 2007; Groppa and Benavides 2008; Gupta et al. 2013). Overproduction of spermidine and spermine in rice enhances salt tolerance (Roy and Wu 2002). Xu et al. (1996) found that *HVA7*, a LEA from barley, when transferred to rice, confers tolerance to drought and salinity stress. Dehydrins, another

LEA protein, were shown to enhance plant tolerance to various stresses (Hanin et al. 2011). Brini et al. (2007b) found positive correlation between wheat *dehydrin DHN-5* and salt tolerance. The study showed that the expression of the wheat *dehydrin DHN-5* in *Arabidopsis* led to an increase in tolerance to salt and osmotic stresses. Some of the transgenics where functional genes overexpressed to impart salt tolerance in different plant species have been shown in Table 8.2.

8.4 Transgenic Breeding: Regulatory Genes Controlling Salinity Stress Responses

8.4.1 Transcription Factors

Plant response to salt stress is a complex process and involves a vast array of genes working in different or overlapping regulatory pathways. As stress response is complex process and regulated by multi-genes, it is very challenging to achieve success in improving plant stress tolerance with the single functional gene approach (Mittler and Blumwald 2010; Varshney et al. 2011). Thus, instead of manipulating single functional gene, engineering regulatory genes or master regulators can be potential strategy for controlling stress responses. Transcription factors (TFs) and signaling proteins are master regulators of many genes involved in stress responses; hence, they are possible candidates for genetic engineering to obtain salinity-tolerant crops (Table 8.3). Transcription factors from various families like *AP2/ERF*, *NAC*, *MYB*, *MYC*, *DREB*, *Cys2/His2 zinc finger*, *bZIP*, and *WRKY* have been reported to be involved salt-stress tolerance (Golldack et al. 2011, 2014).

NAC TF family found to be involved in abiotic stress responses along with other important functions in plants (Nakashima et al. 2012). Hu et al. (2008) reported that overexpression of *SNAC1* and *SNAC2* genes from *NAC* TF family helps the survival of transgenic *Oryza sativa* plants under high-salinity conditions. Similarly, overexpression of TF gene *OsNAC04* leads to drought and salinity stress tolerance in *O. sativa* (Zheng et al. 2009). The AP2/ERF another family of plant-specific TFs which is known to play key role against various abiotic stresses (Mizoi et al. 2012). The transgenic plants for gene *GmDREB2* from soybean showed enhanced salinity and drought tolerance (Chen et al. 2007). Overexpression *Oryza sativa MYB2* (TF from MYB family) exhibited salt tolerance by variation in expression levels of various stress responsive genes (Yang et al. 2012).

Several groups reported a key role of *WRKY* TFs in responses to various abiotic stresses including salinity stress (Banerjee and Roychoudhury 2015). Li et al. (2013) reported *ZmWRKY33* (WRKY TF family) enhanced tolerance to salinity stress in *Arabidopsis* while overexpressing *GmWRKY54* exhibited salt tolerance, probably through the regulation of another TF *DREB2A and STZ/Zat10* (Zhou et al. 2008). *bZIP* is also another family of TFs having an important role in response to various abiotic stresses including salinity stress (Jakoby et al. 2002). Wang et al. (2010) demonstrated that overexpression of *ThbZIP1* gene of *Tamarix hispida* from TF *bZIP*

Table 8.3 Examples of regulatory genes leading to improvement of salt-stress tolerance of crop plants

Enhanced tolerance	Gene(s)- (TF family)	Donor	Transgenic plant	References
Enhanced drought and salinity tolerance	*OsDREB2A - (AP2/ ERFBP)*	*Oryza sativa*	Rice	Mallikarjuna et al. (2011)
Enhanced salinity tolerance	*StDREB1 - AP2/ ERFBP*	*Solanum tuberosum*	Potato	Bouaziz et al. (2013)
Enhanced salinity tolerance	*GmERF7-(AP2/ ERFBP)*	*Glycine max*	Tobacco	Zhai et al. (2013)
Enhanced drought and salinity tolerance	*TaERF3 -(AP2/ ERFBP)*	*Triticum aestivum*	Wheat	Rong et al. (2014)
Enhanced drought and salinity tolerance	*EaDREB2 -(AP2/ ERFBP)*	*Erianthus arundinaceus*	Sugarcane	Augustine et al. (2015)
Enhanced drought and salinity tolerance	*SsDREB - (AP2/ ERFBP)*	*Suaeda salsa*	Tobacco	Zhang et al. (2015)
Enhanced drought and salinity tolerance	*TaPIMP1-(MYB)*	*Triticum aestivum*	Tobacco	Liu et al. (2011)
Enhanced drought, cold, salinity tolerance	*OsMYB2-(MYB)*	*Oryza sativa*	Rice	Yang et al. (2012)
Enhanced drought, cold, salinity tolerance	*MdSIMYB1-(MYB)*	*Malus × domestica*	Apple	Wang et al. (2014)
Enhanced NaCl, ABA, mannitol tolerance	*SbMYB2-(MYB)* *SbMYB7-(MYB)*	*Scutellaria baicalensis*	Tobacco	Qi et al. (2015)
Enhanced salinity tolerance	*OsMYB91-(MYB)*	*Oryza sativa*	Rice	Zhu et al. (2015)
Enhanced drought, cold, salinity tolerance	*SlAREB1 -(bZIP)*	*Solanum lycopersicum*	Tomato	Orellana et al. (2010)
Enhanced drought, cold, salinity tolerance	*ThbZIP1-(bZIP)*	*Tamarix hispida*	Tobacco	Wang et al. (2010)
Enhanced salinity tolerance	*LrbZIP -(bZIP)*	*Nelumbo nucifera*	Tobacco	Cheng et al. (2013)
Enhanced salinity and drought tolerance	*OsbZIP71-(bZIP)*	*Oryza sativa*	Rice	Liu C. et al. (2014)
Enhanced salinity tolerance	*GhWRKY39-(WRKY)*	*Gossypium hirsutum*	Tobacco	Shi et al. (2014)
Enhanced salinity and drought tolerance	*TaWRKY10-(WRKY)*	*Triticum aestivum*	Tobacco	Wang et al. (2013)
Enhanced salinity and drought tolerance	*ZmWRKY58-(WRKY)*	*Zea may*	Rice	Cai et al. (2014)
Enhanced salinity and drought tolerance	*MtWRKY76-(WRKY)*	*Medicago truncatula*	*Medicago truncatula*	Liu et al. (2016)
Enhanced salinity and drought tolerance	*OsNAC04-(NAC)*	*Oryza sativa*	Rice	Zheng et al. (2009)
Enhanced salinity and drought tolerance	*OsNAP -(NAC)*	*Oryza sativa*	Rice	Chen et al. (2014)
Enhanced salinity and drought tolerance	*ONAC022 -(NAC)*	*Oryza sativa*	Rice	Hong et al. (2016)

(continued)

Table 8.3 (continued)

Enhanced tolerance	Gene(s)- (TF family)	Donor	Transgenic plant	References
Enhanced salinity tolerance	*ShCML44*	Rice	Rice	Xu et al. (2013)
Enhanced salinity, cold, and drought tolerance	*ShCML44*	Wild tomato	Tomato	Munir et al. (2016)
Enhanced salinity, cold, and drought tolerance	*OsCDPK7(CDPK)*	*Oryza sativa*	Rice	Saijo et al. (2000)
Enhanced salinity and drought tolerance	*OsCPK4(CDPK)*	*Oryza sativa*	Rice	Campo et al. (2014)
Enhanced salinity tolerance	*CalcineurinA subunit*	Mouse	Rice	Ma et al. (2005)
Enhanced salinity tolerance	*ZmMKK4*	*Zea mays*	*Arabidopsis*	Kong et al. (2011)
Enhanced salinity and drought tolerance	*GhMPK2*	*Gossypium hirsutum*	Tobacco	Zhang et al. (2011)
Enhanced salinity, cold, and drought tolerance	*OsMAPK5*	Rice	Rice	Xiong et al. (2003)
Enhanced salinity tolerance	*OsMKK6*	Rice	Rice	Kumar and Sinha (2013)
Enhanced salinity and drought tolerance	*MKK5*	*Arabidopsis*	*Arabidopsis*	Xing et al. (2015)
Enhanced salinity tolerance	*PtMAPKK4*	*Populus trichocarpa*	Tobacco	Yang et al. (2017)
Enhanced salinity and drought tolerance	*OsSIK1*	Rice	Rice	Ouyang et al. (2010)
Enhanced salinity tolerance	*PtSnRK2*	Poplar	*Arabidopsis*	Song et al. (2016)

Modified after Wang et al. (2016)

contributes to salinity tolerance by enhancing the activity of antioxidant enzymes such as peroxidase and superoxide dismutase and by accumulating compatible osmolytes like soluble sugars and soluble proteins. *SlAREB1* and *SlAREB2* are two members of *bZIP* TF in *Solanum lycopersicum*. Transgenic tomato overexpressing *SlAREB1* plants showed improved salinity and drought tolerance (Orellana et al. 2010).

8.4.2 Signaling Proteins

In addition to TFs, genetic engineering of signaling proteins has also become one of the feasible approaches. Some of the examples of transgenics where regulatory genes are overexpressed to impart salt tolerance are presented in Table 8.3. Several studies have reported that abiotic stresses (cold, high salt, and drought) trigger rapid increase in plant cells calcium (Ca^{2+}) levels (Sanders et al. 2002). Calcium signaling is often coupled with protein phosphorylation and dephosphorylation mediated by protein kinases and phosphatases, respectively. Changes in cellular Ca^{2+} level are being mediated by different Ca^{2+} binding proteins like calmodulin (CaM) and CaM-related proteins (CML), calcium-dependent protein kinases (CDPKs), and calcineurin B-like proteins (CBL) (Bouché et al. 2005). CDPK is one of the best studied protein kinases in the Ca^{2+} signaling pathway. Another family of protein kinases that function in stress tolerance is mitogen-activated protein kinases (MAPKs) (Zhang and Klessig 2001).

Overexpression of *ShCML44*, cold-responsive calmodulin-like gene in tomato, showed enhanced tolerance to salinity stress with higher germination rate and better growth of seedling (Munir et al. 2016). Similarly, transgenic rice overexpressing *OsMSR2*, a novel calmodulin-like gene, enhanced salinity tolerance with altered expression pattern of genes related to stress (Xu et al. 2013). Transgenic rice plant expressing calcineurin A subunit from mouse exhibited a higher level of salinity stress tolerance; also it has been observed that Na^+ content is higher in roots of untransformed wild-type plants than that of transgenic roots (Ma et al. 2005).

The *OsCPK4* gene is a member of calcium-dependent protein kinases in rice. Recently Campo et al. (2014) showed that transgenic rice plants overexpressing of *OsCPK4* significantly enhances salt and drought tolerance. Mitogen-activated protein kinase (MAPK) cascades also play crucial regulatory roles in various stress responses other than plant development processes. Zhang et al. (2011) reported ectopic expression of cotton *GhMPK2* in *Nicotiana tabacum* and found elevated levels of proline and induced expression of several genes related to stress, and as a result transgenic *Nicotiana tabacum* exhibited enhanced drought and salt tolerance. Similarly, overexpression *MAPK* from rice (*OsMAPK5*) exhibited increased kinase activity along with increased tolerance for salinity and other abiotic stresses like drought and cold (Xiong and Yang 2003). Overexpression of mitogen-activated protein kinase kinase 5 (*MKK5*) in *Arabidopsis* wild-type plants improved their tolerance level against various salt treatments (Xing et al. 2015). In another study overexpression of mitogen-activated protein kinase kinase 4 from *Populus trichocarpa* (*PtMAPKK4*) shows improved salt tolerance in tobacco. Specifically, under salt-stress condition, *PtMAPKK4* overexpressing lines showed improved germination and growth and development (Yang et al. 2017). However, some MAPKs can also have contrary effects particularly in case of rice; overexpression of *OsMAPK33* caused increased sensitivity to salinity and drought stress compared to wild-type plants (Lee et al. 2011). Receptor-like kinases (RLKs), another type of kinase, also have an important role in stress responses. Overexpressing *OsSIK1* (*OsSIK1-ox*), one of the putative RLKs, showed greater tolerance to salt stress as compared to control plants, gene-silenced plants by RNA interference (RNAi), knockout mutants *sik1* in rice (Ouyang et al. 2010). Sucrose non-fermenting 1 (SNF1)-related protein kinases (SnRKs) is one type of well-characterized protein kinase involved in stress responses (Halford and Hey 2009). In one study *PtSnRK2.5* and *PtSnRK2.7*genes (SnRKs from Poplar) heterologously overexpressed in *Arabidopsis* and found that overexpression of *PtSnRK2* leads to enhanced tolerance level for salt stress.

8.4.3 *Manipulating the miRNAs*

In the past decade, miRNAs have become major players in the regulation of plant response to environmental abiotic stresses (Zhang 2015, Shriram et al. 2016). There has been great interest in exploring regulatory roles of microRNAs against different stresses including salt for their exploitation in genetic engineering for higher stress tolerance, biomass, and yield (Zhang and Wang 2016, Patel et al. 2018). Transgenic plants overexpressing miR319 showed significantly higher plant tolerance to drought and salinity stress in creeping bentgrass (*Agrostis stolonifera*) (Zhou et al. 2013). In another study, expression of miR408 was shown to improve higher tolerance to salinity, cold, and oxidative stress in *Arabidopsis* seedlings (Ma et al. 2015). A rice microRNA *osa-miR393* was overexpressed in *Arabidopsis* plant resulting in enhanced salt tolerance (Gao et al. 2011a; b). In several other studies, novel microRNAs have been identified suggesting that these well-characterized candidates could become targets for plant genetic engineering investigations as successful in silico predictions could result in finding the target genes involved in pathways of signaling, ion homeostasis besides sustained plant growth under salt stress.

8.5 Halophyte Genes for Improving Salt-Stress Tolerance of Crops

In plants, gene expression and regulation decides the fate of plants from growth and development to stress tolerance. Modification/ manipulation in the regulation of these entities can dramatically change the fate of plant's life. In this sense, stress tolerance of plants can be improved by manipulating particular genes. In terms of stress tolerance, it is proved that the tolerant and sensitive plants possess same set of genes, but their efficient regulation or subtle changes in gene sequence can make one plant sensitive and other plant tolerant to the same environmental condition. This phenomenon is also true for salt-sensitive glycophytes and salt-tolerant halophytic plants. The halophytes are naturally tolerant to high salinity. Their genetic analysis revealed that differences in promoter activities and gene duplication in halophytes as compared to their glycophytic relative is responsible for their high salt tolerance (Nikalje et al. 2017). For example, the *NHX8* showed stress-induced expression in *Arabidopsis* while in *Thellungiella*, it showed constitutive expression. However *Arabidopsis* possess a single copy of *CBL10*, while *T. parvula* contain three copies; such changes make *Thellungiella* more salt tolerant. In addition, the efficient post-translational modifications are highly efficient in halophytes (Bose et al. 2015), and the halophytic gene sequences are more complex with presence of extra transposons and intergenic sequences (Rui et al. 2007). Therefore, for genetic improvement of crops, it may be important to choose genes from halophytic origin. Overexpression of *NHX1* gene of *Aeluropus littoralis* in soybean resulted in less sodium accumulation in aerial parts than underground parts, increased potassium ion content under salt stress and increased salt tolerance up to 150 mM NaCl (Liu et al. 2014). Shabala and Potossin (2014) opined that retention of potassium ions under salt stress is a key factor for salt tolerance in plants and specialty of halophytes. Further, Bose et al. (2015) have confirmed this by showing that maintenance of negative water potential because of high H^+ ATPase activity is important for halotropism. Similarly different halophytic antiporters were overexpressed in rice, and the transgenic plants showed high salt tolerance. *PtNHA1* and *PtNHX* from *Puccinellia tenuiflora* were transformed into rice, and the resulting transgenic rice showed improved tolerance to NaCl and $NaHCO_3$. Transgenic rice harboring *AgNHX1* from *Atriplex gmelinii* increased vacuolar antiporter activity by almost eightfold and improved its tolerance up to 300 mM NaCl (Ohta et al. 2002).

8.6 New Research on the Salt Pan

Genome editing tools have opened up new avenues for specific and targeted modifications in the crop plants (de Wiel et al. 2017). The method enables the introduction of targeted precise genomic changes using customized nucleases (Jain 2015). Genes associated with salt tolerance such as those involved in signaling, ion homeostasis, osmolyte synthesis, and transporters can be the suitable candidates for editing based manipulation. The plasma membrane ATPase plays a critical role in the regulation of ion homeostasis under salt stress and hence has been used as the target gene in a recent study. Osakabe et al. (2016) induced mutation of an abiotic stress tolerance gene encoding OPEN STOMATA 2 (OST2) (AHA1) – a major plasma membrane H + -ATPase via the precise site modification by using truncated gRNAs (tru-gRNAs) in the CRISPR-Cas9 system (Table 8.3).

High-throughput screening methods have advanced our knowledge about the genomes and phenomes. Plant stress biology research depends on robust screening methods for contrasting salt-stress-responsive phenotypes at different levels of tissue, organ, and whole-plant level. This branch of research, plant phenomics, is now being applied to facilitate efficient and reliable evaluation of stress (and salt) tolerant lines. Several such platforms for phenotyping are now available. Some of these include the High Resolution Plant Phenomics Centre (http://www.plantphenomics.org.au/HRPPC), Plant AccelatorTM (http://www.plantaccelerator.org.au/), Jülich Plant Phenotyping Centre- JPPC (http://www.fz-juelich.de/ibg/ibg-2/EN/_organisation/JPPC/JPPC_node.html) and Deep Plant Phenomics (Ubbens and Stavness 2017). Campbell et al. (2015) have developed a novel approach to analyze the dynamic plant responses to salt stress and studied the genetic basis of salt stress associated, genetically determined changes using a longitudinal genome-wide association model. This study highlights the use of image-based phenomics platforms combined with genome-wide association studies (GWAS) for dissecting the plant stress responses and should enable to establish liaison between expressed phenotypes with related genomic regions and environmental conditions. Further research into plant genetic manipulation via precise genetic tools will benefit from efficient phenotyping screens and high-throughput analysis tools.

8.7 Conclusions

Increasing salinity severely affects crop productivity and is becoming threat to world agriculture. The development of several genomics-assisted approaches including genetically modified plants has been advocated to circumvent this problem. Toward this goal, several stress-responsive genes have been identified and successfully introduced into other crops to create transgenic crops with enhanced stress tolerance. The most impressive results were obtained when manipulating transcription and signaling factors, as they control a broad range of downstream events, which results in superior tolerance to multiple stresses. However, challenges still lie ahead before successfully improving crop yield under saline conditions as most methods have been limited by the problem of yield penalty. Salinity tolerance involves a complex of responses at molecular, cellular, metabolic, physiological, and whole-plant levels. The marker-assisted selection as the molecular breeding method has begun to deliver its expected benefits in commercial breeding programs for salinity stress tolerance. For this, in addition to the key loci identified for salt tolerance traits majorly in rice, emphasis should also be given on identification and validation of other new loci in rice and other crops and their pyramiding in elite genetic background for enhanced salt tolerance through molecular marker-assisted breeding. Generation of salinity-tolerant transgenic varieties should necessarily involve gene stacking where multiple genes need to be overexpressed using advanced genetic engineering tools. Furthermore, the critical step is the field trials required to evaluate the transgenic plants, especially focusing on their growth and tolerance in the whole life period. New and novel information is generated through omics methods such as metabolomics and proteomics, and it is expected to develop more understanding of the salt-stress responses. It is also equally important that further understanding how plants perceive stress signals (salt sensors, osmosensors), transmit, and trigger a cascade of genetic mechanisms is necessary to develop crop plants that can tolerate extreme environments. With the current renewed interest in stress genomics, fast-forward approaches of phenomics, allele mining, and stress-metabolite profiling, it is expected to gain thorough understanding of salt-adaptive diversity for use in crop breeding for salt tolerance. Continued research should be aimed at development of salt-tolerant crop germplasm to expand the utilization of saline soils for enhancing agricultural productivity and environmental sustainability.

References

Abebe T, Guenzi AC, Martin B, Cushman JC (2003) Tolerance of mannitol-accumulating transgenic wheat to water stress and salinity. Plant Physiol 131:1748–1755

Ahmad P, Ahanger MA, Alyemeni MN, Wijaya L, Alam P, Ashraf M (2018) Mitigation of sodium chloride toxicity in *Solanum lycopersicum* L. by supplementation of jasmonic acid and nitric oxide. J Plant Inter 13(1):64–72

Apse MP, Aharon GS, Snedden WA, Blumwald E (1999) Salt tolerance conferred by over expression of a vacuolar Na^+/H^+ antiport in *Arabidopsis*. Science 285:1256–1258

Augustine SM, Ashwin Narayan J, Syamaladevi DP, Appunu C, Chakravarthi M, Ravichandran V, Tuteja N, Subramonian N (2015) Overexpression of EaDREB2 and pyramiding of EaDREB2 with the pea DNA helicase gene (PDH45) enhance drought and salinity tolerance in sugarcane (*Saccharum* spp. hybrid). Plant Cell Rep 34:247–263

Babu NN, Krishnan SG, Vinod KK, Krishnamurthy SL, Singh VK, Singh MP, Singh R, Ellur RK, Rai V, Bollinedi H, Bhowmick PK, Yadav AK, Nagarajan M, Singh NK, Prabhu KV, Singh AK (2017a) Marker Aided Incorporation of Saltol, a Major QTL Associated with Seedling Stage Salt Tolerance, into *Oryza sativa* 'Pusa Basmati 1121'. Front Plant Sci 8:41

Babu NN, Vinod KK, Krishnamurthy SL, Gopala Krishnan S, Yadav A, Bhowmick PK, Nagarajan M, Singh NK, Prabhu KV, Singh AK (2017b) Microsatellite based linkage disequilibrium analyses reveal *Saltol* haplotype fragmentation and identify novel QTLs for seedling stage salinity tolerance in rice (*Oryza sativa* L.). J Plant Biochem Biotechnol 26(3):310–320

Badawi GH, Kawano N, Yamauchi Y, Shimada E, Sasaki R, Kubo A, Tanaka K (2004) Overexpression of ascorbate peroxidase in tobacco chloroplasts enhances the tolerance to salt stress and water deficit. Physiol Plant 121:231–238

Banerjee A, Roychoudhury A (2015) WRKY proteins: signaling and regulation of expression during abiotic stress responses. Sci World J 2015:807560

Bertrand A, Dhont C, Bipfubusa M, Chalifour FP, Drouin P, Beauchamp CJ (2015) Improving salt stress responses of the symbiosis in alfalfa using salt-tolerant cultivar and rhizobial strain. Appl Soil Eco 87:108–117

Binzel ML, Hess FD, Bressan RA, Hasegawa PM (1998) Intracellular compartmentation of ions in salt adapted tobacco cells. Plant Physiol 86:607–614

Bizimana JB, Luzi-Kihupi A, Murori RW, Singh RK (2017) Identification of quantitative trait loci for salinity tolerance in rice (*Oryza sativa* L.) using R29/Hasawi mapping population. J Genet 96(4):571–582

Boriboonkaset T, Theerawitaya C, Yamada N, Pichakum A, Supaibulwatana K, Cha-Um S, Takabe T, Kirdmanee C (2013) Regulation of some carbohydrate metabolism-related genes, starch and soluble sugar contents, photosynthetic activities and yield attributes of two contrasting rice genotypes subjected to salt stress. Protoplasma 250:1157–1167

Bose J, Rodrigo-Moreno A, Lai D, Xie Y, Shen W, Shabala S (2015) Rapid regulation of the plasma membrane H+-ATPase activity is essential to salinity tolerance in two halophyte species, Atriplex lentiformis and Chenopodium quinoa. Annals of Botany 115(3):481–494

Bouaziz D, Pirrello J, Charfeddine M, Hammami A, Jbir R, Dhieb A, Bouzayen M, Gargouri-Bouzid R (2013) Overexpression of StDREB1 transcription factor increases tolerance to salt in transgenic potato plants. Mol Biotechnol 54:803–817

Bouché N, Yellin A, Snedden WA, Fromm H (2005) Plant-specific calmodulin-binding proteins. Annu Rev Plant Biol 56:435–466

Brini F, Hanin M, Mezghani I, Berkowitz GA, Masmoudi K (2007a) Overexpression of wheat Na^+/H^+ antiporter TNHX1 and H^+-pyrophosphatase TVP1 improve salt-and drought stress tolerance in *Arabidopsis thaliana* plants. J Exp Bot 58:301–308

Brini F, Hanin M, Lumbreras V, Amara I, Khoudi H, Hassairi A, Pagès M, Masmoudi K (2007b) Overexpression of wheat dehydrin DHN-5 enhances tolerance to salt and osmotic stress in *Arabidopsis thaliana*. Plant Cell Rep 26:2017–2026

Cai R, Zhao Y, Wang Y, Lin Y, Peng X, Li Q, Chang Y, Jiang H, Xiang Y, Cheng B (2014) Overexpression of a maize WRKY58 gene enhances drought and salt tolerance in transgenic rice. Plant Cell Tissue Organ Cult 119:565–577

Campbell MT, Knecht AC, Berger B, Brien CJ, Wang D, Walia H (2015) Integrating image-based phenomics and association analysis to dissect the genetic architecture of temporal salinity responses in rice. Plant Physiol 168(4):1476–1489

Campo S, Baldrich P, Messeguer J, Lalanne E, Coca M, Segundo BS (2014) Overexpression of a calcium-dependent protein kinase confers salt and drought tolerance in Rice by preventing membrane lipid peroxidation. Plant Physiol 165(2):688–704

Checker VG, Chhibbar AK, Khurana P (2012) Stress-inducible expression of barley Hva1 gene in transgenic mulberry displays enhanced tolerance against drought, salinity and cold stress. Transgenic Res 21:939–957

Chen M, Wang QY, Cheng XG, Xu ZS, Li LC, Ye XG, Xia LQ, Ma YZ (2007) GmDREB2, a soybean DRE-binding transcription factor, conferred drought and high salt tolerance in transgenic plants. Biochem Biophys Res Commun 353:299–305

Chen JB, Yang JW, Zhang ZY, Feng XF, Wang SM (2013) Two P5CS genes from common bean exhibiting different tolerance to salt stress in transgenic *Arabidopsis*. J Genet 92(3):461–469

Chen X, Wang Y, Lv B, Li J, Luo L, Lu S, Zhang X, Ma H, Ming F (2014) The NAC family transcription factor OsNAP confers abiotic stress response through the ABA pathway. Plant Cell Physiol 55:604–619

Cheng L, Li S, Hussain J, Xu X, Yin J, Zhang Y, Chen X, Li L (2013) Isolation and functional characterization of a salt responsive transcriptional factor, LrbZIP from lotus root (*Nelumbo nucifera* Gaertn). Mol Biol Rep 40:4033–4045

Choi JY, Seo YS, Kim SJ, Kim WT, Shin JS (2011) Constitutive expression of CaXTH3, a hot pepper xyloglucan endotransglucosylase/hydrolase, enhanced tolerance to salt and drought stresses without phenotypic defects in tomato plants (*Solanum lycopersicum* cv. Dotaerang). Plant Cell Rep 30(5):867–877

Cominelli E, Conti L, Tonelli C, Galbiati M (2013) Challenges and perspectives to improve crop drought and salinity tolerance. Nat Biotechnol 30:355–361

De Leon TB, Linscombe S, Subudhi PK (2017) Identification and validation of QTLs for seedling salinity tolerance in introgression lines of a salt tolerant rice landrace 'Pokkali'. PLoS One 12(4):e0175361

Dietz KJ, Tavakoli N, Kluge C, Mimura T, Sharma SS, Harris GC, Chardonnens AN, Golldack D (2001) Significance of the V type ATPase for the adaptation to stressful growth conditions and its regulation on the molecular and biochemical level. J Exp Bot 52(363):1969–1980

Do TD, Chen H, Vu HTT, Hamwieh A, Yamada T, Sato T, Yan Y, Cong H, Shono M, Suenaga K, Xu D (2016) Ncl synchronously regulates Na+, K+, andCl-in soybean greatly increases the grain yield in saline field conditions. Sci Rep 6:19147

Ellur RK, Khanna A, Yadav A, Pathania S, Rajashekara H, Singh VK et al (2016) Improvement of Basmati rice varieties for resistance to blast and bacterial blight diseases using marker assisted backcross breeding. Plant Sci 242:330–341

Faize M, Burgo SL, Faize L, Piqueras A, Nicolas E, Barba-Espin G, Clemente-Moreno MJ, Alcobendas R, Artlip T, Hernandez JA (2011) Involvement of cytosolic ascorbate peroxidase and Cu/Zn-superoxide dismutase for improved tolerance against drought stress. J Exp Bot 62:2599–2613

Flowers TJ, Yeo AR (1997) Breeding for salt resistance in plants. In: Jaiwal PK, Singh PR, Gulaati A (eds) Strategies for improving salt tolerance in higher plants. Science Publishers Inc, Enfield, pp 247–264

Flowers TJ, Troke PF, Yeo AR (1977) The mechanism of salt tolerance in halophytes. Ann Rev Plant Physiol 28:89–121

Gao P, Bai X, Yang L, Lv D, Pan X, Li Y (2011a) Osa-MIR393: a salinity and alkaline stress-related microRNA gene. Mol Biol Rep 38:237–242

Gao SQ, Chen M, Xu ZS, Zhao CP, Li L, Xu HJ, Tang YM, Zhao X, Ma YZ (2011b) The soybean GmbZIP1 transcription factor enhances multiple abiotic stress tolerances in transgenic plants. Plant Mol Biol 75:537–553

Garg AK, Kim J-K, Owens TG, Ranwala AP, Choi YD, Kochian LV, Wu R (2002) Trehalose accumulation in rice plants confers high tolerance levels to different abiotic stresses. Proc Natl Acad Sci U S A 99:15898–15903

Goel D, Singh AK, Yadav V, Babbar SB, Bansal KC (2010) Overexpression of osmotin gene confers tolerance to salt and drought stresses in transgenic tomato (*Solanum lycopersicum* L.). Protoplasma 245:133–141

Golldack D, Luking I, Yang O (2011) Plant tolerance to drought and salinity: stress regulating transcription factors and their functional significance in the cellular transcriptional network. Plant Cell Rep 30:1383–1391

Golldack D, Li C, Mohan H, Probst N (2014) Tolerance to drought and salt stress in plants: unraveling the signaling networks. Front Plant Sci 5:151

Gouiaa S, Khoudi H (2015) Co-expression of vacuolar Na^+/H^+ antiporter and H^+-pyrophosphatase with an IRES-mediated dicistronic vector improves salinity tolerance and enhances potassium biofortification of tomato. Phytochemistry 117:537–546

Gouiaa S, Khoudi H, Leidi EO, Pardo JM, Masmoudi K (2012) Expression of wheat Na(+)/H(+) antiporter TNHXS1 and H(+)- pyrophosphatase TVP1 genes in tobacco from a bicistronic transcriptional unit improves salt tolerance. Plant Mol Biol 79:137–155

Groppa MD, Benavides MP (2008) Polyamines and abiotic stress: recent advances. Amino Acids 34(1):35–45

Guo Q, Zhang J, Gao Q, Xing S, Li F, Wang W (2008) Drought tolerance through overexpression of monoubiquitin in transgenic tobacco. J Plant Physiol 165(16):1745–1755

Gupta KJ, Stoimenova M, Kaiser WM (2005) In higher plants, only root mitochondria, but not leaf mitochondria reduce nitrite to NO, *in vitro* and *in situ*. J Exp Bot 56(420):2601–2609

Gupta K, Dey A, Gupta B (2013) Polyamines and their role in plant osmotic stress tolerance. In: Tuteja N, Gill SS (eds) Climate change and plant abiotic stress tolerance. Wiley-VCH, Weinheim, pp 1053–1072

Halford NG, Hey SJ (2009) Snf1-related protein kinases (*SnRKs*) act with in an intricate network that links metabolic and stress signaling in plants. Biochem J 419:247–259

Hanin M, Brini F, Ebel C, Toda Y, Takeda S, Masmoudi K (2011) Plant dehydrins and stress tolerance: versatile proteins for complex mechanisms. Plant Signal Behav 6:1503–1509

Hanin M, Ebel C, Ngom M, Laplaze L, Masmoudi K (2016) New insights on plant salt tolerance mechanisms and their potential use for breeding. Front Plant Sci 7

Hasegawa PM (2013) Sodium (Na+) homeostasis and salt tolerance of plants. Environ Exp Bot 92:19–31

Hasegawa PM, Bressan RA, Zhu JK, Bohnert HJ (2000) Plant cellular and molecular responses to high salinity. Ann Rev Plant Biol 51:463–499

He CX, Yan JQ, Shen GX, Fu LH, Holaday AS, Auld D, Blumwald E, Zhang H (2005) Expression of an *Arabidopsis* vacuolar sodium/proton antiporter gene in cotton improves photosynthetic performance under salt conditions and increases fiber yield in the field. Plant Cell Physiol 46:1848–1854

Hong Z, Lakkineni K, Zhang Z, Verma DPS (2000) Removal of feedback inhibition of 1 pyrroline-5-carboxylase synthetase (P5CS) results in increased proline accumulation and protection of plants from osmotic stress. Plant Physiol 122:1129–1136

Hong Y, Zhang H, Huang L, Li D, Song F (2016) Overexpression of a stress-responsive NAC transcription factor gene ONAC022 improves drought and salt tolerance in rice. Front Plant Sci 7:4

Hoque ABMZ, Haque MA, Sarker MRA, Rahman MA (2015) Marker-assisted introgression of saltol locus into genetic background of BRRI Dhan-49. Int J Biosci 6:71–80

Horie T, Hauser F, Schroeder JI (2009) HKT transporter-mediated salinity resistance mechanisms in *Arabidopsis* and monocot crop plants. Trends Plant Sci 14:660–668

Hu H, You J, Fang Y, Zhu X, Qi Z, Xiong L (2008) Characterization of transcription factor gene SNAC2 conferring cold and salt tolerance in rice. Plant Mol Biol 67:169–181

Hu L, Zhang P, Jiang Y, Fu J (2015) Metabolomic analysis revealed differential adaptation to salinity and alkalinity stress in Kentucky bluegrass (*Poa pratensis*). Plant Mol Bio Report 33:56–68

Ishitani M, Liu J, Halfter U, Kim CS, Shi W, Zhu JK (2000) SOS3 function in plant salt tolerance requires N-myristoylation and calcium binding. Plant Cell 12(9):1667–1677

Jain M (2015) Function genomics of abiotic stress tolerance in plants: a CRISPR approach. Front Plant Sci 6

Jakoby M, Weisshaar B, Dröge-Laser W, Vicente-Carbajosa J, Tiedemann J, Kroj T, Parcy F (2002) bZIP transcription factors in Arabidopsis. Trends Plant Sci 7(3):106–111

Jamoussi RJ, Elabbassi MM, Jouira HB, Hanana M, Zoghlami N, Ghorbel A, Mliki A (2014) Physiological responses of transgenic tobacco plants expressing the dehydration responsive RD22 gene of *Vitis vinifera* to salt stress. Turk J Bot 38:268–280

Kere GM, Chen C, Guo Q, Chen J (2017) Genetics of salt tolerance in cucumber (*Cucumis sativus* L.) revealed by quantitative traits loci analysis. Sci Lett 5(1):22–30

Kishor PBK, Hong Z, Miao GH, Hu CAA, Verma DPS (1995) Overexpression of Δ-pyrroline-5-carboxilase synthetase increases proline production and confers osmotolerance in transgenic plants. Plant Physiol 108:1387–1394
Kong X, Pan J, Zhang M, Xing X, Zhou Y, Liu Y, Li D, Li D (2011) ZmMKK4, a novel group C mitogen-activated protein kinase kinase in maize (*Zea mays*), confers salt and cold tolerance in transgenic *Arabidopsis*. Plant Cell Environ 34(8):1291–1303
Kumar K, Sinha AK (2013) Overexpression of constitutively active mitogen activated protein kinase kinase 6 enhances tolerance to salt stress in rice. Rice 6:25
Kumar G, Purty RS, Sharma MP, Singla-Pareek SL, Pareek A (2009) Physiological responses among *Brassica* species under salinity stress show strong correlation with transcript abundance for SOS pathway-related genes. J Plant Physiol 166:507–520
Lee SK, Kim BG, Kwon TR, Jeong MJ, Park SR, Lee JW, Byun MO, Kwon HB, Matthews BF, Hong CB, Park SC (2011) Overexpression of the mitogen-activated protein kinase gene OsMAPK33 enhances sensitivity to salt stress in rice (*Oryza sativa* L.). J Biosci 36(1):139–151
Li HW, Zang BS, Deng XW, Xi-Ping W (2011) Overexpression of the trehalose-6-phosphate synthase gene OsTPS1 enhances abiotic stress tolerance in rice. Planta 234:1007
Li H, Gao Y, Xu H, Dai Y, Deng D, Chen J (2013) ZmWRKY33, a WRKY maize transcription factor conferring enhanced salt stress tolerances in *Arabidopsis*. Plant Growth Regul 70:207–216
Linh LH, Khanh TD, Luanl NV, Cuc DTK, Duc LD, Linh TH, Ismail AM, Ham LH (2012) Application of marker assisted backcrossing to pyramid salinity tolerance (Saltol) into Rice Cultivar- Bac Thom 7. VNU J Sci Nat Sci Technol 28:87–99
Liu J, Ishitani M, Halfter U, Kim CS, Zhu JK (2000) The *Arabidopsis thaliana* SOS2 gene encodes a protein kinase that is required for salt tolerance. Proc Natl Acad Sci U S A 97(7):3730–3734
Liu H, Zhou X, Dong N, Liu X, Zhang H, Zhang Z (2011) Expression of a wheat MYB gene in transgenic tobacco enhances resistance to *Ralstonia solanacearum*, and to drought and salt stresses. Funct Integr Genomics 11:431–443
Liu C, Mao B, Ou S, Wang W, Liu L, Wu Y, Chu C, Wang X (2014) OsbZIP71, abZIP transcription factor, confers salinity and drought tolerance in rice. Plant Mol Biol 84:19–36
Liu L, Zhang Z, Dong J, Wang T (2016) Overexpression of MtWRKY76 increases both salt and drought tolerance in *Medicago truncatula*. Environ Exp Bot 123:50–58
Luo M, Zhao Y, Zhang R, Xing J, Duan M, Li J, Wang N, Wang W, Zhang S, Chen Z, Zhang H, Shi Z, Song W, Zhao J (2017) Mapping of a major QTL for salt tolerance of mature field-grown maize plants based on SNP markers. BMC Plant Biol 17:140
Lyzenga WJ, Stone SL (2012) Abiotic stress tolerance mediated by protein ubiquitination. J Exp Bot 63(2):599–616
Ma X, Qian Q, Zhu D (2005) Expression of a calcineurin gene improves salt stress tolerance in transgenic rice. Plant Mol Bio 58:483–495
Ma C, Burd S, Lers A (2015) miR408 is involved in abiotic stress responses in Arabidopsis. Plant J Cell Mol Biol 84:169–187
Mallikarjuna G, Mallikarjuna K, Reddy MK, Kaul T (2011) Expression of OsDREB2A transcription factor confers enhanced dehydration and salt stress tolerance in rice (*Oryza sativa* L.). Biotechnol Lett 33:1689–1697
Mantri N., Patade V., Pang E. (2014) Recent Advances in Rapid and Sensitive Screening For Abiotic Stress Tolerance. In: Tran LP (ed) Improvement of Crops in the Era of Climate Changes, vol 2 (trans: Ahmad P, Wani MR, Azooz MM). Springer, New York. ISBN 978-1-4614-8823-1
Mickelbart MV, Hasegawa PM, Bailey-Serres J (2015) Genetic mechanisms of abiotic stress tolerance that translate to crop yield stability. Nat Rev Genet 16:237–251
Miller G, Shulaev V, Mittler R (2008) Reactive oxygen signaling and abiotic stress. Physiol Plant 133:481–489
Mittler R, Blumwald E (2010) Genetic engineering for modern agriculture: challenges and perspectives. Ann Rev Plant Biol 61:443–462
Mizoi J, Shinozaki K, Yamaguchi-Shinozaki K (2012) AP2/ERF family transcription factors in plant abiotic stress responses. Biochim Biophys Acta 1819(2):86–96

Moller IS, Gilliham M, Jha D, Mayo GM, Roy SJ, Coates JC, Haseloff J, Tester M (2009) Shoot Na+ exclusion and increased salinity tolerance engineered by cell type-specific alteration of Na+ transport in *Arabidopsis*. Plant Cell 21(7):2163–2178

Munir S, Liu H, Xing Y, Hussain S, Ouyang B, Zhang Y, Li H, Ye Z (2016) Overexpression of calmodulin-like (*ShCML44*) stress-responsive gene from *Solanum habrochaites* enhances tolerance to multiple abiotic stresses. Sci Rep 6:31772

Munns R (2002) Comparative physiology of salt and water stress. Plant Cell Environ 25:239–250

Munns R, Tester M (2008) Mechanisms of salinity tolerance. Ann Rev Plant Biol 59:651–681

Nakashima K, Takasaki H, Mizoi J, Shinozaki K, Yamaguchi-Shinozaki K (2012) NAC transcription factors in plant abiotic stress responses. Biochim Biophys Acta 1819:97–103

Nikalje GC, Nikam TD, Suprasanna P (2017) Looking at halophytic adaptation to high salinity through genomics landscape. Curr Genom 18(6). https://doi.org/10.2174/1389202918666170228143007.

Nongpiur RC, Singla-Pareek SL, Pareek A (2016) Genomics approaches for improving salinity stress tolerance in crop plants. Curr Genomics 17(4):343–357

Ohta M, Hayashi Y, Nakashima A, Hamada A, Tanaka A, Nakamura T, Hayakawa T (2002) Introduction of a Na /H antiporter gene from confers salt tolerance to rice. FEBS Letts 532(3):279–282

Orellana S, Yañez M, Espinoza A, Verdugo I, González E, Ruiz-Lara S, Casaretto JA (2010) The transcription factor SlAREB1 confers drought, salt stress tolerance and regulates biotic and abiotic stress-related genes in tomato. Plant Cell Environ 33:2191–2208

Osakabe Y, Watanabe T, Sugano SS, Ueta R, Ishihara R, Shinozaki K, Osakabe K (2016) Optimization of CRISPR/Cas9 genome editing to modify abiotic stress responses in plants. Sci Rep 6(1)

Otoch MLO, Sobreira ACM, Aragão MEF, Orellano EG, Lima MGS, Melo DF (2001) Salt modulation of vacuolar H+-ATPase and H+-Pyrophosphatase activities in *Vigna unguiculata*. J Plant Physiol 158:545–551

Ouyang SQ, Liu YF, Liu P, Lei G, He SJ, Ma B, Zhang WK, Zhang JS, Chen SY (2010) Receptor-like kinase *OsSIK1* improves drought and salt stress tolerance in rice (*Oryza sativa*) plants. Plant J 62:316–329

Park HY, Seok HY, Park BK, Kim SH, Goh CH, Lee BH, Lee CH, Moon YH (2008) Overexpression of *Arabidopsis* ZEP enhances tolerance to osmotic stress. Biochem Biophys Res Commun 375:80–85

Pasapula V, Shen G, Kuppu S, Paez-Valencia J, Mendoza M, Hou P, Chen J, Qiu X, Zhu L, Zhang X, Auld D, Blumwald E, Zhang H, Gaxiola R, Payton P (2011) Expression of an *Arabidopsis* vacuolar H^+-pyrophosphatase gene (AVP1) in cotton improves drought-and salt tolerance and increases fibre yield in the field conditions. Plant Biotechnol J 9:88–99

Patel P, Yadav K, Ganapathi TR and Suprasanna Penna (2018) Plant miRNAome: cross talk in abiotic stressful times. In: Rajpal VR, Sehgal D, Raina SN (eds) Genomics-assisted breeding for crop improvement: abiotic stress tolerance. Springer (In Press)

Penna S (2003) Building stress tolerance through over-producing trehalose in transgenic plants. Trends Plant Sci 8(8):355–357

Prashanth SR, Sadhasivam V, Parida A (2008) Overexpression of cytosolic copper/zinc superoxide dismutase from a mangrove plant *Avicennia marina* in indica rice var Pusa Basmati-1 confers abiotic stress tolerance. Transgenic Res 17:281–291

Qi L, Yang J, Yuan Y, Huang L, Chen P (2015) Overexpression of twoR2R3-MYB genes from *Scutellaria baicalensis* induces phenylpropanoid accumulation and enhances oxidative stress resistance in transgenic tobacco. Plant Physiol Biochem 94:235–243

Qiu QS, Guo Y, Dietrich MA, Schumaker KS, Zhu J-K (2002) Regulation of SOS1, a plasma membrane Na^+/H^+ exchanger in *Arabidopsis thaliana*, by SOS2 and SOS3. Proc Natl Acad Sci U S A 99:8436–8441

Quan R, Wang J, Hui J, Bai H, Lyu X, Zhu Y, Zhang H, Zhang Z, Li S, Huang R (2018) Improvement of salt tolerance using wild rice genes. Front Plant Sci 8:2269

Ravikiran KT, Krishnamurthy SL, Warraich AS, Sharma PC (2017) Diversity and haplotypes of rice genotypes for seedling stage salinity tolerance analyzed through morpho-physiological and SSR markers. Field Crops Res. Online published. https://doi.org/10.1016/j.fcr.2017.04.006

Reddy PS, Jogeswar G, Rasineni GK, Maheswari M, Reddy AR, Varshney RK, Kishor PBK (2015) Proline over-accumulation alleviates salt stress and protects photosynthetic and antioxidant enzyme activities in transgenic sorghum [*Sorghum bicolor* (L.) Moench]. Plant Physiol Biochem 94:104–113

Rong W, Qi L, Wang A, Ye X, Du L, Liang H, Xin Z, Zhang Z (2014) The ERF transcription factor TaERF3 promotes tolerance to salt and drought stresses in wheat. Plant Biotechnol J 12:468–479

Roxas VP, Smith RK Jr, Allen ER, Allen RD (1997) Overexpression of glutathione S-transferase/glutathione peroxidase enhances the growth of transgenic tobacco seedlings during stress. Nat Biotechnol 15:988–991

Roy M, Wu R (2002) Overexpression of S-adenosyl methionine decarboxylase gene in rice increases polyamine level and enhances sodium chloride-stress tolerance. Plant Sci 163(5):987–992

Roy SJ, Negrão S, Tester M (2014) Salt resistant crop plants. Curr Opin Biotechnol 26:115–124

Rubio F, Gassmann W, Schroeder JI (1995) Sodium-driven potassium uptake by the plant potassium transporter HKT1 and mutations conferring salt tolerance. Science 270:1660–1663

Rui A, Qi-Jun C, Mao-Feng C, Ping-Li L, Zhao S, Zhi-Xiang Q, Jia C, Xue-Chen W (2007) AtNHX8, a member of the monovalent cation:proton antiporter-1 family in Arabidopsis thaliana, encodes a putative Li+/H+ antiporter. Plant J 49(4):718–728

Saijo Y, Hata S, Kyozuka J, Shimamoto K, Izui K (2000) Overexpression of a single Ca2+ dependent protein kinase confers both cold and salt/drought tolerance on rice plants. Plant J 23:319–327

Sakamoto A, Murata A, Murata N (1998) Metabolic engineering of rice leading to biosynthesis of glycine betaine and tolerance to salt and cold. Plant Mol Biol 38:1011–1019

Sanders D (2000) Plant biology: the salty tale of *Arabidopsis*. Curr Biol 10(13):R486–R488

Sanders D, Pelloux J, Brownlee C, Harper JF (2002) Calcium at the crossroads of signaling. Plant Cell 14:401–417

Schachtman DP, Schroeder JI (1994) Structure and transport mechanism of a high-affinity potassium uptake transporter from higher plants. Nature 370:655–658

Schilling RK, Marschner P, Shavrukov Y, Berger B, Tester M, Roy SJ, Plett DC (2013) Expression of the *Arabidopsis* vacuolar H^+-pyrophosphatase gene (AVP1) improves the shoot biomass of transgenic barley and increases grain yield in a saline field. Plant Biotechnol J 12:378–386

Serrano R, Mulet JM, Rios G, Marquez JA, de Larrinoa IF, Leube MP, Mendizabal I, Pascual-Ahuir A, Proft M, Ros R, Montesinos C (1999) A glimpse of the mechanisms of ion homeostasis during salt stress. J Exp Bot 50:1023–1036

Shabala S, Cuin TA (2008) Potassium transport and plant salt tolerance. Physiol Plantarum 133:651–669

Shabala S, Pottosin I (2014) Regulation of potassium transport in plants under hostile conditions: implications for abiotic and biotic stress tolerance. Physiol Plant 151(3):257–279

Shen G, Wei J, Qiu X, Hu R, Kuppu S, Auld D, Blumwald E, Gaxiola R, Payton P, Zhang H (2014) Co-overexpression of AVP1 and AtNHX1 in cotton further improves drought and salt tolerance in transgenic cotton plants. Plant Mol Biol Rep 33:167–177

Sheveleva E, Chmara W, Bohnert HJ, Jensen RG (1997) Increased salt and drought tolerance by D-Ononitol production in transgenic *Nicotiana tabacum*. Plant Physiol 115:1211–1219

Shi W, Liu D, Hao L, Wu CA, Guo X, Li H (2014) GhWRKY39, a member of the WRKY transcription factor family in cotton, has a positive role in disease resistance and salt stress tolerance. Plant Cell Tissue Organ Cult 118:17–32

Shinozaki K, Yamaguchi-Shinozakiy K, Seki M (2003) Regulatory network of gene expression in the drought and cold stress responses. Curr Opin Plant 6(5):410–417

Shriram V, Kumar V, Devarumath RM, Khare TS, Wani SH (2016) MicroRNAs as potential targets for abiotic stress tolerance in plants. Front Plant Sci 7

Singh A, Singh VK, Singh SP, Pandian RTP, Ellur RK, Singh D, et al. (2012) Molecular breeding for the development of multiple disease resistance in Basmati rice. AoB Plants 2012 ls029

Smith CA, Melino VJ, Sweetman C, Soole KL (2009) Manipulation of alternative oxidase can influence salt tolerance in *Arabidopsis thaliana*. Physiol Plant 137:459–447

Song X, Yu X, Hori C, Demura T, Ohtani M, Zhuge Q (2016) Heterologous overexpression of poplar *SnRK2* genes enhanced salt stress tolerance in *Arabidopsis thaliana*. Front Plant Sci 7:612

Sperling O, Lazarovitch N, Schwartz A, Shapira O (2014) Effects of high salinity irrigation on growth, gas-exchange, and photoprotection in date palms (*Phoenix dactylifera* L., cv. Medjool). Environ and Exp Bot. 99:100–109

Su J, Wu R (2004) Stress-inducible synthesis of proline in transgenic rice confers faster growth under stress conditions than that with constitutive synthesis. Plant Sci 166:941–948

Székely G, Abrahám E, Cséplo A, Rigó G, Zsigmond L, Csiszár J, Ayaydin F, Strizhov N, Jásik J, Schmelzer E, Koncz C, Szabados L (2008) Duplicated P5CS genes of Arabidopsis play distinct roles in stress regulation and developmental control of proline biosynthesis. Plant J 53:11–28

Tarczynski MC, Jensen RG, Bohnert HJ (1992) Expression of a bacterial mtlD gene in transgenic tobacco leads to production and accumulation of mannitol. Proc Natl Acad Sci U S A 89:2600–2604

Thomas JC, Sepahi M, Arendall B, Bohnert HJ (1995) Enhancement of seed germination in high salinity by engineering mannitol expression in *Arabidopsis thaliana*. Plant Cell Environ 18(7):801–806

Thomson MJ, Ocampo M, Egdane J, Rahman MA, Sajise AG, Adorada DL et al (2010) Characterizing the Saltol quantitative trait locus for salinity tolerance in rice. Rice 3: 148–160

Tsutsumi K, Yamada N, Cha-Um S, Tanaka Y, Takabe T (2015) Differential accumulation of glycine betaine and choline monooxygenase in bladder hairs and lamina leaves of *Atriplex gmelini*, under high salinity. J Plant Physiol 176:101–107

Turan S, Katrina C, Kumar S (2012) Salinity tolerance in plants: breeding and genetic engineering. Aust J Crop Sci 6(9):1337–1348

Türkan I, Demiral T (2009) Recent developments in understanding salinity tolerance. Environ Exp Bot 67:2–9

Ubbens JR, Stavness I (2017) Deep plant phenomics: a deep learning platform for complex plant phenotyping tasks. Front Plant Sci 8:1190. https://doi.org/10.3389/fpls.2017.01190

van de Wiel CCM, Schaart JG, Lotz LAP, Smulders MJM (2017) New traits in crops produced by genome editing techniques based on deletions. Plant Biotechnol Rep 11(1):1–8

Varshney RK, Bansal KC, Aggarwal PK, Datta SK, Craufurd PQ (2011) Agricultural biotechnology for crop improvement in a variable climate: hope or hype? Trends Plant Sci 16:363–371

Wang B, Lüttge U, Ratajczak R (2001) Effects of salt treatment and osmotic stress on V-ATPase and V-PPase in leaves of the halophyte *Suaeda salsa*. J Exp Bot 52(365):2355–2365

Wang Y, Gao C, Liang Y, Wang C, Yang C, Liu G (2010) A novel bZIP gene from *Tamarix hispida* mediates physiological responses to salt stress in tobacco plants. J Plant Physiol 167:222–230

Wang C, Deng P, Chen L, Wang X, Ma H, Hu W, Yao N, Feng Y, Chai R, Yang G, He G (2013) A wheat WRKY transcription factor TaWRKY10 confers tolerance to multiple abiotic stresses in transgenic tobacco. PLoS One 8(6):e65120

Wang RK, Cao ZH, Hao YJ (2014) Overexpression of a R2R3MYB gene MdSIMYB1 increases tolerance to multiple stresses in transgenic tobacco and apples. Physiol Plant 150:76–87

Wang H, Wang H, Shao H, Tang X (2016) Recent advances in utilizing transcription factors to improve plant abiotic stress tolerance by transgenic technology. Front Plant Sci 7:67. https://doi.org/10.3389/fpls.2016.00067

Xing Y, Chen W-H, Jia W, Zhang J (2015) Mitogen-activated protein kinase kinase 5 (MKK5)-mediated signalling cascade regulates expression of iron superoxide dismutase gene in *Arabidopsis* under salinity stress. J Exp Bot 66(19):5971–5981

Xiong L, Yang Y (2003) Disease resistance and abiotic stress tolerance in rice are inversely modulated by an abscisic acid–inducible mitogen-activated protein kinase. Plant Cell 15:745–759

Xu D, Duan X, Wang B, Hong B, Ho THD, Wu R (1996) Expression of a late embryogenesis abundant protein gene, HVA1, from barley confers tolerance to water deficit and salt stress in transgenic rice. Plant Physiol 110:249–257

Xu G, Cui Y, Li M, Wang M, Yu Y, Zhang B, Huang L, Xia X (2013) *OsMSR2*, a novel rice calmodulin-like gene, confers enhanced salt tolerance in rice (*Oryza sativa* L.). AJCS 7(3):368–373

Xue ZY, Zhi DY, Xue GP, Zhang H, Zhao YX, Xia GM (2004) Enhanced salt tolerance of transgenic wheat (*Triticum aestivum* L.) expressing a vacuolar Na^+/H^+ antiporter gene with improved grain yields in saline soils in the field and a reduced level of leaf Na^+. Plant Sci 167:849–859

Yamaguchi T, Blumwald E (2005) Developing salt-tolerant crop plants: challenges and opportunities. Trends Plant Sci 10:615–620

Yang J, Zhang J, Liu K, Wang Z, Liu L (2007) Involvement of polyamines in the drought resistance of rice. J Exp Bot 58(6):1545–1555

Yang Q, Chen ZZ, Zhoua XF, Yina HB, Lia X, Xina XF, Honga XH, Zhu JK, Gong Z (2009) Over-expression of SOS (salt overly sensitive) genes increases salt tolerance in transgenic *Arabidopsis*. Mol Plant 2:22–31

Yang A, Dai X, Zhang WH (2012) AR2R3-type MYB gene, OsMYB2, is involved in salt, cold, and dehydration tolerance in rice. J Exp Bot 63:2541–2556

Yang C, Wang R, Gou L, Si Y, Guan Q (2017) Overexpression of *Populus trichocarpa* mitogen-activated protein Kinase Kinase4 enhances salt tolerance in tobacco. Int J Mol Sci 18:2090

Yusuf MA, Kumar D, Rajwanshi R, Strasser RJ, Govindjee MTM, Sarin NB (2010) Overexpression of γ-tocopherol methyl transferase gene in transgenic *Brassica juncea* plants alleviates abiotic stress: physiological and chlorophyll a fluorescence measurements. Biochim Biophys Acta 1797:1428–1438

Zhai Y, Wang Y, Li Y, Lei T, Yan F, Su L, Li X, Zhao Y, Sun X, Li J, Wang Q (2013) Isolation and molecular characterization of GmERF7, a soybean ethylene-response factor that increases salt stress tolerance in tobacco. Gene 513:174–183

Zhang B (2015) MicroRNA: a new target for improving plant tolerance to abiotic stress. J Exp Bot 66:1749–1761

Zhang HX, Blumwald E (2001) Transgenic salt tolerant tomato plants accumulate salt in the foliage but not in the fruits. Nat Biotechnol 19:765–768

Zhang S, Klessig DF (2001) MAPK cascades in plant defense signaling. TRENDS Plant Sci 6:520–527

Zhang B, Wang Q (2016) MicroRNA, a new target for engineering new crop cultivars. Bioengineered 7(1):7–10

Zhang HX, Hodson JN, Williams JP, Blumwald E (2001) Engineering salt-tolerant Brassica plants: characterization of yield and seed oil quality in transgenic plants with increased vacuolar sodium accumulation. Proc Natl Acad Sci U S A 98:12832–12836

Zhang L, Xi D, Li S, Gao Z, Zhao S, Shi J, Wu C, Guo X (2011) A cotton group C MAP kinase gene, GhMPK2, positively regulates salt and drought tolerance in tobacco. Plant Mol Biol 77:17–31

Zhang X, Liu X, Wu L, Yu G, Wang X, Ma H (2015) The SsDREB transcription factor from the succulent halophyte *Suaeda salsa* enhances abiotic stress tolerance in transgenic tobacco. Int J Genomics 2015:875497

Zhang X, Shabala S, Koutoulis A, Shabala L, Zhou M (2017) Meta-analysis of major QTL for abiotic stress tolerance in barley and implications for barley breeding. Planta 245(2):283–295

Zhao YL, Wang HM, Shao BX, Chen W, Guo ZJ, Gong HY, Sang XH, Wang JJ, Ye WW (2016) SSR-based association mapping of salt tolerance in cotton (*Gossypium hirsutum* L.). Genet Mol Res 15(2):gmr.15027370

Zheng X, Chen B, Lu G, Han B (2009) Overexpression of a NAC transcription factor enhances rice drought and salt tolerance. Biochem Biophys Res Commun 379:985–989

Zhou QY, Tian AG, Zou HF, Xie ZM, Lei G, Huang J, Wang CM, Wang HW, Zhang JS, Chen SY (2008) Soybean WRKY-type transcription factor genes, GmWRKY13, GmWRKY21, and GmWRKY54, confer differential tolerance to abiotic stresses in transgenic *Arabidopsis* plants. Plant Biotechnol J 6:486–503

Zhou M, Li D, Li Z, Hu Q, Yang C, Zhu L, Luo H (2013) Constitutive expression of a miR319 gene alters plant development and enhances salt and drought tolerance in transgenic creeping bentgrass. Plant Physiol 161:1375–1391

Zhu N, Cheng S, Liu X, Du H, Dai M, Zhou DX, Yang W, Zhao Y (2015) The R2R3-type MYB gene OsMYB91 has a function in coordinating plant growth and salt stress tolerance in rice. Plant Sci 236:146–156

Chapter 9
Advances in Genetics and Breeding of Salt Tolerance in Soybean

Huatao Chen, Heng Ye, Tuyen D. Do, Jianfeng Zhou, Babu Valliyodan, Grover J. Shannon, Pengyin Chen, Xin Chen, and Henry T. Nguyen

Abstract Salt stress is one of the major abiotic factors affecting crop growth and production. In general, soybean is sensitive to salt stress. The success of soybean improvement for salt tolerance depends on discovery and utilization of genetic variation in the germplasm. In this chapter, advance in salt-tolerant research and breeding was summarized by highlighting the genetic diversity, quantitative trait loci (QTL), identification of the major locus (*Glyma03g32900*), and improvement of soybean varieties in salt tolerance. The ion exclusion and tissue tolerance mechanisms regulated by this major locus are discussed. In addition, genomic resources and high-throughput phenotyping platforms that can facilitate a better understanding of phenotype-genotype association and formulate genomic-assisted breeding strategies are prospected.

Keywords Soybean · QTL · QTN · Next-generation sequencing · Marker-assisted selection · Single nucleotide polymorphism

---

H. Chen
Division of Plant Sciences, University of Missouri, Columbia, MO, USA

Institute of Industrial Crops, Jiangsu Academy of Agricultural Sciences, Nanjing, China
e-mail: cht@jaas.ac.cn; chenhuat@missouri.edu

H. Ye · T. D. Do · B. Valliyodan · G. J. Shannon · H. T. Nguyen (✉)
Division of Plant Sciences, University of Missouri, Columbia, MO, USA
e-mail: yehe@missouri.edu; tddqk7@mail.missouri.edu; valliyodanb@missouri.edu; shannong@missouri.edu; nguyenhenry@missouri.edu

J. Zhou
Division of Agricultural Systems Management, University of Missouri, Columbia, MO, USA
e-mail: zhoujianf@missouri.edu

P. Chen
Division of Plant Sciences, University of Missouri, Delta Research Center, Portageville, MO, USA
e-mail: chenpe@missouri.edu

X. Chen (✉)
Institute of Industrial Crops, Jiangsu Academy of Agricultural Sciences, Nanjing, China
e-mail: cx@jaas.ac.cn

V. Kumar et al. (eds.), *Salinity Responses and Tolerance in Plants, Volume 2*,
https://doi.org/10.1007/978-3-319-90318-7_9

Abbreviations

CAX	Cation exchanger
GWAS	Genome-wide association studies
MAS	Marker-assisted selection
NGS	Next-generation sequencing (NGS)
NHX	Na^+/H^+ antiporter
QTL	Quantitative trait loci
QTN	Quantitative trait nucleotides
SNP	Single nucleotide polymorphism

9.1 Introduction

Food security is under serious concern because of the increasing world population (Rosegrant and Cline 2003; Stocking 2003; Schmidhuber and Tubiello 2007; Brown and Funk 2008). Demand for crop production increase up to 100–110% was predicted from 2005 to 2050 (Tilman et al. 2011). Salt stress is one of the major abiotic factors affecting crop growth and production. The saline and sodic soils were estimated as 397 and 434 million ha worldwide, respectively (FAO 2000). It was reported that 19.5% of irrigated land and 2.1% of dry land were affected by salt stress. These salt-affected lands produced about 20% of the world's food consumption (Pimentel et al. 2004). Therefore, improving crop salt tolerance for these lands is a practical way to boost crop yield to meet the increasing food demand.

Soybean [*Glycine max* (L.) Merr.] is the primary crop source for protein and oil. In 2015, 320 million metric tons of soybean were produced, which provide 61% of the oilseed and 71% of protein meal consumption in the world (www.soystats.com). Soybean is regarded as a salt-sensitive crop (Munns and Tester 2008). Soybean yields are negatively affected by the soil salinity with a concentration exceeding 5 dS/m (Ashraf 1994; Phang et al. 2008). Seed yields could be reduced 50% when the electrical conductivity of the saturation extract of soil was 9 millimhos/cm (Abel and MacKenzie 1964). Papiernik et al. (2005) reported that increasing salt stress led to 40% yield reduction in soybean. Toxicity will happen when Na^+ and Cl^- and ions are absorbed and accumulated at high concentrations in soybean plants, and plant death happens with increasing salt concentration in the soil (Pathan et al. 2007; Phang et al. 2008). Therefore, improving the salt tolerance in soybean is needed to ensure food security for the world. The success of such improvement depends largely on discovery and utilization of genetic variation present in the germplasm and characterization of salt-tolerant genes and mechanisms.

Fig. 9.1 Phenotyping of different soybean accessions under salt-stress condition

9.1.1 Salt Stress Inhibits Normal Growth and Development of Soybean

The germination of soybean seeds can be delayed in low salt condition, and the germination rates are significantly decreased in a high salt concentration condition (Abel and MacKenzie 1964). Abel (1969) found the effect of salt on seed germination was more prominent in salt-sensitive genotypes than salt-tolerant ones. However, other reports showed no significant correlation between germination and seedling stages in soybean (Essa 2002). Recent research showed that soybeans are more sensitive to salt stress at the seedling stage compared to the germination stage (Hosseini et al. 2002; Phang et al. 2008). Therefore, salt tolerance at germination and seedling stages could be regulated by different mechanisms.

The growth of soybean plant is inhibited by salt stress. Salt-tolerant soybean varieties demonstrated less injury than salt-sensitive varieties (Fig. 9.1). The salt injury phenotypes include leaf scorching, reduction of plant size, biomass, number of internodes, number of branches, number of pods, weight per plant, and weight of 100 seeds (Abel and MacKenzie 1964; Chang et al. 1994). The quality of soybean seeds is also affected by salt stress, as the oil and protein content of seeds can be affected under salt stress (Abel and MacKenzie 1964; Chang et al. 1994; Wan et al. 2002). Chlorophyll content of soybean leaves significantly declined under salt stress (Hamayun et al. 2015). The gibberellins content (GAs) and free sialic acid (free SA) of soybean plants were also reduced under salt stress (Hamayun et al. 2015). In the contrast, the abscisic acid (ABA) content and jasmonic acid (JA) content were increased by salt stress (Hamayun et al. 2015).

Intensive work has been conducted to investigate salt tolerance at the seedling stage. But little has been focused on the salt tolerance during germination. Therefore, future efforts are needed to address the salt tolerance during seed germination.

Rhizobium colonization of inoculated soybean root surfaces was not affected by salt treatment, but the subsequent nodule formation and development were significantly affected by salt stress (Singleton and Bohlool 1984). Total shoot N content was declined under salt stress, and the decrease in concentration of N in shoots was also observed after salt treatment. The decrease in N content in shoots was correlated by the sharp reduction in nodule number and dry weight. Nodule number and weight were reduced by approximately 50% at a low NaCl concentration condition and by more than 90% at a high NaCl concentration condition. The weakening of the soybean root system by salt stress could be a reason for the reduction in nodule formation and development, but other mechanisms may exist and need to be investigated. The special ion toxicities are different between soybean cultivars and wild soybean accessions under salt stress. The ion Cl^- is the main resource of ion toxicity of soybean cultivars suffered under NaCl stress, and wild soybean accessions were tolerant to Cl^- toxicity. The salt tolerance was regulated mainly by withholding of Cl^- in roots and stems to decrease its accumulation in leaves of soybean cultivars. The salt tolerance was mainly determined by the success of Na^+ in roots and stems to decrease its concentration in leaves in wild soybean accession (Luo et al. 2005b).

9.2 Salt Tolerance in Plants

Salt tolerance in plants involves complex physiological traits, metabolic pathways, and molecular networks (Gupta and Huang 2014). Salinity affects growth by imposing both osmotic and ionic stress. The presence of high concentrations of salt lowers soil water potential, thus making it harder for roots to uptake water. The ionic stress is associated with the gradual accumulation of salts in plant tissues over time (Munns and Tester 2008). Exploring salt-tolerant mechanisms in plants will allow the development of integrative approaches to facilitate breeding for salt-tolerant crop varieties.

Current knowledge points to three major mechanisms of salt tolerance in plants, including ion exclusion in roots, osmotic tolerance, and tissue tolerance (Roy et al. 2014). So far, the information of the mechanism of osmotic tolerance is still limited. Progress has been made in ion exclusion and tissue tolerance mechanisms in plants. Some specific molecular pathways have been identified to regulate salt tolerance in plants through maintaining ion homeostasis.

9.2.1 Maintaining Ion Homeostasis to Tolerate Salt Stress in Plants

Maintaining ion homeostasis by ion uptake and compartmentalization is an essential process for growth during salt stress (Niu et al. 1995; Serrano et al. 1999; Hasegawa 2013). The major form of salt present in the soil is NaCl; thus, the main focus of research is the study of the transport mechanism of Na^+ and Cl^- ions and Na^+ compartmentalization.

9.2.1.1 Ion Exclusion

When plants grow in typical NaCl-dominated saline environments, high Na^+ will be accumulated in the cytosol, which causes high Na^+/K^+ ratios. The high Na^+/K^+ ratios will disrupt enzymatic functions that are normally activated by K^+ in cells (Bhandal and Malik 1988; Munns et al. 2006; Tester and Davenport 2003). Therefore, it is very important for cells to maintain a low concentration of cytosolic Na^+ or to maintain a relatively low Na^+/K^+ ratio in the cytosol to resist salt stress (Maathuis and Amtmann 1999). Therefore, ion exclusion is one strategy for plants to resist salt stress by preventing these ions from entering the stem and leaf tissues. High-affinity K^+ transporters (HKTs) play an important role in maintaining a low Na^+/K^+ ratio in plants (Gierth and Mäser 2007; Hauser and Horie 2010). The first plant HKT was identified as TaHKT1 from wheat, and it works as either a high-affinity Na^+–K^+ symporter or a low-affinity Na^+ transporter depending on the external Na^+–K^+ concentrations (Gassmann et al. 1996; Rubio et al. 1995). The *HKT* gene family can unload Na^+ from xylem to roots to prevent excessive Na^+ transported to aboveground tissues (Berthomieu et al. 2003; Mäser et al. 2002; Møller et al. 2009; Ren et al. 2005; Rus et al. 2004; Sunarpi et al. 2005). These genes usually show cell-type-specific expression, and manipulation of the expression of these genes in the target tissues could be an approach to modify salt tolerance.

9.2.1.2 Tissue Tolerance

Three major mechanisms contributing to tissue tolerance have been identified, including accumulation of sodium in the vacuole, synthesis of compatible solutes, and production of enzymes catalyzing detoxification of reactive oxygen species. ABA and reactive oxygen species (ROS) signaling pathways were found to be involved in tissue tolerance to salt stress (Ren et al. 2012). The tissue tolerance was found to be enhanced by increasing cyclic electron flow, which promotes Na^+ sequestration into vacuoles in soybean (He et al. 2015). A salt-tolerant soybean line (S111-9) was found to accumulate Na^+ in the vacuole, while sensitive line (Melrose) plants accumulated Na^+ in the chloroplast under salt stress (He et al. 2015). This salt-tolerant line showed higher expression of genes associated with Na^+ transport into the vacuole of cells compared with the sensitive line (He et al. 2015). Therefore, manipulation of Na^+ accumulation in vacuoles could be another strategy to improve the performance of soybean under salt stress.

9.2.2 SOS Signaling Pathway Conferring Salt Tolerance

Salt overly sensitive (SOS) signaling pathway has been identified to maintain ion homeostasis through the mediation of cellular signaling (Zhu et al. 1998; Zhu 2000, 2002; Wu et al. 1996). The SOS signaling pathway was identified based on isolation and characterization of *Arabidopsis* mutants showing root growth hypersensitivity after salt treatment (Zhu et al. 1998). The SOS signaling pathway consists of three major members: SOS1, SOS2, and SOS3. *SOS1*, encoding a Na^+/H^+ antiporter, is crucial in modulating Na^+ efflux at the cellular level (Qiu et al. 2002; Shi et al. 2002). Previous reports showed that *SOS1* is required for Na^+ tolerance in plants under salt stress and plays important role in regulating Na^+ compartmentalization in the vacuole and controlling Na^+ loading to the xylem (Shi et al. 2002; Oh et al. 2010). *SOS1* was thought to be the target of the phospholipase D signaling pathway in ion sensing and dynamic equilibrium adjustment under salt stress (Yu et al. 2010). However, overexpression of *SOS1* gene in either *A. thaliana* or *T. salsuginea* did not lead to a significant improvement in Na^+ tolerance (Shi et al. 2003). These findings suggest that tissue-specific expression of *SOS1*, such as stelar cells to unload Na from xylem, is needed to improve salt tolerance in plants.

SOS2 encodes a serine/threonine kinase and can be activated by salt-stress-elicited Ca^+ signals. SOS2 was also shown to interact with ABI2 and nucleoside diphosphate kinase 2, suggesting SOS cross talks with both ABA and ROS signaling pathways (Ohta et al. 2003; Verslues et al. 2007). Furthermore, SOS2 was also found to regulate the activity of the transporters, including Na^+/H^+ exchanger, which are located in tonoplast in plants(Cheng et al. 2004; Qiu et al. 2004; Batelli et al. 2007; Huertas et al. 2012).

SOS3 encodes a myristoylated Ca^+ binding protein. The SOS3 protein contains an N-terminus myristoylation site, which plays a critical role in salt tolerance in *Arabidopsis* (Ishitani et al. 2000). SOS3 protein was identified as a primary calcium sensor to perceive the increase in cytosolic Ca^{2+} triggered by excess Na^+ in cytoplasm and it subsequently activate SOS2 to regulate the activity of ion transporters (Liu and Zhu 1998; Halfter et al. 2000; Liu et al. 2000; Hrabak et al. 2003). The positive roles of SOS3 were identified in the promotion of lateral root growth in *Arabidopsis* under mild salt stress (Zhao et al. 2011).

Overall, SOS signaling pathway is not purely a linear system from an unknown Na^+ sensor through SOS3 and SOS2 to mediate the Na^+/H^+ antiporter activity of SOS1 (Ji et al. 2013). Its additional functions have also been identified as a node point of cross talk with other signaling pathways (Katiyar-Agarwal et al. 2006; Jaspers et al. 2010). Manipulation of the genes in the SOS signaling pathway could be another approach to improve salt tolerance in soybean.

9.3 Natural Genetic Variation, QTLs/QTNs Mapped, and Genes Cloned for Salt Tolerance in Soybean

Stress-tolerant plants have evolved their stress adaptation mechanisms under natural selection. Exotic and wild relatives growing under wild conditions tend to have better stress-tolerant characteristics, including salt-tolerant traits (Mickelbart et al. 2015). Incorporation of these salt-tolerant resource into cultivars by transferring the underlying natural variations is essential to improve the performance of cultivars under salt stress. These key natural variations can be identified through quantitative trait locus (QTL) mapping and genome-wide association studies (GWAS). With the assistance of genomic strategies, these natural variations can be identified and expeditiously incorporated into the cultivars by marker-assisted selection (MAS) and genomic selection. Further characterization of genes underlying the stress tolerance traits will elucidate specific biological mechanisms of stress tolerance, which will be consequently translated to crop breeding.

9.3.1 Genetic Diversity of Salt Tolerance in Soybean

A large amount of work has been done to investigate salt tolerance in soybean. Based on the responses of soybean plants to salt stress, several phenotyping indices were developed to evaluate salt-tolerant levels of soybean plants (Table 9.2). The most commonly used indices are based on visual rating, including leaf scorch score, salt-tolerant rating, and survival rate. With these phenotyping indices, rich genetic variations of salt tolerance were observed in soybean, and salt-tolerant soybean lines were identified over the years (Table 9.1). Initially, ten salt-tolerant soybean cultivars were identified, and the tolerant cultivars can significantly inhibit the accumulation of Cl^- in leaf tissue than the sensitive cultivars (Parker et al. 1983). These tolerant cultivars also showed clear yield advantage of the sensitive cultivars by 40% (Parker et al. 1983). Later on, 19 soybean cultivars were identified as Cl^- excluders from 60 lines (Yang and Blanchar 1993). Recent screening for salt tolerance identified 150 salt-tolerant germplasm lines, such as Lee, Lee 68, and S-100 (www.ars-grin.gov/npgs/searchgrin.html). Genetic variation in salt tolerance was also observed in wild soybean. Recently, five wild soybean accessions were identified to be tolerant to salt stress, including BB52, N23232, JWS156-1, PI 483463,

Table 9.1 Major salt-tolerant soybean lines

Accession name	Sub-species	References
S-100	*Glycine max*	Lee et al. (2004)
Lee	*Glycine max*	Lee et al. (2004)
Lee 68	*Glycine max*	Chen et al. (2008)
Lee 74	*Glycine max*	Lee et al. (2004)
Fengzitianandou	*Glycine max*	Zhang et al. (2014)
Baiqiu 1	*Glycine max*	Zhang et al. (2014)
FT-Abyara	*Glycine max*	Xu et al. (2016)
Jindou 6	*Glycine max*	Xu et al. (2016)
Tiefeng 8	*Glycine max*	Guan et al. (2014)
Fiskeby III	*Glycine max*	Do et al. (2018)
JWS156–1	*Glycine soja*	Hamwieh and Xu (2008)
PI483463	*Glycine soja*	Lee et al. (2009) and Ha et al. (2013)
W05	*Glycine soja*	Qi et al. (2014)

and W05 (Luo et al. 2005b; Chen et al. 2013; Hamwieh and Xu 2008; Lee et al. 2009; Ha et al. 2013; Qi et al. 2014). In a more recent study, a total of 123 soybean accessions (117 cultivated soybean accessions and 6 wild soybean accessions) were identified as tolerant to salt stress using hydroponic culture in the greenhouse (Do et al. 2016). Several new soybean accessions were identified to excellent salt tolerance from 600 soybean accessions, and these lines include a Brazilian soybean cultivar FT-Abyara, a Chinese soybean cultivar Jindou 6, and a Japanese wild soybean accession JWS156–1 (Xu et al. 2016). The existing genetic diversity in salt tolerance in soybean offers genetic resources to breed salt-tolerant varieties.

9.3.2 Genetic Studies of Salt Tolerance in Soybean

The heredity of salt tolerance in soybean was previously analyzed as a quality trait using classical genetics approach, and the gene symbols *Ncl* and *ncl* were proposed as the dominant for tolerance and the recessive for sensitive, respectively (Abel 1969). Inheritance test in F_2, F_3, and BC_1F_1 populations pointed that Cl^- accumulation in the upper part of soybean was controlled by a single locus (*Ncl*) (Abel 1969). Shao et al. (1994) and Guan et al. (2014) further confirmed the result of a single locus controlling salt tolerance in cultivated soybean. The inheritance of salt tolerance in wild soybean was studied using the accession PI483463. The F_2 population from a cross of PI483463 (wild tolerant) × S-100 (cultivated tolerant) segregated into a tolerant to sensitive ratio of 15:1, indicating the wild accession may have additional locus/loci for salt tolerance. Recent study in a recombinant inbred line (RIL) population pointed out that salt tolerance in soybean was controlled by major gene(s) mixed with minor polygenes and salt tolerance in soybean should be considered as a quantitative trait (Chen et al. 2011b).

9.3.3 QTL Mapping for Salt Tolerance During Seedling Stage in Soybean

QTL mapping for salt tolerance has been focused mainly on the seedling stage in soybean (Table 9.2). The first QTL mapping was performed in an $F_{2:5}$ population derived from the cross of S-100 (salt tolerant) and Tokyo (salt sensitive), and two QTL were mapped on linkage groups L and N (Lee et al. 2004). The QTL on linkage group N (Chr.3) showed the major effect in salt tolerance with a phenotypic contribution of up to 60% (Lee et al. 2004). QTL mapping in a RIL population derived from the cross between Kefeng No. 1 and Nannong1138-2 identified eight QTLs associated with salt tolerance and confirmed the QTL on linkage group N (Chr.3) as a major one (Chen et al. 2008). The same major QTL on linkage group N (Chr.3) was confirmed in an F_2 population derived from a cross between the Jackson (sensitive cultivar) and JWS156-1(tolerant wild) (Hamwieh and Xu 2008). Later on,

Table 9.2 Markers associated with salt tolerance in soybean

Tolerance indices	Population	Parents	Linkage group (Chr.)	Markers	R^2 (%)	Donor allele	References
Visual salt-tolerant ratings	$F_{2:5}$	S-100 x Tokyo	N (Chr.3)	Satt237	41.0	S-100	Lee et al. (2004)
Salt-tolerant rating	F_2	PI548657 × JWS156–1	N (Chr.3)	Satt339	68.7	JWS156–1	Hamwieh and Xu (2008)
Plant leaf SPAD value	F_2	PI548657 × JWS156–1	N (Chr.3)	Satt339	49.6	JWS156–1	Hamwieh and Xu (2008)
Salt-tolerant rating	RIL	FT-Abyara × C01	N (Chr.3)	Sat_091	44.0	FT-Abyara	Hamwieh et al. (2011)
Plant leaf SPAD value	RIL	FT-Abyara × C01	N (Chr.3)	Sat_091	38.2	FT-Abyara	Hamwieh et al. (2011)
Salt-tolerant rating	RIL	Jin dou No. 6×0197	N (Chr.3)	Sat_091	47.1	Jin dou No. 6	Hamwieh et al. (2011)
Plant leaf SPAD value	RIL	Jin dou No. 6 × 0197	N (Chr.3)	Sat_091	40.9	Jin dou No. 6	Hamwieh et al. (2011)
Leaf scorch	RIL	PI 483463 × Hutcheson	N (Chr.3)	Satt255	56.5	PI 483463	Ha et al. (2013)
Percentage of plant survival	RIL	Kefeng No.1 × Nannong1138–2	N (Chr.3)	satt237	7.8	Nannong1138–2	Chen et al. (2008)
Plant survival days; salt-tolerant rating	RIL	Kefeng No.1 × Nannong1138–2	G (Chr.18)	Sat_164	17.9	Nannong1138–2	Chen et al. (2008)
Percentage of plant survival; salt-tolerant rating	RIL	Kefeng No.1 × Nannong1138–2	K (Chr.9)	Sat_363	19.2	Nannong1138–2	Chen et al. (2008)
Salt-tolerant rating	RIL	Kefeng No.1 × Nannong1138–2	M (Chr.7)	Satt655	19.7	Nannong1138–2	Chen et al. (2008)
Leaf scorch	$F_{2:3}$	Williams 82 × Fiskeby III	Chr.13	Gm13_37204738	11.5	Fiskeby III	Do et al. (2018)

this major QTL was confirmed for its effect in salt tolerance at the near-isogenic backgrounds (Hamwieh et al. 2011). The genomic region of this major QTL was further narrowed to 209 kb flanked by markers QS08064 and Barcsoyssr_3_1301 on Chr.3 (Guan et al. 2014).

Overall, the major salt tolerance QTL is located on linkage group N (Chr.3), and other important QTLs conferring salt tolerance were also identified on linkage groups G (Chr.18), K (Chr.9), and M (Chr.7) in soybean (Table 9.2). The donor lines and associated DNA markers for these QTLs can be used in genomic-assisted breeding to incorporate the target genomic regions into the breeding lines (Table 9.2). Further work on the effect of these QTLs is still needed to fully understand their value in breeding practice. Recently, a QTL mapping study was performed in a cross of William 82 × Fiskeby III (Do et al. 2018). In this study, two QTLs were mapped with one overlapped with the previously reported one on Chr.3 and a new one with a relatively minor effect (R^2 = 11.5%) on Chr.13. Both of the tolerance donor alleles at the two loci were derived from the salt-tolerant cultivar Fiskeby III, which could explain the supreme salt-tolerant performance of this tolerant cultivar.

9.3.4 QTL Mapping Associated with Salt Tolerance During Germination Stage in Soybean

Seed germination is highly sensitive to salt and severely inhibited as salinity increases (Fredj et al. 2013). The tolerance mechanisms at the germination stage evolved divergently compared with that at the seedling stage in soybean (Pathan et al. 2007; Phang et al. 2008). Relatively less research has been done for salt tolerance at the germination stage. Eleven QTLs were detected to be associated with salt tolerance during germination stage using a soybean RIL population, and each of them explained 4.49–25.94% of phenotypic variations (Kan et al. 2016). In another study, a total of 83 QTL-by-environment interactions and 1 epistatic QTL were detected using epistatic association mapping implemented with an empirical Bayes algorithm (Zhang et al. 2014). So far, the QTL information for salt tolerance in soybean germination stage is still limited. Therefore, more work needs to be carried out to focus on salt tolerance at germination stage in soybean, as salt tolerance during seed germination stage is a critical determinant of uniform stand establishment under salt stress.

9.3.5 Positional Cloning of the Major Salt-Tolerant QTL Located on Chr.3

The major salt-tolerant QTL located on Chr.3 (linkage group N) was identified by several research groups using different soybean mapping populations. A research group from Hong Kong, China, by association mapping assisted by whole-genome resequencing technique (Qi et al. 2014), first cloned this QTL. *Glyma03g32900*, encoding an ion transporter, was identified as the underlying gene for this QTL and named as *GmCHX1* (Qi et al. 2014). The function of this gene in salt tolerance was confirmed using the transgenic hairy root assay and B2Y cells overexpression. *GmCHX1* was identified to exclude Na^+ from aboveground tissues, as lower Na^+ and the Na^+/K^+ ratio were observed in the soybean lines with the wild-type alleles and *GmCHX1* transgenic tobacco lines, suggesting that *GmCHX1* is involved in ion homeostasis under salt stress (Qi et al. 2014). Haplotype analysis showed that *GmCHX1* is highly conserved in the tolerant alleles than that in the sensitive alleles, suggesting the tolerant allele was subjected to selection (Qi et al. 2014). In the same year, the other research group in Beijing, China, also cloned this major QTL through map-based cloning (Guan et al. 2014). The QTL underlying gene was cloned as the same gene of *Glyma03g32900* on Chr.3 and named as *GmSALT3* (Guan et al. 2014). In addition to the work done by the first group, Guan et al. (2014) identified that *GmSALT3* was preferentially expressed in the stelar cells around xylem. Therefore, the mechanism of ion homeostasis regulated by this gene was hypothesized to unload Na^+ from xylem to achieve Na^+ exclusion in aboveground tissues (Fig. 9.2). Later on, this group found that *GmSALT3* can also exclude Cl^- from aboveground tissues, suggesting that this gene could be a cotransporter for Na^+ and Cl^- (Liu et al. 2016). More recently, the same gene was map-based cloned and named as *Ncl* by a Japanese group (Do et al. 2016). The ability of cotransporting Na^+ and Cl^- was confirmed in their studies (Do et al. 2016).

Overall, three research groups cloned the major QTL on Chr.3 as the same gene encoding an ion transporter. The salt-tolerant mechanisms regulated by this gene

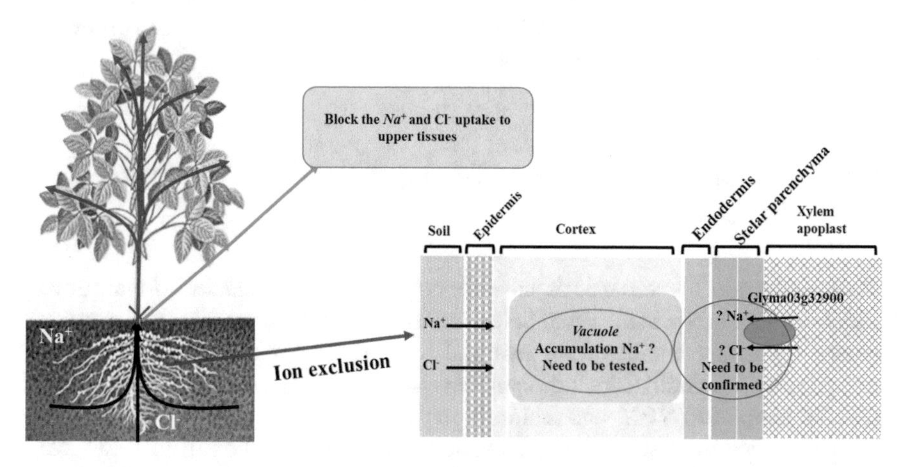

Fig. 9.2 Working model of salt tolerance regulated by *glyma03g32900* in soybean. The figure was modified from Fig. 9.1 in Roy et al. (2014)

can be deduced as improvement tissue tolerance by compartmentalization Na^+ and Cl^- into vacuoles in the root cells and be unloading these ions from xylem back to roots (Fig. 9.2). However, more experiments are needed to confirm this working model.

9.3.6 GWAS for Salt Tolerance in Soybean

Benefited from the advance of high-density SNP array genotyping platform and next-generation sequencing (NGS) techniques, GWAS was broadly used to detect the associated SNPs for many agronomy and stress-resistant traits in soybean (Mamidi et al. 2011; Hao et al. 2012; Hwang et al. 2014; Wen et al. 2014, 2015; Zhang et al. 2015). Recently, GWAS was successfully performed to identify the SNP associated with salt tolerance in soybean. Eight SNPs were detected to be associated with salt-tolerant phenotypes at the germination stage in soybean (Kan et al. 2015). Nine candidate genes were pointed based on the positions of these associated SNPs (Kan et al. 2015). The whole-genome resequencing (WGRS) data of the 106 soybean lines were also used to detect the SNPs associated with salt tolerance at the seedling stage, and a significant SNP within the previously cloned salt-tolerant gene *Glyma03g32900* was identified, explaining 63% of the phenotypic variation (Patil et al. 2016). More recently, a GWAS was performed to detect SNPs associated with salt tolerance in a diverse set of 283 soybean lines using SoySNP50K BeadChip database (Zeng et al. 2017). As a result, 45 SNPs representing 9 genomic regions on 9 chromosomes were identified to be significantly associated with two salt-tolerant indices. Seven out of the nine QTLs were detected for the first time (Zeng et al. 2017). These SNPs associated with salt tolerance can be used to develop DNA markers for marker-assisted selection.

9.3.7 Candidate Genes Contributing to Salt-Tolerant Traits in Soybean

Maintaining ion homeostasis is an essential process to maintain normal growth under salt stress (Niu et al. 1995; Serrano et al. 1999; Hasegawa 2013). Genes from several exchanger families were analyzed conferring K^+, Na^+, and Cl^- transportation. These gene families include cation exchanger (CAX), (high-affinity K^+) HKT, (Na^+/H^+ antiporter) NHX, and (chloride channel) CLC types. These types of genes were identified to regulate the uptake and compartmentalization of ions (K^+, Na^+, and Cl^-) and supposed to be involved in maintaining ion homeostasis (Niu et al. 1995; Serrano et al. 1999; Hasegawa 2013). According to previous reports, several of these ion transport genes were identified conferring salt tolerance in soybean by regulating ion homeostasis under salt stress (Table 9.3).

Table 9.3 Genomic location and function annotation about the candidate genes compared to the major gene (*GmSALT3*, *GmCHX1*, *Ncl*) conferring salt tolerance in soybean

Gene name	Chromosome	Function	Reference
GmSALT3, *GmCHX1*, *Ncl*	3	Natural variation; regulating the accumulation of Na^+, K^+, and Cl^-	Guan et al. (2014), Qi et al. (2014), Do et al. (2016), and Liu et al. (2016)
GmHKT1	1	Affected K^+ and Na^+ transport	Chen et al. (2011a)
GmHKT1;4	6	Regulation Na^+/K^+ ratio	Chen et al. (2014)
GmNHX1	20	Sodium/hydrogen exchanger	Sun et al. (2006) and Li et al. (2006)
GmCLC1	5	Chloride channel	Li et al. (2006) and Wei et al. (2016)
GmCAX1	Not match	Membrane cation/proton antiporter	Luo et al. (2005a)

9.4 Advance in Breeding for Salt Tolerance in Soybean

9.4.1 Breeding Pedigree and Narrow Genetic Base for Salt Tolerance in Soybean

The genetic base for salt tolerance in soybean is narrow in the USA. Hundreds of current elite cultivars can be traced back to a small number of ancestral lines (Gizlice et al. 1994; Sneller 1994). Only 35 ancestral lines were found to contribute more than 95% of all alleles and 5 ancestral lines account for more than 55% genetic background in the publicly released varieties in USA (Gizlice et al. 1994). Research has been conducted to understand the breeding pedigree of the major salt-tolerant gene on Chr.3 in the US soybean varieties. Lee et al. (2004) screened for the major salt-tolerant gene on Chr.3 using two tightly linked DNA markers (Sat_091 and Satt237) in 27 soybean cultivars with pedigrees from the US soybean ancestors S-100 (tolerant) or Tokyo (sensitive). In general, lines with S-100 allele at this locus are tolerant to salt stress, while lines with Tokyo allele are sensitive (Lee et al. 2004). Interestingly, three salt-tolerant lines (Hill, Dyer, and Forrest) have different alleles as the tolerant ancestor (S-100), and one salt-sensitive line (Davis) has a different allele as the sensitive ancestor (Tokyo) (Lee et al. 2004). Another salt-tolerant cultivar (FT-Abyara) was found to be not related to S-100, suggesting the existence of other donor sources of this major tolerance gene other than S-100 (Hamwieh et al. 2011). Recently, haplotype analysis for this major locus using whole-genome resequencing data of 93 US ancestral lines showed that salt-tolerant alleles can be traced back to Lee and S-100 and the sensitive alleles can be traced back to Williams 82, Tokyo, Davis, and Arksoy (Patil et al. 2016). Therefore, the pedigree of this major locus was deduced based on above three reports (Fig. 9.3). This pedigree further confirmed the narrow genetic base in US soybean germplasm.

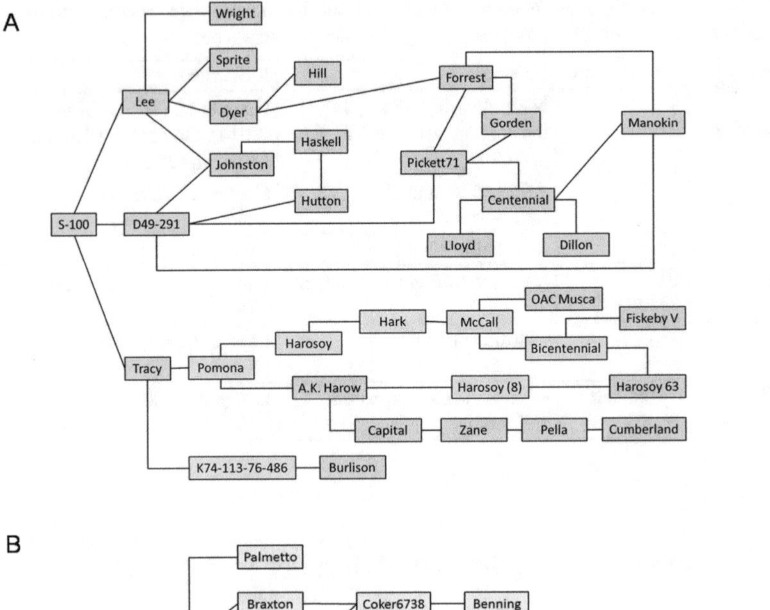

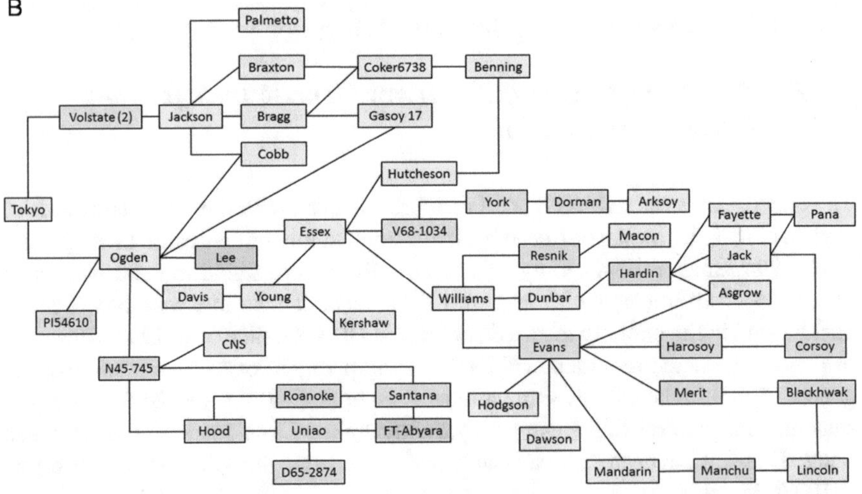

Fig. 9.3 Pedigree tracing of salt-tolerant and salt-sensitive soybean lines. Green represents salt-tolerant line with tolerant alleles at the major salt-tolerant gene on Chr.3; yellow represents salt-sensitive lines with the sensitive alleles at the locus

Similar narrow genetic base for salt tolerance in soybean was also discovered in the rest of the world. The donor sources were all traced back to the cloned major gene on Chr.3. In China, efforts were conducted to understand the domestication of this gene for salt tolerance. The tolerant alleles of this gene were found to be raised and accumulated at the river valley regions known for saline stress in the central part of China (Guan et al. 2014). The significant lower genetic diversity of the tolerant alleles compared with the sensitive alleles suggested that the tolerant alleles could be raised from more recent mutations and supposed to be gain-function mutations. The accumulation of the tolerant alleles at the regions of saline stress suggested the intensive artificial selection during soybean domestication. Currently, the genetic resource for salt tolerance in soybean is still limited to this major gene on Chr.3. Therefore, enriching the current gene pool of soybean is urgent to sustain the breeding success for salt tolerance by the introduction of new genetic resources (new alleles and new loci).

9.4.2 Markers and MAS for Salt Tolerance in Soybean

Marker-assisted selection (MAS) is a powerful genomic tool in breeding by targeting the desired genomic region during the crossing and selection. MAS has been applied in breeding for salt-tolerant soybean varieties, and high selection efficiency has been approved up to 94.2% (Guan et al. 2014). Strong correlations between SNP markers associated with salt-tolerant gene and salt-tolerant phenotypes were detected in different germplasm populations (Patil et al. 2016). These SNP markers showed a high selection rate of more than 91%. The major salt-tolerant gene on Chr.3 was incorporated into a salt-sensitive line (Jackson) by MAS using three DNA markers associated with the gene, and the introgression line (BC_4F_2) showed significantly improve salt tolerance (Do et al. 2016). These results demonstrated that salt-tolerant soybean varieties can be developed through introgression of major salt-tolerant genes using MAS method.

9.5 Future Perspectives

The mechanism regulated by the major salt-tolerant gene on Chr.3 has been discovered and utilized in the soybean breeding; however, the current genetic resources are also limited to this major gene. Hence, the next step is to explore the new loci in different soybean germplasm. Several new QTLs for salt tolerance are being detected other than the major one on Chr.3 (Table 9.2). Validation of the effects of these QTLs on salt tolerance and other agronomic traits is needed. Introduction of these newly detected QTL into elite soybean germplasm is necessary to develop new salt-tolerant varieties and enrich the genetic resources in salt tolerance for sustainable breeding. New important salt-tolerant genes in soybean need to be characterized. The mechanisms for salt tolerance are different between germination and seedling stages (Pathan et al. 2007; Phang et al. 2008). Considering the importance of salt tolerance during seed germination, further research is also needed to address this issue.

Currently, the phenotyping of salt tolerance is still relying on visual scoring. This requires experience, and the results may vary from person to person. The visual

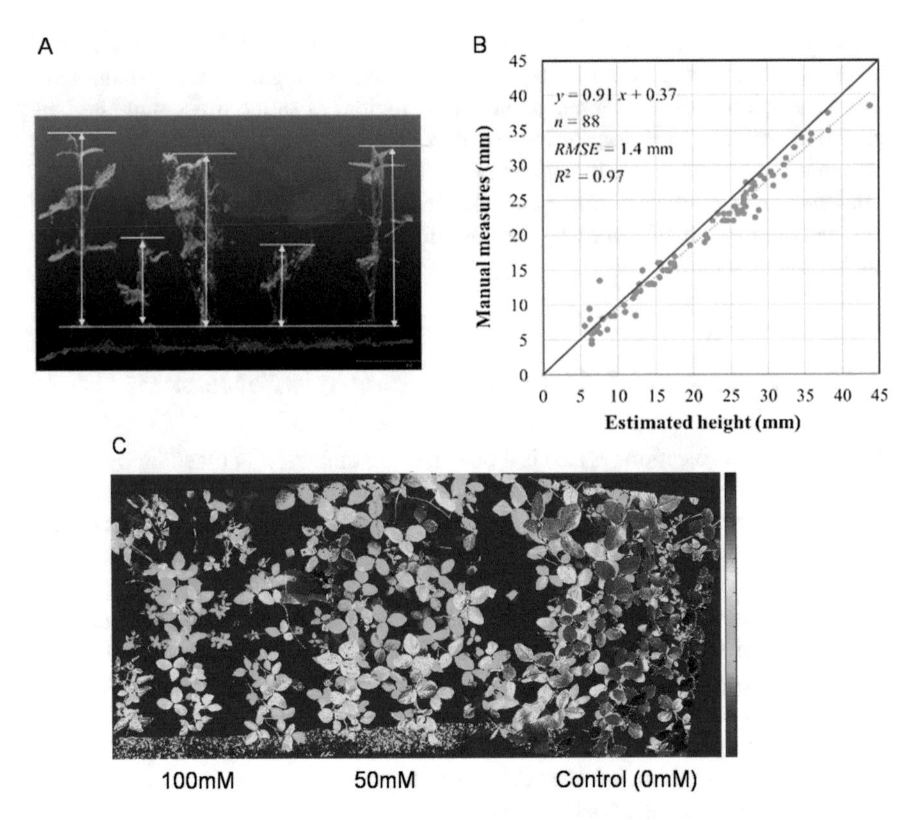

Fig. 9.4 Imaging system was used to evaluate the performances of different soybean accessions under salt-stress and control conditions. (**a**, **b**) Plant height estimated by imaging system and correlation analysis with real measurement data. (**c**) Plant size of soybean accessions measured by imaging system under salt-stress (NaCl) and control conditions

scoring systems usually rate 1–5 or 1–10 scales, which may not fully capture the variation in tolerance performance of different genotypes. High-throughput phenotyping technology with the development of an image-based data analysis system provides a new powerful tool. High-throughput phenotyping platforms that monitor canopy architecture and stress response are useful to advance our knowledge of salt tolerance (Fig. 9.4). It enables quantitative, nondestructive assessment of the responses of soybean to salinity during the seedling stage under salt-stress condition. Canopy spectral reflectance data can be measured with remote sensing, imaging, and digital processing in a high-throughput and cost-effective manner. Some stress-related traits, such as plant height, canopy size, canopy temperature, chlorophyll content, photosynthesis rate, and transpiration rate, can be estimated by analysis canopy spectral reflectance data. Certain wavelengths in the canopy reflectance data, such as the near-infrared wavelengths (730–1305 nm), was found to be highly correlated with agronomic traits, which offers a new source of variation to represent those original agronomic traits including stress tolerance.

Concurrently, next-generation sequencing technologies will facilitate faster understanding of the genetic architecture and characterization of salt-tolerant genes, leading to improved marker-assisted selection methods. Finally, the targeted genome editing by CRISPR/Cas9 system (Jacobs et al. 2015; Sun et al. 2015) can enable modification of these alleles to maximize the performance of soybean under stress conditions. The development of pan genomes and resequencing information of various legume crops can strengthen the comparative legume genomics for discovery and characterization of key genes and gene families involved in stress tolerance. Integration of these genomic resources and technologies into the next-generation breeding strategies will accelerate the genetic improvement of yield under stress conditions for crops, including soybean.

Acknowledgments This chapter is a joint contribution from the University of Missouri (MU), USA, and Jiangsu Academy of Agricultural Sciences (JAAS), China. We thank JAAS for Huatao Chen visiting scholarship at MU.

References

Abel GH (1969) Inheritance of the capacity for chloride inclusion and chloride exclusion by soybeans. Crop Sci 9:697–698

Abel GH, MacKenzie AJ (1964) Salt tolerance of soybean varieties (*Glycine max* L. Merrill) during germination and later growth. Crop Sci 4:157–161

Ashraf M (1994) Breeding for salinity tolerance in plants. Crit. Rev. Plant Sci. 13, 17–42. https://doi.org/10.1080/713608051.

Batelli G, Verslues PE, Agius F et al (2007) SOS2 promotes salt tolerance in part by interacting with the vacuolar H^+-ATPase and upregulating its transport activity. Mol Cell Biol 27:7781–7790

Berthomieu P, Conéjéro G, Nublat A et al (2003) Functional analysis of AtHKT1 in Arabidopsis shows that Na^+ recirculation by the phloem is crucial for salt tolerance. EMBO J 22:2004–2014

Bhandal IS, Malik CP (1988) Potassium estimation, uptake, and its role in the physiology and metabolism of flowering plants. Int Rev Cytol 110:205–254

Brown ME, Funk CC (2008) Food security under climate change. Science 319:580–581

Chang RZ, Chen YW, Shao GH et al (1994) Effect of salt stress on agronomic characters and chemical quality of seeds in soybean. Soybean Sci 13:101–105

Chen H, Cui S, Fu S et al (2008) Identification of quantitative trait loci associated with salt tolerance during seedling growth in soybean (*Glycine max* L.). Aust J Agric Res 59:1086–1091

Chen H, He H, Yu D (2011a) Overexpression of a novel soybean gene modulating Na^+ and K^+ transport enhances salt tolerance in transgenic tobacco plants. Physiol Plant 141(1):11–18

Chen HT, Chen X, Yu DY (2011b) Inheritance analysis and mapping quantitative trait loci (QTLs) associated with salt tolerance during seedling growth in soybean. Chin J Oil Crop Sci 33(3):231–234

Chen P, Yan K, Shao H et al (2013) Physiological mechanisms for high salt tolerance in wild soybean (Glycine soja) from Yellow River Delta, China: Photosynthesis, Osmotic Regulation, Ion Flux and antioxidant Capacity. PLoS One 8(12):e83227

Chen H, Chen X, Gu H et al (2014) GmHKT1;4, a novel soybean gene regulating Na^+/K^+ ratio in roots enhances salt tolerance in transgenic plants. Plant Growth Regul 73:299–308

Cheng NH, Pittman JK, Zhu JK et al (2004) The protein kinase SOS2 activates the Arabidopsis H^+/Ca^{2+} antiporter CAX1 to integrate calcium transport and salt tolerance. J Biol Chem 279:2922–2926

Do TD, Chen H, Hien VT et al (2016) Ncl synchronously regulates Na+, K+, and Cl− in soybean and greatly increases the grain yield in saline field conditions. Scientific Reports, 6, https://doi.org/10.1038/srep19147

Do TD, Vuong TD, Dunn D, Smothers S, Patil G, Yungbluth DC, Chen P, Scaboo A, Xu D, Carter TE, Nguyen HT, Grover Shannon J (2018) Mapping and confirmation of loci for salt tolerance in a novel soybeangermplasm, Fiskeby III. Theor Appl Genet 131(3):513–524. https://doi.org/10.1007/s00122-017-3015-0

Essa TA (2002) Effect of salinity stress on growth and nutrient composition of three soybean (*Glycine max* L. Merrill) cultivars. J Agron Crop Sci 188:86–93

FAO A (2000) Extent and causes of salt affected soils in participating countries. Available from http://www.fao.org/ag/agl/agll/spush/topic2.htm

Fredj MB, Zhani K, Hannachi C, Mehwachi T (2013) Effect of NaCl priming on seed germination of four coriander cultivars (*Coriandrum sativum*). Eurasia J Bio Sci 7:11–29

Gassmann W, Rubio F, Schroeder JI (1996) Alkali cation selectivity of the wheat root high-affinity potassium transporter HKT1. Plant J 10:869–952

Gierth M, Mäser P (2007) Potassium transporters in plants – involvement in K^+ acquisition, redistribution and homeostasis. FEBS Lett 581:2348–2356

Gizlice Z, Carter Jr TE, Burton JW (1994) Genetic base for the North American public soybean cultivars released between 1947 and 1988. Crop Sci 34:1143–1151

Guan R, Qu Y, Guo Y et al (2014) Salinity tolerance in soybean is modulated by natural variation in GmSALT3. Plant J 80:937–950

Gupta B, Huang B (2014) Mechanism of salinity tolerance in plants: physiological, biochemical, and molecular characterization. Int J Genom 2014:1–18. https://doi.org/10.1155/2014/701596

Ha BK, Vuong TD, Velusamy V et al (2013) Genetic mapping of quantitative trait loci conditioning salt tolerance in wild soybean (*Glycine soja*) PI 483463. Euphytica 193:79–88

Halfter U, Ishitani M, Zhu JK (2000) The *Arabidopsis* SOS2 protein kinase physically interacts with and is activated by the calcium-binding protein SOS3. Proc Natl Acad Sci U S A 97:3735–3740

Hamayun M, Hussain A, Khan SA et al (2015) Kinetin modulates physio-hormonal attributes and isoflavone contents of soybean grown under salinity stress. Front Plant Sci 6:377. https://doi.org/10.3389/fpls

Hamwieh A, Xu DH (2008) Conserved salt tolerance quantitative trait locus (QTL) in wild and cultivated soybeans. Breed Sci 58:355–359

Hamwieh A, Do DD, Cong H et al (2011) Identification and validation of a major QTL for salt tolerance in soybean. Euphytica 79:451–459

Hao D, Chao M, Yin Z et al (2012) Genome-wide association analysis detecting significant single nucleotide polymorphisms for chlorophyll and chlorophyll fluorescence parameters in soybean (*Glycine max*) landraces. Euphytica 186:919–931

Hasegawa PM (2013) Sodium (Na^+) homeostasis and salt tolerance of plants. Environ Exp Bot 92:19–31

Hauser F, Horie T (2010) A conserved primary salt tolerance mechanism mediated by HKT transporters: a mechanism for sodium exclusion and maintenance of high K^+/Na^+ ratio in leaves during salinity stress. Plant Cell Environ 33:552–565

He Y, Fu J, Yu C et al (2015) Increasing cyclic electron flow is related to Na^+ sequestration into vacuoles for salt tolerance in soybean. J Exp Bot 66:6877–6889

Hosseini MK, Powell AA, Bingham IJ (2002) Comparison of the seed germination and early seedling growth of soybean in saline conditions. Seed Sci Res 12:165–172

Hrabak EM, Chan CW, Gribskov M et al (2003) The *Arabidopsis* CDPK-SnRK superfamily of protein kinases. Plant Physiol 132:666–680

Huertas R, Olías R, Eljakaoui Z et al (2012) Overexpression of SlSOS2 (SlCIPK24) confers salt tolerance to transgenic tomato. Plant Cell Environ 35:1467–1482

Hwang EY, Song Q, Jia G et al (2014) A genome-wide association study of seed protein and oil content in soybean. BMC Genomics 15:1

Ishitani M, Liu J, Halfter U et al (2000) SOS3 function in plant salt tolerance requires N-myristoylation and calcium binding. Plant Cell 12(9):1667–1678

Jacobs TB, LaFayette PR, Schmitz RJ, Parrott WA (2015) Targeted genome modifications in soybean with CRISPR/Cas9. BMC Biotechnol12;15:16. https://doi.org/10.1186/s12896-015-0131-2

Jaspers P, Brosché M, Overmyer K et al (2010) The transcription factor interacting protein RCD1 contains a novel conserved domain. Plant Signal Behav 5:78–80

Ji H, Pardo JM, Batelli G et al (2013) The salt overly sensitive (SOS) pathway: established and emerging roles. Mol Plant 6:275–286

Kan GZ, Zhang W, Yang W et al (2015) Association mapping of soybean seed germination under salt stress. Mol Gen Genomics 290:2147–2162

Kan G, Ning L, Li Y et al (2016) Identification of novel loci for salt stress at the seed germination stage in soybean. Breed Sci 66(4):530–541

Katiyar-Agarwal S, Zhu J, Kim K et al (2006) The plasma membrane Na^+/H^+ antiporter SOS1 interacts with RCD1 and functions in oxidative stress tolerance in *Arabidopsis*. Proc Natl Acad Sci U S A 103:18816–18821

Lee GJ, Carter TE Jr, Villagarcia MR et al (2004) A major QTL conditioning salt tolerance in S-100 soybean and descendent cultivars. Theor Appl Genet 109:1610–1619

Lee JD, Shannon JG, Vuong TD et al (2009) Inheritance of salt tolerance in wild soybean (*Glycine soja* Sieb. and Zucc.) accession PI483463. J Hered 100:798–801

Liu J, Zhu JK (1998) A calcium sensor homolog required for plant salt tolerance. Science 280:1943–1945

Liu J, Ishitani M, Halfter U et al (2000) The *Arabidopsis thaliana* SOS2 gene encodes a protein kinase that is required for salt tolerance. Proc Natl Acad Sci U S A 97:3730–3734

Liu Y, Yu L, Qu Y et al (2016) *GmSALT3*, which confers improved soybean salt tolerance in the field, increases leaf Cl^- exclusion prior to Na^+ exclusion but does not improve early vigor under salinity. Front Plant Sci 7:1485

Luo GZ, Wang HW, Huang J et al (2005a) A putative plasma membrane cation/proton antiporter from soybean confers salt tolerance in Arabidopsis. Plant Mol Biol 59:809–820

Luo Q, Yu B, Liu Y (2005b) Differential sensitivity to chloride and sodium ions in seedlings of *Glycine max* and *G. soja* under NaCl stress. J Plant Physiol 162:1003–1012

Maathuis FJM, Amtmann A (1999) K^+ nutrition and Na^+ toxicity: the basis of cellular K^+/Na^+ ratios. Ann Bot 84:123–133

Mamidi S, Chikara S, Goos RJ et al (2011) Genome-wide association analysis identifies candidate genes associated with iron deficiency chlorosis in soybean. Plant Genome 4:154–164

Mäser P, Eckelman B, Vaidyanathan R et al (2002) Altered shoot/root Na^+ distribution and bifurcating salt sensitivity in *Arabidopsis* by genetic disruption of the Na^+ transporter AtHKT1. FEBS Lett 531:157–161

Mickelbart MV, Hasegawa PM, Bailey-Serres J (2015) Genetic mechanisms of abiotic stress tolerance that translate to crop yield stability. Nat. Rev. Genet16:237–251. https://doi.org/10.1038/nrg3901

Møller IS, Gilliham M, Jha D et al (2009) Shoot Na^+ exclusion and increased salinity tolerance engineered by cell type-specific alteration of Na^+ transport in *Arabidopsis*. Plant Cell 21:2163–2178

Munns R, Tester M (2008) Mechanisms of salinity tolerance. Annu Rev Plant Biol 59:651–681

Munns R, James AJ, Läuchli A (2006) Approaches to increasing the salt tolerance of wheat and other cereals. J Exp Bot 57:1025–1043

Niu X, Bressan RA, Hasegawa PM et al (1995) Ion homeostasis in NaCl stress environments. Plant Physiol 109(3):735–742

Oh DH, Lee SY, Bressan RA et al (2010) Intracellular consequences of SOS1 deficiency during salt stress. J Exp Bot 61:1205–1213

Ohta M, Guo Y, Halfter U et al (2003) A novel domain in the protein kinase SOS2 mediates interaction with the protein phosphatase 2C ABI2. Proc Natl Acad Sci U S A 100:11771–11776

Papiernik SK, Grieve CM, Lesch SM et al (2005) Effects of salinity, imazethapyr, and chlorimuron application on soybean growth and yield. Commun Soil Sci Plant Anal 36:951–967

Parker MB, Gascho GJ, Gains TP (1983) Chloride toxicity of soybeans grown on Atlantic Coast flatwoods soils. Agron J 75:439–443

Pathan MS, Lee JD, Shannon JG et al (2007) Recent advances in breeding for drought and salt stress tolerance in soybean. In: Jenks MA, Hasegawa PM, Jain SM (eds) Advances in molecular-breeding toward drought and salt tolerant crops. Springer, Dordrecht, pp 739–773

Patil G, Do T, Vuong TD et al (2016) Genomic-assisted haplotype analysis and the development of high-throughput SNP markers for salinity tolerance in soybean. Sci Rep 6:19199

Phang TH, Shao GH, Lam HM (2008) Salt tolerance in soybean. J Integr Plant Biol 50(10):1196–1212

Pimentel D, Berger B, Filiberto D et al (2004) Water resources: agricultural and environmental issues. Bioscience 54:909–918

Qi X, Li MW, Xie M et al (2014) Identification of a novel salt tolerance gene in wild soybean by whole-genome sequencing. Nat Commun 5:4340

Qiu QS, Guo Y, Dietrich MA et al (2002) Regulation of SOS1, a plasma membrane Na^+/H^+ exchanger in *Arabidopsis thaliana*, by SOS2 and SOS3. Proc Natl Acad Sci U S A 99:8436–8441

Qiu QS, Guo Y, Quintero FJ et al (2004) Regulation of vacuolar Na^+/H^+ exchange in *Arabidopsis thaliana* by the salt-overly-sensitive (SOS) pathway. J Biol Chem 279:207–215

Ren ZH, Gao JP, Li LG et al (2005) A rice quantitative trait locus for salt tolerance encodes a sodium transporter. Nat Genet 37:1141–1146

Ren S, Weeda S, Li H et al (2012) Salt tolerance in soybean WF-7 is partially regulated by ABA and ROS signaling and involves withholding toxic Cl^- ions from aerial tissues. Plant Cell Rep 31:1527–1533

Rosegrant MW, Cline SA (2003) Global food security: challenges and policies. Science 302:1917–1919

Roy SJ, Negrao S, Tester M (2014) Salt resistant crop plants. Curr. Opin. Biotechnol. 26:115–124.https://doi.org/10.1016/j.copbio.2013.12.004

Rubio F, Gassmann W, Schroeder JI (1995) Sodium-driven potassium uptake by the plant potassium transporter HKT1 and mutations conferring salt tolerance. Science 270:1660–1663

Rus A, Lee BH, Muñoz-Mayor A et al (2004) AtHKT1 facilitates Na^+ homeostasis and K^+ nutrition in planta. Plant Physiol 136:2500–2511

Schmidhuber J, Tubiello FN (2007) Global food security under climate change. Proc Natl Acad Sci U S A 104:19703–19708

Serrano R, Mulet JM, Rios G et al (1999) A glimpse of the mechanisms of ion homeostasis during salt stress. J Exp Bot 50:1023–1036

Shao GH, Chang RZ, Chen YW et al (1994) Study on inheritance of salt tolerance in soybean. Acta Agron Sin 20:721–726

Shi HZ, Quintero FJ, Pardo JM et al (2002) The putative plasma membrane Na^+/H^+ antiporter SOS1 controls long-distance Na^+ transport in plants. Plant Cell 14:465–477

Shi H, Lee BH, Wu SJ et al (2003) Overexpression of a plasma membrane Na^+/H^+ antiporter gene improves salt tolerance in *Arabidopsis thaliana*. Nat Biotechnol 21:81–85

Singleton PW, Bohlool BB (1984) Effect of salinity on nodule formation by soybean. Plant Physiol 74:72–76

Sneller CH (1994) Pedigree analysis of elite soybean cultivars. Crop Sci. 34:1515–1522

Stocking MA (2003) Tropical soils and food security: the next 50 years. Science 302:1356–1359

Sun YX, Wang D, Bai YL et al (2006) Studies on the overexpression of the soybean *GmNHX1* in *Lotus corniculatus*: the reduced Na^+ level is the basis of the increased salt tolerance. Chin Sci Bull 51:1306–1315

Sun X, Hu Z, Chen R, Jiang Q, Song G, Zhang H, Xi, Y (2015) Targeted mutagenesis in soybean using the CRISPR-Cas9 system. Sci Rep 5: 10342.

Sunarpi TH, Motoda J, Kubo M et al (2005) Enhanced salt tolerance mediated by AtHKT1 transporter-induced Na^+ unloading from xylem parenchyma cells. Plant J 44:928–938

Tester M, Davenport R (2003) Na^+ tolerance and Na^+ transport in higher plants. Ann Bot 91:503–527
Tilman D, Balzer C, Hill J et al (2011) Global food demand and the sustainable intensification of agriculture. Proc Natl Acad Sci U S A 108:20260–20264
Verslues PE, Batelli G, Grillo S et al (2007) Interaction of SOS2 with Nucleoside Diphosphate Kinase 2 and catalases reveals a point of connection between salt stress and H_2O_2 signaling in *Arabidopsis thaliana*. Mol Cell Biol 27:7771–7780
Wan C, Shao G, Chen Y et al (2002) Relationship between salt tolerance and chemical quality of soybean under salt stress. Chin J Oil Crop Sci 24:67–72
Wei P, Wang L, Liu A et al (2016) GmCLC1 confers enhanced salt tolerance through regulating chloride accumulation in soybean. Front Plant Sci 7:1082
Wen ZX, Tan R, Yuan J et al (2014) Genome-wide association mapping of quantitative resistance to sudden death syndrome in soybean. BMC Genomics 15:1
Wen ZX, Boyse JF, Song Q et al (2015) Genomic consequences of selection and genome-wide association mapping in soybean. BMC Genomics 16:671
Wu SJ, Lei D, Zhu JK (1996) SOS1, a genetic locus essential for salt tolerance and potassium acquisition. Plant Cell 8:617–627
Xu DH, Do TD, Chen HT et al (2016) Genetic analysis of salt tolerance in soybean. Plant & animal genome conference XXIV, P0983. (https://pag.confex.com/pag/xxiv/webprogram/Paper20285.html)
Yang J, Blanchar RW (1993) Differentiating chloride susceptibility in soybean cultivars. Agron J 85:880–885
Yu L, Nie J, Cao C et al (2010) Phosphatidic acid mediates salt stress response by regulation of MPK6 in *Arabidopsis thaliana*. New Phytol 188:762–773
Zeng A, Chen P, Korth K et al (2017) Genome-wide association study (GWAS) of salt tolerance in worldwide soybean germplasm lines. Mol Breed 37:1–14. https://doi.org/10.1007/s11032-017-0634-8
Zhang WJ, Niu Y, Bu SH et al (2014) Epistatic association mapping for alkaline and salinity tolerance traits in the soybean germination stage. PLoS One 9(1):e84750
Zhang J, Song Q, Cregan PB et al (2015) Genome-wide association study for flowering time, maturity dates and plant height in early maturing soybean (*Glycine max*) germplasm. BMC Genomics 16:217
Zhao Y, Wang T, Zhang W et al (2011) SOS3 mediates lateral root development under low salt stress through regulation of auxin redistribution and maxima in *Arabidopsis*. New Phytol 189:1122–1134
Zhu JK (2000) Genetic analysis of plant salt tolerance using Arabidopsis. Plant Physiol 124:941–948
Zhu JK (2002) Salt and drought stress signal transduction in plants. Annu Rev Plant Biol 53:247–273
Zhu JK, Liu J, Xiong L (1998) Genetic analysis of salt tolerance in *Arabidopsis*: evidence for a critical role of potassium nutrition. Plant Cell 10:1181–1191

Chapter 10
Proteomics Perspectives in Post-Genomic Era for Producing Salinity Stress-Tolerant Crops

Pannaga Krishnamurthy, Lin Qingsong, and Prakash P. Kumar

Abstract Plant growth and productivity are affected by both biotic and abiotic stress factors. Among the abiotic stresses, salt stress is the most prevalent and deleterious environmental factor which limits crop yield globally. Combined with the increasing population and food demands, this poses a great challenge to humanity. Currently, salinity affects more than 20% of the irrigated land. This is estimated to increase drastically in the near future due to the excessive irrigation practices. These factors have necessitated the researchers to understand the salt tolerance mechanisms in plants in order to use various approaches to generate salt-tolerant crops. Due to their sessile nature, plants cannot evade the stressful environment, and therefore, some species have evolved various adaptive strategies to grow and reproduce under unfavorable environments. Salt stress imparts both osmotic and ionic stress to the plants, affecting their metabolism and ion homeostasis, thereby leading to reduced growth and productivity and death in some cases. Salt tolerance is a complex phenomenon involving changes in the biochemical, molecular, and physiological processes of the plant. These changes consisting of a readjustment in the genomic and proteomic complement of the plants are imperative in unraveling the tolerance mechanisms. Recent advances in the omics research have shed more light on a range of promising candidate genes and proteins that render salt tolerance to plants. In this chapter, we describe the general effects of salt stress, the tolerance mechanisms of plants, and how recent advances in the field of proteomics can be utilized to enhance salt tolerance of crop plants.

Keywords Proteomics · Salt stress · Salt tolerance · Crop plants

P. Krishnamurthy (✉) · L. Qingsong · P. P. Kumar
Department of Biological Sciences, and NUS Environmental Research Institute (NERI), National University of Singapore, Singapore, Singapore
e-mail: eripk@nus.edu.sg

V. Kumar et al. (eds.), *Salinity Responses and Tolerance in Plants, Volume 2*,
https://doi.org/10.1007/978-3-319-90318-7_10

Abbreviations

1DE	One-dimensional gel electrophoresis
2DE	Two-dimensional gel electrophoresis
ABA	Abscisic acid
APX	Ascorbate peroxidase
bHLH	Basic helix-loop-helix
CAT	Catalase
CCOMT	CoA O-methyltransferase
CNGCs	Cyclic nucleotide-gated channels
DHAR	Dehydroascorbate reductase
DIGE	Difference gel electrophoresis
EC	Electric conductivity
GPX	Glutathione peroxidase
HKT1	High-affinity potassium transporter
ICAT	Isotope-coded affinity tags
iTRAQ	Isobaric tags for relative and absolute quantitation
JA	Jasmonic acid
LRR	Leucine-rich repeat
MAPK	Mitogen-activated protein kinase
MDAR	Monodehydroascorbate reductase
MRM	Multiple reaction monitoring
MS	Mass spectrometry
MudPIT	Multidimensional protein identification technology
NaCl	Sodium chloride
NHX	Sodium/hydrogen exchanger
NSCC	Nonselective cation channels
PIP	Plasma membrane intrinsic proteins
POD	Peroxidases
PTMs	Posttranslational modifications
ROS	Reactive oxygen species
SA	Salicylic acid
SAM	S-adenosyl methionine
SILAC	Stable isotope labeling by amino acids in cell culture
SOD	Superoxide dismutase
SOS1	Salt overly sensitive1
SRM	Selective reaction monitoring
SUMO	Small ubiquitin-like modifiers
SWATH MS	Sequential window acquisition of all theoretical mass spectra
TIP	Tonoplast intrinsic proteins
VDAC	Voltage-dependent anion channel
VPPase	Vacuolar pyrophosphatase

10.1 Salt Tolerance in Plants

Salinity is one of the most deleterious environmental stress factors affecting agricultural productivity worldwide. More than 6% of land area is affected by salinity, mainly due to the increasing irrigation in addition to inundation by seawater (Flowers and Yeo 1995; Tester and Davenport 2003). When irrigated water evaporates, it leaves behind the dissolved salts, and continual irrigation of a given field progressively leads to accumulation of excessive quantities of salts such as sodium chloride and sodium sulfate. The soil is considered to be saline when the electric conductivity (EC) of the soil solution reaches 4 dS m^{-1} (equivalent to 40 mM NaCl), which generates an osmotic pressure of ~ 0.2 MPa (Munns and Tester 2008) (dS m^{-1} = deci Siemens per meter; for reference, EC of most fresh drinking water will be less than 0.1 dS m^{-1}, and EC of seawater is around 54 dS m^{-1}). Sodium and chloride, the two major components of saline soils, impart both osmotic and ionic stress to the plants. In the initial stage, plants experience an osmotic stress due to the decreased osmotic potential in the soil which is reversible (Munns and Tester 2008). With the increasing Na^+ and Cl^-, ionic stress is imposed to the plants leading to impairment of metabolic processes and photosynthetic efficiency in the second phase (Deinlein et al. 2014; Flowers and Yeo 1995). As a consequence, plant growth and development are negatively affected leading to plant death in some cases.

Plants exhibit variable tolerance to salt depending on species and climatic and soil conditions (Tang et al. 2015). Genetic variability exists in salt tolerance of plants, and the degree of tolerance varies with species as well as within the varieties of species. Based on the ability of the plants to grow in saline environments, they are classified as glycophytes and halophytes. Glycophytes include most crop plants that cannot grow in the presence of high salt concentrations (100–200 mM) (Acosta-Motos et al. 2017). Among the crop plants, barley (*Hordeum vulgare*) is shown to be more tolerant compared to rice (*Oryza sativa*) and wheat (*Triticum aestivum*) (Gupta and Huang 2014; Munns and Tester 2008), whereas halophytes such as the mangrove plants can survive and complete their life cycle in high salt concentrations (300–500 mM) as they have developed better adaptive mechanisms to grow in saline conditions (Flowers and Colmer 2015; Parida and Das 2005).

Some of the adaptive strategies exhibited by plants to cope with salinity are (a) salt exclusion at the roots, (b) compartmentalization of salt, (c) accumulation of osmolytes, and (d) salt secretion. Since roots are in direct contact with the saline soil, they need to provide the first line of defense. Roots of both glycophytes and halophytes exclude unwanted salt at the roots with the aid of apoplastic barriers (suberin lamellae and Casparian bands) in the endodermis and exodermis (Fig. 10.1). The salt tolerance of plants has been related to the extent of barrier deposition in the roots (Ma and Peterson 2003; Parida and Jha 2010; Peterson 1988; Ranathunge et al. 2003; Reinhardt and Rost 1995; Scholander 1968; Schreiber and Franke 2011; Schreiber et al. 1999; Krishnamurthy et al. 2011, 2014a; Kim et al. 2016). However, the salt that manages to enter the roots and leaves of such tolerant plants is still toxic to them because the enzymes involved in metabolic processes are sensitive to salt

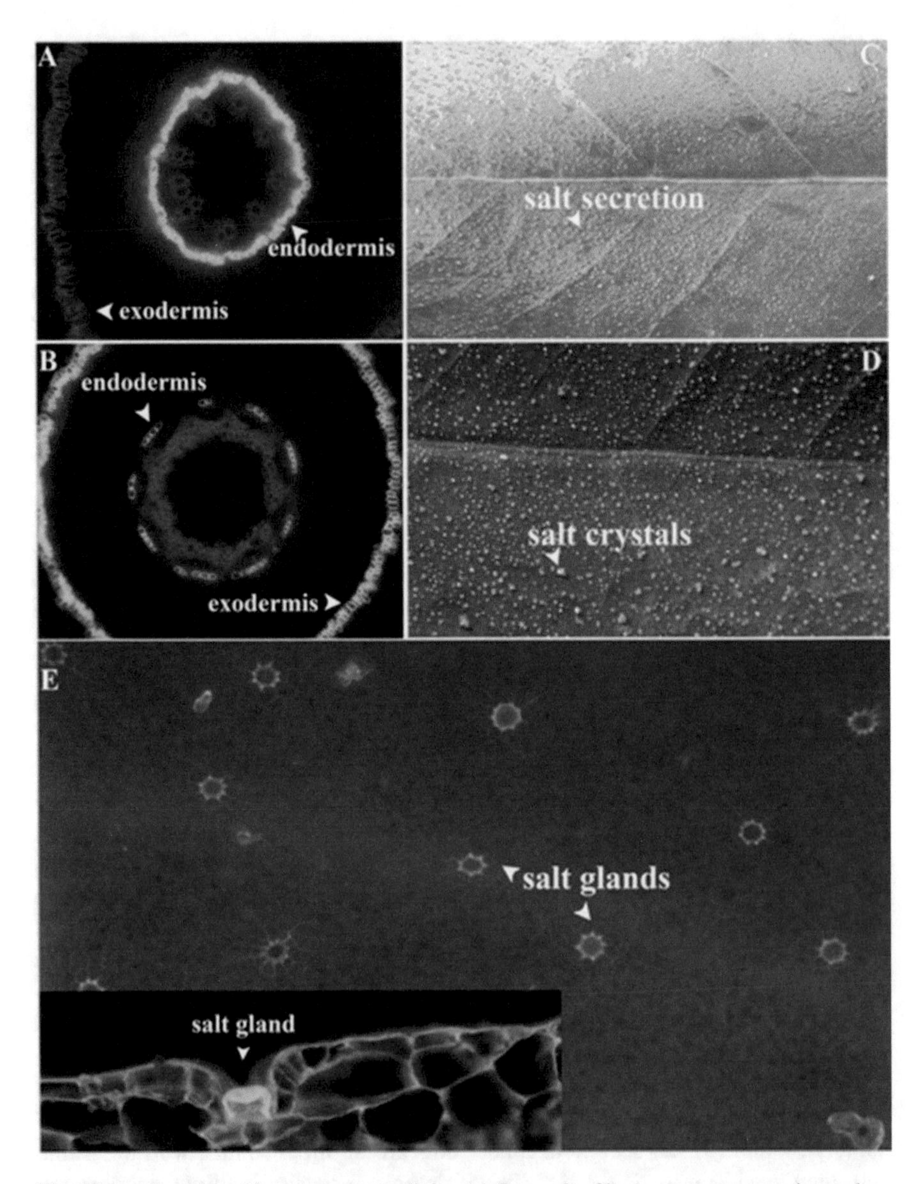

Fig. 10.1 Adaptations of mangroves to salinity. (**a**) Some ultrafiltrator mangrove species such as *Bruguiera cylindrica* filter ~99% of salt from the seawater at their roots. This is facilitated by deposition of two sets of complete rings of suberized hydrophobic barriers at the root, namely, the endodermis and multilayered exodermis. The largely apoplastic pathway of movement for water and dissolved ions gets routed through the cell membranes (symplastic route) by the exodermis and endodermis layers. (**b**) However, due to the formation of incomplete suberized cell walls (broken ring) of the endodermis toward tips of the roots, salt-secreting mangrove species such as *Avicennia officinalis* can only filter ~95% of salt from the seawater. (**c**) To compensate for the inefficient ultrafiltration, the latter species secrete excess salt at the leaf surface. (**d**) The secreted salt solution can dry up leaving behind salt crystals on the leaves. (**e**) Salt secretion is carried out via specialized salt glands (*inset*) present largely on the adaxial surface of the leaves

both in glycophytes and halophytes (Parida and Das 2005; Parida and Jha 2010; Zhu 2003). Sodium/hydrogen exchangers (NHXs) are involved in compartmentalizing the excess cytosolic Na^+ into the vacuoles using the hydrogen gradient generated by two electrogenic H^+ pumps, the vacuolar type H^+-ATPase (V-ATPase) and vacuolar pyrophosphatase (VPPase) (Apse et al. 1999; Parida and Das 2005; Parida and Jha 2010). Some plants also exhibit compartmentalization of toxic NaCl into the older leaves which are eventually shed, protecting the young leaves (Cheeseman 1988). To maintain the ionic balance in the vacuoles, the cytoplasm accumulates compatible solutes (osmolytes) which are neutral, low-molecular mass compounds and do not interfere with the general metabolic processes (Hasegawa et al. 2000; Parida and Das 2005; Parida and Jha 2010). The major compatible solutes found in plants are proline, glycine betaine, polyols (mannitol, pinitol, inositol, etc.), and sugars (glucose, fructose, sucrose) (Parida and Das 2005; Parida and Jha 2010; Tsuzuki et al. 2011). Osmolytes also function as chaperones and as scavengers of stress-induced oxygen radicals in many plants (Bohnert et al. 1995; Parida and Das 2005; Parida and Jha 2010). While all these three strategies can be found in most of the plant species, salt secretion is predominantly an adaptation of halophytes and more so in mangrove plants (Fig. 10.1). The excess salt from the photosynthetic tissue is actively secreted through microscopic foliar salt glands/salt hairs (bladders) (Bohnert et al. 1995; Parida and Jha 2010; Flowers and Colmer 2015). These structures are commonly found in halophytes such as *Frankenia*, *Cressa*, *Spartina*, *Aeluropus*, *Atriplex*, *Chloris*, *Statice*, *Plumbago*, *Tamarix*, *Limonium*, *Armeria*, *Glaux*, and *Reamuria*, as well as in mangrove species such as *Avicennia*, *Aegiceras*, *Aegialitis*, and *Acanthus* (Hasanuzzaman et al. 2013; Popp 1995). In order to be able to generate salt-tolerant crop plants, it is critical to understand the salt tolerance mechanisms in both glycophytes and halophytes. Several decades of salinity research have shed light on most of these processes, although the underlying mechanisms of many are still not understood completely and pose a great challenge to the researchers.

10.2 Importance of Proteomic Research

Recent advances in plant omics research have led to the opening up of new opportunities and perspectives for improving crop productivity under various abiotic stresses. Salt tolerance is a complex process brought about by the interplay of multiple genes, which involves many physiological, biochemical, and molecular processes (Flowers 2004; Munns and Tester 2008). With the aid of transcriptomics and microarray technology, researchers have profiled the global gene expression patterns in various plant species such as *Arabidopsis*, rice, maize, *Medicago*, *Mesembryanthemum*, *Thellungiella*, and *Avicennia* that have helped to identify and characterize a number of salt-responsive genes, many of which have been used for developing salt-tolerant crops (Jiang and Deyholos 2006; Mishra and Tanna 2017; Qing et al. 2009; Rabbani et al. 2003; Seki et al. 2002; Szittya et al. 2008; Taji et al.

2004; Walia et al. 2007; Wang et al. 2003; Krishnamurthy et al. 2017). Despite the advantages of transcriptomic approach, functional gene expression profiles can only be achieved by proteomic analysis. It has been established that the changes in gene expression levels need not necessarily correspond with the changes at protein level (Gygi et al. 1999; Kosava et al. 2011). This is due to the fact that posttranscriptional and posttranslational regulations occur in cells. Further, because of the combined effects of specific temporal and spatial expression patterns of genes, studies that focus on transcriptomics alone are not sufficient to elucidate the overall mechanism of stress remediation in plants. Some transcripts might not be translated into proteins in selected tissues/cells. Therefore, if we only rely on transcriptomics, such posttranscriptional regulation will be missed. In addition, genes and their regulatory elements may have complex 3D relationships involving long-range chromatin interactions which have emerged as a new level of complexity (Sanyal et al. 2012). Moreover, posttranslational modifications such as phosphorylation, glycosylation, ubiquitination, SUMOylation, carbonylation, and protein-protein interactions are important for many biological processes (Kosava et al. 2011; Wu et al. 2016). Proteins are also dynamic in cells which get targeted for proteolysis in addition to the modifications. Hence quantitative analysis of gene expression at protein level is necessary for understanding the mechanism of salt tolerance (Sobhanian et al. 2011). Investigation of changes in plant proteome is highly important since proteins, unlike transcripts, are direct players of plant stress response which play key roles in growth and development of a plant as well as in response to various stress conditions. Proteomics, which helps to analyze the profiles and functions of proteins on a large scale, is considered one of the next-generation research tools in life sciences along with genomics and transcriptomics (Eldakak et al. 2013; Zhuang et al. 2014). Since the genome sequences of several plant species have been deciphered and made publicly available, researchers are focusing to link the genomic, transcriptomic, and metabolomic profiles to the expression and biological functions of proteins. Combined omics studies are being proven to be valuable to address these limitations and provide additional perspectives on the molecular regulatory mechanisms (Gupta and Huang 2014; Zhuang et al. 2014).Thus, in the post genome era, proteomics has emerged as an important complementary tool to other omics approaches such as transcriptomics and metabolomics.

10.3 Advances in Plant Proteomic Tools

In the recent past, there have been rapid developments and advances in protein quantitative methods. Expression profiling of proteins is the core of proteomics approaches (Parker et al. 2006), and several techniques have been developed for the differential analysis of protein expression. Major steps critical for any proteomic approach include sample (tissue/cell or organelle) preparation, protein extraction, purification and fractionation, labeling and modification, analysis by mass spectrometry (MS), identification of proteins, analysis of data, and validation (Hakeem

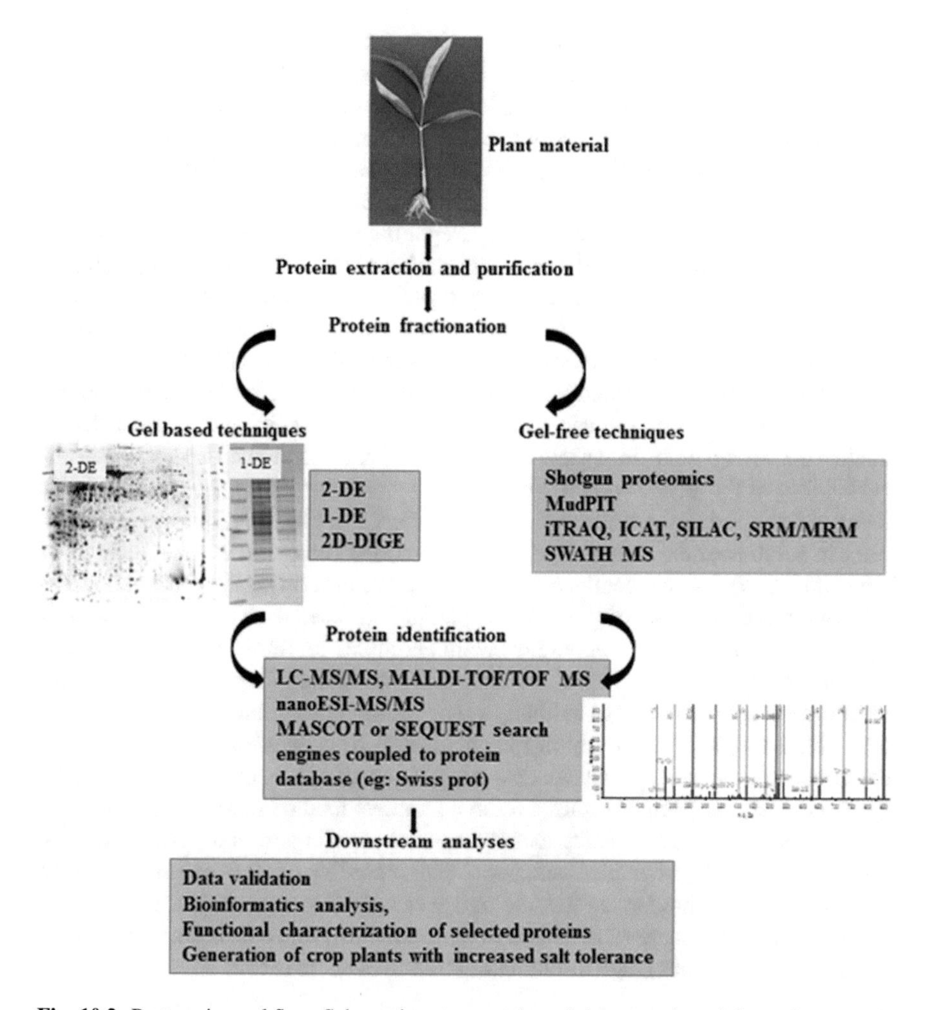

Fig. 10.2 Proteomic workflow. Schematic representation of the general workflow of proteomic analysis. The protein extracted and purified from the plant material is fractionated using gel-based and gel-free techniques. The fractionated peptides are identified using various mass spectrometry analyses. Refer to the text for the abbreviations for the various techniques listed

et al. 2013; Jorrin-Novo et al. 2009) (Fig. 10.2). Protein extraction protocols widely used for plant materials include tissue homogenization in buffers or inorganic solvents (phenol, TCA-acetone). Sample preparation is a major challenge in achieving good proteomic data, and developing efficient extraction procedures has therefore gained importance (Rossignol et al. 2006). Gel-based or gel-free techniques are the two most common protein fractionation techniques used. One-dimensional gel electrophoresis (1-DE) and two-dimensional gel electrophoresis (2-DE) are the predominantly used gel-based techniques, although 2-DE is the most preferred in combination with MS. With the development of "difference gel electrophoresis"

(2D-DIGE), quantitative proteomics can be carried out with a relatively small amount of protein (0.025–0.050 mg) (Dunkley et al. 2006). Gel-free technique includes shotgun liquid chromatography coupled with tandem mass spectrometry (LC-MS/MS) and is popular lately because we can identify more number of proteins from a complex sample using this approach compared to gel-based techniques. In addition, the gel-free multidimensional protein identification technology (MudPIT) is used for the detection of hydrophobic proteins (Gorg et al. 2009). The recent development in second-generation proteomic technologies for quantitative proteomics such as isotope-coded affinity tags (ICAT) (Shiio and Aebersold 2006), stable isotope labeling by amino acids in cell culture (SILAC) (Palmblad et al. 2008), and isobaric tags for relative and absolute quantitation (iTRAQ) (Gan et al. 2007) has led to the identification of several salt-responsive proteins from various plant species. Some of the salt tolerance-related proteomic studies are listed in Table 10.1 along with the species involved and the technique used. Despite the availability of the advanced gel-free quantitative proteomic approaches, 2-DE even today is preferred as the detergents and buffers used in this technique are quite efficient in solubilization of more hydrophobic proteins compared to other techniques (Hu et al. 2015). More recently, multiple reaction monitoring (MRM, also known as selective reaction monitoring – SRM) technique has been used in mass spectrometry to selectively quantify compounds within complex mixtures. Using MRM, salt-responsive proteins from *Brassica napus* were quantified (Luo et al. 2015). This technique is an alternative to antibody-based immunodetection assays owing to its high sensitivity and selectivity in quantitation of low-abundance proteins from a complex mixture. Sequential window acquisition of all theoretical mass spectra (SWATH MS) has been lately used as an alternative method to the traditional mass spectrometry-based proteomics techniques such as shotgun and MRM methods. This method allows a complete and permanent recording of all fragment ions of the detectable peptide precursors in a biological sample (Hu et al. 2015). Therefore, it combines the advantages of MRM (high reproducibility and consistency) with those of shotgun MS studies (high throughput) (Huang et al. 2015b).

10.4 Proteomic Studies in Salt Tolerance Research

Salt tolerance is a complex process brought about by the interaction of various hormonal, biochemical, physiological, and molecular processes. Proteomic research has been carried out in various plant species comprising both glycophytes and halophytes to identify the key salt tolerance-related proteins and understand the mechanism of salt tolerance. Some of the studies in glycophytes involve *Arabidopsis thaliana* (Jiang et al. 2007; Ndimba et al. 2005), tobacco (Dani et al. 2005; Razavizadeh et al. 2009), grass pea (Chattopadhyay et al. 2011; Mastrobuoni et al. 2012), *Agrostis stolonifera* (Xu et al. 2010), *Populus cathayana* (Chen et al. 2011), rice (Abbasi and Komatsu 2004; Chitteti and Peng 2007; Dooki et al. 2006; Kim et al. 2005; Yan et al. 2005; Liu et al. 2014), barley (Fatehi et al. 2012; Rasoulnia et al. 2011; Witzel et al. 2009),

Table 10.1 Examples of studies carried out using various proteomic techniques in both glycophytes and halophytes

Species	Stress condition	Technique used	No. of proteins identified
Wheat	0%, 1.5%, 2.0%, and 2.5% NaCl (2 days)	2D-DIGE (Gao et al. 2011)	52 (leaves)
		2D MALDI-TOF (Caruso et al. 2008)	38 (leaves)
	100 mM NaCl (2 days)	iTRAQ (Jiang et al. 2017)	121 (roots)
	350 mM NaCl (4 days)		
Rice	250 mM NaCl (30 min)	2DE Nano-LC/MS/MS (Liu et al. 2014)	104 (roots),
	150 mM NaCl (24, 48 and 72 h)	2-DE MALDI-TOF/MS (Yan et al. 2005)	59 (leaves)
			12 (roots)
Soybean	40 mM NaCl(1 week)	2-DE Edman sequencing	19 (leaves)
	200 mM NaCl(12 h)	MALDI/TOF (Sobhanian et al. 2010b)	22 (hypocotyls)
		iTRAQ (Ji et al. 2016)	44 (roots)
			278 (leaves),
			440 (roots)
Date palm	3.6 M (1 month)	2D-DIGE (El Rabey et al. 2015)	9 (leaves)
Brassica napus	200 mM NaCl (24 h, 48 h and 72 h)	2-DE MALDI-TOF/TOFMS (Jia et al. 2015)	42 (leaves)
Barley	0, 50, 100, 150, 200, and 250 mM NaCl (13 days)	2-DEnanoLC-ESI-Q-TOF MS/MS (Witzel et al. 2009)	26 (roots)
	300 mM NaCl (3 weeks)	2DE MALDI-TOF-TOF MS (Fatehi et al. 2012) SDS PAGE NanoLC-ESI-Q-TOF MS (Rasoulnia et al. 2011)	44 (leaves)
	300 mM NaCl (1 week)		44 (crown tissue)
Potato	150 mM NaCl (3, 8 days)	2D-DIGE MALDI-TOF/TOF (Evers et al. 2012)	90 (leaves)
Sorghum bicolor	200 mM NaCl(96 h)	2DE MALDI-TOD/TOF (Swami et al. 2011)	21 (leaves)
Cotton	200 mM NaCl(24 h)	iTRAQ (Li et al. 2015)	128 (roots)
	240 mM NaCl (1 week)	iTRAQ (Chen et al. 2016)	611 (roots),
			1477 (leaves)
Peanut	0-300 mM NaCl (12 days)	2-DE ESI-Q-TOF MS/MS (Jain et al. 2006)	24 (callus)
A. thaliana	50 mM and 150 mM NaCl (5 days)	2-DE iTRAQ LC-MS/MS (Pang et al. 2010)	119 (leaves)
	200 mM NaCl (6 h)	2D-DIGE MALDI-TOF/TOF MS (Ndimba et al. 2005)	75 (cell culture)
Tobacco (*Nicotiana tabacum*)	100 mM NaCl (20 days)	2-DE LC-MS/MS (Dani et al. 2005)2-DE (Razavizadeh et al. 2009)	20 (apoplast)
	150 mM NaCl (2 days)		18 (leaves)

(continued)

Table 10.1 (continued)

Species	Stress condition	Technique used	No. of proteins identified
Populus cathayana	70 and 150 mM NaCl (1 week)	2-DE Q-TOF/MS/MS (Chen et al. 2011)	73 (leaves)
Grass pea	500 mM NaCl (12, 24 and 36 h)	2-DE LC-tandem MS/MS (Chattopadhyay et al. 2011)	48 (leaves)
Bent grass	10dS^{-1} (28 days)	2-DE MALDI-TOF/TOF (Xu et al. 2010)	106 (leaves)
			24 (roots)
Canola	175 and 350 mM NaCl (3 weeks)	2-DE MALDI-TOF/TOF (Bandehagh et al. 2011)	75 (leaves)
Maize	25 mM NaCl (1 h)	2-DE MALDI-TOF/TOF MS (Zorb et al. 2010)	14 (roots)
Tomato	120 mM NaCl (7 days)	2-DE LC-ESI-MS/MS (Chen et al. 2009)	23 (seedlings)
Cucumber	50 mM NaCl (3 days)	2-DEMALDI-TOF/MS (Du et al. 2010)	29 (roots)
Halogeton glomeratus	200 mM NaCl (24 h, 72 h and 7 days)	2DE MALDI TOF/TOF MS/MS (Wang et al. 2015)	49 (leaves)
Chlamydomonas reinhardtii	100 mM and 150 mM NaCl (1, 3, 8 and 24 h)	SILAC GC-MS/MS (Mastrobuoni et al. 2012)	3115 (cells)
Thellungiella halophila	50 mM and 150 mM NaCl (5 days)	2-DE iTRAQ LC-MS/MS (Pang et al. 2010)	69 (leaves)
Bruguiera gymnorrhiza	500 mM NaCl (1, 3, 6, 12, 24 h and 12 days)	2DE (Tada and Kashimura 2009)	2 (root)
			3 (lateral root)
			1 (leaf)
Dunaliella salina	0.75 M and 3 M NaCl	2D-DIGE MALDI-TOF/TOF MS (Jia et al. 2016)	24 (cell culture)
Avicennia officinalis	Field samples (~500 mM NaCl)	2-DE LC-MS/MS (Krishnamurthy et al. 2014b)	419 (leaves)
	Field samples (~500 mM NaCl)	Shotgun LC-MS/MS (Tan et al. 2015)	2188 (leaf epidermal peels)
Salicornia europaea	0, 200, 600, and 800 mM NaCl (21 days)	2DE-MALDI/TOF/TOF MS (Wang et al. 2009)	111 (leaves)
Suaeda aegyptiaca	0, 150, 300, 450, and 600 mM NaCl (30 days)	2DE-LC/MS/MS (Askari et al. 2006)	27 (leaves)
Suaeda salsa	0,100,200mMNaCl (3 weeks)	2-DE MALDI-TOF/MS (Li et al. 2011)	57 (leaves)
Aeluropus lagopoides	450 mM NaCl (10 days)	2-DE Edman sequencing Nano LC-MS/MS (Sobhanian et al. 2010a)	83 (leaves)
Tangut nitaria	200 mM NaCl (1, 3, 5 and 7 days)	iTRAQ (Cheng et al. 2015)	71 (leaves)

(continued)

Table 10.1 (continued)

Species	Stress condition	Technique used	No. of proteins identified
Kandelia candel	300, 450 and 600 mM NaCl (3 days)	2-DE MALDI-TOF-TOF MS (Wang et al. 2014)	48 (leaves)
Puccinellia tenuiflora	50 and 150 mM NaCl (7 days)	2-DE ESI-Q-TOF MS/MS (Yu et al. 2011)	107 (leaves)
Mesembryanthemum crystallinum	400 mM NaCl (12 days)	SDS-PAGE LC-MS/MS (Cosentino et al. 2013)	14 (leaves)

The specific salt treatment and the techniques used along with the key references [in brackets] for the cited examples are provided

wheat (Caruso et al. 2008; Gao et al. 2011; Jiang et al. 2017), canola (Bandehagh et al. 2011), soybean (Ji et al. 2016; Sobhanian et al. 2010b), sorghum (Swami et al. 2011), peanut (Jain et al. 2006), date palm (El Rabey et al. 2015), maize (Zorb et al. 2009; Zorb et al. 2010), cucumber (Du et al. 2010), tomato (Chen et al. 2009), potato (Evers et al. 2012), and cotton (Chen et al. 2016; Li et al. 2015) (Table 10.1). A summary of the proteomic work carried out in major glycophytic crops can be found in earlier reports (Salekdeh et al. 2002; Sobhanian et al. 2011). Similarly, proteomic studies involving halophytes are *Suaeda salsa* (Li et al. 2011), *M. crystallinum* (Cosentino et al. 2013), *Suaeda aegyptiaca* (Askari et al. 2006), *Puccinellia tenuiflora* (Yu et al. 2011), *Bruguiera gymnorrhiza* (Tada and Kashimura 2009), *Avicennia officinalis* (Krishnamurthy et al. 2014b; Tan et al. 2015), *Salicornia europaea* (Wang et al. 2009), *Thellungiella halophila* (Pang et al. 2010), *Dunaliella salina* (Jia et al. 2016), *Kandelia candel* (Wang et al. 2014), *Aeluropus lagopoides* (Sobhanian et al. 2010a), *Tangut nitaria* (Cheng et al. 2015), and *Chlamydomonas reinhardtii* (Mastrobuoni et al. 2012) (Table 10.1). Based on these proteomic studies, several similarities and differences were revealed in the responses of glycophytes and halophytes to salt stress. The overall responses have been grouped into the following categories, and the major proteins affected by salt are discussed in a greater detail.

10.5 Salt Exclusion

Salt exclusion at the roots under salinity conditions is a well-known process that occurs both in glycophytes and halophytes (Faiyue et al. 2010; Krishnamurthy et al. 2011; Yeo et al. 1987). This involves the deposition of physical hydrophobic barriers (suberin lamellae and Casparian bands), mainly made up of suberin and lignin which block the leakage of ions and water into the xylem through an apoplast bypass flow (Enstone et al. 2003; Schreiber et al. 1999). In a rice comparative proteomic study between salt-tolerant Pokkali and the sensitive cultivar IR29, caffeoyl-CoA O-methyltransferase (CCOMT) was markedly upregulated by salt stress. CCOMT is involved in suberin and lignin biosynthesis. In other studies, an increase in suberin and lignin upon salt treatment leading to reduced bypass flow has been shown in Pokkali (Ranjithakumari and Radhakrishnan 2008). In a proteomic study of halophyte plant *Salicornia europaea*, increased lignification of the xylem vessels associated with an increased biosynthesis of S-adenosyl methionine (SAM, enzyme involved in biosynthesis of lignin) was found to reduce the apoplast bypass flow. In addition, increased expression of several plasma membrane ion transporters such as SOS1, sodium/hydrogen antiporters (Nhx), voltage-dependent anion channel (VDAC), and calcium-mediated sensors was seen suggesting that the salt exclusion process is coupled to the salt ion translocation through membrane-bound transporters (Wang et al. 2009).

10.6 Signaling and Regulation of Gene Expression

Plants require numerous signaling molecules to regulate the complex stress responses. Several key signaling-related proteins reported to be involved in salt stress response include proteins related to Ca^{2+} and phytohormone signaling, GTP-binding proteins, 14-3-3-like proteins, and various protein kinases. Salt stress-induced Ca^{2+}-dependent signaling has been reported to transduce stress signal from plasma membrane to the nucleus leading to Na^+ homeostasis and salt tolerance (Mahajan et al. 2008). Proteomic studies have revealed several proteins involved in Ca^{2+} signaling pathway which have been reported to be involved in salt tolerance. In maize, a Ca^{2+}-sensing receptor and a Na^+-sensing element were induced by salt (Zorb et al. 2010). In plants like rice, *Arabidopsis*, *S. salsa*, and *A. officinalis*, calcium-binding proteins have been reported to be regulated by salinity (Jiang et al. 2007; Krishnamurthy et al. 2014b; Li et al. 2010b, 2011; Pang et al. 2010). In addition, levels of calmodulin, calcineurin-like phosphoesterases, calreticulin, and annexin were affected by salt in various plants (Jiang et al. 2007; Li et al. 2010b; Zorb et al. 2010). These changes may modulate the levels of Ca^{2+} in the intracellular stores, thereby inducing several phosphatases and kinases (Jiang et al. 2007). Salinity-related works have shown that the Ca^{2+} signaling activates the SOS pathway which is important for regulation of cellular Na^+/K^+ homeostasis (Zhu 2002). However, not many proteomic reports can be found related to SOS1, which could be due to the low abundance of signaling proteins compared to the cytosolic proteins making it difficult for their identification. The phytohormones ABA, JA, ethylene, and SA play major role in plant abiotic stress tolerance. In several instances, the cross talk between these hormones leads to the tolerance to a particular stress. ABA-responsive proteins and ABA/stress-inducible proteins are shown to be increased by salt in pea and rice (Nat et al. 2004; Salekdeh et al. 2002). The mitogen-activated protein kinase (MAPK), which negatively regulates SA and positively regulates JA, and several pathogenesis-related proteins (PR1, PR10, PR17) that are regulated by JA/ethylene signaling were all induced by salt (Zhang et al. 2012a; Chattopadhyay et al. 2011; Jain et al. 2006; Li et al. 2010b; Pang et al. 2010). GTP-binding proteins such as RAb-GTPases and RAC-GTPases, which regulate vesicle trafficking and endocytosis, have been reported from several studies (Jiang et al. 2007; Krishnamurthy et al. 2014b; Ndimba et al. 2005; Pang et al. 2010; Wang et al. 2008a). 14-3-3-like proteins are an important class of proteins known to interact with MAPKs as well as CDPKs and regulate cellular ion homeostasis by activating plasma membrane H^+-ATPases. This class of proteins has been found in most of the proteomic studies related to salt stress (Chattopadhyay et al. 2011; Cheng et al. 2015; Ji et al. 2016; Krishnamurthy et al. 2014b; Wang et al. 2008a; Zorb et al. 2010). Several studies have reported a large number of kinases that regulate key salt-responsive processes via phosphorylation and dephosphosphorylation. The receptor-like protein kinase plays an important role in salt tolerance via the activation of MAPK pathway (Krishnamurthy et al. 2014b). Members of the large family of leucine-rich repeat (LRR) protein kinases act as important signaling components in plant disease and stress responses (Cheng et al. 2009; Krishnamurthy et al. 2014b). Various signal transduction pathways activated by salt would alter the expression of several genes by regulating the transcription factors and transcription-related proteins that are important for salinity tolerance of plants. Several transcription factors (TF) such as basic helix-loop-helix (bHLH), basic transcription factor 3 (BTF3), and NAC family proteins have been reported to be induced by salt in proteomic studies (Aghaei and Komatsu 2013; Chen et al. 2009; Fatehi et al. 2012; Vincent et al. 2007; Yan et al. 2005). The bHLH and NAC TFs have been shown to render salt tolerance in rice and *Arabidopsis* (Zhou et al. 2009; He et al. 2005).

10.7 Metabolic Processes

Several metabolic pathways including carbohydrates, proteins, nucleic acids, and lipids are perturbed by salt stress. Plants differ in regulating different metabolic processes in response to salt stress. As mentioned earlier, plants accumulate high levels of organic solutes such as glycine betaine, proline, polyols, and sugars along with several osmoprotectants in order to overcome the salinity-induced osmotic stress. Accordingly, proteomic studies of salt-stressed plants revealed an increase in the levels of enzymes involved in the biosynthesis of these osmolytes. Glutamine synthase and Δ-pyrroline-5-carboxylate synthase (P5CS) involved in proline biosynthesis were increased, while proline dehydrogenase, which degrades proline, was decreased in *Thellungiella* (Kosova et al. 2013). Several studies have shown that glutamine synthase is induced by salt stress and plays a major role in salt tolerance (Pang et al. 2010). Increase in the levels of SAM synthase (SAMS), betaine aldehyde dehydrogenase, glycine dehydrogenase, and choline monooxygenase, which are involved in the biosynthesis of another important osmolyte glycine betaine, was reported in several proteomic (Askari et al. 2006; Veeranagamallaiah et al. 2008; Wang et al. 2015) studies as well as transcriptomic studies (Gharat et al. 2016; Mishra and Tanna 2017). Some of the osmoprotectants such as LEA proteins, dehydrins, and osmotins were also expressed in response to salt stress (Jyothi-Prakash et al. 2014; Kosova et al. 2013; Pang et al. 2010). These are hydrophilic proteins that play an important protective role during cellular dehydration occurring under abiotic stresses.

Salt stress alters the energy metabolism of plants by affecting photosynthesis. Several metabolic processes are severely decreased under salt stress due to the closure of stomata leading to reduced CO_2 availability, which in turn decreases the photosynthetic rate (Kosava et al. 2013). A decreased abundance of RuBisCO, oxygen-evolving complex protein (OEC proteins), carbonic anhydrase, chlorophyll a/b binding protein, and RuBisCO activase was observed in many glycophytic plants under salt treatment (Aghaei and Komatsu 2013; Bandehagh et al. 2011; Chattopadhyay et al. 2011; Pang et al. 2010; Sobhanian et al. 2010b; Kosava et al. 2013; Kosova et al. 2013) indicating reduced assimilation of CO_2 under stressed conditions. On the contrary, increased degradation of RuBisCO along with increased expression of several other enzymes involved in catabolic processes such as Calvin cycle and glycolysis was observed (Pang et al. 2010; Rasoulnia et al. 2011). However, proteomic studies with halophytes have shown that the net photosynthetic rate is either maintained or increased under salt stress in these plants. Salt-induced increase in the abundance of RuBisCO subunits, chlorophyll a/b binding proteins, and D2 protein (a core protein of photosystem II) has been reported (Cheng et al. 2015; Wang et al. 2008b, 2015). This shows how halophytes, unlike glycophytes, are successful in maintaining the plant growth and productivity under saline conditions. Other key metabolism-related proteins reported in proteomic studies in response to salt stress are fructose 1, 6-bisphosphate aldolases (FBP aldolase), fructose 1, 6-bisphosphatase (FBPase), alcohol dehydrogenase, enolase, and short-chain dehydrogenase reductase (Abbasi and Komatsu 2004; Chattopadhyay et al. 2011; Krishnamurthy et al. 2014b; Sobhanian et al. 2010b; Tada and Kashimura 2009; Kim et al. 2005). FBP aldolases increase the production of amino acids and sugars through an increase in glycolysis, and they are also known to increase the production of osmolytes leading to increased salt tolerance (Tada and Kashimura 2009). In addition, aldolases and enolases interact directly and regulate the function of vacuolar H^+-ATPase, which activates the Na^+ compartmentalization into vacuoles by Na^+/H^+ exchangers, rendering salt tolerance (Barkla et al. 2009). Short-chain dehydrogenase reductase is involved in nutrient signaling, hormone biosynthesis, and synthesis of secondary metabolites contributing to salt tolerance (Cheng et al. 2002). Under salt stress, demand for energy production increases in order to maintain ionic balance and plant growth. An increase in the levels of proteins involved in metabolic processes leading to energy production such as TCA cycle, glycolysis, pentose phosphate pathway, and photorespiratory pathway has been found in both halophytes and glycophytes (Kosova et al. 2013). Some of the reported proteins involved in these processes are glyceraldehyde-3-phosphate dehydrogenase, succinyl-CoA ligase, carbonic anhydrase, malate dehydrogenase, pyruvate dehydrogenase, cytochrome c oxidase, and glucose-6-phosphate isomerase (Ndimba et al. 2005; Pang et al. 2010; Sobhanian et al. 2011; Yan et al. 2005; Krishnamurthy et al. 2014b). In addition, several subunits of ATP synthases that provide energy for various cellular processes were increased in many salt-treated plants (Krishnamurthy et al. 2014b; Li et al. 2011; Pang et al. 2010; Veeranagamallaiah et al. 2008). Increase in the levels of all these proteins shows that plants respond to salt by producing more energy, osmolytes, and sugars, which are shown to be important for stress tolerance in both glycophytes and halophytes (Parida et al. 2004a).

10.8 Stress Responsive

Another important class of differentially expressed proteins identified in proteomic studies under salt treatment is stress-responsive proteins. Along with the alterations in the metabolic processes, salt stress induces oxidative damage to the plants, the damage being higher in glycophytes compared to halophytes (Kosova et al. 2013). The oxidative stress leads to accumulation of reactive oxygen species (ROS) such as superoxide, hydrogen peroxide, hydroxyl radical, and singlet oxygen that are cytotoxic and can seriously damage lipids, proteins, and nucleic acids, which would disturb the normal metabolism (Parida et al. 2004b).Plants with high levels of antioxidants have greater resistance to this oxidative damage (Parida et al. 2004b; Parida and Jha 2010). Several ROS-scavenging enzymes (antioxidants) such as superoxide dismutase (SOD), catalase (CAT), ascorbate peroxidase (APX), dehydroascorbate reductase (DHAR), monodehydroascorbate reductase (MDAR), glutathione peroxidase (GPX), peroxidases (POD), thioredoxin h, and cytochrome P450 monooxygenase were found in response to salt in proteomic studies (Abbasi and Komatsu 2004; Aghaei and Komatsu 2013; Cheng et al. 2015; Dooki et al. 2006; Ji et al. 2016; Jiang et al. 2017; Kim et al. 2005; Li et al. 2015; Pang et al. 2010; Swami et al. 2011; Wang et al. 2009; Witzel et al. 2009). SODs are enhanced by salt and carry out dismutation of superoxide into oxygen and H_2O_2 which is later removed by CATs, APX, and GPX in various pathways. In addition to antioxidants, heat-shock proteins (HSPs), chaperones, cold-shock domain-containing proteins (CSPs), and AAA-ATPases are some of the stress-responsive proteins found in proteomic studies (Ruan et al. 2011; Wang et al. 2009; Yu et al. 2011; Zhang et al. 2012a). HSPs, CSPs, and chaperones are known to be involved in protein structure stabilization through folding, repair, and renaturation of the damaged proteins due to stress (Zhang et al. 2012a). AAA-ATPases form a large protein family and play key roles in proteolysis, protein folding, membrane trafficking, microtubule regulation, and cytoskeleton regulation. Under salt stress, targeted degradation of several proteins via the proteasome pathway involving ubiquitin/polyubiquitin, several peptidases and proteases have been shown to be important for stress remediation and salt tolerance. Many of these proteins have been found in several proteomic studies to be upregulated by salt (Chitteti and Peng 2007; Jiang et al. 2007; Li et al. 2010b; Ndimba et al. 2005; Pang et al. 2010). Thus, these stress-responsive proteins play a vital role in salt tolerance of plants.

10.9 Ion Transport

Salt stress alters ion balance in plants due to the high levels of Na^+ and Cl^- entry into the cells. This necessitates the plants to reestablish ion homeostasis in the cells by means of ion uptake, exclusion, and compartmentation. This is important for plant survival as the cytosolic enzymes in both halophytes and glycophytes are sensitive to Na^+ (Munns and Tester 2008). In most of the plants, Na^+ enters the plant cells through nonselective cation channels such as NSCCs and cyclic nucleotide-gated channels (CNGCs) as well as high-affinity potassium transporter (HKT1). Ion exclusion from the cytosol into the external apoplast is usually carried out by the plasma membrane sodium/hydrogen antiporters (SOS1) which function by utilizing the H^+ gradient generated by the plasma membrane H^+-ATPases. Similarly, the Na^+ ion compartmentation occurs by the action of vacuolar sodium/hydrogen exchangers (NHXs) which function with the help of vacuolar-H^+-ATPases and PPases. Increased levels of SOS1 were found in a relatively salt-tolerant sugar beet (Wakeel et al. 2011), whereas NHX, V-ATPases, and VPPases have been found to be upregulated by salt in many studies involving various plant species (Du et al. 2010; Jiang et al. 2007; Krishnamurthy et al. 2014b; Pang et al. 2010; Tan et al. 2015; Wang et al. 2008a; Xu et al. 2010). In several halophytes, Na^+ is sequestered into the vacuoles as a cheap osmolyte to maintain the osmotic balance under high saline conditions (Parida and Jha 2010). Several other stress-responsive transporters/channels found in plant proteomic studies are K^+ transporters, ABC transporters, aquaporins, chloride channels, and voltage-dependent anion channels (VDAC). K^+ transporters have been identified in several plants and are shown to maintain the ionic balance under stressed conditions. ABC transporters were induced by salt in *P. patens* (Wang et al. 2008b) and wheat (Peng et al. 2009; Wang et al. 2008a) and found in the mangrove plant *A. officinalis* (Krishnamurthy et al. 2014b; Tan et al. 2015). These transporters are involved in various cellular functions such as detoxification and compartmentalization of heavy metals, micronutrient homeostasis, Cd^{2+} and heavy metal efflux, transport of stress-related secondary metabolites, and osmolyte transport (Yazaki 2006). Overexpression studies of ABC transporters in *Arabidopsis* have shown that ABC transporters contribute to salt and drought tolerance by reducing Na^+ levels in transgenic plants (Kim et al. 2007, 2010). It is a challenging task for plants to maintain the water status under salt stress due to the increased ion accumulation. Although the phospholipid bilayer exhibits substantial permeability to water, aquaporins are recruited for water flux across the plasma membrane (plasma membrane intrinsic proteins, PIP) and tonoplast (tonoplast intrinsic proteins, TIP) (Maurel and Chrispeels 2001) and play a key role in water transport under drought and salt stress in various plant species (Boursiac et al. 2005; Maurel et al. 2008). In a study involving mangrove, PIPs were shown to be involved in salt secretion through the leaf salt glands (Tan et al. 2015). Chloride channel protein was found to be reduced in salt-treated *P. patens* and was suggested to limit the transport of Cl^- into the cells and increase salt tolerance (Wang et al. 2008b). VDACs are porins present on the outer membranes of the mitochondria and have been shown to be induced upon salt stress in several species such as *Z. mays*, *B. vulgaris*, *A. lagopoides*, and *A. officinalis* (Krishnamurthy et al. 2014b; Sobhanian et al. 2010a; Wakeel et al. 2011; Wang et al. 2009; Zorb et al. 2010). VDAC2 was shown to be involved in rendering salt tolerance by inducing the expression of SOS3, which would later lead to the efflux of Na^+ through SOS1, a PM Na^{+}/H^{+} antiporter (Wen et al. 2011). In addition, the levels of a calcium-permeable channel, annexin, were upregulated by salt stress in several plant species (Aghaei and Komatsu 2013; Lee et al. 2004; Li et al. 2010b; Manaa et al. 2011; Pang et al. 2010; Peng et al. 2009; Sobhanian et al. 2010b; Wang et al. 2009). Increase in this channel may help in osmotic adjustment, cell expansion, and exocytosis during salt stress (Lee et al. 2004; Zhang et al. 2012a). All these studies strengthen the vital role of ion transporters in salt tolerance.

10.10 Posttranslational Modification

Posttranslational modifications (PTMs) play a key role in plant salt tolerance because they are involved in fine-tuning of protein function, half-life, localization, and interactions to mitigate several damages caused by environmental stresses (Hashiguchi and Komatsu 2016; Xiong et al. 2016). Among these, protein phosphorylation is one of the major mechanisms for transmission of stress signals, while other modifications such as S-nitrosylation, carbonylation, ubiquitination, and SUMOylation have gained more importance recently as major posttranslational regulatory processes in eukaryotes (Guerra et al. 2015). These modifications may occur at different levels on the same protein target during the translational process, thereby affecting the amount of target and the corresponding phenotypic trait. In a rice root phosphoproteomic study, 17 phosphoproteins were found to be increased, while 11 were found to be decreased upon salt treatment (Chitteti and Peng 2007). Some of the increased proteins include HSP70, small GTP-binding protein, mannose-binding rice lectin, and OsRac2, while protein kinases and ATP synthase were among the decreased proteins. Similarly, in maize several proteins were phosphorylated (fructokinase, UDP-glucosyl transferase, 2-Cys peroxiredoxin, etc.), while others were found to be dephosphorylated (CaM, isocitrate dehydrogenase, 40S ribosomal proteins, etc.) by salt stress (Zorb et al. 2010). Similarly, phosphorylation of dehydrin (DHN5) in wheat and several PR10 proteins from peanut upon salt treatment were reported (Brini et al. 2007; Jain et al. 2006). In another study involving citrus, several carbonylated proteins (such as chaperonin 60 subunit alpha, ADH, HSP70, glycolytic enzymes, and mitochondrial processing peptidases) and nitrosylated proteins (such as GST, SOD, GAPDH, GST, RuBisCO activase, PGK, peroxiredoxin, and tubulin) were found (Tanou et al. 2009). The newly emerging studies on PTMs suggest that these could be important determinants of salt tolerance in plants.

10.11 Selected Examples of Enhanced Salt Tolerance in Plants Achieved Using Information Derived from *Omics* Studies

As described in the earlier sections, it is evident that the omics research has helped researchers to identify the key genes and proteins responsible for salt tolerance. Because of the multigenic property of salt tolerance trait, researchers have achieved limited success in improving the salt tolerance of crop plants using conventional breeding programs so far. However, several salt-tolerant crops with appreciable productivity have been developed over the past decade using the transgenic approaches. Ion transporters have gained more importance in the past decade in transgenic studies for their role in alleviation of Na^+-induced toxicity in plant cells. Increase in salinity tolerance has been achieved in various plant species (e.g., *Arabidopsis thaliana*, *Nicotiana tabacum*, *Oryza sativa*, and *Medicago sativa*) by overexpressing NHX genes from plants such as *Agropyron elongatum*, *Pennisetum glaucum*, *Hordeum vulgare*, and *Triticum aestivum*, indicating that these genes are involved in regulating salt tolerance in plants (Bayat et al. 2011; Qiao et al. 2007; Verma et al. 2007; Zhang et al. 2012b). Transgenic poplar plants overexpressing *PtSOS2* showed increased tolerance to salt stress, which was associated with a decreased Na^+ accumulation in the leaves of transgenic plants (Yang et al. 2015). The chloride transporter *GmCLC1* enhanced salt tolerance through regulating chloride accumulation in soybean (Wei et al. 2016), while overexpression of the high-affinity potassium transporter, *HvHKT2;1*, increased salt tolerance of barley (Mian et al. 2011). Transgenic *Arabidopsis* plants overexpressing an ABC transporter gene *AtABCG36* showed increased salt and drought resistance with concomitant reduction of Na^+ contents in the transgenic plants. Similarly using vacuolar H^+ ATPase subunit C1 (*SaVHAC1*), enhanced salinity tolerance in rice was achieved (Baisakh et al. 2012). Overexpression of dehydration-responsive element-binding transcription factor (*LcDREB3a*), MYB-related transcription factor (*LcMYB-1*), NAC transcription factor*TaNAC29*, and basic helix-loop-helix (bHLH) protein gene (*OrbHLH001*) all led to enhanced salt tolerance in *Arabidopsis thaliana* (Cheng et al. 2013; Li et al. 2010a; Xianjun et al. 2011; Huang et al. 2015a). Transgenic mulberry plants overexpressing *HVA1* that encodes a group of three late embryogenesis abundant (LEA) proteins increased accumulation of proline leading to increased salt tolerance (Lal et al. 2008). Overexpression of antioxidant enzymes SOD and APX enhanced salt stress tolerance in sweet potato by reducing the oxidative damage to the plants (Yan et al. 2016).These are only a few examples showing how the information obtained by omics research involving both transcriptomic and proteomic approaches are helping to advance in the field of crop improvement. It has been observed that enhanced tolerance to one of the abiotic stresses conferred by the above approaches often results in enhanced tolerance to other abiotic stresses as well (e.g., salinity and drought, salinity and heat stresses), which is a collateral benefit.

10.12 Concluding Remarks and Future Perspectives

Salinity induces both osmotic and ionic stress to the plants and perturbs plant metabolism and ionic balance. Under such conditions, plants need to evolve with adaptive strategies that help them survive and reproduce. Glycophytes and halophytes exhibit differences in their extent of salt tolerance and the mechanisms that occur in halophytes could be exploited in order to generate salt-tolerant crop plants which are mainly glycophytes. Proteins play an important role in salt tolerance as they are the direct players in most of the processes such as metabolism and growth. Plants respond to stress by readjusting the cellular metabolism and ionic balance, increasing the signaling, and expressing stress remediation-related proteins. Recent advances in proteomic techniques such as iTRAQ, SWATH, MudPIT, and MRM MS have opened new avenues for identifying low-abundance proteins expressed in response to salt stress that were difficult to identify earlier (Hu et al. 2015). Despite the decades of salinity research, there are still many challenges in utilizing all these information to generate salt-tolerant crop plants. Integrating newly emerging proteomic data with genomic, transcriptomic, and metabolomic data would be highly useful in unraveling the complex mechanism of salt tolerance, which could offer novel strategies to engineer crop plants that can tolerate high-salinity conditions in the future.

Acknowledgments The research work in our laboratory is supported by the Singapore National Research Foundation under its Environment and Water Research Programme and administered by PUB, Singapore's National Water Agency, Singapore, NRF-EWI-IRIS (R-706-000-010-272 and R-706-000-040-279).

References

Abbasi FM, Komatsu S (2004) A proteomic approach to analyze salt-responsive proteins in rice leaf sheath. Proteomics 4(7):2072–2081. https://doi.org/10.1002/pmic.200300741

Acosta-Motos JR, Ortuno MF, Bernal-Vicente A, Diaz-Vivancos P, Sanchez-Blanco MJ, Hernandez JA (2017) Plant responses to salt stress: adaptive mechanisms. Agronomy 7(18):1–38. https://doi.org/10.3390/agronomy7010018

Aghaei K, Komatsu S (2013) Crop and medicinal plants proteomics in response to salt stress. Front Plant Sci 4:8. https://doi.org/10.3389/fpls.2013.00008

Apse MP, Aharon GS, Snedden WA, Blumwald E (1999) Salt tolerance conferred by overexpression of a vacuolar Na^+/H^+ antiport in *Arabidopsis*. Science 285(5431):1256–1258

Askari H, Edqvist J, Hajheidari M, Kafi M, Salekdeh GH (2006) Effects of salinity levels on proteome of *Suaeda aegyptiaca* leaves. Proteomics 6(8):2542–2554. https://doi.org/10.1002/pmic.200500328

Baisakh N, RamanaRao MV, Rajasekaran K, Subudhi P, Janda J, Galbraith D, Vanier C, Pereira A (2012) Enhanced salt stress tolerance of rice plants expressing a vacuolar H^+ -ATPase subunit c1 (SaVHAc1) gene from the halophyte grass Spartina alterniflora Loisel. Plant Biotechnol J 10(4):453–464. https://doi.org/10.1111/j.1467-7652.2012.00678.x

Bandehagh A, Salekdeh GH, Toorchi M, Mohammadi A, Komatsu S (2011) Comparative proteomic analysis of canola leaves under salinity stress. Proteomics 11(10):1965–1975. https://doi.org/10.1002/pmic.201000564

Barkla BJ, Vera-Estrella R, Hernandez-Coronado M, Pantoja O (2009) Quantitative proteomics of the tonoplast reveals a role for glycolytic enzymes in salt tolerance. Plant Cell 21(12):4044–4058. https://doi.org/10.1105/tpc.109.069211

Bayat F, Shiran B, Belyaev DV (2011) Overexpression of *HvNHX2*, a vacuolar Na^+/H^+ antiporter gene from barley, improves salt tolerance in *Arabidopsis thaliana*. Aust J Crop Sci 5:428–432

Bohnert HJ, Nelson DE, Jensen RG (1995) Adaptations to environmental stresses. Plant Cell 7(7):1099–1111. https://doi.org/10.1105/tpc.7.7.1099

Boursiac Y, Chen S, Luu DT, Sorieul M, van den Dries N, Maurel C (2005) Early effects of salinity on water transport in *Arabidopsis* roots. Molecular and cellular features of aquaporin expression. Plant Physiol 139(2):790–805. https://doi.org/10.1104/pp.105.065029

Brini F, Hanin M, Lumbreras V, Irar S, Pages M, Masmoudi K (2007) Functional characterization of DHN-5, a dehydrin showing a differential phosphorylation pattern in two Tunisian durum wheat (*Triticum durum* Desf.) varieties with marked differences in salt and drought tolerance. Plant Sci 172:20–28

Caruso G, Cavaliere C, Guarino C, Gubbiotti R, Foglia P, Lagana A (2008) Identification of changes in *Triticum durum* L. leaf proteome in response to salt stress by two-dimensional electrophoresis and MALDI-TOF mass spectrometry. Anal Bioanal Chem 391(1):381–390. https://doi.org/10.1007/s00216-008-2008-x

Chattopadhyay A, Subba P, Pandey A, Bhushan D, Kumar R, Datta A, Chakraborty S, Chakraborty N (2011) Analysis of the grasspea proteome and identification of stress-responsive proteins upon exposure to high salinity, low temperature, and abscisic acid treatment. Phytochemistry 72(10):1293–1307. https://doi.org/10.1016/j.phytochem.2011.01.024

Cheeseman JM (1988) Mechanisms of salinity tolerance in plants. Plant Physiol 87(3):547–550

Chen S, Gollop N, Heuer B (2009) Proteomic analysis of salt-stressed tomato (*Solanum lycopersicum*) seedlings: effect of genotype and exogenous application of glycinebetaine. J Exp Bot 60(7):2005–2019. https://doi.org/10.1093/jxb/erp075

Chen F, Zhang S, Jiang H, Ma W, Korpelainen H, Li C (2011) Comparative proteomics analysis of salt response reveals sex-related photosynthetic inhibition by salinity in Populus cathayana cuttings. J Proteome Res 10(9):3944–3958. https://doi.org/10.1021/pr200535r

Chen T, Zhang L, Shang H, Liu S, Peng J, Gong W, Shi Y, Zhang S, Li J, Gong J, Ge Q, Liu A, Ma H, Zhao X, Yuan Y (2016) iTRAQ-based quantitative proteomic analysis of cotton roots and leaves reveals pathways associated with salt stress. PLoS One 11(2):e0148487. https://doi.org/10.1371/journal.pone.0148487

Cheng WH, Endo A, Zhou L, Penney J, Chen HC, Arroyo A, Leon P, Nambara E, Asami T, Seo M, Koshiba T, Sheen J (2002) A unique short-chain dehydrogenase/reductase in *Arabidopsis* glucose signaling and abscisic acid biosynthesis and functions. Plant Cell 14(11):2723–2743

Cheng Y, Qi Y, Zhu Q, Chen X, Wang N, Zhao X, Chen H, Cui X, Xu L, Zhang W (2009) New changes in the plasma-membrane-associated proteome of rice roots under salt stress. Proteomics 9(11):3100–3114. https://doi.org/10.1002/pmic.200800340

Cheng L, Li X, Huang X, Ma T, Liang Y, Ma X, Peng X, KJia J, Chen S, Chen Y, Deng B, Liu G (2013) Overexpression of sheep grass R1-MYB transcription factor LcMYB1 confers salt tolerance in transgenic *Arabidopsis*. Plant Physiol Biochem 70:252–260

Cheng T, Chen J, Zhang J, Shi S, Zhou Y, Lu L, Wang P, Jiang Z, Yang J, Zhang S, Shi J (2015) Physiological and proteomic analyses of leaves from the halophyte *Tangut Nitraria* reveals diverse response pathways critical for high salinity tolerance. Front Plant Sci 6:30. https://doi.org/10.3389/fpls.2015.00030

Chitteti BR, Peng Z (2007) Proteome and phosphoproteome differential expression under salinity stress in rice (*Oryza sativa*) roots. J Proteome Res 6(5):1718–1727. https://doi.org/10.1021/pr060678z

Cosentino C, Di Silvestre D, Fischer-Schliebs E, Homann U, De Palma A, Comunian C, Mauri PL, Thiel G (2013) Proteomic analysis of *Mesembryanthemum crystallinum* leaf microsomal fractions finds an imbalance in V-ATPase stoichiometry during the salt-induced transition from C3 to CAM. Biochem J 450(2):407–415. https://doi.org/10.1042/BJ20121087

Dani V, Simon WJ, Duranti M, Croy RR (2005) Changes in the tobacco leaf apoplast proteome in response to salt stress. Proteomics 5(3):737–745. https://doi.org/10.1002/pmic.200401119

Deinlein U, Stephan AB, Horie T, Luo W, Xu G, Schroeder JI (2014) Plant salt-tolerance mechanisms. Trends Plant Sci 19(6):371–379. https://doi.org/10.1016/j.tplants.2014.02.001

Dooki AD, Mayer-Posner FJ, Askari H, Zaiee AA, Salekdeh GH (2006) Proteomic responses of rice young panicles to salinity. Proteomics 6(24):6498–6507. https://doi.org/10.1002/pmic.200600367

Du CX, Fan HF, Guo SR, Tezuka T, Li J (2010) Proteomic analysis of cucumber seedling roots subjected to salt stress. Phytochemistry 71(13):1450–1459. https://doi.org/10.1016/j.phytochem.2010.05.020

Dunkley TP, Hester S, Shadforth IP, Runions J, Weimar T, Hanton SL, Griffin JL, Bessant C, Brandizzi F, Hawes C, Watson RB, Dupree P, Lilley KS (2006) Mapping the *Arabidopsis* organelle proteome. Proc Natl Acad Sci 103(17):6518–6523. https://doi.org/10.1073/pnas.0506958103

El Rabey HA, Al-Malki AL, Abulnaja KO, Rohde W (2015) Proteome analysis for understanding abiotic stress (salinity and drought) tolerance in date palm (*Phoenix dactylifera* L.). Int J Genomics 2015:407165. https://doi.org/10.1155/2015/407165

Eldakak M, Milad SI, Nawar AI, Rohila JS (2013) Proteomics: a biotechnology tool for crop improvement. Front Plant Sci 4:35. https://doi.org/10.3389/fpls.2013.00035

Enstone JE, Perterson CA, Ma FS (2003) Root endodermis and exodermis: structure, function and response to the environment. J Plant Growth Regul 21:335–351

Evers D, Legay S, Lamoureux D, Hausman JF, Hoffmann L, Renaut J (2012) Towards a synthetic view of potato cold and salt stress response by transcriptomic and proteomic analyses. Plant Mol Biol 78(4–5):503–514. https://doi.org/10.1007/s11103-012-9879-0

Faiyue B, Al-Azzawi MJ, Flowers TJ (2010) The role of lateral roots in bypass flow in rice (*Oryza sativa* L.). Plant Cell Environ 33(5):702–716. https://doi.org/10.1111/j.1365-3040.2009.02078.x

Fatehi F, Hosseinzadeh A, Alizadeh H, Brimavandi T, Struik PC (2012) The proteome response of salt-resistant and salt-sensitive barley genotypes to long-term salinity stress. Mol Biol Rep 39(5):6387–6397. https://doi.org/10.1007/s11033-012-1460-z

Flowers TJ (2004) Improving crop salt tolerance. J Exp Biol 55(396):307–319. https://doi.org/10.1093/jxb/erh003

Flowers TJ, Colmer TD (2015) Plant salt tolerance: adaptations in halophytes. Ann Bot 115(3):327–331

Flowers TJ, Yeo AR (1995) Breeding for salinity resistance in crop plants: where next? Australian. J Plant Physiol 22:875–884

Gan CS, Chong PK, Pham TK, Wright PC (2007) Technical, experimental, and biological variations in isobaric tags for relative and absolute quantitation (iTRAQ). J Proteome Res 6(2):821–827. https://doi.org/10.1021/pr060474i

Gao L, Yan X, Li X, Guo G, Hu Y, Ma W, Yan Y (2011) Proteome analysis of wheat leaf under salt stress by two-dimensional difference gel electrophoresis (2D-DIGE). Phytochemistry 72(10):1180–1191. https://doi.org/10.1016/j.phytochem.2010.12.008

Gharat SA, Parmar S, Tambat S, Vasudevan M, Shaw BP (2016) Transcriptome analysis of the response to NaCl in *Suaeda maritima* provides an insight into salt tolerance mechanisms in halophytes. PLoS One 11(9):e0163485. https://doi.org/10.1371/journal.pone.0163485

Gorg A, Drews O, Luck C, Weiland F, Weiss W (2009) 2-DE with IPGs. Electrophoresis 30(Suppl 1):S122–S132. https://doi.org/10.1002/elps.200900051

Guerra D, Crosatti C, Khoshro HH, Mastrangelo AM, Mica E, Mazzucotelli E (2015) Post-transcriptional and post-translational regulations of drought and heat response in plants: a spider's web of mechanisms. Front Plant Sci 6:57. https://doi.org/10.3389/fpls.2015.00057

Gupta B, Huang B (2014) Mechanism of salinity tolerance in plants: physiological, biochemical, and molecular characterization. Int J Genomics 2014:701596. https://doi.org/10.1155/2014/701596

Gygi SP, Rochon Y, Franza BR, Aebersold R (1999) Correlation between protein and mRNA abundance in yeast. Mol Cell Biol 19(3):1720–1730

Hakeem KR, Chandan R, Rehman RU, Tahir I, Sabir M, Iqbal M (eds) (2013) Salt stress in plants: signalling, omics and adaptations enhancing plant productivity under salt stress: relevance of poly-omics. Springer, New York. https://doi.org/10.1007/978-1-4614-6108-1_3

Hasanuzzaman M, Nahar K, Fujita M, Ahmad P, Chandna R, Prasad MNV, Ozturk M (eds) (2013) Salt stress in plants: signalling, omics and adaptations Enhancing plant productivity under salt stress: relevance of poly-omics. Springer, New York. https://doi.org/10.1007/978-1-4614-6108-1_6

Hasegawa PM, Bressan RA, Zhu JK, Bohnert HJ (2000) Plant cellular and molecular responses to high salinity. Annu Rev Plant Physiol Plant Mol Biol 51:463–499. https://doi.org/10.1146/annurev.arplant.51.1.463

Hashiguchi A, Komatsu S (2016) Impact of post-translational modifications of crop proteins under abiotic stress. Proteomes 4(4):42. https://doi.org/10.3390/proteomes4040042

He XJ, Mu RL, Cao WH, Zhang ZG, Zhang JS, Chen SY (2005) AtNAC2, a transcription factor downstream of ethylene and auxin signaling pathways, is involved in salt stress response and lateral root development. Plant J 44(6):903–916. https://doi.org/10.1111/j.1365-313X.2005.02575.x

Hu J, Rampitsch C, Bykova NV (2015) Advances in plant proteomics toward improvement of crop productivity and stress resistancex. Front Plant Sci 6:209. https://doi.org/10.3389/fpls.2015.00209

Huang Q, Wang Y, Li B, Chang J, Chen M, Li K, Yang G, He G (2015a) TaNAC29, a NAC transcription factor from wheat, enhances salt and drought tolerance in transgenic Arabidopsis. BMC Plant Biol 15:268. https://doi.org/10.1186/s12870-015-0644-9

Huang Q, Yang L, Luo J, Guo L, Wang Z, Yang X, Jin W, Fang Y, Ye J, Shan B, Zhang Y (2015b) SWATH enables precise label-free quantification on proteome scale. Proteomics 15(7):1215–1223. https://doi.org/10.1002/pmic.201400270

Jain S, Srivastava S, Sarin NB, Kav NN (2006) Proteomics reveals elevated levels of PR 10 proteins in saline-tolerant peanut (*Arachis hypogaea*) calli. Plant Physiol Biochem 44(4):253–259. https://doi.org/10.1016/j.plaphy.2006.04.006

Ji W, Cong R, Li S, Li R, Qin Z, Li Y, Zhou X, Chen S, Li J (2016) Comparative proteomic analysis of soybean leaves and roots by iTRAQ provides insights into response mechanisms to short-term salt stress. Front Plant Sci 7:573. https://doi.org/10.3389/fpls.2016.00573

Jia H, Shao M, He Y, Guan R, Chu P, Jiang H (2015) Proteome dynamics and physiological responses to short-term salt stress in *Brassica napus* leaves. PLoS One 10(12):e0144808. https://doi.org/10.1371/journal.pone.0144808

Jia YL, Chen H, Zhang C, Gao LJ, Wang XC, Qiu LL, Wu JF (2016) Proteomic analysis of halotolerant proteins under high and low salt stress in *Dunaliella salina* using two-dimensional differential in-gel electrophoresis. Genet Mol Biol 39(2):239–247. https://doi.org/10.1590/1678-4685-GMB-2015-0108

Jiang Y, Deyholos MK (2006) Comprehensive transcriptional profiling of NaCl-stressed *Arabidopsis* roots reveals novel classes of responsive genes. BMC Plant Biol 6:25. https://doi.org/10.1186/1471-2229-6-25

Jiang Y, Yang B, Harris NS, Deyholos MK (2007) Comparative proteomic analysis of NaCl stress-responsive proteins in *Arabidopsis* roots. J Exp Bot 58(13):3591–3607. https://doi.org/10.1093/jxb/erm207

Jiang Q, Li X, Niu F, Sun X, Hu Z, Zhang H (2017) iTRAQ-based quantitative proteomic analysis of wheat roots in response to salt stress. Proteomics 17:1–13. https://doi.org/10.1002/pmic.201600265

Jorrin-Novo JV, Maldonado AM, Echevarria-Zomeno S, Valledor L, Castillejo MA, Curto M, Valero J, Sghaier B, Donoso G, Redondo I (2009) Plant proteomics update (2007–2008): second-generation proteomic techniques, an appropriate experimental design, and data analysis to fulfill MIAPE standards, increase plant proteome coverage and expand biological knowledge. J Proteome 72(3):285–314

Jyothi-Prakash PA, Mohanty B, Wijaya E, Lim TM, Lin Q, Loh CS, Kumar PP (2014) Identification of salt gland-associated genes and characterization of a dehydrin from the salt secretor mangrove Avicennia officinalis. BMC Plant Biol 14:291. https://doi.org/10.1186/s12870-014-0291-6

Kim DW, Rakwal R, Agrawal GK, Jung YH, Shibato J, Jwa NS, Iwahashi Y, Iwahashi H, Kim DH, Shim Ie S, Usui K (2005) A hydroponic rice seedling culture model system for investigating proteome of salt stress in rice leaf. Electrophoresis 26(23):4521–4539. https://doi.org/10.1002/elps.200500334

Kim DY, Bovet L, Maeshima M, Martinoia E, Lee Y (2007) The ABC transporter AtPDR8 is a cadmium extrusion pump conferring heavy metal resistance. Plant J 50(2):207–218. https://doi.org/10.1111/j.1365-313X.2007.03044.x

Kim DY, Jin JY, Alejandro S, Martinoia E, Lee Y (2010) Overexpression of AtABCG36 improves drought and salt stress resistance in *Arabidopsis*. Physiol Plant 139(2):170–180. https://doi.org/10.1111/j.1399-3054.2010.01353.x

Kim K, Seo E, Chang SK, Park TJ, Lee SJ (2016) Novel water filtration of saline water in the outermost layer of mangrove roots. Sci Rep 6:20426. https://doi.org/10.1038/srep20426

Kosava K, Vitamvas P, Prasil IT, TRenaut J (2011) Plant proteome changes under abiotic stress- condition of proteomics studies to understanding plant stress response. J Proteome 74:1301–1322

Kosava K, Vitamvas P, Urban MO, Prail IT (2013) Plant proteome responses to salinity stress-comparison of glycophytes and halophytes. Funct Plant Biol 40:775–786

Kosova K, Prail IT, Vitamvas P (2013) Protein contribution to plant salinity response and tolerance acquisition. Int J Mol Sci 14(4):6757–6789. https://doi.org/10.3390/ijms14046757

Krishnamurthy P, Ranathunge K, Nayak S, Schreiber L, Mathew MK (2011) Root apoplastic barriers block Na^+ transport to shoots in rice (*Oryza sativa* L.). J Exp Bot 62(12):4215–4228. https://doi.org/10.1093/jxb/err135

Krishnamurthy P, Jyothi-Prakash PA, Qin L, He J, Lin Q, Loh CS, Kumar PP (2014a) Role of root hydrophobic barriers in salt exclusion of a mangrove plant *Avicennia officinalis*. Plant Cell Environ 37(7):1656–1671. https://doi.org/10.1111/pce.12272

Krishnamurthy P, Tan XF, Lim TK, Lim TM, Kumar PP, Loh CS, Lin Q (2014b) Proteomic analysis of plasma membrane and tonoplast from the leaves of mangrove plant *Avicennia officinalis*. Proteomics 14(21–22):2545–2557. https://doi.org/10.1002/pmic.201300527

Krishnamurthy P, Mohanty B, Wijaya E, Lee DY, Lim TM, Lin Q, Xu J, Loh CS, Kumar P (2017) Transcriptomics analysis of salt stress tolerance in the roots of the mangrove *Avicennia officinalis*, Scientific Reports 7:10031

Lal S, Gulyani V, Khurana P (2008) Overexpression of HVA1 gene from barley generates tolerance to salinity and water stress in transgenic mulberry (Morus indica). Transgenic Res 17(4):651–663. https://doi.org/10.1007/s11248-007-9145-4

Lee S, Lee EJ, Yang EJ, Lee JE, Park AR, Song WH, Park OK (2004) Proteomic identification of annexins, calcium-dependent membrane binding proteins that mediate osmotic stress and abscisic acid signal transduction in *Arabidopsis*. Plant Cell 16(6):1378–1391. https://doi.org/10.1105/tpc.021683

Li F, Guo S, Zhao Y, Chen D, Chong K, Xu Y (2010a) Overexpression of a homopeptide repeat-containing bHLH protein gene (OrbHLH001) from Dongxiang Wild Rice confers freezing and salt tolerance in transgenic Arabidopsis. Plant Cell Rep 29(9):977–986. https://doi.org/10.1007/s00299-010-0883-z

Li XJ, Yang MF, Chen H, Qu LQ, Chen F, Shen SH (2010b) Abscisic acid pretreatment enhances salt tolerance of rice seedlings: proteomic evidence. Biochim Biophys Acta 1804(4):929–940. https://doi.org/10.1016/j.bbapap.2010.01.004

Li W, Zhang C, Lu Q, Wen X, Lu C (2011) The combined effect of salt stress and heat shock on proteome profiling in *Suaeda salsa*. J Plant Physiol 168(15):1743–1752. https://doi.org/10.1016/j.jplph.2011.03.018

Li W, Zhao F, Fang W, Xie D, Hou J, Yang X, Zhao Y, Tang Z, Nie L, Lv S (2015) Identification of early salt stress responsive proteins in seedling roots of upland cotton (*Gossypium hirsutum* L.) employing iTRAQ-based proteomic technique. Front Plant Sci 6:732. https://doi.org/10.3389/fpls.2015.00732

Liu CW, Chang TS, Hsu YK, Wang AZ, Yen HC, Wu YP, Wang CS, Lai CC (2014) Comparative proteomic analysis of early salt stress responsive proteins in roots and leaves of rice. Proteomics 14(15):1759–1775. https://doi.org/10.1002/pmic.201300276

Luo J, Tang S, Peng X, Yan X, Zeng X, Li J, Li X, Wu G (2015) Elucidation of cross-talk and specificity of early response mechanisms to salt and PEG-simulated drought stresses in *Brassica napus* using comparative proteomic analysis. PLoS One 10(10):e0138974. https://doi.org/10.1371/journal.pone.0138974

Ma FS, Peterson CA (2003) Current insights into the development, structure and chemistry of the endodermis and exodermis of roots. Can J Bot 81:405–421

Mahajan S, Pandey GK, Tuteja N (2008) Calcium- and salt-stress signaling in plants: shedding light on SOS pathway. Arch Biochem Biophys 471(2):146–158. https://doi.org/10.1016/j.abb.2008.01.010

Manaa A, Ahmed HB, Smiti S, Faurobert M (2011) Salt-stress induced physiological and proteomic changes in tomato (*Solanum lycopersicum*) seedlings. OMICS 15(11):801–809. https://doi.org/10.1089/omi.2011.0045

Mastrobuoni G, Irgang S, Pietzke M, Assmus HE, Wenzel M, Schulze WX, Kempa S (2012) Proteome dynamics and early salt stress response of the photosynthetic organism *Chlamydomonas reinhardtii*. BMC Genomics 13:215. https://doi.org/10.1186/1471-2164-13-215

Maurel C, Chrispeels MJ (2001) Aquaporins. A molecular entry into plant water relations. Plant Physiol 125(1):135–138

Maurel C, Verdoucq L, Luu DT, Santoni V (2008) Plant aquaporins: membrane channels with multiple integrated functions. Annu Rev Plant Biol 59:595–624. https://doi.org/10.1146/annurev.arplant.59.032607.092734

Mian A, Oomen RJ, Isayenkov S, Sentenac H, Maathuis FJ, Very AA (2011) Over-expression of an Na^+-and K^+-permeable HKT transporter in barley improves salt tolerance. Plant J 68(3):468–479. https://doi.org/10.1111/j.1365-313X.2011.04701.x

Mishra A, Tanna B (2017) Halophytes: potential resources for salt stress tolerance genes and promoters. Front Plant Sci 8:829. https://doi.org/10.3389/fpls.2017.00829

Munns R, Tester M (2008) Mechanisms of salinity tolerance. Annu Rev Plant Biol 59:651–681. https://doi.org/10.1146/annurev.arplant.59.032607.092911

Nat NVK, Sanjeeva S, Laksiri G, Stanford FB (2004) Proteome-level changes in the roots of *Pisum sativum* in response to salinity. Ann Appl Biol 145:217–230

Ndimba BK, Chivasa S, Simon WJ, Slabas AR (2005) Identification of Arabidopsis salt and osmotic stress responsive proteins using two-dimensional difference gel electrophoresis and mass spectrometry. Proteomics 5(16):4185–4196. https://doi.org/10.1002/pmic.200401282

Palmblad M, Mills DJ, Bindschedler LV (2008) Heat-shock response in *Arabidopsis thaliana* explored by multiplexed quantitative proteomics using differential metabolic labeling. J Proteome Res 7(2):780–785. https://doi.org/10.1021/pr0705340

Pang Q, Chen S, Dai S, Chen Y, Wang Y, Yan X (2010) Comparative proteomics of salt tolerance in Arabidopsis thaliana and *Thellungiella halophila*. J Proteome Res 9(5):2584–2599. https://doi.org/10.1021/pr100034f

Parida AK, Das AB (2005) Salt tolerance and salinity effects on plants: a review. Ecotoxicol Environ Saf 60(3):324–349. https://doi.org/10.1016/j.ecoenv.2004.06.010

Parida AK, Jha B (2010) Salt tolerance mechanisms in mangroves: a review. Trees 24:199–217

Parida AK, Das AB, Mittra B, Mohanty P (2004a) Salt-stress induced alterations in protein profile and protease activity in the mangrove *Bruguiera parviflora*. Z Naturforsch C 59(5–6):408–414

Parida AK, Das AB, Mohanty P (2004b) Defense potentials to NaCl in a mangrove, *Bruguiera parviflora*: differential changes of isoforms of some antioxidative enzymes. J Plant Physiol 161(5):531–542. https://doi.org/10.1078/0176-1617-01084

Parker R, Flowers TJ, Moore AL, Harpham NV (2006) An accurate and reproducible method for proteome profiling of the effects of salt stress in the rice leaf lamina. J Exp Bot 57(5):1109–1118. https://doi.org/10.1093/jxb/erj134

Peng Z, Wang M, Li F, Lv H, Li C, Xia G (2009) A proteomic study of the response to salinity and drought stress in an introgression strain of bread wheat. Mol Cell Proteomics 8(12):2676–2686. https://doi.org/10.1074/mcp.M900052-MCP200

Peterson CA (1988) Exodermal Casparian bands: their significance for ion uptake by roots. Physiol Plant 72:204–208

Popp M (1995) Salt resistance in herbaceous halophytes and mangroves. Prog Bot 56:415–429

Qiao WH, Zhao XY, Li W, Luo Y, Zhang XS (2007) Overexpression of AeNHX1, a root-specific vacuolar Na^+/H^+ antiporter from Agropyron elongatum, confers salt tolerance to Arabidopsis and Festuca plants. Plant Cell Rep 26(9):1663–1672. https://doi.org/10.1007/s00299-007-0354-3

Qing DJ, Lu HF, Li N, Dong HT, Dong DF, Li YZ (2009) Comparative profiles of gene expression in leaves and roots of maize seedlings under conditions of salt stress and the removal of salt stress. Plant Cell Physiol 50(4):889–903. https://doi.org/10.1093/pcp/pcp038

Rabbani MA, Maruyama K, Abe H, Khan MA, Katsura K, Ito Y, Yoshiwara K, Seki M, Shinozaki K, Yamaguchi-Shinozaki K (2003) Monitoring expression profiles of rice genes under cold, drought, and high-salinity stresses and abscisic acid application using cDNA microarray and RNA gel-blot analyses. Plant Physiol 133(4):1755–1767. https://doi.org/10.1104/pp.103.025742

Ranathunge K, Steudle E, Lafitte R (2003) Control of water uptake by rice (*Oryza sativa* L.): role of the outer part of the root. Planta 217(2):193–205. https://doi.org/10.1007/s00425-003-0984-9

Ranjithakumari BD, Radhakrishnan (2008) Plant proteomics. In: Nangia SB (ed) Plant with biotic and abiotic factors interaction proteomics. A. P. H. Publishing Corporation, New Delhi

Rasoulnia A, Bihamta MR, Peyghambari SA, Alizadeh H, Rahnama A (2011) Proteomic response of barley leaves to salinity. Mol Biol Rep 38(8):5055–5063. https://doi.org/10.1007/s11033-010-0651-8

Razavizadeh R, Ehsanpour AA, Ahsan N, Komatsu S (2009) Proteome analysis of tobacco leaves under salt stress. Peptides 30(9):1651–1659. https://doi.org/10.1016/j.peptides.2009.06.023

Reinhardt DH, Rost TL (1995) Salinity accelerates endodermal development and induces an exodermis in cotton seedling roots. Environ Exp Bot 35:563–574

Rossignol M, Peltier JB, Mock HP, Matros A, Maldonado AM, Jorrin JV (2006) Plant proteome analysis: a 2004–2006 update. Proteomics 6(20):5529–5548. https://doi.org/10.1002/pmic.200600260

Ruan SL, Ma HS, Wang SH, Fu YP, Xin Y, Liu WZ, Wang F, Tong JX, Wang SZ, Chen HZ (2011) Proteomic identification of OsCYP2, a rice cyclophilin that confers salt tolerance in rice (*Oryza sativa* L.) seedlings when overexpressed. BMC Plant Biol 11:34. https://doi.org/10.1186/1471-2229-11-34

Salekdeh GH, Siopongco J, Wade LJ, Ghareyazie B, Bennett J (2002) Proteomic analysis of rice leaves during drought stress and recovery. Proteomics 2(9):1131–1145. https://doi.org/10.1002/1615-9861(200209)2:9<1131::AID-PROT1131>3.0.CO;2-1

Sanyal A, Lajoie BR, Jain G, Dekker J (2012) The long-range interaction landscape of gene promoters. Nature 489(7414):109–113. https://doi.org/10.1038/nature11279

Scholander PF (1968) How mangroves desalinate water. Physiol Plant 21:251–261

Schreiber L, Franke BR (2011) Endodermis and exodermis in roots. Wiley, Chichester

Schreiber L, Hartmann K, Skrabs M, Zeier J (1999) Apoplastic barriers in roots: chemical composition of endodermal and hypodermal cell walls. J Exp Bot 50:1267–1280

Seki M, Narusaka M, Ishida J, Nanjo T, Fujita M, Oono Y, Kamiya A, Nakajima M, Enju A, Sakurai T, Satou M, Akiyama K, Taji T, Yamaguchi-Shinozaki K, Carninci P, Kawai J, Hayashizaki Y, Shinozaki K (2002) Monitoring the expression profiles of 7000 *Arabidopsis* genes under drought, cold and high-salinity stresses using a full-length cDNA microarray. Plant J 31(3):279–292

Shiio Y, Aebersold R (2006) Quantitative proteome analysis using isotope-coded affinity tags and mass spectrometry. Nat Protoc 1(1):139–145. https://doi.org/10.1038/nprot.2006.22

Sobhanian H, Motamed N, Jazii FR, Nakamura T, Komatsu S (2010a) Salt stress induced differential proteome and metabolome response in the shoots of *Aeluropus lagopoides* (Poaceae), a halophyte C(4) plant. J Proteome Res 9(6):2882–2897. https://doi.org/10.1021/pr900974k

Sobhanian H, Razavizadeh R, Nanjo Y, Ehsanpour AA, Jazii FR, Motamed N, Komatsu S (2010b) Proteome analysis of soybean leaves, hypocotyls and roots under salt stress. Proteome Sci 8:19. https://doi.org/10.1186/1477-5956-8-19

Sobhanian H, Aghaei K, Komatsu S (2011) Changes in the plant proteome resulting from salt stress: toward the creation of salt-tolerant crops? J Proteome 74(8):1323–1337. https://doi.org/10.1016/j.jprot.2011.03.018

Swami AK, Alam SI, Sengupta N, Sarin R (2011) Differential proteomic analysis of salt stress response in *Sorghum bicolor* leaves. Environ Exp Bot 71:321–328

Szittya G, Moxon S, Santos DM, Jing R, Fevereiro MP, Moulton V, Dalmay T (2008) High-throughput sequencing of *Medicago truncatula* short RNAs identifies eight new miRNA families. BMC Genomics 9:593. https://doi.org/10.1186/1471-2164-9-593

Tada Y, Kashimura T (2009) Proteomic analysis of salt-responsive proteins in the mangrove plant, Bruguiera gymnorhiza. Plant Cell Physiol 50(3):439–446. https://doi.org/10.1093/pcp/pcp002

Taji T, Seki M, Satou M, Sakurai T, Kobayashi M, Ishiyama K, Narusaka Y, Narusaka M, Zhu JK, Shinozaki K (2004) Comparative genomics in salt tolerance between *Arabidopsis* and *Arabidopsis*-related halophyte salt cress using *Arabidopsis* microarray. Plant Physiol 135(3):1697–1709. https://doi.org/10.1104/pp.104.039909

Tan WK, Lim TK, Loh CS, Kumar P, Lin Q (2015) Proteomic characterisation of the salt gland-enriched tissues of the mangrove tree species *Avicennia officinalis*. PLoS One 10(7):e0133386. https://doi.org/10.1371/journal.pone.0133386

Tang X, Mu X, Shao H, Wang H, Brestic M (2015) Global plant-responding mechanisms to salt stress: physiological and molecular levels and implications in biotechnology. Crit Rev Biotechnol 35(4):425–437. https://doi.org/10.3109/07388551.2014.889080

Tanou G, Job C, Rajjou L, Arc E, Belghazi M, Diamantidis G, Molassiotis A, Job D (2009) Proteomics reveals the overlapping roles of hydrogen peroxide and nitric oxide in the acclimation of citrus plants to salinity. Plant J 60(5):795–804. https://doi.org/10.1111/j.1365-313X.2009.04000.x

Tester M, Davenport R (2003) Na^+ tolerance and Na^+ transport in higher plants. Ann Bot 91(5):503–527

Tsuzuki M, Moskvin OV, Kuribayashi M, Sato K, Retamal S, Abo M, Zeilstra-Ryalls J, Gomelsky M (2011) Salt stress-induced changes in the transcriptome, compatible solutes, and membrane lipids in the facultatively phototrophic bacterium *Rhodobacter sphaeroides*. Appl Environ Microbiol 77(21):7551–7559. https://doi.org/10.1128/AEM.05463-11

Veeranagamallaiah G, Jyothsnakumari G, Thippeswamy M, Reddy PCO, Surabhi GK, Sriranganayakulu G, Mahesh Y, Rajasekhar B, Madhurarekha C, Sudhakar C (2008) Proteomic analysis of salt stress responses in foxtail millet (*Setaria italica* L. cv. Prasad) seedlings. Plant Sci 175:631–641

Verma D, Singla-Pareek SL, Rajagopal D, Reddy MK, Sopory SK (2007) Functional validation of a novel isoform of Na^+/H^+ antiporter from Pennisetum glaucum for enhancing salinity tolerance in rice. J Biosci 32(3):621–628

Vincent D, Ergul A, Bohlman MC, Tattersall EA, Tillett RL, Wheatley MD, Woolsey R, Quilici DR, Joets J, Schlauch K, Schooley DA, Cushman JC, Cramer GR (2007) Proteomic analysis reveals differences between *Vitis vinifera* L. cv. Chardonnay and cv. *Cabernet Sauvignon* and their responses to water deficit and salinity. J Exp Bot 58(7):1873–1892. https://doi.org/10.1093/jxb/erm012

Wakeel A, Asif AR, Pitann B, Schubert S (2011) Proteome analysis of sugar beet (*Beta vulgaris* L.) elucidates constitutive adaptation during the first phase of salt stress. J Plant Physiol 168(6):519–526. https://doi.org/10.1016/j.jplph.2010.08.016

Walia H, Wilson C, Zeng L, Ismail AM, Condamine P, Close TJ (2007) Genome-wide transcriptional analysis of salinity stressed japonica and indica rice genotypes during panicle initiation stage. Plant Mol Biol 63(5):609–623. https://doi.org/10.1007/s11103-006-9112-0

Wang H, Miyazaki S, Kawai K, Deyholos M, Galbraith DW, Bohnert HJ (2003) Temporal progression of gene expression responses to salt shock in maize roots. Plant Mol Biol 52(4):873–891

Wang MC, Peng ZY, Li CL, Li F, Liu C, Xia GM (2008a) Proteomic analysis on a high salt tolerance introgression strain of *Triticum aestivum/Thinopyrum ponticum*. Proteomics 8(7):1470–1489. https://doi.org/10.1002/pmic.200700569

Wang X, Yang P, Gao Q, Liu X, Kuang T, Shen S, He Y (2008b) Proteomic analysis of the response to high-salinity stress in *Physcomitrella patens*. Planta 228(1):167–177. https://doi.org/10.1007/s00425-008-0727-z

Wang X, Fan P, Song H, Chen X, Li X, Li Y (2009) Comparative proteomic analysis of differentially expressed proteins in shoots of *Salicornia europaea* under different salinity. J Proteome Res 8(7):3331–3345. https://doi.org/10.1021/pr801083a

Wang L, Liu X, Liang M, Tan F, Liang W, Chen Y, Lin Y, Huang L, Xing J, Chen W (2014) Proteomic analysis of salt-responsive proteins in the leaves of mangrove *Kandelia candel* during short-term stress. PLoS One 9(1):e83141. https://doi.org/10.1371/journal.pone.0083141

Wang J, Meng Y, Li B, Ma X, Lai Y, Si E, Yang K, Xu X, Shang X, Wang H, Wang D (2015) Physiological and proteomic analyses of salt stress response in the halophyte *Halogeton glomeratus*. Plant Cell Environ 38(4):655–669. https://doi.org/10.1111/pce.12428

Wei P, Wang L, Liu A, Yu B, Lam HM (2016) GmCLC1 confers enhanced salt tolerance through regulating chloride accumulation in soybean. Front Plant Sci 7:1082. https://doi.org/10.3389/fpls.2016.01082

Wen G, Cai L, Liu Z, Li DK, Lou Q, Li XF, Wan J, Yang Y (2011) *Arabidopsis thaliana VDAC2* involvement in salt stress response pathway. Afr J Biotechnol 10:11588–11593

Witzel K, Weidner A, Surabhi GK, Borner A, Mock HP (2009) Salt stress-induced alterations in the root proteome of barley genotypes with contrasting response towards salinity. J Exp Bot 60(12):3545–3557. https://doi.org/10.1093/jxb/erp198

Wu X, Gong F, Cao D, Hu X, Wang W (2016) Advances in crop proteomics: PTMs of proteins under abiotic stress. Proteomics 16(5):847–865. https://doi.org/10.1002/pmic.201500301

Xianjun P, Xingyong M, Weihong F, Man S, Liqin C, Alam I, Lee BH, Dongmei Q, Shihua S, Gongshe L (2011) Improved drought and salt tolerance of Arabidopsis thaliana by transgenic expression of a novel DREB gene from Leymus chinensis. Plant Cell Rep 30(8):1493–1502. https://doi.org/10.1007/s00299-011-1058-2

Xiong Y, Peng X, Cheng Z, Liu W, Wang GL (2016) A comprehensive catalog of the lysine-acetylation targets in rice (*Oryza sativa*) based on proteomic analyses. J Proteome 138:20–29. https://doi.org/10.1016/j.jprot.2016.01.019

Xu C, Sibicky T, Huang B (2010) Protein profile analysis of salt-responsive proteins in leaves and roots in two cultivars of creeping bentgrass differing in salinity tolerance. Plant Cell Rep 29(6):595–615. https://doi.org/10.1007/s00299-010-0847-3

Yan S, Tang Z, Su W, Sun W (2005) Proteomic analysis of salt stress-responsive proteins in rice root. Proteomics 5(1):235–244. https://doi.org/10.1002/pmic.200400853

Yan H, Li Q, Park SC, Wang X, Liu YJ, Zhang YG, Tang W, Kou M, Ma DF (2016) Overexpression of CuZnSOD and APX enhance salt stress tolerance in sweet potato. Plant Physiol Biochem 109:20–27. https://doi.org/10.1016/j.plaphy.2016.09.003

Yang Y, Tang RJ, Jiang CM, Li B, Kang T, Liu H, Zhao N, Ma XJ, Yang L, Chen SL, Zhang HX (2015) Overexpression of the PtSOS2 gene improves tolerance to salt stress in transgenic poplar plants. Plant Biotechnol J 13(7):962–973. https://doi.org/10.1111/pbi.12335

Yazaki K (2006) ABC transporters involved in the transport of plant secondary metabolites. FEBS Lett 580(4):1183–1191. https://doi.org/10.1016/j.febslet.2005.12.009

Yeo AR, Yeo ME, Flowers TJ (1987) The contribution of an apoplastic pathway to sodium uptake by rice roots in saline conditions. J Exp Bot 38:1141–1153

Yu J, Chen S, Zhao Q, Wang T, Yang C, Diaz C, Sun G, Dai S (2011) Physiological and proteomic analysis of salinity tolerance in *Puccinellia tenuiflora*. J Proteome Res 10(9):3852–3870. https://doi.org/10.1021/pr101102p

Zhang H, Han B, Wang T, Chen S, Li H, Zhang Y, Dai S (2012a) Mechanisms of plant salt response: insights from proteomics. J Proteome Res 11(1):49–67. https://doi.org/10.1021/pr200861w

Zhang YM, Liu ZH, Wen ZY, Zgang HM, Yang F, Guo XL (2012b) The vacuolar Na^+–H^+ antiport gene *TaNHX2* confers salt tolerance on transgenic alfalfa (*Medicago sativa*). Funct Plant Biol 39:708–716

Zhou J, Li F, Wang JL, Ma Y, Chong K, Xu YY (2009) Basic helix-loop-helix transcription factor from wild rice (OrbHLH2) improves tolerance to salt- and osmotic stress in *Arabidopsis*. J Plant Physiol 166(12):1296–1306. https://doi.org/10.1016/j.jplph.2009.02.007

Zhu JK (2002) Salt and drought stress signal transduction in plants. Annu Rev Plant Biol 53:247–273. https://doi.org/10.1146/annurev.arplant.53.091401.143329

Zhu JK (2003) Regulation of ion homeostasis under salt stress. Curr Opin Plant Biol 6(5):441–445

Zhuang J, Zhang J, Hou XL, Wang F, Xiong AS (2014) Transcriptomic, proteomic, metabolomic and functional genomic approaches for the study of abiotiuc stress in vegetable crops. Crit Rev Plant Sci 33:225–237

Zorb C, Herbst R, Forreiter C, Schubert S (2009) Short-term effects of salt exposure on the maize chloroplast protein pattern. Proteomics 9(17):4209–4220. https://doi.org/10.1002/pmic.200800791

Zorb C, Schmitt S, Muhling KH (2010) Proteomic changes in maize roots after short-term adjustment to saline growth conditions. Proteomics 10(24):4441–4449. https://doi.org/10.1002/pmic.201000231

Chapter 11
Metabolomics for Crop Improvement Against Salinity Stress

Luisa D'Amelia, Emilia Dell'Aversana, Pasqualina Woodrow, Loredana F. Ciarmiello, and Petronia Carillo

Abstract In the post-genomic era, increasing efforts have been done to describe the relationship between genome and phenotype in plants. It has become clear that even a complete understanding of the state of the genes, messages, and proteins in a living system does not reveal its phenotype. Metabolites are the main readouts of gene vs environment interactions and represent the sum of all the levels of regulation in between gene and enzyme. Therefore, metabolome can be considered as the final recipient of biological information flow. Some metabolites have a very short lifetime and are indicators of specific metabolic reaction and of plant status. Indeed, it is well known that many of them are transformed during specific stresses and involved in plant stress response and resistance.

Salinity provides an important example of the effectiveness of metabolic changes in response to stress. In fact, exposure to salinity triggers specific strategies for cell osmotic adjustment and control of ion and water homeostasis to minimize stress damage and to reestablish growth. A ubiquitous mechanism that plants have evolved to adapt to salinity involves sodium sequestration in the vacuole, as a cheap osmoticum, and synthesis and accumulation of compatible compounds, both for osmotic adjustment and oxidative stress protection in the cytosol.

Metabolomics has been utilized for the study of plants in response to salinity stress in order to dissect particular patterns associated with stress tolerance. These studies have proven that certain metabolites are present in case of salt-induced metabolic dysfunction and can act as effectors of osmotic readjustment or antioxidant response. Thus, the presence of particular metabolite patterns can be associated with stress tolerance and could serve as accurate markers for salt-tolerant crop selection in breeding programs.

Keywords Osmotic adjustment · Glycine betaine · Asparagine · Asparagine synthetase · P5CS · Nitrate reductase

L. D'Amelia · E. Dell'Aversana · P. Woodrow · L. F. Ciarmiello · P. Carillo (✉)
Dipartimento di Scienze e Tecnologie Ambientali, Biologiche e Farmaceutiche, Università degli Studi della Campania "Luigi Vanvitelli", Caserta, Italy
e-mail: petronia.carillo@unicampania.it

V. Kumar et al. (eds.), *Salinity Responses and Tolerance in Plants, Volume 2*,
https://doi.org/10.1007/978-3-319-90318-7_11

Abbreviations

BMRB	BioMagResBank
CE	capillary electrophoresis
EI-MS	electron impact ionization MS
ELS	evaporative light scattering
ESI	electrospray ionization
FT-ICR	Fourier transform ion cyclotron resonance
GABA	γ-aminobutyric acid
GB	glycine betaine
GC	gas chromatography
HILIC	hydrophilic interaction liquid chromatography
HPAEC-PAD	high-performance anion-exchange chromatography with pulsed amperometric detection
HPLC	high-performance liquid chromatography
IE	ion-exchange
IT	ion trap
LC	liquid chromatography
LT	linear trap
MMCD	Madison Metabolomics Consortium Database
MS	mass spectrometry
m/z	mass-to-charge ratio
PAs	polyamines
Q	quadrupole
RI	refractive index
SEC	size-exclusion chromatography
SPE	solid-phase extraction
TOF	time of flight
UPLC	ultra-performance liquid chromatography

11.1 Introduction

Worldwide, more than 20% of total cultivated and 33% of irrigated agricultural lands have been damaged by salt (Negrão et al. 2017; Shrivastava and Kumar 2015). 1.5 million hectares are taken out of production each year as a result of high-salinity levels in the soil due to low precipitation, high surface evaporation, weathering of native rocks, and poor cultural practices (Zhang and Blumwald 2001). More than 50% of the arable land could be salinized by the year 2050 (Jamil et al. 2011). Irrigation with brackish water and seawater intrusion into freshwater aquifers highly contributes to soil salinization in irrigated areas (Rana and Katerji 2000). Only in the Mediterranean Basin, more than 40% of soils are prone to salinity (Nedjimi 2014). This stress has detrimental effects on agricultural crop productivity impairing germination, plant vigor, and crop yield (Munns and Tester 2008) (Fig. 11.1).

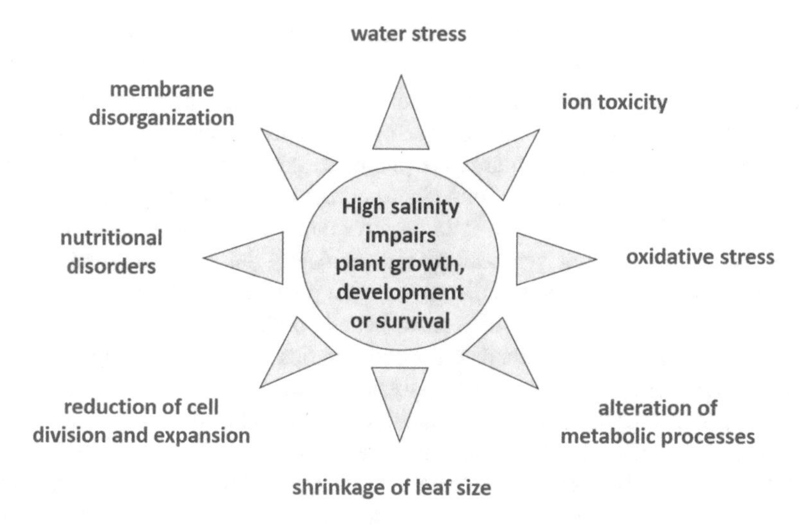

Fig. 11.1 High salinity causes pleiotropic effects

Osmotic stress and ion toxicity are the main causes of plant growth restriction in salinized soils (Gorham et al. 2010; Munns and Tester 2008). High levels of soil salinity reduce the capacity of roots to extract water from soil, and high concentrations of salts inside the plant itself turn out to be toxic, causing plant nutritional imbalance and oxidative stress (Hasegawa et al. 2000; Munns 2002). These effects impair plant growth, development, and survival, also in dependence on the concurrent salt toxicity levels and phenological stage sensitivity to salt stress (Hasegawa et al. 2000; Lutts et al. 1995). In particular, sodium affects cytosol and organelles metabolism because it can replace potassium in key enzymatic reactions, and potassium to sodium ratio is critical for the cell performance under salinity (Cuin et al. 2009). However, plant cells adopt specific strategies under salinity in order to get cell osmotic adjustment and control of ion and water homeostasis to attenuate stress and to restore plant growth (Hasegawa et al. 2000; Woodrow et al. 2017). Plants manage salinity stress injuries through a plethora of molecular, biochemical, and physiological changes resulting in a remodulation of metabolic pathways to reach a new homeostatic equilibrium. Among the different strategies, they have evolved a conserved and ubiquitous mechanism for the adaptation to salinity, which consists in sodium sequestration in the vacuole, as a cheap osmoticum, and synthesis and accumulation of osmolytes for turgor maintenance and protection of macromolecular structures against oxidative stress in the cytosol (Sairam and Tyagi 2004; Shabala and Munns 2012). These metabolites, nontoxic at high concentrations, are rapidly synthesized under salinity and rapidly removed (degraded or reallocated) when no longer required (Carillo et al. 2005). A wide range of osmoprotective compounds has been identified. Most of compatible solutes are nitrogen-containing compounds and sugars (Ashraf and Foolad 2007; Mansour 2000; Rhodes and Hanson 1993). The synthesis of these osmolytes has a very high energy cost (50–70 moles ATP for mole) (Raven 1985), and their accumulation must be tightly regulated to ward off any negative effects in cells and ensure plant survival (Shabala and Munns 2012).

Therefore, particular patterns of compatible metabolites accumulation can be associated with species-specific stress tolerance (Slama et al. 2015) and considered reliable stress markers. In addition to their role in maintaining osmotic balance within the cell, osmolytes may play other unique protective metabolic roles, such as acting as antioxidants, buffering cellular redox potential, stabilizing membranes and macromolecules, and functioning as immediate sources of energy during recovery from stress (Yancey 2005). Therefore it is an advantageous mechanism accumulating osmolytes as they, being synthesized independently of developmental stage in response to environmental signals, keep the main physiological functions of the cell active (Slama et al. 2015). Recently, it has been hypothesized that salinity can activate alternative gene expression patterns coordinating a reprogramming of metabolites, through the generation of new polymorphisms. This latter could be due to retrotransposon mobilization upon induction of salt-induced transcription factors able to bind the promoter of specific retrotransposons (Woodrow et al. 2010, 2011). Retrotransposition bursts could be critical for the remodeling of the gene regulatory networks and the creation of new metabolite patterns for salinity stress tolerance and adaptation (Mirouze and Paszkowski 2011; Woodrow et al. 2012).

The measurement of these patterns of metabolites is crucial for the comprehension of plant molecular and physiological responses to salinity and instrumental to unravel the functions of genes as a tool in functional genomics and systems biology in order to develop new breeding and selection strategies for improving salt tolerance in agricultural crops (Kumar and Khare 2016; Obata and Fernie 2012).

11.2 Plant Metabolome

Metabolome has been described as the final recipient of biological information flow. The word metabolome was created in analogy with transcriptome and proteome and includes all the small-molecule metabolites (such as metabolic intermediates, hormones and other signaling molecules, and secondary metabolites) present in biological samples. A metabolite is usually defined as any molecule less than 1.5 kDa in size (Wakayama et al. 2015) (Fig. 11.2). Metabolites are the main readouts of gene versus environment interactions and represent the sum of all the levels of regulation in between gene and enzyme. The response of plants to stresses is directly dependent on the changes of metabolites content, and therefore it would be useful to detect as much different metabolites as possible. However, we are still far from being able to fully detect the entire set of plant small molecules. It has been showed that for complex organisms, such as plants, there are more metabolites than genes, whereas for microorganisms there are generally fewer metabolites than genes (Schwab 2003). For instance, the yeast *Saccharomyces cerevisiae* has fewer than 600 metabolites and more than 6000 genes.

In 1993, the total number of metabolites produced by plants was estimated in a range between 100,000 and 200,000 (Mazza and Miniati 1993). However, this number seems to increase with improved analytical methods (Last et al. 2007), and still there are no reliable methods to assay all plant compounds. Actually it is con-

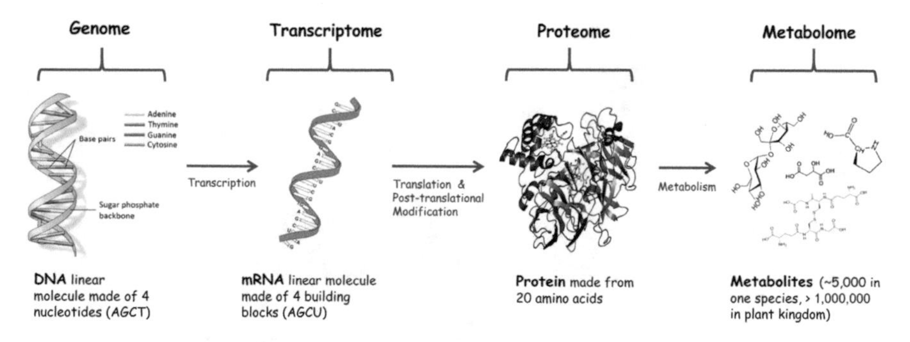

Fig. 11.2 Biological information flow

sidered that in one plant species between 5000 and 25,000 different compounds may exist (Trethewey 2004), while more than 1,000,000 compounds are present in the 400,000 existing plant species (Benkeblia 2014).

Metabolites result from a very complex network involving many enzymes and regulatory interactions, where products from one enzymatic reaction are substrates for other chemical reactions. Some of these metabolites, indicators of specific stresses, have a very short lifetime, and for this reason it is difficult or impossible to establish a direct link between stress-induced genes and metabolites.

11.3 Metabolomics, Metabolic Profiling and Target Analysis

Metabolomics is an effective tool that allows a comprehensive understanding of the fluctuations of metabolism in response to environmental changes and has been recently applied by many researchers to assess the effect of abiotic stresses on agricultural crops (Arbona et al. 2013; Guo et al. 2015; Kumari et al. 2015; Obata and Fernie 2012; Sanchez et al. 2008; Shulaev et al. 2008; Zhang et al. 2011). The increase of the accuracy and specificity of high-throughput analytical methods combined to new processing database mining strategies has broaden our horizons on understanding the molecular basis of stress response and resistance (Hall 2006) (Fig. 11.3).

However, it is much more difficult to measure the metabolome than the transcriptome or proteome. The genome, transcriptome, and proteome elucidations are based on target chemical analyses of biopolymers composed of 4 different nucleotides (genome and transcriptome) or about 20 amino acids (proteome). These compounds are highly similar chemically and facilitate high-throughput analytical approaches. Within the metabolome, there is, however, a large variance in chemical structures and properties. Metabolites differ in size, chemical composition, hydrophobicity, and stability, and there are added complications from various types of isomers, and also matrix effects can interfere with the analyses. Because of their complexity, it is virtually impossible to simultaneously determine with only one analytical technique the complete metabolome (Villas-Bôas et al. 2005) (Fig. 11.4).

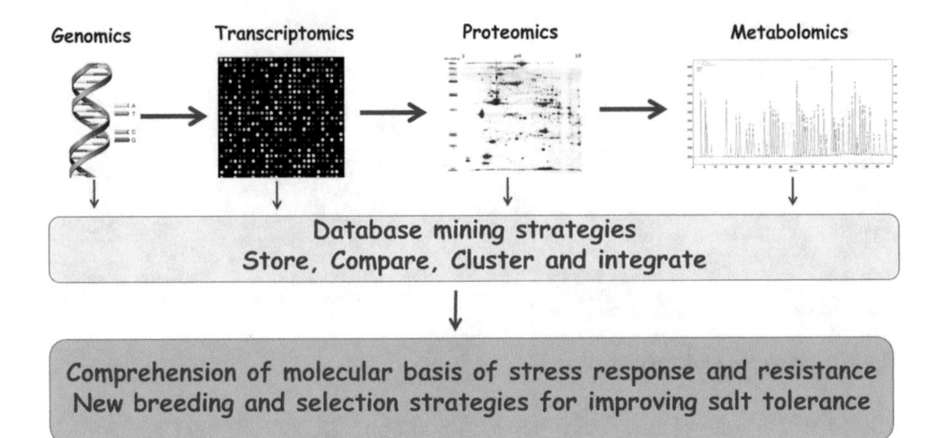

Fig. 11.3 Integrated omics data collection and data mining strategies

- Transcripts (mRNA)
  - Only need to identify and sequence 4 Nucleic Acids
  - Organism specific
- Proteins
  - Only need to identify and sequence about 20 Amino Acids
  - Organism specific
- Metabolites
  - Labile
  - Chemical diversity
  - Wide dynamic range
  - Not organism specific

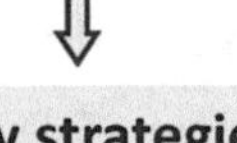

Many strategies needed for this

Fig. 11.4 Different strategies for measuring the *omics*

Over the last decade, several methods offering the possibility for the analysis of highly complex mixtures of compounds have been developed and implemented offering the possibility of high-throughput experiments. The two main technical approaches for the metabolomic analyses are nuclear magnetic resonance (NMR) spectroscopy- and mass spectrometry (MS)-based techniques (Fuhrer and Zamboni 2015).

NMR is based on atomic interactions. Atoms with non-zero magnetic moment absorb and re-emit electromagnetic radiation in a strong magnetic field. These atoms can be ^{1}H, ^{13}C, ^{14}N, ^{15}N, and ^{31}P. Atoms in a molecule have a specific spectrum of radiation characterized by their intensity, frequency, and magnetic properties and can be used for identification and quantification of metabolites in complex biological extracts. It allows not only the quantification of the concentration of metabolites but also provides information about the chemical structure of unknown compounds (Alonso et al. 2015). NMR advantages are the potential for high-throughput analysis, the fast and easy sample preparation, and the non-discriminating and nondestructive nature of the technique. Moreover, there are central repository for experimental NMR spectral data like the BioMagResBank (BMRB) and the Madison Metabolomics Consortium Database (MMCD). However, in complex mixtures the identification of single metabolites based on chemical shift signals is challenging, and only highly concentrated metabolites can be detected (Dettmer et al. 2007). Often 1H NMR spectrum signals are broadened beyond detection, in particular when metabolites contain functional groups deficient in protons or when protons can chemically exchange with the solvent.

MS is an analytical method that allows the separation of mixture of ions according to their mass-to-charge ratio (m/z) generally through static or oscillating magnetic fields. This mixture is obtained by sample molecules ionization through an energy note electron beam. The ionized molecules are unstable and fragment into lighter ions according to typical patterns as a function of their chemical structure. Different peak patterns characterize the fingerprint of the original molecule and their relative intensity. Combination of MS with a separation technique like gas chromatography (GC), liquid chromatography (LC), or capillary electrophoresis (CE) decreases the speed of analysis but simplifies the mass spectra due to different retention times of metabolites and makes feasible the separation and quantitative determination of small metabolites in highly complex biological samples avoiding matrix effects (Dettmer et al. 2007).

If GC and electron impact ionization MS (EI-MS) are combined, a highly reproducible fragmentation pattern occurs and the metabolite resolution improves, as well as their specific detection and quantification. Quadrupole (Q) instruments or time of flight (TOF)-MS further improves the MS detection performance by increasing the deconvolution and the mass accuracy and reducing the run time (Obata and Fernie 2012).

GC-MS protocols, having being used since long time and therefore accurately set up, are characterized by robustness, facility of chromatogram evaluation, and interpretation and existence of several comprehensive reference library of electron-ionization mass spectra (e.g., NIST, Golm Metabolome Database, and Fiehn GC-MS Database) (Kind et al. 2009; Kopka et al. 2005; Kumari et al. 2011, 2015). These features together with the short running time and relatively low running costs make GC-MS particularly advantageous. The preparation of samples for the GC-MS analysis often requires only a lyophilization step, even if the complex mixtures must be not too concentrated for avoiding problems with the analysis. GS-MS is the most popular technique for the analyses of low-molecular weight compounds (smaller than 1 kDa), in particular sugars, sugar alcohols, amino acids, organic acids, and polyamines. The functional groups of these polar metabolites need to be derivatized to reduce polarity and increase thermal stability and volatility of the molecules. If additional steps for derivatization are required, these are time-consuming and can cause metabolite losses associated with the sample handling or formation of undesired derivatization products (Kopka 2006).

LC-MS is a good technique for the reliable analyses of a wide variety of not thermostable plant polar metabolites with low or high molecular weight. It separates the metabolites in a liquid phase by a conventional high-performance liquid chromatography (HPLC) or an ultra-performance liquid chromatography (UPLC), which has a better resolution and sensitivity also in very complex mixtures. Both

LC-MS techniques do not require volatilization of samples. If LC and electrospray ionization MS (ESI-MS) are combined to Q, TOF, triple Q (Q3), ion trap (IT), linear trap Q (LTQ)-Orbitrap, and Fourier transform ion cyclotron resonance (FT-ICR)-MS, the metabolite deconvolution, resolution, detection, and quantification can highly improve while reducing the run time (Dettmer et al. 2007; Obata and Fernie 2012). LC reverse phase column coupled to these MS techniques can even be used to simultaneously assay many charged metabolites with similar structure, such as phosphorylated intermediates (Arrivault et al. 2009; Lunn et al. 2006), having low interference from other compounds in the plant extracts. The relatively simple sample preparation and flexibility of assay allow high throughput; however, these characteristics can even cause low discovery rates and false identifications, due to the presence of potential mass isomers and instrumental noise. This drawback makes difficult to establish unique LC-MS comprehensive reference spectral libraries, representing unequivocally assigned plant metabolites, even if recently reference libraries do exist like WEIZMASS spectral library (Shahaf et al. 2016). For confirming the peak identity with high confidence and the reliability of metabolite annotation, recovery experiments and isotope labelling of authentic chemical standards are recommended; moreover purification of new compounds should be followed by (NMR) analysis (Giavalisco et al. 2009; Shahaf et al. 2016).

CE-MS is a technique for the analyses of polar and charged compounds on the basis of their intrinsic electrophoretic mobility, depending on their charge-to-mass ratio. If CE is combined to ESI, for sample ionization, and coupled to TOF-MS, the metabolite resolution highly improves. Moreover, only small volumes (nanoliters) of samples are required for CE which are separated with high electric fields in a run time of seconds (Obata and Fernie 2012). On the other hand, CE-MS lacks concentration sensitivity, migration time stability, method robustness, and reference spectra libraries. These drawbacks limit the use of CE-MS in the field of plant metabolomics; however because it constitutes a completely different analytical approach, it is often use in combination to LC-MS to provide a complementary and wider view of plant metabolome (Ramautar et al. 2017).

11.4 Salt Stress Compatible Compound Profiling

Salt stress induces a molecular reprogramming that plays a pivotal role in plants under salt stress. A wide range of osmoprotective compounds protects enzymes from denaturation and stabilizes membrane or macromolecules in addition to the principal role of osmotic adjustment (Ashraf and Foolad 2007). In particular, they protect cellular structures against ROS and thanks to their hydrophilicity are able to replace water at the surface of proteins or membranes, thus acting as low molecular weight chaperones (Hasegawa et al. 2000). Compatible compounds include nitrogen-containing metabolites such as amino acids, amines, and betaines but also molecules like sugars and polyols (Carillo et al. 2011a). For the analysis of these different compatible compounds in complex biological samples, different techniques can be used singularly or in combination (Table 11.1).

Table 11.1 Comparison of various analytical techniques used for compatible compounds profiling

Metabolites	Analytical methods	Advantages	Limitation
Amino acids	RP-HPLC	Quickness and precision in quantitative analysis, automated analysis, high sensitivity, suitable for different types of samples, reproducible, cheap costs	Proline not resolved with OPA derivatization, Ile and Leu overlapping with FMOC derivatization, very long run times
	GC-MS	High sensitivity, excellent resolution	Laborious manual derivatization procedure, failure in determination of Arg
	CE-MS	Analyses of all nonvolatile polar amino acids, no derivatization, short run time, almost no matrix interference	Complex methodology and quantification
	LC-MS	Analyses of all nonvolatile polar amino acids without derivatization	Low discovery rates and false identifications
	HPLC-MS/MS	High rapidity (1.5 min versus 20–90 min for standard methods)	Slow preparation of samples, Ile and Leu overlapping, Gln and Lys peaks interference
	^{1}H NMR	High-throughput analysis, fast and easy sample preparation, the non-discriminating and nondestructive nature of the technique	Detection of only highly concentrated metabolites
Proline	Ninhydrin-based method	High-throughput and harmless spectrophotometric method	Interference of high contents of anthocyanins, pro quantification
Glycine betaine	HPLC	Highly sensitive procedure	Need of tedious anion exchange resin filtration of aqueous extracts
Polyamines	LC-MS, GC-MS, and CE-MS	The most powerful methods	Wasteful extraction and pretreatment of samples, long run times
Carbohydrates and polyols	HPLC	High run time, analysis of large carbohydrates in a single run by RI and ELS detectors	Some detection mode requires a derivatization step
	HPAEC-PAD	Good sensitivity and selective separations at high pH, no derivatization	Lack of commercially available standards, impossibility to put HPAEC-PAD in line with MS
	GC-MS	Good resolution	A derivatization step always required, difficult in high-throughput metabolomics studies
	^{1}H NMR	Analysis of simple and complex carbohydrates	Expensive, need of considerable amount of pure samples and standards
Trehalose	Fluorometric method	Fast, simple, and accurate method	Need of expensive recombinant trehalase

11.4.1 Amino Acids

The analysis of the salt-induced changes in amino acid levels is of pivotal importance and requires fast and reliable analytical methods. This analysis is commonly performed by pre-column derivatization with ortho-phthalaldehyde (OPA) and/or 9-fluorenylmethyl chloroformate (FMOC) followed by reversed-phase HPLC and fluorescence detection (Carillo et al. 2005; Nasir et al. 2010). The problem with OPA derivatization is that proline is not resolved, while with FMOC derivatization there can be the overlapping of amino acids like isoleucine and leucine. The combination of the two methods can resolve all amino acids. The run time of this method can vary between 20 and 90 min; the longer is the chromatographic run time, the better is the peak resolution. Therefore, this technique is accurate, precise, robust, and reproducible, and above all the maintenance costs are relatively cheap, but the drawback is that the time of analysis for each sample can be very long. However, while HPLC is unable to carry out the simultaneous determination of ^{14}N- and ^{15}N-labelled amino acids, essential for the metabolic flux analysis experiments, MS-based methods can be used to perform it (Thiele et al. 2012). GC-MS usually requires derivatization of the amino acids which can be performed for obtaining fluorinated and non-fluorinated chloroformate and anhydride derivatives of amino acids (Waldhier et al. 2010).

A method commonly used is a methoxyamination followed by trimethylsilylation, which converts metabolites with polar groups into their trimethylsilyl (TMS) derivatives (Weckwerth et al. 2004). With this method, a comprehensive study of six salt-tolerant or salt-sensitive *Lotus* species has revealed that 50% of all metabolites identified with GC-TOF-MS give similar responses to salinity with the exception of amino acids (Sanchez et al. 2011). Although this GC method shows high sensitivity and excellent resolution, it has the drawback that derivatization procedure is too laborious, and the method fails in the determination of some amino acids, like arginine. CE-MS and LC-MS have the advantage that all nonvolatile polar amino acids can be assayed without derivatization. In particular, short analysis times, selectivity, and almost no matrix interferences have been reported for a CE-ESI-MS method developed for the determination of free amino acids (Thiele et al. 2012). Hydrophilic interaction liquid chromatography (HILIC)-ESI-MS/MS has been successfully used for separating and quantifying amino acids in different matrices (Iwasaki et al. 2012). A further alternative is the separation on ion exchange columns which had been carried out only with electrochemical detectors without derivatization (Casella and Contursi 2003). ^{1}H NMR spectroscopy has also been used to analyze amino acids in maize plants as affected by salinity (Gavaghan et al. 2011).

Plant responses to salinity determine large changes in amino acid levels. In fact, under salinity, there is a reorganization of amino acids; in particular proline, glutamate and amides, alanine, and minor amino acids increase, while other amino acids are stable or decrease (Annunziata et al. 2017; Cramer et al. 2007; Jorge et al. 2016; Woodrow et al. 2017; Zuther et al. 2007). Glutamate can also decrease, functioning as nitrogen donor in biosynthetic transamination, for the production of other amino acids which highly increase under salinity (e.g., proline, glutamine, and asparagine)

(Forde and Lea 2007; Lea et al. 2007). A strong increase of glutamine and asparagine in wheat can be seen under salinity (Annunziata et al. 2017; Carillo et al. 2005; Wang et al. 2005; Woodrow et al. 2017), for their possible role in osmotic adjustment and macromolecule protection (Mansour 2000). In particular, asparagine accumulation appears necessary to prevent ammonium toxicity (Herrera-Rodríguez et al. 2006), by re-assimilating the nitrogen released during photorespiration or protein degradation under salinity (Carillo et al. 2011c; Masclaux-Daubresse et al. 2010; Sanchez et al. 2011).

Minor amino acids, occurring at low concentration in control plants, increase under salinity and in particular branched-chain amino acids (BCAAs) which can serve both as compatible compounds and as alternative electron donor for the mitochondrial electron transport chain (Pires et al. 2016). The increase of aromatic amino acid biosynthesis during the first 12 h of salt stress in *Arabidopsis* could be prodromal for the induction of lignin biosynthesis for cell wall strengthening (Kim et al. 2007).

γ-Aminobutyric acid (GABA) is a nonprotein amino acid deriving from glutamate decarboxylation that can be accumulated under salt stress (Akçay et al. 2012; Carillo 2018; Renault et al. 2013; Wang et al. 2017; Woodrow et al. 2017). Its proton-consuming synthesis buffers cytosolic acidosis, contributing to the biochemical pH-stat, and decreases, together with asparagine, the excess of ammonium accumulated under salt stress acting as temporary nitrogen store (Fait et al. 2008; Molina-Rueda et al. 2015). For its nature as zwitterion, it can function as compatible compound being able to accumulate in high concentration in cells without toxic effects (Signorelli et al. 2015). Moreover, it has radical scavenging activity (Molina-Rueda et al. 2015).

11.4.2 Proline and Glycine Betaine

Overproduction of proline and glycine betaine (GB) is a widespread response observed in plants experiencing salt stress. These two metabolites can be determined by using the same methods applied for the amino acid analysis; however for both of them, targeted high-throughput extraction and/or determination methods have been set up. In particular, proline can be assayed by a fast and harmless ninhydrin-based spectrophotometric method at low pH using a microplate reader (Woodrow et al. 2017), which is a modification of the method of Bates 1973 (Bates et al. 1973). The drawback is that in some species submitted to prolonged stresses, high contents of pigments such as anthocyanins can interfere with the proline quantification. In order to decrease background noise, the pigment chromogen can be extracted using harmful solvents such as benzene (Troll and Lindsley 1955) or toluene (Bates et al. 1973). GB assay can be performed by a highly sensitive HPLC procedure described by (Bessieres et al. 1999; Carillo et al. 2011b), which requires an anion exchange resin (AG1 8X resin, 200–400 mesh, OH^- form, Bio-Rad) filtration of aqueous extracts. This procedure allows the proline to be completely retained, and it is essential for avoiding that this amino acid could interfere with GB quantification (GB and proline show similar retention time of 4–5 min in the HPLC run). The drawback is that the filtration is tedious and time-consuming.

Proline is synthesized both from glutamate, which is considered to be the main pathway of accumulation under stress, and from arginine/ornithine (Kishor et al. 2014). It is still not clear if proline accumulation under stress represents a symptom of stress, a stress response, or an adaptive strategy (Woodrow et al. 2017). It can also be seen as a response to a decreased nitrogen requirement due to stress-induced growth retardation (Sanchez et al. 2008). However, proline, in addition to its role as an osmolyte, has many important roles under stress conditions among which stabilizing membranes and proteins, scavenging ROS, or buffering cellular redox potential (Carillo et al. 2008, 2011a). Proline can also induce the expression of salt stress-responsive genes, which have proline-responsive elements (e.g., PRE, ACTCAT) in their promoters (Carillo et al. 2011c; Chinnusamy et al. 2005). Wild barley, which has a higher osmoregulation ability compared to cultivated barley, accumulates more proline (Wu et al. 2013). In high-nitrate conditions, proline accounts for about 40% of the osmotic adjustment in the cytoplasmic compartments of durum wheat old leaves. Its nitrogen-dependent accumulation may offer the important advantage that it can be metabolized to allow reallocation of energy, carbon, and nitrogen from the older leaves to the younger tissues (Carillo et al. 2008).

GB preferentially accumulates at later stages of the stress and in younger tissues (Carillo et al. 2008). GB is an amphoteric compound extremely soluble in water and electrical neutral in a wide range of physiological pH values. It not only act as osmoregulator but also stabilizes protein complexes and enzymes by interacting with their hydrophilic and hydrophobic domains, protects membranes by the damaging effects of salinity, and can play a role as ROS scavenger (Carillo et al. 2011c; Chen and Murata 2011). In addition, ethanolamine, a precursor of GB that has strong osmoprotective properties, accumulates under short-term salt stress and can confer stress tolerance (Sakamoto and Murata 2002).

11.5 Polyamines

Free, conjugated, and bound polyamines (PAs) can be extracted in perchloric acid, neutralized, hydrolyzed, and dansylated before undergoing fluorimetric HPLC analysis (Quinet et al. 2010). Several MS-based analytical methods for PAs determination have also been developed: LC-MS, GC-MS, and CE-MS techniques are among the most powerful methods for the analysis of these metabolites (Magnes et al. 2014). There are however drawbacks: in particular, the extraction and pretreatment of samples require many time-consuming steps, methods are species-specific or matrix-specific, chromatographic run times are too long, and there are limits of detection. Ion-pairing reagents such as perfluorinated carboxylic acids had been presented to improve the separation of polyamines on C18 columns and can favor the coupling of LC with ESI-MS (Häkkinen et al. 2008). Recently a higher efficient solid-phase extraction (SPE)-LC/MS/MS technique has been developed and could have solved most of these problems (Magnes et al. 2014).

Putrescine, spermidine, and spermine are the main PAs present in plants (Liu et al. 2015). They play an important role in modulating the defense response of plants to diverse environmental stresses among which salinity (Gill and Tuteja 2010). PAs have anti-senescence and anti-stress effects due to their acid neutralizing and antioxidant properties, as well as for their membrane and cell wall stabilizing abilities. Plant overexpressing PAs biosynthetic enzymes or exogenous application of PAs, such as putescine, has been successfully used to enhance salinity tolerance. PAs are also able to bind anion macromolecules, such as nucleic acids and proteins, regulating transcription and translation (Gill and Tuteja 2010).

11.6 Carbohydrates and Polyols

Carbohydrates are difficult to analyze because they have high polarity and similar structural characteristics and of the absence of a suitable chromophores or fluorophores making useless ultraviolet and fluorescence detectors (Zhao et al. 2016). Refractive index (RI) detector or an evaporative light scattering (ELS) detection system is also used coupled to ion exchange (IE)/size-exclusion chromatography (SEC) columns (Raessler 2011; Yin et al. 2010), but they are poorly sensitive, lack specificity of detection, and suffer from drift (Mechri et al. 2015). Moreover, ELS and SEC columns are often optimized for single class of sugars (e.g., monosaccharides or oligosaccharides), but not for all those found in plant tissues together (Raessler 2011). Therefore, different analytical procedures have been applied, among which HILIC, that is, suitable for analysis of polar compounds in complex systems (Buszewski and Noga 2012), though it requires derivatization and high organic content mobile phases (50–80% acetonitrile) which could cause sample solubility problems at high concentrations. High-performance anion exchange chromatography with pulsed amperometric detection (HPAEC-PAD) permits better sensitivity and higher selective separations of carbohydrates at high pH, thanks to their weakly acidic nature using a strong anion exchange stationary phase. Moreover, the use of PAD allows direct quantification of non-derivatized sugars (Corradini et al. 2012). However, it suffers for the lack of commercially available standards; therefore it is not possible to identify all peaks; in addition, the mobile phase from HPAEC-PAD cannot be directly put online with MS because the eluent high salt concentration could interfere with the metabolite identification (Adamo et al. 2009). An alternative method is 1H NMR, used to determine the structures of carbohydrates present in plants under abiotic stress (Zhao et al. 2016). This technique is not cheap, and a considerable amount of pure samples are required (at least 1 g of lyophilized sample or 3 mg of purified oligosaccharide) (Tuomivaara et al. 2015). LC-MS technique has utilized Cl^- in negative electrospray mode or Li^+, Na^+, Cs^+, and $NH4^+$ in positive mode for carbohydrates and polyol measurement because these anion and cations can form detectable metastable complexes with sugars

(Mechri et al. 2015). Alternatively, GC-MS provides both high sensitivity and specificity resolving and quantifying different classes of soluble sugars. When GC-MS is coupled to EI, it becomes very efficient to detect uncharged compounds like monosaccharides (Martínez Montero et al. 2004). However, since carbohydrates are nonvolatile and thermally stable, because of the presence of multiple -OH groups, their analysis by GC-MS is difficult to employ in high-throughput metabolomics studies or in studies where large amounts of samples are not available, because a chemical derivatization of metabolites in a multi-step procedure is required (Ruiz-Matute et al. 2011). As an example, it is difficult to apply GC-MS to the high-throughput detection of trehalose in response to abiotic stress, for which a targeted fluorometric method using recombinant *Escherichia coli* cytoplasmic trehalase (treF) was developed (Carillo et al. 2013).

Carbohydrates play a vital role in a variety of biological functions, including salt stress defense mechanisms since they can function as osmolytes to maintain cell homeostasis. Generally salinity increases the soluble sugar content (Rosa et al. 2009). Rice cell cultures in the initial phase of salt stress (100 mM NaCl for 1–24 h) show accumulation of galactose, glucose, fructose, as well as phosphorylated sugars (Liu et al. 2013), while *Arabidopsis* cell cultures under long-term salt stress exposure show co-induction of glycolytic metabolites and sucrose (Jorge et al. 2016). *Thellungiella* and *Arabidopsis* both accumulate soluble sugars (fructose, glucose, sucrose, raffinose) and other unknown putative complex glycans under salinity, even if the extremophile one has much higher concentrations (Gong et al. 2005). Hexoses and sucrose contribute to the osmolality of shoot and root tissues in durum wheat seedlings under salinity (Annunziata et al. 2017; Woodrow et al. 2017). In rice, exogenous application of trehalose highly reduces salt stress damages (Fernandez et al. 2010). Polyol accumulation (e.g., mannitol, sorbitol, cyclic polyols, myo-inositol and its methylated derivatives) is correlated with tolerance to salinity (Bohnert et al. 1995). It has been suggested that both trehalose (Luo et al. 2008) and polyols (with their water-like hydroxyl groups) (Mechri et al. 2015) can create a protective hydration shell structure around macromolecules and also function as scavengers of ROS, preventing salt-related peroxidation of lipids and cell damage.

11.7 Conclusions

The diverse compatible compound accumulation in different species under salinity confirms that osmoprotection is not controlled by one specific pathway (Jorge et al. 2016) and that plant tolerance to salinity is a complex trait difficult to interpret that still needs to be studied (Kumar et al. 2015). However, the methods applied for the study of the complex chemical pattern present under salinity are often not quantitative, not sensitive enough, and not specific for all the metabolites. The huge chemical complexity and diversity of these compounds requires multiple analytical platforms/MS-based techniques to profile the wide range of metabolites. Applying one general extraction and determination system can have the consequence that

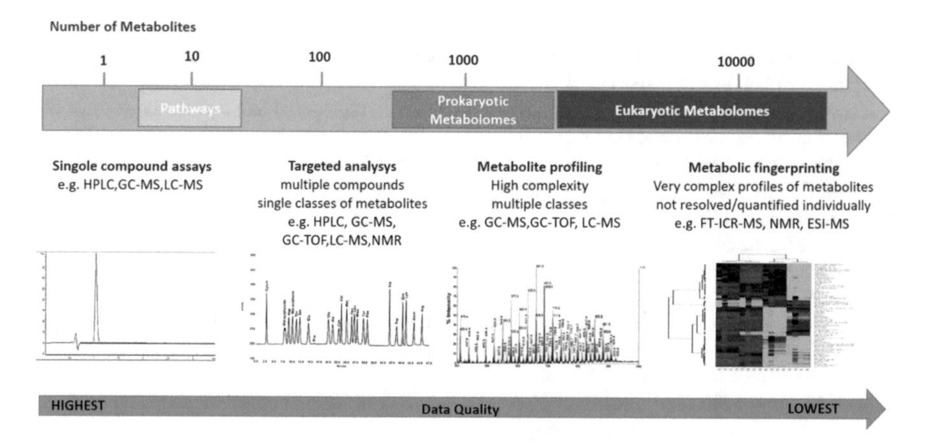

Fig. 11.5 Choice of metabolomics techniques vs. data quality. (Modified from Fernie et al. 2004)

many metabolites cannot be correctly measured. However, the more accurate the system is, the fewer metabolites are analyzed (Oksman-Caldentey et al. 2004) (Fig. 11.5).

Today, there is still need for specialized protocols to analyze specific and/or low concentration classes of metabolites (Stitt and Fernie 2003). Specific metabolites or classes of compounds can be selected for analysis and quantitatively profiled by cheap and routine metabolic profiling methods or by a quantitative, specific, and sensitive approach. In this view, the quantitative analysis of groups or classes of specific metabolites, e.g., sugars or amino acids, can be done by metabolic profiling. Instead, specific markers of salt stress exposure, such as proline, glycine betaine, and trehalose, can be analyzed by target analysis, a more directed approach that is used for the measurement of selected metabolites (Fiehn et al. 2000).

It is clear that when more than one analytical method is applied, the datasets deriving from different metabolomics platforms have to be controlled and correctly integrated (Schuhmacher et al. 2013). The quantitative results obtained by metabolic profiling using different biochemical techniques are independent of the technology used for their acquisition, and they can be put together to produce an integrated study of the entire metabolome (Dettmer et al. 2007). Only in this way can the biochemical strategies that enable salt tolerance be unraveled and the large gaps in our understanding of the biochemical pathways and regulation of metabolites synthesis in plants under salinity bridged.

References

Adamo M, Qiu D, Dick LW Jr, Zeng M, Lee A-H, Cheng K-C (2009) Evaluation of oligosaccharide methods for carbohydrate analysis in a fully human monoclonal antibody and comparison of the results to the monosaccharide composition determination by a novel calculation. J Pharm Biomed Anal 49:181–192

Akçay N, Bor M, Karabudak T, Özdemir F, Türkan İ (2012) Contribution of Gamma amino butyric acid (GABA) to salt stress responses of *Nicotiana sylvestris* CMSII mutant and wild type plants. J Plant Physiol 169:452–458

Alonso A, Marsal S, Julià A (2015) Analytical methods in untargeted metabolomics: state of the art in 2015. Front Bioeng Biotechnol 3:23

Annunziata MG, Ciarmiello LF, Woodrow P, Maximova E, Fuggi A, Carillo P (2017) Durum wheat roots adapt to salinity remodeling the cellular content of nitrogen metabolites and sucrose. Front Plant Sci 7:2035

Arbona V, Manzi M, de Ollas C, Gómez-Cadenas A (2013) Metabolomics as a tool to investigate abiotic stress tolerance in plants. Int J Mol Sci 14:4885–4911

Arrivault S, Guenther M, Ivakov A, Feil R, Vosloh D, Van Dongen JT, Sulpice R, Stitt M (2009) Use of reverse-phase liquid chromatography, linked to tandem mass spectrometry, to profile the Calvin cycle and other metabolic intermediates in Arabidopsis rosettes at different carbon dioxide concentrations. Plant J 59:826–839

Ashraf M, Foolad MR (2007) Roles of glycine betaine and proline in improving plant abiotic stress resistance. Environ Exp Bot 59:206–216

Bates LS, Waldren RP, Teare ID (1973) Rapid determination of free proline for water-stress studies. Plant Soil 39:205–207

Benkeblia N (2014) Metabolomics and postharvest sciences: challenges and perspecitives. Acta Hortic 1047:303–308

Bessieres M-A, Gibon Y, Lefeuvre JC, Larher F (1999) A single-step purification for glycine betaine determination in plant extracts by isocratic HPLC. J Agric Food Chem 47:3718–3722

Bohnert HJ, Nelson DE, Jensen RG (1995) Adaptations to environmental stresses. Plant Cell 7:1099–1111

Buszewski B, Noga S (2012) Hydrophilic interaction liquid chromatography (HILIC)—a powerful separation technique. Anal Bioanal Chem 402:231–247

Carillo P (2018) GABA Shunt in durum wheat. Front Plant Sci 9:1–9

Carillo P, Mastrolonardo G, Nacca F, Fuggi A (2005) Nitrate reductase in durum wheat seedlings as affected by nitrate nutrition and salinity. Funct Plant Biol 32:209–219

Carillo P, Mastrolonardo G, Nacca F, Parisi D, Verlotta A, Fuggi A (2008) Nitrogen metabolism in durum wheat under salinity: accumulation of proline and glycine betaine. Funct Plant Biol 35:412–426

Carillo P, Annunziata MG, Pontecorvo G, Fuggi A, Woodrow P (2011a) Salinity stress and salt tolerance, abiotic stress in plants – mechanisms and adaptations. In: Shanker A (ed) InTech. https://doi.org/10.5772/22331. Available from: https://www.intechopen.com/books/abiotic-stress-in-plants-mechanisms-and-adaptations/salinity-stress-and-salt-tolerance

Carillo P, Gibon Y, PrometheusWiki contributors (2011b) Extraction and determination of glycine betaine. PrometheusWiki, http://www.publish.csiro.au/prometheuswiki/tiki-pagehistory.php?page=Extraction and determination of glycine betaine&preview=10. Accessed 9 May 2017

Carillo P, Parisi D, Woodrow P, Pontecorvo G, Massaro G, Annunziata MG, Fuggi A, Sulpice R (2011c) Salt-induced accumulation of glycine betaine is inhibited by high light in durum wheat. Funct Plant Biol 38:139–150

Carillo P, Feil R, Gibon Y, Satoh-Nagasawa N, Jackson D, Bläsing OE, Stitt M, Lunn JE (2013) A fluorometric assay for trehalose in the picomole range. Plant Methods 9:21

Casella IG, Contursi M (2003) Isocratic ion chromatographic determination of underivatized amino acids by electrochemical detection. Anal Chim Acta 478:179–189

Chen THH, Murata N (2011) Glycinebetaine protects plants against abiotic stress: mechanisms and biotechnological applications. Plant Cell Environ 34:1–20

Chinnusamy V, Jagendorf A, Zhu J-K (2005) Understanding and improving salt tolerance in plants. Crop Sci 45:437–448

Corradini C, Cavazza A, Bignardi C (2012) High-performance anion-exchange chromatography coupled with pulsed electrochemical detection as a powerful tool to evaluate carbohydrates of food interest: principles and applications. Int J Carbohydr Chem 2012:13

Cramer GR, Ergül A, Grimplet J, Tillett RL, Tattersall EAR, Bohlman MC, Vincent D, Sonderegger J, Evans J, Osborne C, Quilici D, Schlauch KA, Schooley DA, Cushman JC (2007) Water and salinity stress in grapevines: early and late changes in transcript and metabolite profiles. Funct Integr Genomics 7:111–134

Cuin TA, Tian Y, Betts SA, Chalmandrier R, Shabala S (2009) Ionic relations and osmotic adjustment in durum and bread wheat under saline conditions. Funct Plant Biol 36:110–119

Dettmer K, Aronov PA, Hammock BD (2007) Mass spectrometry-based metabolomics. Mass Spectrom Rev 26:51–78

Fait A, Fromm H, Walter D, Galili G, Fernie AR (2008) Highway or byway: the metabolic role of the GABA shunt in plants. Trends Plant Sci 13:14–19

Fernandez O, Béthencourt L, Quero A, Sangwan RS, Clément C (2010) Trehalose and plant stress responses: friend or foe? Trends Plant Sci 15:409–417

Fernie AR, Trethewey RN, Krotzky AJ, Willmitzer L (2004) Metabolite profiling: from diagnostics to systems biology. Nat Rev Mol Cell Biol 5:763–769

Fiehn O, Kopka J, Dormann P, Altmann T, Trethewey RN, Willmitzer L (2000) Metabolite profiling for plant functional genomics. Nat Biotechnol 18:1157–1161

Forde BG, Lea PJ (2007) Glutamate in plants: metabolism, regulation, and signalling. J Exp Bot 58:2339–2358

Fuhrer T, Zamboni N (2015) High-throughput discovery metabolomics. Curr Opin Biotechnol 31:73–78

Gavaghan CL, Li JV, Hadfield ST, Hole S, Nicholson JK, Wilson ID, Howe PWA, Stanley PD, Holmes E (2011) Application of NMR-based metabolomics to the investigation of salt stress in maize (*Zea mays*). Phytochem Anal 22:214–224

Giavalisco P, Kohl K, Hummel J, Seiwert B, Willmitzer L (2009) 13C isotope-labeled metabolomes allowing for improved compound annotation and relative quantification in liquid chromatography–mass spectrometry-based metabolomic research. Anal Chem 81:6546–6551

Gill SS, Tuteja N (2010) Polyamines and abiotic stress tolerance in plants. Plant Signal Behav 5:26–33

Gong Q, Li P, Ma S, Indu Rupassara S, Bohnert HJ (2005) Salinity stress adaptation competence in the extremophile Thellungiella halophila in comparison with its relative *Arabidopsis thaliana*. Plant J 44:826–839

Gorham J, Läuchli A, Leidi EO (2010) Plant responses to salinity. In: Stewart JM, Oosterhuis DM, Heitholt JJ, Mauney JR (eds) Physiology of cotton. Springer, Dordrecht, pp 129–141

Guo R, Yang Z, Li F, Yan C, Zhong X, Liu Q, Xia X, Li H, Zhao L (2015) Comparative metabolic responses and adaptive strategies of wheat (*Triticum aestivum*) to salt and alkali stress. BMC Plant Biol 15:170

Häkkinen MR, Keinänen TA, Vepsäläinen J, Khomutov AR, Alhonen L, Jänne J, Auriola S (2008) Quantitative determination of underivatized polyamines by using isotope dilution RP-LC-ESI-MS/MS. J Pharm Biomed Anal 48:414–421

Hall RD (2006) Plant metabolomics: from holistic hope, to hype, to hot topic. New Phytol 169:453–468

Hasegawa PM, Bressan RA, Zhu J-K, Bohnert HJ (2000) Plant cellular and molecular responses to high salinity. Annu Rev Plant Physiol Plant Mol Biol 51:463–499

Herrera-Rodríguez MB, Maldonado JM, Pérez-Vicente R (2006) Role of asparagine and asparagine synthetase genes in sunflower (*Helianthus annuus*) germination and natural senescence. J Plant Physiol 163:1061–1070

Iwasaki Y, Sawada T, Hatayama K, Ohyagi A, Tsukuda Y, Namekawa K, Ito R, Saito K, Nakazawa H (2012) Separation technique for the determination of highly polar metabolites in biological samples. Metabolites 2:496–515

Jamil A, Riaz S, Ashraf M, Foolad MR (2011) Gene expression profiling of plants under salt stress. Crit Rev Plant Sci 30:435–458

Jorge TF, Rodrigues JA, Caldana C, Schmidt R, van Dongen JT, Thomas-Oates J, António C (2016) Mass spectrometry-based plant metabolomics: metabolite responses to abiotic stress. Mass Spectrom Rev 35:620–649

Kim JK, Bamba T, Harada K, Fukusaki E, Kobayashi A (2007) Time-course metabolic profiling in *Arabidopsis thaliana* cell cultures after salt stress treatment*. J Exp Bot 58:415–424
Kind T, Wohlgemuth G, Lee Do Y, Lu Y, Palazoglu M, Shahbaz S, Fiehn O (2009) FiehnLib: mass spectral and retention index libraries for metabolomics based on quadrupole and time-of-flight gas chromatography/mass spectrometry. Anal Chem 81:10038–10048
Kishor PBK, Sreenivasulu N (2014) Is proline accumulation per se correlated with stress tolerance or is proline homeostasis a more critical issue? Plant, Cell Environ 37:300–311
Kopka J (2006) Current challenges and developments in GC–MS based metabolite profiling technology. J Biotechnol 124:312–322
Kopka J, Schauer N, Krueger S, Birkemeyer C, Usadel B, Bergmüller E, Dörmann P, Weckwerth W, Gibon Y, Stitt M, Willmitzer L, Fernie AR, Steinhauser D (2005) GMD@CSB.DB: the Golm Metabolome Database. Bioinformatics 21:1635–1638
Kumar V, Khare T (2016) Differential growth and yield responses of salt-tolerant and susceptible rice cultivars to individual (Na+ and Cl−) and additive stress effects of NaCl. Acta Physiol Plant 38:170
Kumar V, Singh A, Mithra SVA, Krishnamurthy SL, Parida SK, Jain S, Tiwari KK, Kumar P, Rao AR, Sharma SK, Khurana JP, Singh NK, Mohapatra T (2015) Genome-wide association mapping of salinity tolerance in rice (*Oryza sativa*). DNA Res: Int J Rapid Publ Rep Genes Genomes 22:133–145
Kumari S, Stevens D, Kind T, Denkert C, Fiehn O (2011) Applying in-silico retention index and mass spectra matching for identification of unknown metabolites in accurate mass GC-TOF mass spectrometry. Anal Chem 83:5895–5902
Kumari A, Das P, Parida AK, Agarwal PK (2015) Proteomics, metabolomics, and ionomics perspectives of salinity tolerance in halophytes. Front Plant Sci 6:537
Last RL, Jones AD, Shachar-Hill Y (2007) Towards the plant metabolome and beyond. Nat Rev Mol Cell Biol 8:167
Lea PJ, Sodek L, Parry MAJ, Shewry PR, Halford NG (2007) Asparagine in plants. Ann Appl Biol 150:1–26
Liu D, Ford KL, Roessner U, Natera S, Cassin AM, Patterson JH, Bacic A (2013) Rice suspension cultured cells are evaluated as a model system to study salt responsive networks in plants using a combined proteomic and metabolomic profiling approach. Proteomics 13:2046–2062
Liu J-H, Wang W, Wu H, Gong X, Moriguchi T (2015) Polyamines function in stress tolerance: from synthesis to regulation. Front Plant Sci 6:827
Lunn John E, Feil R, Hendriks Janneke HM, Gibon Y, Morcuende R, Osuna D, Scheible W-R, Carillo P, Hajirezaei M-R, Stitt M (2006) Sugar-induced increases in trehalose 6-phosphate are correlated with redox activation of ADPglucose pyrophosphorylase and higher rates of starch synthesis in <em>*Arabidopsis thaliana*</em>. Biochem J 397:139–148
Luo Y, Li W-M, Wang W (2008) Trehalose: protector of antioxidant enzymes or reactive oxygen species scavenger under heat stress? Environ Exp Bot 63:378–384
Lutts S, Kinet JM, Bouharmont J (1995) Changes in plant response to NaCl during development of rice (*Oryza sativa* L.) varieties differing in salinity resistance. J Exp Bot 46:1843–1852
Magnes C, Fauland A, Gander E, Narath S, Ratzer M, Eisenberg T, Madeo F, Pieber T, Sinner F (2014) Polyamines in biological samples: rapid and robust quantification by solid-phase extraction online-coupled to liquid chromatography–tandem mass spectrometry. J Chromatogr A 1331:44–51
Mansour MMF (2000) Nitrogen containing compounds and adaptation of plants to salinity stress. Biol Plant 43:491–500
Martínez Montero C, Rodríguez Dodero MC, Guillén Sánchez DA, Barroso CG (2004) Analysis of low molecular weight carbohydrates in food and beverages: a review. Chromatographia 59:15–30
Masclaux-Daubresse C, Daniel-Vedele F, Dechorgnat J, Chardon F, Gaufichon L, Suzuki A (2010) Nitrogen uptake, assimilation and remobilization in plants: challenges for sustainable and productive agriculture. Ann Bot 105:1141–1157
Mazza G, Miniati E (1993) Anthocyanins in fruits, vegetables and grains. CRC Press, London

Mechri B, Tekaya M, Cheheb H, Hammami M (2015) Determination of mannitol sorbitol and myo-inositol in olive tree roots and rhizospheric soil by gas chromatography and effect of severe drought conditions on their profiles. J Chromatogr Sci 53:1631–1638
Mirouze M, Paszkowski J (2011) Epigenetic contribution to stress adaptation in plants. Curr Opin Plant Biol 14:267–274
Molina-Rueda J, Garrido-Aranda A, Gallardo F (2015) Glutamate decarboxylase. In: D' Mello J (ed) Amino acids in higher plants. CAB International, Wallingford, pp 129–141
Munns R (2002) Comparative physiology of salt and water stress. Plant Cell Environ 25:239–250
Munns R, Tester M (2008) Mechanisms of salinity tolerance. Annu Rev Plant Biol 59:651–681
Nasir FA, Batarseh M, Abdel-Ghani AH, Jiries A (2010) Free amino acids content in some halophytes under salinity stress in arid environment, Jordan. CLEAN Soil Air Water 38:592–600
Nedjimi B (2014) Effects of salinity on growth, membrane permeability and root hydraulic conductivity in three saltbush species. Biochem Syst Ecol 52:4–13
Negrão S, Schmöckel SM, Tester M (2017) Evaluating physiological responses of plants to salinity stress. Ann Bot 119:1–11
Obata T, Fernie AR (2012) The use of metabolomics to dissect plant responses to abiotic stresses. Cell Mol Life Sci 69:3225–3243
Oksman-Caldentey K-M, Inzé D, Orešič M (2004) Connecting genes to metabolites by a systems biology approach. Proc Natl Acad Sci U S A 101:9949–9950
Pires MV, Pereira Júnior AA, Medeiros DB, Daloso DM, Pham PA, Barros KA, Engqvist MKM, Florian A, Krahnert I, Maurino VG, Araújo WL, Fernie AR (2016) The influence of alternative pathways of respiration that utilize branched-chain amino acids following water shortage in Arabidopsis. Plant Cell Environ 39:1304–1319
Quinet M, Ndayiragije A, Lefèvre I, Lambillotte B, Dupont-Gillain CC, Lutts S (2010) Putrescine differently influences the effect of salt stress on polyamine metabolism and ethylene synthesis in rice cultivars differing in salt resistance. J Exp Bot 61:2719–2733
Raessler M (2011) Sample preparation and current applications of liquid chromatography for the determination of non-structural carbohydrates in plants. Trends Anal Chem 30:1833–1843
Ramautar R, Somsen GW, de Jong GJ (2017) CE–MS for metabolomics: developments and applications in the period 2014–2016. Electrophoresis 38:190–202
Rana G, Katerji N (2000) Measurement and estimation of actual evapotranspiration in the field under Mediterranean climate: a review. Eur J Agron 13:125–153
Raven JA (1985) Tansley review no. 2. Regulation of pH and generation of osmolarity in vascular plants: a cost-benefit analysis in relation to efficiency of use of energy, nitrogen and water. New Phytol 101:25–77
Renault H, El Amrani A, Berger A, Mouille G, Soubigou-Taconnat L, Bouchereau A, Deleu C (2013) γ-aminobutyric acid transaminase deficiency impairs central carbon metabolism and leads to cell wall defects during salt stress in Arabidopsis roots. Plant Cell Environ 36:1009–1018
Rhodes D, Hanson A (1993) Quaternary ammonium and tertiary sulfonium compounds in higher plants. Annu Rev Plant Physiol Plant Mol Biol 44:357–384
Rosa M, Prado C, Podazza G, Interdonato R, González JA, Hilal M, Prado FE (2009) Soluble sugars—metabolism, sensing and abiotic stress: a complex network in the life of plants. Plant Signal Behav 4:388–393
Ruiz-Matute AI, Hernández-Hernández O, Rodríguez-Sánchez S, Sanz ML, Martínez-Castro I (2011) Derivatization of carbohydrates for GC and GC–MS analyses. J Chromatogr B 879:1226–1240
Sairam R, Tyagi A (2004) Physiology and molecular biology of salinity stress tolerance in plants. Curr Sci 86:407–421
Sakamoto A, Murata N (2002) The role of glycine betaine in the protection of plants from stress: clues from transgenic plants. Plant Cell Environ 25:163–171
Sanchez DH, Siahpoosh MR, Roessner U, Udvardi M, Kopka J (2008) Plant metabolomics reveals conserved and divergent metabolic responses to salinity. Physiol Plant 132:209–219

Sanchez DH, Pieckenstain FL, Szymanski J, Erban A, Bromke M, Hannah MA, Kraemer U, Kopka J, Udvardi MK (2011) Comparative functional genomics of salt stress in related model and cultivated plants identifies and overcomes limitations to translational genomics. PLoS One 6:e17094

Schuhmacher R, Krska R, Weckwerth W, Goodacre R (2013) Metabolomics and metabolite profiling. Anal Bioanal Chem 405:5003–5004

Schwab W (2003) Metabolome diversity: too few genes, too many metabolites? Phytochemistry 62:837–849

Shabala S, Munns R (2012) Salinity stress: physiological constraints and adaptive mechanisms. CAB, UK

Shahaf N, Rogachev I, Heinig U, Meir S, Malitsky S, Battat M, Wyner H, Zheng S, Wehrens R, Aharoni A (2016) The WEIZMASS spectral library for high-confidence metabolite identification. Nat Commun 7:12423

Shrivastava P, Kumar R (2015) Soil salinity: a serious environmental issue and plant growth promoting bacteria as one of the tools for its alleviation. Saudi J Biol Sci 22:123–131

Shulaev V, Cortes D, Miller G, Mittler R (2008) Metabolomics for plant stress response. Physiol Plant 132:199–208

Signorelli S, Dans PD, Coitiño EL, Borsani O, Monza J (2015) Connecting proline and γ-aminobutyric acid in stressed plants through non-enzymatic reactions. PLoS One 10:e0115349

Slama I, Abdelly C, Bouchereau A, Flowers T, Savouré A (2015) Diversity, distribution and roles of osmoprotective compounds accumulated in halophytes under abiotic stress. Ann Bot 115:433–447

Stitt M, Fernie AR (2003) From measurements of metabolites to metabolomics: an 'on the fly' perspective illustrated by recent studies of carbon–nitrogen interactions. Curr Opin Biotechnol 14:136–144

Thiele B, Stein N, Oldiges M, Hofmann D (2012) Direct analysis of underivatized amino acids in plant extracts by LC-MS-MS. In: Alterman MA, Hunziker P (eds) Amino acid analysis: methods and protocols. Humana Press, Totowa, pp 317–328

Trethewey RN (2004) Metabolite profiling as an aid to metabolic engineering in plants. Curr Opin Plant Biol 7:196–201

Troll W, Lindsley J (1955) A photometric method for the determination of proline. J Biol Chem 215:655–660

Tuomivaara ST, Yaoi K, O'Neill MA, York WS (2015) Generation and structural validation of a library of diverse xyloglucan-derived oligosaccharides, including an update on xyloglucan nomenclature. Carbohydr Res 402:56–66

Villas-Bôas SG, Mas S, Åkesson M, Smedsgaard J, Nielsen J (2005) Mass spectrometry in metabolome analysis. Mass Spectrom Rev 24:613–646

Wakayama M, Hirayama A, Soga T (2015) Capillary electrophoresis-mass spectrometry. In: Bjerrum JT (ed) Metabonomics: methods and protocols. Springer, New York, pp 113–122

Waldhier MC, Dettmer K, Gruber MA, Oefner PJ (2010) Comparison of derivatization and chromatographic methods for GC–MS analysis of amino acid enantiomers in physiological samples. J Chromatogr B 878:1103–1112

Wang H, Liu D, Sun J, Zhang A (2005) Asparagine synthetase gene TaASN1 from wheat is upregulated by salt stress, osmotic stress and ABA. J Plant Physiol 162:81–89

Wang Y, Gu W, Meng Y, Xie T, Li L, Li J, Wei S (2017) γ-aminobutyric acid imparts partial protection from salt stress injury to maize seedlings by improving photosynthesis and upregulating osmoprotectants and antioxidants. Sci Rep 7:43609

Weckwerth W, Wenzel K, Fiehn O (2004) Process for the integrated extraction, identification and quantification of metabolites, proteins and RNA to reveal their co-regulation in biochemical networks. Proteomics 4:78–83

Woodrow P, Pontecorvo G, Fantaccione S, Fuggi A, Kafantaris I, Parisi D, Carillo P (2010) Polymorphism of a new Ty1-copia retrotransposon in durum wheat under salt and light stresses. Theor Appl Genet 121:311–322

Woodrow P, Pontecorvo G, Ciarmiello LF, Fuggi A, Carillo P (2011) Ttd1a promoter is involved in DNA–protein binding by salt and light stresses. Mol Biol Rep 38:3787–3794

Woodrow P, Pontecorvo G, Ciarmiello LF, Annunziata MG, Fuggi A, Carillo P (2012) Transcription factors and genes in abiotic stress. In: Venkateswarlu B, Shanker AK, Shanker C, Maheswari M (eds) Crop stress and its management: perspectives and strategies. Springer, Dordrecht, pp 317–357

Woodrow P, Ciarmiello LF, Annunziata MG, Pacifico S, Iannuzzi F, Mirto A, D'Amelia L, Dell'Aversana E, Piccolella S, Fuggi A, Carillo P (2017) Durum wheat seedling responses to simultaneous high light and salinity involve a fine reconfiguration of amino acids and carbohydrate metabolism. Physiol Plant 159:290–312

Wu D, Cai S, Chen M, Ye L, Chen Z, Zhang H, Dai F, Wu F, Zhang G (2013) Tissue metabolic responses to salt stress in wild and cultivated barley. PLoS One 8:e55431

Yancey PH (2005) Organic osmolytes as compatible, metabolic and counteracting cytoprotectants in high osmolarity and other stresses. J Exp Biol 208:2819–2830

Yin Y-G, Kobayashi Y, Sanuki A, Kondo S, Fukuda N, Ezura H, Sugaya S, Matsukura C (2010) Salinity induces carbohydrate accumulation and sugar-regulated starch biosynthetic genes in tomato (*Solanum lycopersicum* L. cv. 'Micro-Tom') fruits in an ABA- and osmotic stress-independent manner. J Exp Bot 61:563–574

Zhang H, Blumwald E (2001) Transgenic salt-tolerant tomato plants accumulate salt in foliage but not in fruit. Nat Biotechnol 19:765–768

Zhang JT, Zhang Y, Du YY, Chen SY, Tang HR (2011) Dynamic metabonomic responses of tobacco (*Nicotiana tabacum*) plants to salt stress. J Proteome Res 10:1904–1914

Zhao L, Chanon AM, Chattopadhyay N, Dami IE, Blakeslee JJ (2016) Quantification of carbohydrates in grape tissues using capillary zone electrophoresis. Front Plant Sci 7:818

Zuther E, Koehl K, Kopka J (2007) Comparative metabolome analysis of the salt response in breeding cultivars of rice. In: Jenks MA, Hasegawa PM, Jain SM (eds) Advances in molecular breeding toward drought and salt tolerance crops. Springer, Berlin, pp 285–315

Chapter 12
Enhancing Crop Productivity in Saline Environment Using Nanobiotechnology

Pradeep Kumar Shukla, Saumya Shukla, Preeti Rajoriya, and Pragati Misra

Abstract Abiotic stresses are the main factor negatively affecting crop growth and productivity worldwide. Salinity is one of the most important environmental stresses, limiting crop production in arid and semiarid areas of the world, and in the saline areas which is three times larger than the land used for agriculture. It is estimated that more than 6% of the world's total land and approximately 20% of irrigated land are affected by salinity, and therefore, is a serious concern in agriculture. Nanoparticles are nowadays gaining importance in improvement of plant systems as they help in reducing adverse effects of stresses, imposing a positive impact on the plant. Applications of nanomaterials can enhance seed germination, improve plant's resistance against abiotic and biotic stress, augment nutrient utilization efficiency, ultimately improving plant growth and developmental processes, with reduced environmental impact compared to traditional approaches. Various reports have evidenced the positive effect of nanoparticles in mitigating the harmful effects of salt stress. This chapter presents a brief glimpse on the effect of various nanoparticles to mitigate the salt-induced damage in the plants.

Keywords Nanoparticles · Salt stress · Agriculture · Gene expression

Abbreviations

AgNPs	Silver nanoparticles
APX	Ascorbate peroxidase

P. K. Shukla · S. Shukla
Department of Biological Sciences, Faculty of Science, Sam Higginbottom University of Agriculture, Technology and Sciences, Allahabad, Uttar Pradesh, India

P. Rajoriya · P. Misra (✉)
Department of Molecular and Cellular Engineering, Jacob Institute of Biotechnology and Bioengineering, Sam Higginbottom University of Agriculture, Technology and Sciences, Allahabad, Uttar Pradesh, India
e-mail: pragati.misra@shiats.edu.in

V. Kumar et al. (eds.), *Salinity Responses and Tolerance in Plants, Volume 2*,
https://doi.org/10.1007/978-3-319-90318-7_12

CAT	Catalase
CeO_2NPs	Cerium oxide nanoparticles
Chl	Chlorophyll
CuNPs	Copper (II) oxide nanoparticles
GPX	Glutathione peroxidase
NPs	Nanoparticles
NSPs	Nanoscale particles
OP	Osmotic potential
PCD	Programmed cell death
POD	Peroxidase
SOD	Superoxide dismutase
ZnONPs	Zinc oxide nanoparticles

12.1 Introduction

Global food demands are likely to rise by 70–110% by 2050 (Tilman et al. 2011), while arable lands are diminishing due to land degradation, urbanization, and seawater intrusion (Munns et al. 2012). Farmers are forced to use salty water for irrigation in arid and semiarid regions due to the variation in global climate (Qadir et al. 2007; Rengasamy 2010; Sanoubar et al. 2016). Salinity is affecting more than 6% of the world's total land and approximately 20% of irrigated land (Munns and Tester 2008). According to Fisher and Turner (1978), arid and semiarid lands embody 40% of the earth's area.

Germination is vital for determining the final plant density if planted seeds completely and vigorously germinate (Baalbaki et al. 1990). The capability of plant to sprout well is controlled by the collaboration between environmental factors and the internal mechanisms of seeds (Ni and Bradford 1992). The first phase of the growth response results from the effects of salt outside the plants. Leaf and root growth is reduced in soils containing salt (Munns 1993; Farooq et al. 2015). Molecular control mechanisms for salt-stress tolerance depend on the regulation of the expression of certain stress genes (Wang et al. 2003). In the arid areas, salt stress is considered as the most serious limiting factor for crop growth and production. Out of the world's total cultivated lands, 23% is saline and 37% is sodic (Khan and Duke 2001). The yields of cultivated crops are adversely affected by high salt concentrations (Rahimi et al. 2012).

12.2 Salt Stress

Soil or water salinization is one of the world's most serious environmental problems in agriculture. For higher yield and improved quality of medicinal and aromatic plants, one should recognize the environmental factors essential for them. The problem of salinity is characterized by an excess of inorganic salts and, is common in, the arid and semiarid lands, where it has been naturally formed under the prevailing climatic conditions, due to higher rates of evapotranspiration and lack of leaching water. Although more frequent in arid lands, salt-affected soils are also present in areas where salinity is caused by poor quality of irrigation water (Jouyban 2012).

In the upcoming 25 years, we are expected to lose 30% of land due to the increasing salinization of arable land. Salinity stress is predictable to have global effects on plant growth and crop plant production which is a threat to agriculture, and so the tolerance of crop plants against salinity stress should be augmented and must be given primacy in agricultural practices (Massoud 1977; Jafari 1994; Wang et al. 2003).

The ionic and osmotic balance of cells in plants is regulated by a number of genes which are liable for salinity resistance and are involved in the salt tolerance process as these genes confines the amount of salt uptake from soil and its transport throughout plants as a result plants do no grow well (Munns 1993; Liang et al. 2006).

Plants are encountering constant difficulty in arid and semiarid regions, due to saline soils, consecutively restraining the spread of plants in their natural habitats (Shanon 1986). Expression of plant under salinity cannot only be determined by its morphological appearance, but also through physiological and biochemical dynamics, including toxic ions status, osmotic potential behaviour, lack of essential mineral elements, etc. along with its interaction with various stresses (Munns 1993, 2002; Neumann 1997; Yao 1998; Hasegawa et al. 2000).

12.3 Adverse Effects of Salt Stress

Salinity adversely affect seed germination, survival percentage, morphological characteristics, development, yield, total carbohydrate content, fatty acid, and protein content of plants. It prompts physiological and metabolic instabilities falling photosynthesis and respiration rate of plants, on the contrary, it augmented the level of amino acids, particularly proline. However, plants grown under salt stress exhibited alleviated content of some secondary plant products. Interactions among salt stress and other environmental factors govern the plant's response and tolerance against salinity (Jouyban 2012; Ashraf and McNeilly 2004).

Plants endure many biotic and abiotic stresses during their lifetime, one such abiotic stress is salinity, and it reduces growth and production of many crops, and so plants adapt various methods and strategies to overcome this stress (Ashraf and Foolad 2007).

In many plant species, soil salinity is known to reduce growth and development through imposing osmotic stress, ion toxicity, mineral deficiencies, and induced physiological and biochemical disorders in metabolic processes (Hasegawa et al. 2000). Salinity stress is often associated with nutritional imbalance.

Proline content and activities of ascorbate peroxidase (APX), catalase (CAT), and peroxidase (POD) in leaves of plants improved due to the application of NSi, whereas Chl a, Chl b, carotenoids, and activity of superoxide dismutase (SOD) were reduced.

This upsurge in antioxidant enzymes activity lessened the oxidative damage produced by salinity. Improvement in membrane stability, chloroplast formation, and sugar accumulation came as a result from NSi and Si treatment (Qados 2015).

From the results of the many studies, it was concluded that increase in the concentration of sodium chloride decreased the plant length which in turn effected the growth of plants (Beltagi et al. 2006; Mustard and Renault 2006; Gama et al. 2007; Jamil et al. 2007; Houimli et al. 2008; Rui et al. 2009; Memon et al. 2010). Different concentrations of NaCl negatively affected the leaf area (Raul et al. 2003; Netondo et al. 2004; Mathur et al. 2006; Chen et al. 2007; Zhao et al. 2007; Yilmaz and Kina 2008; Rui et al. 2009). According to the studies conducted by Raul et al. (2003), Jamil et al. (2005), Gama et al. (2007), and Ha et al. (2008), leaf number declined with the upsurge in the concentration of salt. Fresh and dry weights of the shoot system are affected, both negatively and positively, by changes in salinity concentration, type of salt present, or type of plant species (Bayuelo Jimenez et al. 2002; Jamil et al. 2005; Niaz et al. 2005; Saqib et al. 2006; Turan et al. 2007; Saffan 2008; Rui et al. 2009; Taffouo et al. 2009, 2010; Memon et al. 2010).

Vicia faba L. seedlings were treated with different concentrations (0, 60, 120, 240 mM) of NaCl, and its effect on growth, osmotic potential, chlorophyll content, and protein content was studied. NaCl treatment showed positive responses for plant height with low and medium concentrations however, decrease with the highest concentration, in both measurement periods. Number of leaves or leaf area remains unaffected with low concentration, while a decline was noted for each, with two higher concentrations and in both measurement periods. Both fresh and dry weights of the shoot increased in the two measurement periods. With the duration of the stress, period fall in osmotic potential (OP) was observed with the upsurge in its concentration. Salinity significantly reduced chlorophyll "a," "b," total chl, and carotenoid contents in both measurement periods after 10 days of treatment. An upsurge was observed in the protein content in the two measurement periods due to impact of the salinity stress. A directly proportional relationship was found between protein content and the increase in salt concentrations in the first measurement period, while it was inversely proportional in the second (Qados 2011).

Plant growth and development is deleteriously affected by salinity as it imposes osmotic stress on plants, causes specific ion (Na^+) toxicity, disturbs the major cytosolic enzymes by upsetting intracellular potassium homeostasis, and causes oxidative stress in plant cells (Marschner 1995; Sairam and Srivastava 2002; Chen et al. 2007; Cuin and Shabala 2007). At different spans these activities affect both root and leaf tissues. After weeks of the onset of salinity treatment, Na^+ toxicity in leaves becomes critical (Munns and Tester 2008). Within a couple of hours after exposure to NaCl, enormous reduction of cytosolic K^+ in plant roots (Shabala et al. 2006; Shabala and Cuin 2008) and accretion of reactive oxygen species in root cells (Demidchik et al. 2003, 2007) take place which leads to programmed cell death (PCD) in root cells (Shabala 2009).

12.4 Use of Nanotechnology for Improving Salt Stress of Plants

Nanoparticles (nanoscale particles = NSPs/NPs) can drastically alter their physical and chemical properties in comparison with their respective bulk materials, as they are atomic or molecular aggregates of about 1–100 nm (Ball 2002; Roco 2003). At the global level, the use of nanotechnology in agriculture is at a nascent stage, yet it is increasing. Plant growth is enhanced by wide range of applications developed by nanoscience (Nair et al. 2010). Nanomaterials are of very small size and so exhibit unique features; because of their great surface area than bulk materials, they can alter their physicochemical characteristics, and their solubility and surface reactivity are boosted (Ruffini and Cremonini 2009; Haghighi et al. 2012; Rajoriya et al. 2016, 2017).

At the nanoscale, matter shows extraordinary properties that are not shown by bulk materials. For example, surface area, cation exchange capacity, ion adsorption, complexation, and many more functions of clays would multiply if they are brought to nanoscale. One of the principal ways in which a nanoparticle differs from bulk material is that a high proportion of the atoms in a nanoparticle are present on the surface. Compared with particles of macrosize, NPs may have different surface compositions, different types and densities of sites, and different reactivity with respect to processes such as adsorption and redox reactions, which could be gainfully used in synthesizing nanomaterials for use in agriculture.

12.5 Iron Oxide NPs

A study was conducted on *Mentha piperita* to explore the combined effect of 11 treatments of salinity and iron oxide NPs (FeONPs) on its essential oil composition. After 90 days mature plants were collected, and from their aerial parts, essential oils were extracted using Clevenger-type apparatus, and with the help of gas chromatography and gas chromatography/mass spectrometry, the oil was analyzed. Results confirmed that essential oil amount increased by the application of FeONPs whereas decreased with salinity stress. A strong effect on essential oil production and its composition was observed by the treatment of NPs and salt (Askary et al. 2016).

12.6 Silver NPs

Currently, an important environmental problem is the significant reduction in growth of plants due to salinity. In basil seedlings, application of silver NPs (AgNPs) during germination process augmented germination traits, plant growth, and resistance to salinity conditions and in numerous plant species too, whereas no resistance in tomato plants was observed; however, its role in ameliorating salinity effect and associated mechanisms is still unidentified (Ekhtiyari et al. 2011; Ekhtiyari and Moraghebi 2011; Darvishzadeh et al. 2015).

AgNPs can regulate various plant functional and developmental aspects in a concentration dependent manner, including, enhancement in plant growth, seed germination, photosynthetic quantum competence and chlorophyll content. Further, AgNPs are also used as antimicrobial agent to manage plant diseases (Davies 2009; Lamsal et al. 2011; Salama 2012; Hatami and Ghorbanpour 2013; Kaveh et al. 2013; Vannini et al. 2013).

The effect of AgNPs dose (0.05, 0.5, 1.5, 2, and 2.5 mg/L) on the salt (150 and 100 mM) tolerance of tomato (*Solanum lycopersicum* L.) plants during germination was studied. Salt stress inhibited seed germination and seedling growth, but the effect was alleviated by exposure to AgNPs. Exposure of AgNPs under NaCl stress enhanced germination percentage, germination rate, root length and seedling fresh and dry weight of tomato. The expression of salt-stress genes was investigated by semiquantitative RT-PCR. Of the examined salt-stress genes, four genes, AREB, MAPK2, P5CS, and CRK1, were upregulated by AgNPs under salt stress, and three genes, TAS14, DDF2 and ZFHD1, were downregulated in response to AgNPs. The gene expression patterns associated with AgNPs exposure also suggest the potential involvement of AgNPs in response to stress, indicating that they might be useful for improving plant tolerance to salinity (Almutairi 2016). Application of AgNPs has been found quite effective in improving resistance against salinity during germination of *Foeniculum vulgare* Mill. (Ekhtiyari et al. 2011) and *Cuminum cyminum* L. (Ekhtiyari and Moraghebi 2011). Salt stress was improved by the effect of nano-silicon (N-Si) and nano-zinc oxide (Haghighi et al. 2012; Sedghi et al. 2013; Kalteh et al. 2014). A number of defense mechanisms are adapted by plants which increase tolerance to adverse conditions. Stress conditions trigger a large array of genes that produce a number of proteins to combine in pathways leading to the synergistic augmentation of stress tolerance (Wang et al. 2003).

Effect of different concentrations (0% (control plots were soaked in distilled water), 10, 20, 40 ppm) of AgNPs was studied on salt (0% (control plots were irrigated with deionized water), 30, 60, 90, 120 mM/L) tolerance capacity of tomato plant. No significant effect of AgNPs was observed on the salt tolerance of tomato plant. Fruit number per plant, fruit diameter, average fruit weight, number of branches per plant, and plant height were negatively affected (Younes and Nassef 2015).

12.7 Silica NPs

Silica offers resistance against many biotic and abiotic stresses (Ma and Yamaji 2006; Liang et al. 2007). NPs due to their smaller diameter easily penetrate the pores of cell wall stomata or the base of hairs, and from there they can be easily transported to different organs (Nair et al. 2010). Tolerance against salt stress (Zhu et al. 2004), manganese toxicity (Shi et al. 2005), boron toxicity (Gunes et al. 2007), and cadmium toxicity (Vaculik et al. 2012; Shi et al. 2010) is provided by silica via changing the activity of antioxidant enzymes. Harmful effects of salinity on germination, root length, and plant dry weight on tomato seeds and seedlings were reduced by the application of nanosilica (Haghighi et al. 2012).

Silicon (Si) is the second most abundant element in the earth's crust can potentially increases stress tolerance in plants (Richmond and Sussman 2003; Ma 2004; Ma and Yamaji 2004; Currie and Perry 2007). Si reduces the harmful effects of oxidative stress and delivers resistance against some abiotic and biotic plant stressors (Ma 2004; Liang et al. 2007; Pei et al. 2010). Si augments photosynthesis, increases plant growth, and advances plant resistance to ailment (biotic stress) and numerous abiotic stresses, viz., cold, heat, drought, salinity, and heavy metals (Ma and Yamaji 2004). These effects have been predicted in a wide variety of plants species for their growth and yield (Ma and Yamaji 2004; Liang et al. 2007). Penetration of pesticides or pathogens into the plant cell is prohibited by Si as it acts as a physico-mechanical barrier.

An experiment was conducted to study the effect of SiNPs (without silicon, normal silicon fertilizer, and silicon NPs) on physiological and morphological traits of basil under salinity stress (1, 3, and 6 ds/m). Results revealed a noteworthy decline in growth and development indices due to the salinity stress. Increased concentration of NaCl reduced leaf dry, fresh weight, and chlorophyll content which was significantly augmented by the application of silicon NPs. On the contrary, salinity stress increased the proline content which was a response to stress. Moreover, proline increased by silicon NPs which was due to tolerance induction in plant. Silicon NPs application reduced the pollution effects originated from salinity in basil (Kalteh et al. 2014). *Agropyron elongatum* was treated with SiO_2NPs, and its effect on early seedling growth of plant was investigated by Azimi et al. (2014). Siddiqui et al. (2014) and Sabaghnia and Janmohammadi (2015) separately evaluated the role of nano-silicon dioxide (nano-SiO_2) in plant resistance to salt stress.

12.8 Copper NPs

Copper occurs in multiple oxidative states (Cu^{2+}, Cu^{+}) and is considered as a crucial micronutrient necessary for usual growth of plants and various physiological processes. Cu^{2+} is an important structural constituent of regulatory proteins and is involved in photosynthetic electron transport, mitochondrial respiration, oxidative stress responses, cell wall metabolism, and hormone signaling and plays an essential role in signaling of transcription, protein trafficking machinery, oxidative phosphorylation, and iron mobilization (Marschner 1995; Raven et al. 1999; Yruela 2005; Solymosi and Bertrand 2012). The redox property of Cu can also contribute to its toxicity, as free ions catalyze the production of damaging radicals (Manceau et al. 2008).

Exposure to 1000 mg/L concentration of CuNPs (50 nm) reduced 90% biomass in *Cucurbita pepo* (zucchini) (Stampoulis et al. 2009) and was noxious to the growth of *Phaseolus radiatus* (mung bean) and *Triticum aestivum* (wheat) (Lee et al. 2008). Musante and White (2012) treated *Cucurbita pepo* with CuO NPs and stated a decline in growth and transpiration by 60–70% as compared to control.

Rice (*Oryza sativa*, var. Jyoti) was treated with copper (II) oxide NPs (CuONPs), and its physiological and biochemical behavior was studied. The investigation revealed that germination rate, root and shoot length, biomass, photosynthetic rate, transpiration rate, stomatal conductance, maximal quantum yield of PSII photochemistry, and photosynthetic pigment contents declined at 1000 mg (CuONPs)/L, while uptake of Cu in the roots and shoots augmented at high concentrations of CuONPs. Lower number of thylakoids per granum was found in chloroplasts due to accretion of CuONPs in the cells. Antioxidants, viz., malondialdehyde, proline, ascorbate peroxidase, and superoxide dismutase, augmented as a consequence of oxidative and osmotic stress. This revealed the toxic effect of Cu accumulation in roots and shoots that resulted in loss of photosynthesis (Da Costa and Sharma 2016).

12.9 Zinc Oxide NPs

Zinc is a vital microelement captivated in the form of Zn^{++}. It plays a significant role in the production of biomass (Kaya and Higgs 2002; Cakmak 2008) and acts as a stabilizer of proteins, membranes, and DNA-binding proteins such as Zn fingers (Aravind and Prasad 2003). It is necessary for chlorophyll synthesis, pollen function and fertilization (Pandey et al. 2006), biosynthesis of the plant growth regulator such as indole-3-acetic acid (IAA) (Fang et al. 2008), and carbohydrate and N metabolism which leads to high yield and yield components. Its deficiency also adversely affects carbohydrate metabolism, damages pollen structure, and decreases the yield (Fang et al. 2008). Bybordi and Malakouti (2007) found that application of zinc had a significant effect on seed yield, seed oil content, and 1000-seed weight. So, it is very important to apply zinc fertilizer for increasing crop yields and improving crop quality. However, little attention has been paid to observe the roles of ZnO nanoparticles (ZnONPs) in plants grown under salinity stress.

A study was conducted on five sunflower cultivars (*Helianthus annuus* L. cvs. Alstar, Olsion, Yourflor, Hysun36, and Hysun33) to investigate the effect of foliar application of normal and NPs of zinc oxide (control, ZnO normal and NPs at a rate of 2 g/L) along with different salinity levels (0 and 100 mM NaCl) on growth, proline content, and some antioxidant enzyme activities. The highest proline content and superoxide dismutase activity (SOD) were exhibited by Olsion under saline condition. SOD activity and shoot dry weight were enhanced by foliar spray of ZnO. Zinc oxide NPs (ZnONPs) had positive effect on biomass production of sunflower plants compared to the normal form. According to the result, Olsion and Hysun33 cultivars were appropriate for saline conditions, whereas Hysun36 was suitable for normal condition (Torabian et al. 2016).

Effects of ZnONPs in regulating physiological and biochemical processes of plants in response to salt-induced stress is not much investigated. Calli of five tomato cultivars were exposed to higher concentrations of salt (3.0 and 6.0 g/L NaCl), in the presence of zinc oxide NPs (15 and 30 mg/L) and callus growth traits, rate of regeneration in plants, mineral element (sodium, potassium, phosphorous, and nitrogen) contents, and alterations in the activity of superoxide dismutase (SOD) and glutathione peroxidase (GPX) were evaluated. NaCl exposure augmented callus growth rate and higher sodium content and SOD and GPX activities. Lower concentration of ZnONPs was more effective in alleviating the effects of NaCl than its higher concentration. These results indicated that ZnONPs can be used as a potential anti-stress agent in crop production. Different tomato cultivars showed different degrees of tolerance to salinity in the presence of ZnONPs (Alharby et al. 2016).

In a recent seed priming study, possible roles of ZnO nanoparticles in mitigating harmful effects of salinity stress in lupine (*Lupinus termis*) were studied. Lupine plants were exposed to 150 mM NaCl with and without different concentrations of ZnO (20 mg L^{-1}, 40 mg L^{-1}, and 60 mg L^{-1} for 20 days.

Reduction in plant growth parameters (root length, shoot length, fresh weight, and dry weight) and in the contents of photosynthetic pigments (chlorophyll a and b and carotenoids), as well as in the activity of catalase (CAT), was observed in salinized plants against control plants. On the other side, enhanced contents of organic solutes (soluble sugar, soluble protein, total free amino acids, and proline), total phenols, ascorbic acid, malondialdehyde (MDA), and Na, as well as the activities of superoxide dismutase (SOD), peroxidase (POD), and ascorbate peroxidase (APX), were found in stressed plants over control plants. However, seed priming with ZnNPs mostly stimulated growth of stressed plants, which was accompanied by reinforcement in the levels of photosynthetic pigments, organic solutes, total phenols, ascorbic acid, and Zn, as well as in the activities of SOD, CAT, POD, and APX enzymes over stressed plants alone. On the contrary, priming with ZnNPs caused a decrement in the contents of MDA and Na in stressed plants relative to salinized plants alone. It is worth to mention that this improvement in salt tolerance of plants primed with ZnNPs was more obvious in plants primed with ZnNPs (60 mgL^{-1}) and grown both in unstressed and stressed regimes. It was suggested that seed priming with ZnNPs, especially 60 mg L^{-1} ZnO, is an effective strategy that can be used to enhance salt tolerance of lupine plants (Latef et al. 2017).

12.10 Cerium Oxide NPs

CeO_2NPs exert significant impact on plant growth and production. *Brassica napus* L. (canola) cv. "Dwarf Essex" was taken to study the physiological and biochemical changes under synergistic salt stress (control and 100 mM NaCl) and CeO_2NP effects. Plant growth and the physiological processes of canola were negatively affected by 100 mM NaCl. Under fresh and saline water irrigation conditions,

CeO_2NP-treated plants exhibited higher efficiency of the photosynthetic apparatus and plant biomass and less stress. The results revealed that CeO_2NPs led to variations in canola growth and physiology which enhanced the plant salt-stress response but did not completely improve the salt stress of canola (Rossi et al. 2016).

CeO_2NPs displayed strong effects on plant health both positively and negatively, depending upon the plant species, exposure concentration, and exposure duration and plant growth conditions (Lopez-Moreno et al. 2010b; Ma et al. 2010, 2015, 2016; Wang et al. 2012; Zhao et al. 2014). 2000 mg/L concentration of CeO_2NPs inhibited root elongation of lettuce, while tomato, radish, rape, wheat, cucumber, and cabbage were largely unaffected at the same exposure condition (Ma et al. 2010). The DNA of soybean seedlings was damaged by high concentrations of CeO_2NPs (2000 and 4000 mg/L), while no such effect was observed at lower concentrations (Lopez-Moreno et al. 2010a). 10 mg/L of CeO_2NPs slightly improved tomato growth and yield (Wang et al. 2012). Ce due to dual valance states (Ce^{3+} and Ce^{4+}) is considered as an oxidative stress inducer at other conditions (Ma et al. 2015, 2016) and behaves as an antioxidant at certain conditions (Wang et al. 2012) (Table 12.1).

12.11 Conclusion and Future Perspectives

Salinity stress is a menace to agriculture and a major environmental factor that adversely affects crops. Crop plants are generally vulnerably susceptible to salinity and the extent of sensitivity, and the mechanism adopted to cope up with the same depends on the categorical magnification stage of the plant. Among all other salts, NaCl is the most soluble and widespread salt, it is not surprising that all plants have evolved mechanisms to regulate its accumulation and to cull against it, in favor of other nutrients commonly present in low concentrations. Measures of ionic balances during soil-salinization cycle are still unexplored. Nanotechnology in a naïve sense may appear as a paradigm shift in virtually all branches of science. Nanotechnology promises a breakthrough and sensitized our construal in amending presently abysmal salt tolerance processes through nanoparticles vis-a-vis regenerating soil fertility via reclamation of salt-affected soils. However, constrained data and experimental evidence are available to expound the role and comportment of NPs in salt tolerance/resistance/avoidance mechanisms in planta, and most of the studies are predicated on phenological events, morphological analysis, and antioxidant status. Tasks of metal-predicated and carbon-predicated NPs under salinity are diverse. NPs alleviated the deleterious effects of salt stress and showed regulatory roles in salt-responsive genes. Applications of different NPs shown incremented seed germination, shoot biomass, etc., in salt affected plants. Nevertheless the refined mechanism of actions of NPs is poorly understood in salt-stressed plant systems.

However, in nascent stage, utilization of NPs is revisiting our construal of the theoretical substructures of the agricultural engenderment system along the pedosphere-biosphere-atmosphere continuum. Further research needs to address questions about roles of different NPs on membrane conveyors and candidate genes

Table 12.1 Effect of various NPs in mitigating salt stress in different plants

S. no.	NPs	Plant	NPs conc.	Salt conc.	Reference
1.	FeNPs and potassium silicate	*Vitis vinifera* (grape)	0.08, and 0.8 ppm and 0, 1, 2 mM	50, and 100 mM	Mozafari et al. (2017)
2.	Fe_2O_3 NPs and EDTA iron	*Arachis hypogaea* (peanut)	2, 10, 50, 250, 1000 mg/kg and 45.83 mg/kg	----------	Rui et al. (2016)
3.	AgNPs	*Solanum lycopersicum* (tomato)	0.05, 0.5, 1.5, 2, 2.5 mg/L	150 and 100 mM	Almutairi (2015)
4.	AgNPs	*Solanum lycopersicum* (tomato)	10, 20, 40 ppm	30, 60, 90, and 120 mM/L	Younes and Nassef (2015)
5.	AgNPs	*Triticum aestivum* (wheat)	2, 5, and 10 mM	150 mM	Mohamed et al. (2017)
6.	Si NPs	*Vicia faba* (beans)	----------	1.5, and 3 mM (sodium silicate)	Roohizadeh et al. (2015)
7.	Si NPs	*Ocimum basilicum* (basil)	Without silica, normal silica, and silicon NPs	1, 3, and 6 ds/m	Kalteh et al. (2014)
8.	Nanosilica (NSi)	*Solanum lycopersicum* (tomato)	0.5, 1, 2, 3 mM	150, and 200 mM	Almutairi (2016)
9.	NSi and Si	*Vicia faba* (beans)	1, 2, and 3 mM	50, 100, and 200 mM	Qados (2015)
10.	Nano-Si and Si	*Capsicum annuum* (sweet pepper)	1 or 2 cm^3/L and 4 or 5 cm^3/L	----------	Tantawy et al. (2015)
11.	SiO_2NPs	*Lens culinaris* (lentil)	Distilled water (control), 100 mM NaCl, 1 mM nano-silicon dioxide conc. + 100 mM NaCl		Sabaghnia and Janmohammadi (2015)
12.	CeO_2NPs	*Brassica napus* (canola)	200, and 1000 mg/kg	100 mM	Rossi et al. (2016)
13.	CeO_2 NPs	*Brassica napus* (canola)	500 mg/kg	50 mM	Rossi et al. (2017)
14.	ZnONPs	*Solanum lycopersicum* (tomato)	15 and 30 mg/L	3.0 and 6.0 g/L	Alharby et al. (2016)
15.	ZnONPs	*Helianthus annuus* (sunflower)	None-sprayed ZnO normal and NPs at rate of 2 g/L	100 mM	Torabian et al. (2016)
16.	ZnONPs	*Lupinus termis* (lupine)	20 mg/L, 40 mg/L, 60 mg/L	150 mM	Latef et al. (2017)
17.	ZnO and Fe_3O_4 NPs	*Moringa peregrina* (wild drumstick)	30, 60, 90 mg/L	3000, 6000, 9000 ppm	Soliman et al. (2015)
18.	Chelated calcium and nano-calcium	*Solanum lycopersicum* (tomato)	2 and 3 g/L and 0.5 and 1 g/L	----------	Tantawy et al. (2014)
19.	MWCNTs	*Brassica oleracea* (broccoli)	10, 20, 40, 60 mg/L	80, 100, and 120 mM	Martinez-Ballesta et al. (2016)

conferring salt tolerance (viz., *SOS1*, *SOS3*, *ERA1*, *AAPK*, *PKS3*, *HKT*, *NHX*, *OTS*, *MT1D*). NPs interceded adaptation mechanisms of cell organelles and macromolecular systems to excess of salt are again skeptical. Furthermore, a holistic view of extenuating the restorative effects of NPs on plant stress biology is the future line of action in the area of nanotechnology and additionally needs to be answered.

Acknowledgments The authors are heartily thankful to Hon'ble Vice Chancellor Prof. Rajendra B. Lal, Sam Higginbottom University of Agriculture, Technology and Sciences, Allahabad, for providing essential facilities and valuable suggestions during the course of the investigation.

References

Alharby HF, Metwali EMR, Fuller MP, Aldhebiani AY (2016) Impact of application of zinc oxide nanoparticles on callus induction, plant regeneration, element content and antioxidant enzyme activity in tomato (*Solanum lycopersicum* mill.) under salt stress. Arch Biol Sci 68:723–735

Almutairi ZM (2015) Influence of silver nano-particles on the salt resistance of tomato (*Solanum lycopersicum* L.) during germination. Int J Agric Biol 18:449–457

Almutairi ZM (2016) Effect of nano-silicon application on the expression of salt tolerance genes in germinating tomato (*Solanum lycopersicum* L.) seedlings under salt stress. POJ 9:106–114

Aravind P, Prasad MNV (2003) Zinc alleviates cadmium-induced oxidative stress in *Ceratophyllum demersum* L.: a free floating freshwater macrophyte. Plant Physiol Biochem 41:391–397

Ashraf M, Foolad M (2007) Roles of glycine betaine and proline in improving plant abiotic stress resistance. Environ Exp Bot 59:206–216

Ashraf M, McNeilly T (2004) Salinity tolerance in brassica oilseeds. Crit Rev Plant Sci 23:157–174

Askary M, Talebi SM, Amini F, Bangan ADB (2016) Effect of NaCl and iron oxide nanoparticles on Mentha piperita essential oil composition. EEB 14:27–32

Azimi R, Borzelabad MJ, Feizi H, Azimi A (2014) Interaction of SiO_2 nanoparticles with seed prechilling on germination and early seedling growth of tall wheat grass (*Agropyron elongatum* L.). Pol J Chem Technol 16:25–29

Baalbaki RZ, Zurayk RA, Bleik SN, Talhuk A (1990) Germination and seedling development of drought susceptible wheat under moisture stress. Seed Sci Technol 17:291–302

Ball P (2002) Natural strategies for the molecular engineer. Nanotechnology 13:15–28

Bayuelo Jimenez JS, Debouk DG, Lynch JP (2002) Salinity tolerance in phaseolus species during early vegetative growth. Crop Sci 42:2184–2192

Beltagi MS, Ismail MA, Mohamed FH (2006) Induced salt tolerance in common bean (*Phaseolus vulgaris* L.) by gamma irradiation. Pak J Biol Sci 6:1143–1148

Bybordi A, Malakouti MJ (2007) Effects of zinc fertilizer on the yield and quality of two winter varieties of canola. Zinc crops; improving crop production and human health 24–26 May. Istanbul

Cakmak I (2008) Enrichment of cereal grains with zinc: agronomic or genetic biofortification? Plant Soil 302:1–17

Chen Z, Pottosin II, Cuin TA, Fuglsang AT, Tester M, Jha D, Zepeda-Jazo I, Zhou M, Palmgren MG, Newman IA, Shabala S (2007) Root plasma mem-brane transporters controlling K+/Na+ homeostasis in salt-stressed barley. Plant Physiol 145:1714–1725

Cuin TA, Shabala S (2007) Compatible solutes reduce ROS-induced potassium efflux in Arabidopsis roots. Plant Cell Environ 30:875–885

Currie HA, Perry C (2007) Silica in plants: biological, biochemical and chemical studies. Ann Bot 100:1383–1389

Da Costa MVJ, Sharma PK (2016) Effect of copper oxide nanoparticles on growth, morphology, photosynthesis, and antioxidant response in *Oryza sativa*. Photosynthetica 54:110–119
Darvishzadeh F, Nejatzadeh F, Iranbakhsh AR (2015) Effects of silver nanoparticles on salinity tolerance in basil plant (Ocimum basilicum L.) during germination in vitro. NCMBJ 15:63–70
Davies JC (2009) Nanotechnology oversight: an agenda for the new administration. project on emerging technologies. Woodrow Wilson International Center for Scholars, Washington, DC
Demidchik V, Shabala SN, Coutts KB, Tester M, Davies JM (2003) Free oxygen radicals regulate plasma membrane Ca2+ and K+ permeable channels in plant root cells. J Cell Sci 116:81–88
Demidchik V, Shabala SN, Davies JM (2007) Spatial variation in H_2O_2 response of *Arabidopsis thaliana* root epidermal Ca2+ flux and plasma membrane Ca2+ channels. Plant J 49:377–386
Ekhtiyari R, Moraghebi F (2011) The study of the effects of nano silver technology on salinity tolerance of cumin seed (*Cuminum cyminum* L.). Plant Ecosyst 7:99–107
Ekhtiyari R, Mohebbi H, Mansouri M (2011) The study of the effects of nano silver technology on salinity tolerance of (*Foeniculum vulgare* mill.). Plant Ecosyst 7:55–62
Fang YL, Wang Z, Xin L, Zhao X, Hu Q (2008) Effect of foliar application of zinc, selenium, and iron fertilizers on nutrients concentration and yield of rice grain in China. J Agric Food Chem 56:2079–2084
Farooq M, Hussain M, Wakeel A, Siddique KHM (2015) Salt stress in maize: effects, resistance mechanisms and management. A review. Agron Sustain Dev 35:461–481
Fisher RA, Turner NC (1978) Plant productivity, in arid and semi arid zones. Ann Rev Plant Physiol 29:897–912
Gama PBS, Inanaga S, Tanaka K, Nakazawa R (2007) Physiological response of common bean (*Phaseolus vulgaris* L.) seedlings to salinity stress. Afr J Biotechnol 6:79–88
Gunes A, Inal A, Bagci EG, Coban S, Sahin O (2007) Silicon increases boron tolerance and reduces oxidative damage of wheat grown in soil with excess boron. Biol Plant 51:571–574
Ha E, Ikhajiagba B, Bamidele JF, Ogic-odia E (2008) Salinity effects on young healthy seedling of kyllingia peruviana collected from escravos, Delta state. Glob J Environ Res 2:74–88
Haghighi M, Afifipour Z, Mozafarian M (2012) The effect of N-Si on tomato seed germination under salinity levels. J Biol Environ Sci 6:87–90
Hasegawa PM, Bressen RA, Zhu JK, Bohnert HJ (2000) Plant cellular and molecular responses to high salinity. Ann Rev Plant Physiol 51:463–499
Hatami M, Ghorbanpour M (2013) Effect of nanosilver on physiological performance of pelargonium plants exposed to dark storage. J Hortic Res 21:15–20
Houimli SIM, Denden M, Elhadj SB (2008) Induction of salt tolerance in pepper (B0 or Hbt 10mmtt1) by 24-epibrassinolide. Eurasia J Biol Sci 2:83–90
Jafari M (1994) Salinity and halophytes, Bulletin No. 90. Research Institute of Forests and Rangelands, Tehran
Jamil M, Lee CC, Rehman SU, Lee DB, Ashraf M, Rha ES (2005) Salinity (NaCl) tolerance of brassica species at germination and early seedling growth. Electron J Environ Agric Food Chem 4:970–976
Jamil M, Lee KB, Jung KY, Lee DB, Han MS, Rha ES (2007) Salt stress inhibits germination and early seedling growth in cabbage (*Brassica oleracea* L.). Pak J Biol Sci 10:910–914
Jouyban Z (2012) The effects of salt stress on plant growth. Tech J Eng App Sci 2:7–10
Kalteh M, Alipour ZT, Ashraf S, Aliabadi MM, Nosratabadi AF (2014) Effect of silica nanoparticles on basil (*Ocimum basilicum*) under salinity stress. J Chem Health Risk 4:49–55
Kaveh R, Li YS, Ranjbar S, Tehrani R, Brueck CL, Van Aken B (2013) Changes in *Arabidopsis thaliana* gene expression in response to silver nanoparticles and silver ions. Environ Sci Technol 47:10637–10644
Kaya C, Higgs D (2002) Response of tomato (*Lycopersicon esculentum* L.) cultivars to foliar application of zinc when grown in sand culture at low zinc. Sci Hortic 93:53–64
Khan MA, Duke NC (2001) Halophytes- a resource for the future. Wetl Ecol Manag 6:455–456
Lamsal K, Kim SW, Jung JH, Kim YS, Kim KS, Lee YS (2011) Application of silver nanoparticles for the control of *Colletotrichum* species in vitro and pepper anthracnose disease in field. Mycobiology 39:194–199

Latef AAHA, Alhmad MFA, Abdelfattah KE (2017) The possible roles of priming with ZnO nanoparticles in mitigation of salinity stress in lupine (Lupinus termis) plants. J Plant Growth Regul 36:60–70

Lee WM, An YJ, Yoon H, Kweon HS (2008) Toxicity and bioavailability of copper nanoparticles to the terrestrial plants mung bean (Phaseolus radiatus) and wheat (*Triticum aestivum*): plant agar test for water-insoluble nanoparticles. Environ Toxicol Chem 27:1915–1921

Liang Y, Zhang W, Qin C, Youliang L, Ruixing D (2006) Effect of exogenous silicon (Si) on H+-ATPase activity, phospholipids and fluidity of plasma membrane in leaves of salt-stressed barley (*Hordeum vulgare* L.). Environ Exp Bot 57:212–219

Liang Y, Sun W, Zhu YG, Christie P (2007) Mechanisms of silicon- mediated alleviation of abiotic stresses in higher plants: a review. Environ Pollut 147:422–428

Lopez-Moreno ML, de la Rosa G, Hernandez-Viezcas JA, Castillo-Michel H, Botez CE, Peralta-Videa JR, Gardea-Torresdey JL (2010a) Evidence of the differential biotransformation and genotoxicity of ZnO and CeO2 nanoparticles on soybean (*Glycine max*) plants. Environ Sci Technol 44:7315–7320

Lopez-Moreno ML, de la Rosa G, Hernandez-Viezcas JA, Peralta-Videa JR, Gardea-Torresdey JL (2010b) X-ray absorption spectroscopy (XAS) corroboration of the uptake and storage of CeO2 nanoparticles and assessment of their differential toxicity in four edible plant species. J Agric Food Chem 58:3689–3693

Ma JF (2004) Role of silicon in enhancing the resistance of plants to biotic and abiotic stresses. Soil Sci Plant Nutr 50:11–18

Ma JF, Yamaji N (2004) Role of silicon in enhancing the resistance of plants to biotic and abiotic stresses. Soil Sci Plant Nutr 50:11–18

Ma JF, Yamaji N (2006) Silicon uptake and accumulation in higher plants. Trends Plant Sci 11:392–397

Ma X, Geiser-Lee J, Deng Y, Kolmakov A (2010) Interactions between engineered nanoparticles (ENPs) and plants: phytotoxicity, uptake and accumulation. Sci Total Environ 408:30533061

Ma X, Wang Q, Rossi L, Zhang W (2015) Cerium oxide nanoparticles and bulk cerium oxide leading to different physiological and biochemical responses in *Brassica rapa*. Environ Sci Technol 5:6793–6802

Ma X, Wang Q, Rossi L, Ebbs SD, White JC (2016) Multigenerational exposure to cerium oxide nanoparticles: physiological and biochemical analysis reveals transmissible changes in rapid cycling *Brassica rapa*. NanoImpact 1:46–54

Manceau A, Nagy KL, Marcus MA et al (2008) Formation of metallic copper nanoparticles at the soil-root interface. Environ Sci Technol 42:1766–1772

Marschner H (1995) Mineral nutrition of higher plants. Academic Press, Hartcourt Brace and Company, New York

Martinez-Ballesta MC, Zapata L, Chalbi N, Carvajal M (2016) Multiwalled carbon nanotubes enter broccoli cells enhancing growth and water uptake of plants exposed to salinity. J Nanobiotechnol 14:42

Massoud FI (1977) The use of satellite imagery in detecting and delineating salt affected soils. Pedologie teledetection, AISS-ISSS, Roma

Mathur N, Singh J, Bohra S, Bohra A, Vyas A (2006) Biomass production, productivity and physiological changes in moth bean genotypes at different salinity levels. Am J Plant Physiol 1:210–213

Memon SA, Hou X, Wang LJ (2010) Morphological analysis of salt stress response of pak Choi. EJEAFChe 9:248–254

Mohamed AKSH, Qayyum MF, Abdel-Hadi AM, Rehman RA, Ali S, Rizwan M (2017) Interactive effect of salinity and silver nanoparticles on photosynthetic and biochemical parameters of wheat. Arch Agron Soil Sci 63:1736–1747

Mozafari AA, Asl AG, Ghaderi N (2017) Grape response to salinity stress and role of iron nanoparticle and potassium silicate to mitigate salt induced damage under in vitro conditions. Physiol Mol Biol Plants 24:25–35

Munns R (1993) Physiological processes limiting plant growth in saline soil: some dogmas and hypotheses. Plant Cell Environ 16:15–24
Munns R (2002) Comparative physiology of salt and water stress. Plant Cell Environ 25:239–250
Munns R, Tester M (2008) Mechanisms of salinity tolerance. Ann Rev Plant Biol 59:651–681
Munns R, James RA, Xu B, Athman A, Conn SJ, Jordans C, Byrt CS, Hare RA, Tyerman SD, Tester M, Plett D, Gilliham M (2012) Wheat grain yield on saline soils is improved by an ancestral Naþ transporter gene. Nat Biotechnol 30:360–364
Musante C, White JC (2012) Toxicity of silver and copper to *Cucurbita pepo*: differential effects of nano and bulk-size particles. Environ Toxicol 27:510–517
Mustard J, Renault S (2006) Response of red-osier dogwood (*Cornus sericea*) seedling to NaCl during the onset of bud break. Can J Bot 84:844–851
Nair R, Varghese SH, Nair BG, Maekawa T, Yoshida Y, Kumar DS (2010) Nanoparticulate material delivery to plants. Plant Sci 179:154–163
Netondo GW, Onyango JC, Beck E (2004) Sorghum and salinity. II. Gas exchange and chlorophyll fluorescence of sorghum under salt stress. Crop Sci 44:806–811
Neumann P (1997) Salinity resistance and plant growth revisited. Plant Cell Environ 20:1193–1198
Ni BR, Bradford KJ (1992) Quantitative models characterizing seed germination responses to abscisic acid and osmoticum. Plant Physiol 98:1057–1068
Niaz BH, Athar M, Salim M, Rozema J (2005) Growth and ionic relations of fodder beet and sea beet under saline. CEERS 2:113–120
Pandey N, Pathak GC, Sharma CP (2006) Zinc is critically required for pollen function and fertilisation in lentil. J Trace Elem Med Biol 20:89–96
Pei ZF, Ming DF, Liu D, Wan GL, Geng XX, Gong HJ, Zhou WJ (2010) Silicon improves the tolerance to water-deficit stress induced by polyethylene glycol in wheat (*Triticum aestivum* L.) seedlings. J Plant Growth Regul 29:106–115
Qadir M, Sharma BR, Bruggeman A, Choukr-Allah R, Karajeh F (2007) Nonconventional water resources and opportunities for water augmentation to achieve food security in water scarce countries. Agric Water Manag 87:2–22
Qados AMSA (2011) Effect of salt stress on plant growth and metabolism of bean plant *Vicia faba* (L.). J Saudi Soc Agric Sci 10:7–15
Qados AMSA (2015) Mechanism of nanosilicon-mediated alleviation of salinity stress in faba bean (Vicia faba L.) plants. AJEA 7:78–95
Rahimi R, Mohammakhani A, Roohi V, Armand N (2012) Effects of salt stress and silicon nutrition on cholorophyll content, yield and yield components in fennel (*Foeniculum vulgar* Mill.). Int J Agric Crop Sci 4:1591–1595
Rajoriya P, Misra P, Shukla PK, Ramteke PW (2016) Light regulatory effect on the phytosynthesis of silver nanoparticles using aqueous extract of garlic (*Allium sativum)* and onion (*Allium cepa)* bulb. Curr Sci 111:1364–1368
Rajoriya P, Misra P, Singh VK, Shukla PK, Ramteke PW (2017) Green synthesis of silver nanoparticles. Biotech Today 7:7–20
Raul L, Andres O, Armado L, Bernardo M, Enrique T (2003) Response to salinity of three grain legumes for potential cultivation in arid areas (plant nutrition). Soil Sci Plant Nutr 49:329–336
Raven JA, Evans MC, Korb RE (1999) The role of trace metals in photosynthetic electron transport in O_2 – evolving organisms. Photosynth Res 60:111–150
Rengasamy P (2010) Soil processes affecting crop production in salt-affected soils. Funct Plant Biol 37:613–620
Richmond KE, Sussman M (2003) Got silicon? The non-essential beneficial plant nutrient. Curr Opin Plant Biol 6:268–272
Roco MC (2003) Nanotechnology convergence with modern biology and medicine. Curr Opin Biotechnol 14:337–346
Roohizadeh G, Majd A, Arbabian S (2015) The effect of sodium silicate and silica nanoparticles on seed germination and growth in the Vicia faba L. STPR 2:85–89
Rossi L, Zhang W, Lombardini L, Ma X (2016) The impact of cerium oxide nanoparticles on the salt stress responses of Brassica napus L. Environ Pollut 219:28–36

Rossi L, Zhang W, Ma X (2017) Cerium oxide nanoparticles alter the salt stress tolerance of Brassica napus L. by modifying the formation of root apoplastic barriers. Environ Pollut 229:132–138
Ruffini CM, Cremonini R (2009) Nanoparticles and higher plants. Caryologia 62:161–165
Rui L, Wei S, Mu-Xiang C, Cheng-Jun J, Min W, Bo-Ping Y (2009) Leaf anatomical changes of Bruguiera gymnorrhiza seedlings under salt stress. J Trop Subtrop Bot 17:169–175
Rui M, Ma C, Hao Y, Guo J, Rui Y, Tang X, Zhao Q, Fan X, Zhang Z, Hou T, Zhu S (2016) Iron oxide nanoparticles as a potential iron fertilizer for peanut (Arachis hypogaea). Front Plant Sci 7:815
Sabaghnia N, Janmohammadi M (2015) Effect of nano silicon particles application on salinity tolerance in early growth of some lentil genotypes. Ann UMCS Biol 69:39–55
Saffan SE (2008) Effect of salinity and osmotic stresses on some economic plants. Res J Agric Biol Sci 4:159–166
Sairam RK, Srivastava GC (2002) Changes in antioxidant activity in subcellular fraction of tolerant and susceptible wheat genotypes in response to long term salt stress. Plant Sci 162:897–904
Salama HMH (2012) Effects of silver nanoparticles in some crop plants, common bean (*Phaseolus vulgaris* L.) and corn (*Zea mays* L.). Int Res J Biotech 3:190–197
Sanoubar R, Cellini A, Veroni AM, Spinelli F, Masia A, Vittori Antisari L, Orsini F, Gianquinto G (2016) Salinity thresholds and genotypic variability of cabbage (Brassica oleracea L.) grown under saline stress. J Sci Food Agric 96:319–330
Saqib M, Zorb C, Schubert S (2006) Salt resistant and salt-sensitive wheat genotypes show similar biochemical reaction at protein level in the first phase of salt stress. J Plant Nutr Soil Sci 169:542–548
Sedghi M, Hadi M, Toluie SG (2013) Effect of nano zinc oxide on the germination parameters of soybean seeds under drought stress. Ann WUT Ser Biol XVI:73–78
Shabala S (2009) Salinity and programmed cell death: unravelling mechanisms for ion specific signalling. J Exp Bot 60:709–712
Shabala S, Cuin TA (2008) Potassium transport and plant salt tolerance. Physiol Plant 133:651–669
Shabala S, Demidchik V, Shabala L, Cuin TA, Smith SJ, Miller AJ, Davies JM, Newman IA (2006) Extracellular Ca2+ ameliorates NaCl-induced K+ loss from Arabidopsis root and leaf cells by controlling plasma membrane K+-permeable channels. Plant Physiol 141:1653–1665
Shanon MC (1986) New insights in plant breeding efforts for improved salt tolerance. Hortic Technol 6:96–99
Shi XH, Zhang CC, Wang H, Zhang FS (2005) Effect of Si on the distribution of Cd in rice seedling. Plant Soil 273:53–60
Shi G, Cai Q, Liu C, Wu L (2010) Silicon alleviates cadmium toxicity in peanut plants in relation to cadmium distribution and stimulation of antioxidative enzymes. J Plant Growth Regul 61:45–52
Siddiqui MH, Al-Whaibi MH, Faisal M, Al Sahli AA (2014) Nano silicon dioxide mitigates the adverse effects of salt stress on *Cucurbita pepo* L. Environ Toxicol Chem 33:2429–2437
Soliman AS, El-feky SA, Darwish E (2015) Alleviation of salt stress on Moringa peregrine using foliar application of nanofertilizers. JHF 7:36–47
Solymosi K, Bertrand M (2012) Soil metals, chloroplasts, and secure crop production: a review. Agron Sustain Dev 32:245–272
Stampoulis D, Sinha SK, White JC (2009) Assay-Dependent Phytotoxicity of Nanoparticles to Plants. Environmental Science & Technology 43 (24):9473–9479
Taffouo VD, Kouamou JK, Ngalangue LMT, Ndjeudji BAN, Akoa A (2009) Effects of salinity stress on growth, ions partitioning and yield of some cowpea (Vigna unguiculata L., walp) cultivars. Int J Bot 5:135–143
Taffouo VD, Wamba OF, Yombi E, Nono GV, Akoe A (2010) Growth, yield, water status and ionic distribution response of three bambara groundnut (Vigna subterranean (L.) verdc.) landraces grown under saline conditions. Int J Bot 6:53–58

Tantawy AS, Salama YAM, El-Nemr MA, Abdel-Mawgoud AMR, Ghoname AA (2014) Comparison of chelated calcium with nano calcium on alleviation of salinity negative effects on tomato plants. Middle East J Agric Res 3:912–916

Tantawy AS, Salama YAM, El-Nemr MA, Abdel-Mawgoud AMR (2015) Nano silicon application improves salinity tolerance of sweet pepper plants. Int J ChemTech Res 8:11–17

Tilman D, Balzer C, Hill J, Befort BL (2011) Global food demand and the sustainable intensification of agriculture. Proc Natl Acad Sci U S A 108:20260–20264

Torabian S, Zahedi M, Khoshgoftarmanesh A (2016) Effect of foliar spray of zinc oxide on some antioxidant enzymes activity of Sunflower under salt stress. J Agric Sci Technol 18:1013–1025

Turan MA, Kalkat V, Taban S (2007) Salinity-induced stomatal resistance, proline, chlorophyll and Ion concentrations of bean. Int J Agric Res 2:483–488

Vaculik M, Landberg T, Greger M, Luxová M, Stoláriková M, Lux A (2012) Silicon modifies root anatomy, and uptake and subcellular distribution of cadmium in young maize plants. Ann Bot 110:433–443

Vannini C, Domingo G, Onelli E, Prinsi B, Marsoni M, Espen L, Bracale M (2013) Morphological and proteomic responses of *Eruca sativa* exposed to silver nanoparticles or silver nitrate. PLoS One 8:e68752

Wang W, Vinocur B, Altman A (2003) Plant responses to drought, salinity and extreme temperatures: towards genetic engineering for stress tolerance. Planta 218:1–14

Wang Q, Ma X, Zhang W, Pei H, Chen Y (2012) The impact of cerium oxide nanoparticles on tomato (*Solanum lycopersicum* L.) and its implications for food safety. Metallomics 4:1105e–1112e

Yao AR (1998) Molecular biology of salt tolerance in the context of whole plant physiology. J Exp Bot 49:915–929

Yilmaz H, Kina A (2008) The influence of NaCl salinity on some vegetative and chemical changes of strawberries (*Fragaria x ananassa* L.). Afr J Biotechnol 7:3299–3305

Younes NA, Nassef DMT (2015) Effect of silver nanoparticles on salt tolerancy of tomato transplants (*Solanum lycopersicom* L. Mill.). Assiut J Agric Sci 46:76–85

Yruela I (2005) Copper in plants. Braz J Plant Physiol 17:145–156

Zhao GQ, Ma BL, Ren CZ (2007) Growth, gas exchange, chlorophyll fluorescence and ion content of naked oat in response to salinity. Crop Sci 47:123–131

Zhao L, Peralta-Videa JR, Rico CM, Hernandez-Viezcas JA, Sun Y, Niu G, Servin A, Nunez JE, Duarte-Gardea M, Gardea-Torresdey JL (2014) CeO(2) and ZnO nanoparticles change the nutritional qualities of cucumber (*Cucumis sativus*). J Agric Food Chem 62:2752–2759

Zhu ZJ, Wei GQ, Li J, Qian QQ, Yu JQ (2004) Silicon alleviates salt stress and increases antioxidant enzymes activity in leaves of salt-stressed cucumber (*Cucumis sativus* L.). Plant Sci 167:527–533

Chapter 13
Systems Biology Approach for Elucidation of Plant Responses to Salinity Stress

Amrita Srivastav, Tushar Khare, and Vinay Kumar

Abstract Salinity is one of the most detrimental abiotic stress factors responsible for qualitative and quantitative deterioration in global crop production. By considering the inclusive food requirements, deciphering the complex salinity-mediated responses in plants is significant to develop salt-tolerant crops with higher yields. The recent advancements in analytical technologies have made possible to investigate and generate huge amount of data. Various tools have been successfully employed to check salt-induced alterations at genome, transcriptome, proteome, and metabolome levels of plants. Omics tools in combination with computational biological analysis have provided important information, which can be used to generate biological databases as well as in silico tools. Hence this combinatorial approach of omics-systems biology is currently considered as ultimate path in plant stress physiology analysis, which aims to improve agronomic characters. The present chapter is therefore focused on omics-mediated systems biology-based investigations targeting salinity responses in plants. This chapter also highlights precise computational tools and databases to analyze transcription factors, quantitative trait loci, and small RNAs. Collectively, the chapter illustrates the system biology approach as an essential and convenient tool in salinity studies.

Keywords Salinity · Systems biology · Genomics · Proteomics · Transcriptomics · Metabolomics · Transcription factors · QTLs · Small RNAs

---

Amrita Srivastav and Tushar Khare contributed equally to this work.

A. Srivastav · T. Khare
Department of Biotechnology, Modern College of Arts, Science and Commerce, Savitribai Phule Pune University, Pune, India

V. Kumar (✉)
Department of Biotechnology, Modern College of Arts, Science and Commerce, Savitribai Phule Pune University, Pune, India

Department of Environmental Science, Savitribai Phule Pune University, Pune, India

V. Kumar et al. (eds.), *Salinity Responses and Tolerance in Plants, Volume 2*,
https://doi.org/10.1007/978-3-319-90318-7_13

Abbreviations

ABA	Abscisic acid
AP2	APETALA2
APX	Ascorbate peroxidase
bZIP	Basic region/leucine zipper motif
CAT	Catalase
CE	Capillary electrophoresis
DREB	Dehydration-responsive element-binding
dS cm^{-1}	Deci Siemens per centimeter
EC_e	Electrical conductivity of the saturated extract
ERF	Ethylene response factor
ESTs	Expressed sequence tag
FT-ICR	Fourier-transform ion cyclotron resonance
GC	Gas chromatography
GR	Glutathione reductase
GSH	Glutathione (reduced form)
ICK1	Inhibitor 1 of Cdc2 kinase
IT	Ion trap
LC	Liquid chromatography
MALDI	Matrix-assisted laser desorption
miRNA	MicroRNA
MS	Mass spectroscopy
MYB	Myeloblast
NADPH	Nicotinamide adenine dinucleotide phosphate (reduced form)
NGS	Next-generation sequencing
NMR	Nuclear magnetic resonance spectroscopy
POX	Peroxidase
QTL	Quantitative trait loci
ROI	Reactive oxygen intermediates
ROS	Reactive oxygen species
SAGE	Serial analysis of gene expression
siRNA	Small interfering RNA
SOD	Superoxide dismutase
SOS	Salt overly sensitive
TILLING	Target-induced local lesions in genome
ZAT	Zinc finger transcription factors

13.1 Introduction

Systems biology is the computational and mathematical modeling of complex biological systems. The dynamics of the genome, metabolome, transcriptome, and proteome under stress in plants can be understood well with the help of systems biology

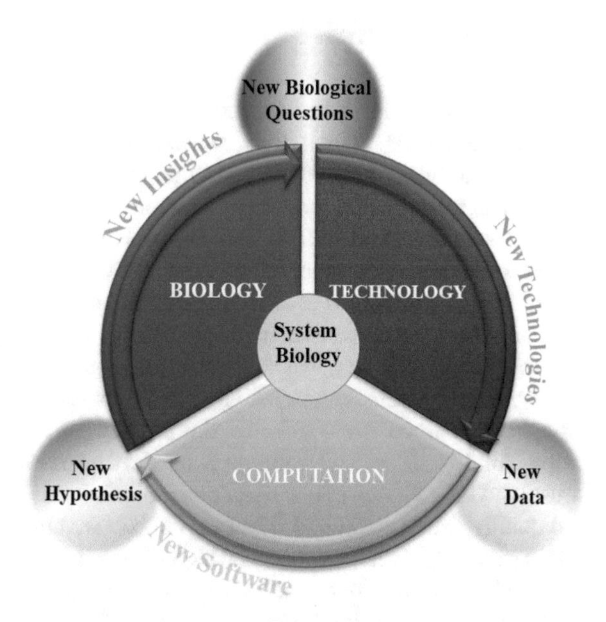

Fig. 13.1 Interfaces between biology, technology, and computation in systems biology approach

and omics (the field of studies in biology ending in -omics) together. In plants, abiotic stresses trigger specific networks of pathways involved in redox and ionic homeostasis as well as osmotic balance. These complicated networks are proposed by unifying the approaches of omics and stress systems biology. Furthermore, cross talk among these pathways is monitored by the regulation and streaming of transcripts and genes. In this chapter, we discuss systems biology and omics as a promising tool to study responses to oxidative and salinity stress in plants.

Salt stress is one of the most important factors limiting plant productivity, with salinity affecting plant physiology and metabolism at multiple levels. Systems biology has been responsible for some of the most important developments in the science of environmental sustainability. It is a holistic approach to deciphering the complexity of biological systems that starts from the understanding that the networks that form the whole of living organisms are more than the sum of their parts. It is collaborative, integrating scientific disciplines – biology, computer science, engineering, bioinformatics, physics, and others – to predict how these systems change over time and under varying conditions and to develop solutions to the world's most pressing abiotic and biotic issues.

Systems biology, ultimately, creates the potential for entirely new kinds of exploration and drives constant innovation in biology-based technology and computation (Fig. 13.1). A fundamental tenet of systems biology is in solving challenging biological problems and always requires the development of new technologies in order to explore new dimension of data space. New data types require novel analytical tools. This virtuous cycle of biology driving technology driving computation can exist only in a cross-disciplinary environment where biologists, chemists, computer scientists, engineers, mathematicians, physicists, physicians, and others can come together in teams to tackle grand challenges.

13.2 Salinity Stress

Salinity is one of agriculture's most crucial problems in large parts of the world (Gupta and Huang 2014). Rice, wheat, and pulses (legumes) are major food resources for about half of the global population. Soil salinity is a major abiotic stress, which limits rice production in about 30% of the rice-growing area worldwide (Srivastava et al. 2014). Increasing salinity progressively decreased plant height, spike length, number of spikelets, grain weight, and yield (straw and grain) in wheat plants (Kalhoro et al. 2016; Kumar and Khare 2016). Adverse effects of salts on wheat plants were associated with the accumulation of less K^+ and more Na^+ and Cl^- in their flag leaf sap, grains, and straw (Kalhoro et al. 2016). Various environmental stresses, viz., high winds, extreme temperatures, soil salinity, drought, and flood, have affected the production and cultivation of agricultural crops. Among these, soil salinity is one of the most devastating environmental stresses, which causes major reductions in cultivated land area, crop productivity, and quality (Yamaguchi and Blumwald 2005; Shahbaz and Ashraf 2013). A saline soil is generally defined as one in which the electrical conductivity of the saturated extract (ECe) in the root zone exceeds 4 dS m^{-1} (approximately 40 mM NaCl) at 25 °C and has an exchangeable sodium of 15%. The yield of most crop plants is reduced at this ECe, though many crops exhibit yield reduction at lower ECe (Jamil et al. 2012). Under the heavy pressure of global population explosion and global climate change, studying salt tolerance is of high importance.

With years of continuous exploration, some general molecular mechanisms of salt tolerance in plants have been revealed. The high-salinity environment mainly disrupts the ionic and osmotic equilibrium of cells, and as a result, genes in several pathways are activated in response to high-sodium concentration. Pathways related to ion pumps (Pons et al. 2011), calcium (White and Broadley 2003), SOS pathway (Mahajan and Tuteja 2005), ABA (abscisic acid) (Teige et al. 2004), mitogen-activated protein kinases (Zhang et al. 2006), glycine betaine (Rhodes and Hanson 1993), proline (Thiery et al. 2004), reactive oxygen species (Miller et al. 2008), and DEAD-box helicases (Vashisht and Tuteja 2006) are of significance in high-salinity environment. They play different roles in maintaining high K^+/Na^+ ratio, synthesizing and segregating ions (Tuteja et al. 2013) and controlling ion concentration (Tuteja 2007). The genes and transcription factors that encode or regulate these components often demonstrate irregular activities in a high-salinity environment. At cellular level, the most significant activities dealing with excessive salt in plants is pumping ions out of a cell to keep the ion equilibrium, while the vacuole located in the cell helps store some ions. In salt-resistant detoxifying mechanisms, especially sequestration by vacuole (Zhu 2003), many salt-tolerant genes with high level of activities in a high-salinity environment are related to vesicle, membrane, and ion transport. For example, H^+-ATPase as a proton pump on cytoplasmic vesicle maintains the ion equilibrium of the cell by pumping H^+ to the vacuole to retain pH and transmembrane proton gradient (Na^+ transporter plays an important role in maintaining high Na^+/K^+ ratio in various tissues (Ren et al. 2005; Cotsaftis et al. 2012). However, the global picture of salt-tolerant mechanisms is still unclear; for example, how ABA induces hydrogen peroxide regulation and how a plant transduces signals in response to salt stress are largely unknown. Multiple sources of data can enhance the understanding of salt tolerance. The genetic variations of different rice responses to salt stress may shed some light on the roles of various genes in salt tolerance.

The availability of plant genome sequence (Yu et al. 2002, 2005) paved the way for in-depth study of salt stress responses and tolerance mechanisms in plants. Microarray gene expression data have provided information on regulatory networks of salinity response. Kawasaki et al. (2001) analyzed the initial phase of salt stress in rice based on gene expression profiles. Huang et al. (2009) identified a zinc finger protein named DST that regulates drought and salt tolerance in rice. The expression of DREB- and ZAT- related genes might be involved in the salt tolerance of the *AtMYB102* chimeric repressor line (Mito et al. 2011). Schmidt et al. (2012) examined TFs like heat-shock factors (HSFs) in response to salinity environment, and they characterized *OsHsfC1b* as playing a role in ABA-mediated salt stress tolerance in rice. Nevertheless, these studies were mainly explored single-gene or some isolated genes approach and lack systems-level understanding of the global molecular mechanism of salt tolerance given that salt resistance reacts in a coordinated and effective manner.

In view of these findings, systems-level study is an apparent choice to fill the gap between isolated genes and the global (whole-genome) mechanism of salt tolerance. Among tens of thousands of genes in microarray data, it is challenging to choose the set of genes that are most relevant to salt tolerance (Blum and Langley 1997; Guoyon and Elisseeff 2003). Biologists often use a volcano plot method, which reflects both fold of change and its statistical significance at the same time in a heuristic fashion (Cui and Churchill 2003). However, such methods may not be sufficient to discover complex relationships between genes and a certain phenotype, trait, or condition (Liang et al. 2011).

13.3 The Omics Technologies and Systems Biology

There is a noteworthy advancement in the accuracy as well as scale of the analysis in the field of molecular systems biology in the past two decades (El-Metwally et al. 2014). Though the basic perspectives of molecular biology (reductionistic) and system biology (holistic) are conflicting, the current advanced research settings allow both the wings to complement each other which provides an easy path to study stress-mediated responses in plants (Duque et al. 2013). The analysis of every component playing vital role during fundamental procedure of the cellular metabolism (replication, transcription, and translation) provides the exhaustive knowledge about every aspect of cellular functioning. Therefore, cumulative genomics, proteomics, transcriptomics, and metabolomics studies can be proven as an ideal source of information to decipher the queries using systems biology approaches. Hence these omics approaches are considered as prime path to interpret the complex plant responses in stress environment.

13.3.1 Genomics

Structural and functional genomics addresses the issues related with basic structure and the respective function of the genes and/or their products. The structural genomic analysis provides the high-resolution characterization of the genome (up to full gene length) which can be subsequently used to develop tools for genomic predictions and annotations. Information related to the structure, location, and feature of the genome offers valuable insights to manipulate the gene of interest to develop plants with desirable agronomic traits. The current trend to obtain the information regarding genomics is whole-genome sequence, which makes it pretty much possible to gain knowledge about the total sequence of any plant genome. The completion of *Arabidopsis* genome is considered as a milestone in genomic research (The Arabidopsis Genome Initiative 2000). Since then, genomic sequence of many plants including economically important glycophytes as well as halophytes (Table 13.1) is available, providing valuable information about their genome. In this manner, structural genomics along with modern next-generation sequencing (NGS)-mediated shotgun sequencing approach greatly contributes in development of molecular markers which are applicable in polymorphism studies. On the other hand, functional genomics describes gene on the basis of its function by retrieving information from sequence data, genome mapping, and functional characterization of gene. The tools such as serial analysis of gene expression (SAGE), target-induced local lesions in genome (TILLING), and array-based hybridization are used in functional genomics studies. Another branch of genomics, known as comparative genomics, uses the information from conserved regions between two closely related species. The comparative mapping and analysis of the genomes provide genomic similarities and dissimilarities providing vital information about genetic constitution of plant (Mohanta et al. 2015a, b, c).

13.3.2 Proteomics

Proteome defines total profile of the expressed proteins with respect to the provided environment in the organism. The proteomics involves techniques for qualitative and quantitative characterization of protein profiles (Tyers and Mann 2003). Apart from the several types of available proteome, whole proteome and phosphoproteome are the choices by the researchers in plant stress response studies (Helmy et al. 2011, 2012a, b; Kumar et al. 2017a). The tools of choice to obtain protein profile commonly involve mass spectrometry coupled with many advancements such as matrix-assisted laser desorption (MALDI), liquid chromatography (LC), ion trap (IT), etc. Such analysis provides the possible peptide fingerprint referred as MS spectra of a protein of interest. Multiple-condition proteome profile reveals quantitative data informing the differentially expressed stress-associated proteins pointing toward their contribution in stress-mediated responses (Liu et al. 2015a, b).

Table 13.1 Details about the genome of some selected plants

Plant	Genome size	GC (%)	Genes	Proteins
Eleusine coracana	1.19 Gb	44.80	85,243	126,312
Hordeum vulgare	5.3 Gb	–	39,734	248,180
Oryza sativa subsp. indica	426.337 Mb	43.73	40,745	88,438
Oryza sativa subsp. japonica	374.423 Mb	43.58	35,679	97,751
Sorghum bicolor	732.2 Mb	–	34,129	47,121
Triticum aestivum	17 Gb	–	217,907	273,739
Zea mays	2.13 Gb	46.91	47,800	58,291
Arabidopsis thaliana	135 Mb	36.05	27,655	48,456
Beta vulgaris	566.55 Mb	37.30	27,429	32,874
Brassica napus	848.2 Mb	37.80	101,040	101,040
Brassica oleracea	488.954 Mb	37.33	53,670	56,687
Cajanus cajan	529.971 Mb	33.71	31,549	31,549
Cicer arietinum	530.894 Mb	32.67	27,758	33,107
Glycine max	978.972 Mb	35.12	58,843	88,647
Glycine soja	863.568 Mb	34.50	50,399	50,399
Gossypium hirsutum	2.18 Gb	34.90	78,043	90,927
Helianthus annuus	3.02 Gb	38.89	57,708	52,191
Hordeum vulgare	4.6 Gb	–	39,734	248,180
Nicotiana tabacum	3.64 Gb	39.20	69,595	84,255
Phaseolus vulgaris	521.077 Mb	36.22	28,134	32,720
Raphanus sativus	426.614 Mb	37.20	58,031	61,216
Solanum lycopersicum	900 Mb	34.76	34,727	34,727
Solanum tuberosum	723 Mb	35.60	39,028	56,215
Theobroma cacao	346 Mb	34.99	29,452	–
Vigna radiata	463.638 Mb	34.40	29,100	35,143
Vitis vinifera	486.197 Mb	35.03	29,863	41,208

Due to specificity of the phosphorylation event as per the associated abiotic stress condition, phosphoproteins provide crucial information pathways activated during specific environment. Hence, via phosphoproteomics, signaling pathways and associated (phospho) proteins can be recognized (Zhang et al. 2014a, b; Kumar et al. 2017a).

13.3.3 *Transcriptomics*

The RNA expression profile of an organism under a specific condition is recognized as transcriptome. Compared to the genome, transcriptome is considered as highly dynamic in nature. Therefore, capturing the transcriptome in spatial and temporal bases is the main aim of the transcriptomic analysis (Duque et al. 2013; El-Metwally et al. 2014). This is achieved through microarrays, NGS-mediated high-throughput

RNA sequencing, serial analysis of gene expression (SAGE), etc. (Kawahara et al. 2012; De Cremer et al. 2013). Comparative transcriptome helps finding candidate genes contributing to stress tolerance. The availability of the online databases and tools hence allows now to achieve novel genome-wide analysis of plant stress responses and tolerance (Mochida and Shinozaki 2011; Jogaiah et al. 2013).

13.3.4 Metabolomics

A large number of plant metabolites with diverse structures and abundance are formed by plants throughout their life cycle. These entities are categorized as primary and secondary metabolites depending on their function. The metabolic analysis of the plants hence involves comprehensive identification, quantitation, as well as localization of various metabolites (Hong et al. 2016). To achieve this goal, series of integrated techniques such as nondestructive NMR (nuclear magnetic resonance spectroscopy), mass spectroscopy (MS) combined with gas chromatography (GC), LC, capillary electrophoresis (CE), and Fourier-transform ion cyclotron resonance (FT-ICR) are applied (Okazaki and Saito 2012; Khakimov et al. 2014). The ultimate information generated through metabolomics provides understandings about metabolic fluxes and metabolite concentrations in the plants present during the specific abiotic stress condition (Toubiana et al. 2013; Kleessen and Nikoloski 2012).

Collectively these four technologies, constituting the principal omics technologies, generate humongous data with a great speed. These all techniques are highly reliant on bioinformatics and other computational tools. Therefore mutually, all techniques are applied during systems biology approach to draw a suitable inference against the abiotic stress-mediated responses in plants (Fig. 13.2).

13.4 Analysis of Key Factors Involved in Salinity: Systems Biology Approach

13.4.1 Transcription Factors

Crop productivity is affected significantly by abiotic stress like salinity. However water stress responses are multigenic and quantitative and alter the phenotypic patterns of plants. Salinity stress occurs in conjunction with high temperature and high light intensities. Systems biology is a multidisciplinary approach with an integration of omics technology which along with defining the system also provides new gene targets for plant and agricultural biotechnology. Transcription factors are major players in water stress signaling; salinity stress and some constitute major hubs in the signaling webs. The main TFs in this network include NAC, MYB, bZIP, bHLH, eRF, and WRKY transcription factors. WRKY and AP2 transcription factors play an important role in abiotic stress signaling networks, and systems biology

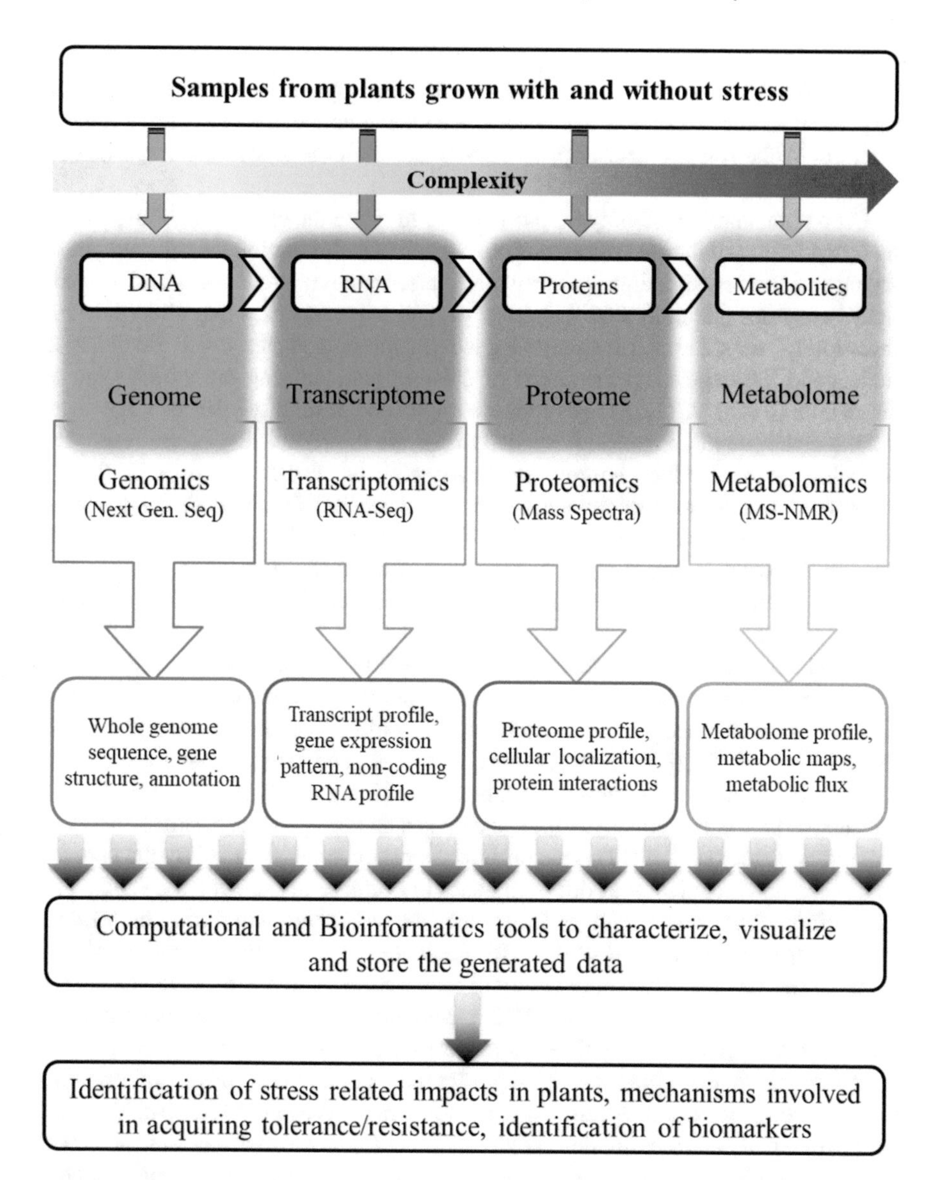

Fig. 13.2 Schematic summery representing a general flow for analysis of stress-mediated changes in plants using omics-systems biology approaches

approaches are getting significant to define their role in the signaling network. Applying systems biology approaches, there are now many transcriptomic analyses and promoter analyses that involve various transcription factors in abiotic stress defense. In addition, reports on nuclear proteomics have elucidated transcription factor proteins that are upregulated at the protein level by subjecting plants to water stress. Interactomics identifies different classes of transcription factor-interacting proteins. Systems biology is trying to make connections between metabolomics, transcription factors, promoters, biosynthetic pathways, fluxes, and downstream responses. As more levels of the systems are characterized, a more detailed understanding of the roles of transcription factors in salinity responses in crops will be obtained.

Crop plants utilize phenotypic plasticity to mitigate the effects of salinity stress by upregulating of various stress-responsive genes including ion transporters, transcriptional factors, signaling pathway modifiers, osmolyte production, and antioxidative enzymes (Ge et al. 2013). Numerous pathway responses that altered due to the salinity mark the salt-responsive genes in tolerant plants which facilitate to understand the expression prototype of existing genes during the whole span of stress (Darko et al. 2017). Many genes are implicated in salinity tolerance; however, a more elaborate investigation is needed to resolve the complexity of the responses to salinity stress at the genomic level (Abogadallah 2010; Darko et al. 2017). Unification of systems biology and omics could specifically elucidate the genomic and metabolic responses of cells in a precise manner, providing better insights into various interconnecting signaling processes that regulate cellular homeostatic machinery during stress. High-salinity level creates ionic imbalance and hypertonic effects, which inhibit crop yield at the molecular, biochemical, and physiological levels, either directly or indirectly (Abogadallah 2010). Moreover, salinity stress is predicted to hamper photosynthesis, enhance photorespiration, deactivate enzymes, increase ROS damage, and ultimately lead to chloroplast damage (Kumar et al. 2017c; Khare et al. 2015). Hence, plants have developed various processes, including salt exclusion and compartmentalization (Zhang et al. 2016), to affect the successive biological and physiological changes that mitigate the harmful effects of salt stress (Darko et al. 2017). This phenotypic plasticity is governed by the upregulation and downregulation of different genes to decrease or protect from ROS damage, reformulate osmotic and ionic balance, and resume growth during high levels of salinity stress (Liu et al. 2014). To elucidate the molecular dynamics of salt tolerance and increase the productivity of crops, substantial efforts have been made to develop genetic model systems. ROS scavenging is an efficient means to mitigate oxidative damage and manipulate the expression of associated genes such as those encoding superoxide dismutase (SOD), ascorbate peroxidase (APX), and glutathione reductases (GRs), thus providing salt tolerance (Abogadallah 2010). Vacuolar compartmentalization maintains Na^+/K^+ and thereby also enhances salt tolerance. Zhang et al. (2016) reported three pathways governing the salinity counter mechanisms in cotton. A salinity counter model for crop plants like wheat demonstrates how various genetic determinants are regulated via different pathways, ultimately leading to cellular homeostasis (Shah et al. 2017). The cell membrane is equipped with proteins serving as Na^+ receptors, which receive stress signals and elicit the production of signaling entities like Ca^{2+}, ROS, and hormones (Liu et al. 2014). The elevated levels of these entities trigger three pathways that ultimately activate salt overly sensitive 1 (SOS1) pathway to pump out Na^+ from the cytosol. All the genes determining these pathways are upregulated. However, in this salinity counter mechanism, the photosynthesis process is inhibited, as RuBisCO and NADPH are deactivated (Zhao et al. 2014. However, an excess of Na^+ hinders the uptake of K^+ and cytosolic enzymes. The activity of Na^+ and K^+ transporters and H^+ pumps and SOS2 and SOS3 protein kinase pathways controls the secretion of toxic Na^+ in the cell. Therefore, salt-tolerant genotypes resume growth at a slow rate when subjected to salt stress, owing to regulation by hormones and cell division-related genes. Increased deposition of ABA in response to salt stress is thought to upregulate cyclin-dependent protein kinase inhibitor (ICK1), which inhibits cell division (Wilkinson and Davies 2010; Lee and Luan 2012). Hence, these interconnected features constitute a breeding target for breeders to improve the potential range of adaptability of their germplasm to salt stress. The salinity tolerance of crop plants such as wheat is a multigenic trait, which is more complicated than in the model plant *Arabidopsis*, in addition to a high sensitivity to salinity. Therefore, it is logical to conclude that wheat employs a more complicated system in response to salinity than *Arabidopsis*. Various genes have been reported to play a significant role in response to salt stress in wheat. For example, SRO (Similar to Red-One) mediates ROS deposition and scavenging by regulating the expression prototype of NADPH dehydrogenase and NADPH oxidase, together with GSH peroxidase and ascorbate-GSH. Dynamic expression of these genes authenticates their inevitability and sufficiency in enhancing salt tolerance (Liu et al. 2014; Zhao et al. 2014). Nevertheless, the processes mediating the genome-wide gene expression in wheat to control the deleterious effects imposed by salinity are still not completely understood. Moreover, it has been reported, using a microarray approach, that out of 32,000 detected ESTs in wheat, 19% were either up-or downregulated (Kawaura et al. 2006, 2008). The adaptability of plants to unfavorable environments has been also explained through polyploidization (Dubcovsky and Dvorak 2007). For instance, tetraploid *Arabidopsis* has a greater tolerance to salt stress, via the homeostasis of K^+ and Na^+, than diploid *Arabidopsis*. However, the molecular mechanisms determining adaptability to environmental stresses via this route are still poorly understood. It has recently been hypothesized that the expression of homologous genes is responsible for increased tolerance to salt stresses in polyploid plants. For example, in allopolyploid cotton (*Gossypium hirsutum*), one copy of the alcohol dehydrogenase A gene (AdhA) is upregulated under cold conditions, while the other responds to water stress (Liu and Adams 2007). Moreover, transcriptomic studies have reported that allohexaploid wheat manifests intensive partitioned expression of homologs in response to drought and heat stress (Liu et al. 2015a, b).

The informatics analysis and computational tools as well as associated methods and algorithms in biology make the analysis of biological data accurate, fast, human error-free, and easily reproducible (Orozco et al. 2013). Therefore, many bioinformatics tasks became indispensable in biological research in general and plant stress multi-omics research in particular which involves tasks like genome sequence assembly (El-Metwally et al. 2013, 2014), sequence alignment, and gene prediction (Stanke and Morgenstern 2005); peptides carry out sequence identification (MS-spectra interpretation tools), gene and protein function prediction (Falda et al. 2012; Yachdav et al. 2014), DNA-protein and protein-protein interaction prediction

(Franceschini et al. 2013), interaction and regulatory networks analysis (Chaouiya 2012), and several other essential tasks (Polpitiya et al. 2008; Henry et al. 2014).

Special types of visualization tools are required for the analysis and processing of data generated by modern analytical and experimental instruments such as genome sequencers and mass spectrometers. This resulted in development of several tools to visualize the biological data and results in a manner that would maximize the utility of the data. The genomic data visualizing tools include GBrowse, UCSC Genome Browser, and Integrated Genome Viewer (IGV) (Pang et al. 2014); proteomics data visualization tools such as PRIDE Inspector and ConPath (Kim et al. 2008; Wang et al. 2012); proteogenomics data or multi-omics data visualization tools such as PGFeval, 3Omics, and Peppy (Helmy et al. 2011, 2012a; Kuo et al. 2013; Risk et al. 2013); metabolomics visualization tools such as MultiExperiment Viewer (MeV) and network visualization tools such as Cytoscape and its associated web versions Cytoscape.js and Cytoscape web.

More knowledge can be gathered through high-throughput data by combining different datasets or by applying various types of analyses into one large-scale comparative analysis. However, this requires the data and results to be sustainably available and accessible to the researchers. Thus, for depositing and storing the biological data and results, several types of databases are available online. The databases are stored in different forms, ranging from those that store plant information such as classification, growth, production, and geographical distribution, plant genomic information (Zhao et al. 2014), plant transcriptomic information, plant proteomic information, plant proteogenomic information (Helmy et al. 2011, 2012a), and plant metabolomic information. Furthermore, some databases are specialized in storing and maintaining plant stress resistance and tolerance information such as STIFDB2, the *Arabidopsis* stress-responsive gene database, QlicRice, and the fungal stress-response database (FSRD). Several tools provide more than one of the abovementioned features such as data analysis and visualization (Helmy et al. 2011) or data visualization and storage (Helmy et al. 2012a). Many tools also provide these services for multi-omics data and results (Kuo et al. 2013).

13.4.2 QTLs

Susceptible and resistant varieties of a plant can be compared with the Gene Expression Omnibus (GEO) and volcano plots analyses, using microarray data. The genes responsive to stress like salt tolerance can be enlisted, and overexpressed genes can be characterized using agriGO (Du et al. 2010), a plant-specific GO term enrichment analysis. During salinity stress molecular alterations at genomic level take place. The overrepresented genes are mainly responsive to iron binding, cation binding, ion binding, and heme binding. The significant behavior of these genes in oxidoreductase activity is related to electron transport in complex chemical reactions that balances the charges during ion transport. The oxidoreductase activity is related to reactive oxygen intermediates (ROI) that are produced in response to oxidative stress because of water deficit during salinity stress (Miller et al. 2008). ROI seriously disrupt the normal metabolism through oxidative damage on lipids, proteins, and nucleic acids. The increased oxidoreductase activity is depicted with enhanced levels of the antioxidative enzymes like catalase (CAT), APX, guaiacol peroxidase (POD), GR, and SOD under salt stress in plants. Biochemicals like proline and glycine betaine also accumulate under stress; they are correlated with osmotic adjustment to endure plant salinity tolerance (Vinocur and Altman 2005). Proline also plays an important role in scavenging free radicals, stabilizing subcellular structures, and buffering cellular redox potential under stresses. Another nitrogen compounds like polyamines are also synthesized under salt tolerance (Gupta et al. 2012a, b). Therefore, it is speculated that these nitrogen-containing compounds are synthesized in cellular nitrogen compound metabolic process. In cellular components, according to the gene enrichment analysis, most of the chosen genes are related to the vesicle and membrane, which are involved with the detoxifying mechanism of salt-resistant genotypes, especially in sequestration by vacuole. It is plausible to infer that some of these chosen proteins on membranes act as transporting ions outside the cell or to the vacuole to maintain pH, transmembrane proton gradient (Gupta and Huang 2014), and high K^+/Na^+ ratio (Ren et al. 2005). In order to assess the performance of improved volcano plot, Microarray-QTL test by using the QTL information is developed as a criterion to evaluate the reliability of chosen genes.

Different tissues of a plant are studied to understand the salt-tolerant mechanism. There is a great challenge in determining feature selection on small sample but high-dimension data. Classical statistical feature selection methods, such as t-test, assume the samples follow some specific distribution as its hypothesis; however, the limited number of samples narrows the usage of these statistical methods. From the feature selection prospective, the volcano plot method uses two dimensions of fold change and t-test p-value to select genes in microarray analysis. This is a fast, simple, and widely used method. Although, the fold change of each differential expression gene does not necessarily reveal the nature of biological meaning. In cases, using machine-learning methods could be a good alternative in microarray analysis. The improved volcano plot method applying specific criteria like MergeValue based on an SVM-RFE procedure could enhance the performance. The improved method also used a bootstraps approach to make the feature selection more robust. The QTL information can be obtained from QTLs zone in the genome using Gramene QTL database (http://www.gramene.org/qtl/), which further helps to identify genes related to salinity response. The QTL information in cooperation with transcription profiles to identify genes related to salinity response. For a given QTL, there may be 25–30 genes per cM (270 kbp in rice) (Khurana and Gaikwad 2005). Given so many possible genes in a QTL region associated to a phenotype, the proposed Microarray-QTL test, which used the same mechanism as GO term enrichment analysis, could help evaluate the relative relevance of these QTL genes to the phenotype quantitatively. Furthermore, predicted protein-protein interactions, protein structure prediction, and gene regulatory motif analysis in studying potential genes related to salt tolerance can also be studied. Such a systems approach is powerful in providing high-confidence predictions of salt-tolerant genes.

The study may provide richer and more concise predictions than a study done by Cotsaftis et al. (2011), which only used the expression level of the gene probes in transcript profiling to predict salt-tolerant genes. Walia et al. (2006) summarized the following components in the salinity response based on their microarray study: (1) adaptive response, (2) nonadaptive response, (3) response to salt injury, and (4) heritable responses conferring tolerance. The involvement of phosphorylation in abiotic stress has been actively studied in recent years in addition to its relationship to salinity stress (Chitteti and Peng 2007), cold stress (Komatsu et al. 1999), and heat stress (Chen et al. 2011). Key proteins can be predicted and mapped to QTLs, which means researchers could conduct experiments to clone and validate these genes. Computational studies can be applied to study other traits and several other species of plants under abiotic stress.

13.4.3 Small RNA Analysis

The role of small-noncoding RNA, including microRNAs (miRNAs) and short-interfering RNAs (siRNAs), is well described in the context of abiotic stress interactions in plants (Khare et al. 2018). The previous method for identification and characterization of small RNAs was via direct cloning strategy. Such identification procedures are tedious tasks due to the multiple existence, smaller size, and limitless molecular-processing mechanisms (Shriram et al. 2016). The cost reduction for the sequencing techniques, along with advancements in sequencing techniques, has enormously assisted in rapid growth of small RNA-related identification and functional annotations (Kumar et al. 2017b). Therefore, currently numerous databases and analytical tools regarding the same are available for small RNA analysis (Rosewick et al. 2013; Tripathi et al. 2015; Shriram et al. 2016).

RNAcentral (http://rnacentral.org/), an inclusive database, has compiled noncoding RNA sequences providing a single port to access noncoding RNA information from many organisms. The database facilitates the information accession by providing browsers and searching tools, in association with other compatible databases (RNAcentral Consortium 2017). The instructive database, especially representing noncoding RNAs, *NONCODE*, (http://www.noncode.org/) incorporated material from 18 unlike organisms including *Arabidopsis* (Zhao et al. 2016). The database provides additional advantages including literature support, high-quality data selection through interface, and conservation annotation. PNRD (http://structuralbiology.cau.edu.cn/PNRD/index.php) represents a collective database for noncoding RNA archives including plant sources. This set includes information for lncRNAs, tRNAs, rRNAs, ratiRNAs, snRNAs, and snoRNAs (Yi et al. 2015). Kozomara and Griffith-Jones (2014) have developed a depository of microRNA sequences, *miRBase*. This collection harbors miRNA sequences along with their annotations from 223 organisms (including 73 plants) (http://www.mirbase.org/). Other small RNA-related databases includes *tasiRNAdb* (http://bioinfo.jit.edu.cn/tasiRNADatabase/) with 583 tasiRNAS regulatory pathways from 18 plants and *siRNAmod* (http://crdd.osdd.net/servers/sirnamod) harboring experimentally authentic chemically altered siRNAs and their biological activities (Zhang et al. 2014a, b; Dar et al. 2016).

Along with these databases, many tools for analysis of small-noncoding RNAs are available. User-friendly, algorithm-based software *the UEA sRNA workbench* had been created by Stocks et al. (2012) (http://srna-workbench.cmp.uea.ac.uk/). It is a downloadable toolkit for analysis of single/multiple samples of small RNAs with available facilities for ta-si predictions, *PAREsnip* for target prediction, and *miRCat* and *miRProf* for expression level predictions. *TAPIR* is also a target prediction tool used to analyze plant miRNAs (http://bioinformatics.psb.ugent.be/webtools/tapir/) (Bonnet et al. 2010). Another tool *plantDARIO* (http://plantdario.bioinfo.uni-leipzig.de/) can be applied for mapping quality control analysis for small RNA data and expression analysis and prediction of small-noncoding RNAs (Patra et al. 2014). A complete database incorporating all the tools for small RNA analysis is developed by Chen et al. (2017), *miRToolsGallery* (http://www.mirtoolsgallery.org/miRToolsGallery/), which is a free database used to search various available tools for small RNA analysis. This versatile database harbors more than 900 tools which can be effectively utilized for specific analytical purposes.

13.5 Conclusion

Extensive studies are underway since many years to decipher the various plant pathways arbitrated to salinity. Due to the latest technological advancements, there is positive increment in speed of analysis as well as quality and quantity of the generated data. The systems biology-based approach to study plants behavior under saline conditions has been proven effective, as the approach has successfully generated ample valuable information from many plants. The systems biology provides an incessant analytical cycle as new data as well as tools are continuously generated. Systems biology has been successfully applied for analysis of major molecular entities with vital role in abiotic stress responses. The main provinces of systems biology, namely, omics, data integration, and data analysis, must be addressed well to decrypt the problems in plant stress biology. This fact also indicates the necessity of constant advancement in developing the bioinformatics tools. Though systems biology is currently serving as an ideal way to study plant stress physiology, there is still a vast scope of improvement. Still this approach has accelerated the process of crop adaptation to improve quality and quantity of agronomic products.

Acknowledgments VK gratefully acknowledge the financial support from Science and Engineering Research Board (SERB), Department of Science and Technology (DST), Government of India (grant number EMR/2016/003896). The authors acknowledge the use of facilities created under DBT Star College and DST-FIST Schemes implemented at Modern College, Ganeshkhind, Pune.

References

Abogadallah GM (2010) Insights into the significance of antioxidative defense under salt stress. Plant Signal Behav 5:369–374. https://doi.org/10.4161/psb.5.4.10873

Blum A, Langley P (1997) Selection of relevant feature and examples in machine learning. Artif Intell 97:245–271

Bonnet E, He Y, Billiau K, Vande PY (2010) TAPIR, a web server for the prediction of plant microRNA targets, including target mimics. Bioinformatics 26:1566–1568. https://doi.org/10.1093/bioinformatics/btq233

Chaouiya C (2012) Logical modelling of gene regulatory networks with GINsim in methods in molecular biology. In: Clifton NJ (ed). Humana Press, Print ISBN 1940-6029

Chen X, Zhang W, Zhang B, Zhou J, Wang Y, Yang Q, Ke Y, He H (2011) Phosphoproteins regulated by heat stress in rice leaves. Proteome Sci 9:37

Chen L, Heikkinen L, Wang CL, Yang Y, Knott KE, Wong G (2017) miRToolsGallery: a microRNA bioinformatics resources database portal. Available at http://www.mirtoolsgallery.org/miRToolsGallery/

Chitteti BR, Peng Z (2007) Proteome and phosphoproteome differential expression under salinity stress in rice (*Oryza sativa* L.) roots. J Proteome Res 6:1718–1727

Cotsaftis O, Plett D, Johnson A, Walia H, Wilson C, Ismail AS, Close TJ, Tester M, Baumann U (2011) Root-specific transcript profiling of contrasting rice genotypes in response to salinity stress. Mol Plant 4:25–41

Cotsaftis O, Plett D, Shirley N, Tester M, Hrmova M (2012) A two-staged model of Na^+ exclusion in rice explained by 3D modeling of HKT transporters and alternative splicing. PLoS One 7(7):e39865

Cui X, Churchill G (2003) Statistical tests for differential expression in cDNA microarray experiments. Genome Biol 4:210

Dar SA, Thakur A, Qureshi A, Kumar M (2016) siRNAmod: a database of experimentally validated chemically modified siRNAs. Sci Rep 6:20031. https://doi.org/10.1038/srep20031

Darko E, Gierczik K, Hudák O, Forgó P, Pál M, Türkösi E et al (2017) Differing metabolic responses to salt stress in wheat-barley addition lines containing different 7H chromosomal fragments. PLoS One 12:e0174170. https://doi.org/10.1371/journal.pone.0174170

De Cremer K, Mathys J, Vos C, Froenicke L, Michelmore RW, Cammue B, De Coninck B (2013) RNAseq-based transcriptome analysis of *Lactuca sativa* infected by the fungal necrotroph Botrytis cinerea. Plant Cell Environ 36:1992–2007. https://doi.org/10.1111/pce.12106

Du Z, Zhou X, Ling Y, Zhang Z, Su Z (2010) agriGO: a GO analysis toolkit for the agricultural community. Nucleic Acids Res 38:W64–W70

Dubcovsky J, Dvorak J (2007) Genome plasticity a key factor in the success of polyploid wheat under domestication. Science 316:1862–1866. https://doi.org/10.1126/science.1143986

Duque AS, de Almeida AM, da Silva AB, da Silva JM, Farinha AP, Santos D, Fevereiro P, de Sousa AS (2013) Abiotic stress responses in plants: unraveling the complexity of genes and networks to survive. In: Abiotic stress-plant responses and applications in agriculture 2013. InTech. https://doi.org/10.5772/45842

El-Metwally S, Hamza T, Zakaria M, Helmy M (2013) Next-generation sequence assembly: four stages of data processing and computational challenges. PLoS Comput Biol 9:e1003345. https://doi.org/10.1371/journal.pcbi.1003345

El-Metwally S, Ouda OM, Helmy M (2014) Next generation sequencing technologies and challenges in sequence assembly. Springer, New York

Falda M, Toppo S, Pescarolo A et al (2012) Argot2: a large scale function prediction tool relying on semantic similarity of weighted Gene Ontology terms. BMC Bioinformatics 13(Suppl 4):S14. https://doi.org/10.1186/1471-2105-13-S4-S14

Franceschini A, Szklarczyk D, Frankild S et al (2013) STRING v9.1: protein-protein interaction networks, with increased coverage and integration. Nucleic Acids Res 41:D808–D815. https://doi.org/10.1093/nar/gks1094

Ge P, Hao P, Cao M, Guo G, Lv D, Subburaj S et al (2013) iTRAQ based quantitative proteomic analysis reveals new metabolic pathways of wheat seedling growth under hydrogen peroxide stress. Proteomics 13:3046–3058. https://doi.org/10.1002/pmic.201300042

Guoyon I, Elisseeff A (2003) An introduction to variable and feature selection. J Mach Learn Res 3:1157–1182

Gupta B, Huang B (2014) Mechanism of salinity tolerance in plants: physiological, biochemical, and molecular characterization. Int J Genomics 2014:701596. https://doi.org/10.1155/2014/701596

Gupta K, Dey A, Gupta B (2012a) Polyamines and their role in plant osmotic stress tolerance. In: Tuteja N, Gill SS (eds) Climate change and plant abiotic stress tolerance. Wiley-VCH, Weinheim, pp 1053–1072

Gupta K, Gupta B, Ghosh B, Sengupta DN (2012b) Spermidine and abscisic acid-mediated phosphorylation of a cytoplasmic protein from rice root in response to salinity stress. Acta Physciol Plant 34:29–40

Helmy M, Tomita M, Ishihama Y (2011) OryzaPG-DB: rice proteome database based on shotgun proteogenomics. BMC Plant Biol 11:63. https://doi.org/10.1186/1471-2229-11-63

Helmy M, Sugiyama N, Tomita M, Ishihama Y (2012a) The rice proteogenomics database oryza PG-DB: development, expansion, and new features. Front Plant Sci 3:65. https://doi.org/10.3389/fpls.2012.00065

Helmy M, Tomita M, Ishihama Y (2012b) Peptide identification by searching large-scale tandem mass spectra against large databases: bioinformatics methods in proteogenomics. Gene Genome Genomics 6:76–85

Henry VJ, Bandrowski AE, Pepin A-S et al (2014) OMICtools: an informative directory for multi-omic data analysis. Database (Oxford) 2014:bau069. https://doi.org/10.1093/database/bau069

Hong J, Yang L, Zhang D, Shi J (2016) Plant metabolomics: an indispensable system biology tool for plant science. Int J Mol Sci 17:767. https://doi.org/10.3390/ijms17060767

Huang X, Chao D, Gao J, Zhu M, Shi M, Lin H (2009) A previously unknown zinc finger protein, DST, regulates drought and salt tolerance in rice via stomatal aperture control. Genes Dev 23:1805–1817

Jamil M, Bashir S, Anwar S, Bibi S, Bangash A, Ullah F, Rha ES (2012) Effect of salinity on physiological and biochemical characteristics of different varieties of rice. Pak J Bot 44:7–13

Jogaiah S, Govind SR, Tran L-SP (2013) Systems biology-based approaches toward understanding drought tolerance in food crops. Crit Rev Biotechnol 33:23–39. https://doi.org/10.3109/07388551.2012.659174

Kalhoro NA, Rajpar I, Kalhoro SA, Ali A, Raza S, Ahmed M, Kalhoro FA, Ramzan M, Wahid F (2016) Effect of salts stress on the growth and yield of wheat (*Triticum aestivum* L.). Am J Plant Sci 7:2257

Kawahara Y, Oono Y, Kanamori H, Matsumoto T, Itoh T, Minami E (2012) Simultaneous RNA-seq analysis of a mixed transcriptome of rice and blast fungus interaction. PLoS One 7:e49423. https://doi.org/10.1371/journal.pone.0049423

Kawasaki S, Borchert C, Deyholos M, Wang H, Brazille S et al (2001) Gene expression profiles during the initial phase of salt stress in rice. Plant Cell 13:889–905

Kawaura K, Mochida K, Yamazaki Y (2006) Transcriptome analysis of salinity stress responses in common wheat using a 22k oligo-DNA microarray. Funct Integr Genomics 6:132–142. https://doi.org/10.1007/s10142-0050010-3

Kawaura K, Mochida L, Ogihara Y (2008) Genome-wide analysis for identification of salt-responsive genes in common wheat. Funct Integr Genomics 8:277–286. https://doi.org/10.1007/s10142-008-0076-9

Khakimov B, Bak S, Engelsen SB (2014) High-throughput cereal metabolomics: current analytical technologies, challenges and perspectives. J Cereal Sci 59:393–418

Khare T, Kumar V, Kishor PK (2015) Na^+ and Cl^- ions show additive effects under NaCl stress on induction of oxidative stress and the responsive antioxidative defense in rice. Protoplasma 252:1149–1165

Khare T, Shriram V, Kumar V (2018) RNAi technology: the role in development of abiotic stress tolerant plants. In: Wani SH (ed) Biochemical, physiological and molecular aveniques for combating abiotic stress tolerance in plants. Elsevier. https://doi.org/10.1016/B978-0-12-813066-7.00008-5

Khurana P, Gaikwad K (2005) The map based sequence of the rice genome. Nature 436:793–800

Kim P-G, Cho H-G, Park K (2008) A scaffold analysis tool using mate-pair information in genome sequencing. J Biomed Biotechnol 2008:675741. https://doi.org/10.1155/2008/675741

Kleessen S, Nikoloski Z (2012) Dynamic regulatory on/off minimization for biological systems under internal temporal perturbations. BMC Syst Biol 6:16

Komatsu S, Karibe H, Hamada T, Rakwal R (1999) Phosphorylation upon cold stress in rice (*Oryza sativa* L.) seedlings. Theor Appl Genet 98:1304–1310

Kozomara A, Griffiths-Jones S (2014) miRBase: annotating high confidence microRNAs using deep sequencing data. Nucleic Acids Res 42:D68–D73. https://doi.org/10.1093/nar/gkt1181

Kumar V, Khare T (2016) Differential growth and yield responses of salt-tolerant and susceptible rice cultivars to individual (Na^+ and Cl^-) and additive stress effects of NaCl. Acta Physiol Plant 38:170

Kumar V, Khare T, Sharma M, Wani SH (2017a) Engineering crops for future: a phosphoproteomics approach. Curr Protein Pept Sci 19:413–426. https://doi.org/10.2174/1389203718666170209152222

Kumar V, Khare T, Shriram V, Wani SH (2017b) Plant small RNAs: the essential epigenetic regulators of gene expression for salt-stress responses and tolerance. Plant Cell Rep 37:61–75. https://doi.org/10.1007/s00299-017-2210-4

Kumar V, Khare T, Sharma M, Wani SH (2017c) ROS-induced signaling and gene expression in crops under salinity stress. In: Reactive oxygen species and antioxidant systems in plants: role and regulation under abiotic stress. Springer, Singapore, p 159–184

Kuo T-C, Tian T-F, Tseng YJ (2013) 3Omics: a web-based systems biology tool for analysis, integration and visualization of human transcriptomic, proteomic and metabolomic data. BMC Syst Biol 7:64. https://doi.org/10.1186/1752-0509-7-64

Lee SC, Luan S (2012) ABA signal transduction at the crossroad of biotic and abiotic stress responses. Plant Cell Environ 35:53–60. https://doi.org/10.1111/j.13653040.2011.02426.x

Liang Y, Zhang F, Wang J, Joshi T, Wang Y, Xu D (2011) Prediction of drought-resistant genes in *Arabidopsis thaliana* using SVM-RFE. PLoS One 6:e21750

Liu Z, Adams KL (2007) Expression partitioning between genes duplicated by polyploidy under abiotic stress and during organ development. Curr Biol 17:1669–1674. https://doi.org/10.1016/j.cub.2007.08.030

Liu S, Liu S, Wang M, Wei T, Meng C, Wang M et al (2014) A wheat SIMILAR TO RCD-ONE gene enhances seedling growth and abiotic stress resistance by modulating redox homeostasis and maintaining genomic integrity. Plant Cell 26:164–180. https://doi.org/10.1105/tpc.113.118687

Liu B, Zhang N, Zhao S et al (2015a) Proteomic changes during tuber dormancy release process revealed by iTRAQ quantitative proteomics in potato. Plant Physiol Biochem 86:181–190. https://doi.org/10.1016/j.plaphy.2014.12.003

Liu W, Xu L, Wang Y, Shen H, Zhu X, Zhang K et al (2015b) Transcriptome-wide analysis of chromium-stress responsive microRNAs to explore miRNA-mediated regulatory networks in radish (*Raphanus sativus* L.). Sci Rep 5:14024. https://doi.org/10.1038/srep14024

Mahajan S, Tuteja N (2005) Cold, salinity and drought stresses: an overview. Arch Biochem Biophys 444:139–158

Miller G, Shulaev V, Mittler R (2008) Reactive oxygen signaling and abiotic stress. Physiol Plant 133:481–489

Mito T, Seki M, Shinozaki K, Ohme-Takagi M, Matsui K (2011) Generation of chimeric repressors that confer salt tolerance in Arabidopsis and rice. Plant Biotechnol J 9:736–746

Mochida K, Shinozaki K (2011) Advances in omics and bioinformatics tools for systems analyses of plant functions. Plant Cell Physiol 52:2017–2038. https://doi.org/10.1093/pcp/pcr153

Mohanta T, Mohanta N, Mohanta Y, Bae H (2015a) Genome-wide identification of calcium dependent protein kinase gene family in plant lineage shows presence of novel d-x-d and d-e-L motifs in EF-hand domain. Front Plant Sci 6:1146. https://doi.org/10.3389/fpls.2015.01146

Mohanta TK, Arora PK, Mohanta N, Parida P, Bae H (2015b) Identification of new members of the MAPK gene family in plants shows diverse conserved domains and novel activation loop variants. BMC Genomics 16:58. https://doi.org/10.1186/s12864-015-1244-7

Mohanta TK, Mohanta N, Mohanta Y, Parida P, Bae H (2015c) Genome-wide identification of Calcineurin B-Like (CBL) gene family of plants reveals novel conserved motifs and evolutionary aspects in calcium signaling events. BMC Plant Biol 15:189. https://doi.org/10.1186/s12870-015-0543-0

Okazaki Y, Saito K (2012) Recent advances of metabolomics in plant biotechnology. Plant Biotechnol Rep 6:1–15

Orozco A, Morera J, Jiménez S, Boza R (2013) A review of bioinformatics training applied to research in molecular medicine, agriculture and biodiversity in Costa Rica and Central America. Brief Bioinform 14:661–670. https://doi.org/10.1093/bib/bbt033

Pang CNI, Tay AP, Aya C et al (2014) Tools to covisualize and coanalyze proteomic data with genomes and transcriptomes: validation of genes and alternative mRNA splicing. J Proteome Res 13:84–98. https://doi.org/10.1021/pr400820p

Patra D, Fasold M, Langenberger D, Steger G, Grosse I, Stadler PF (2014) plantDARIO: web based quantitative and qualitative analysis of small RNA-seq data in plants. Front Plant Sci 5:708. https://doi.org/10.3389/fpls.2014.00708

Polpitiya AD, Qian W-J, Jaitly N et al (2008) DAnTE: a statistical tool for quantitative analysis of -omics data. Bioinformatics 24:1556–1558. https://doi.org/10.1093/bioinformatics/btn217

Pons R, Cornejo M, Sanz A (2011) Differential salinity-induced variations in the activity of H^+-pumps and Na^+/H^+ antiporters that are involved in cytoplasm ion homeostasis as a function of genotype and tolerance level in rice cell lines. Plant Physiol Biochem 49:1399–1409

Ren Z, Gao J, Li L, Cai X, Huang W, Chao D, Zhu M, Wang Z, Luan S, Lin H (2005) A rice quantitative trait locus for salt tolerance encodes a sodium transporter. Nat Genet 37:1141–1146

Rhodes D, Hanson AD (1993) Quaternary ammonium and tertiary sulfonium compounds in higher-plants. Annu Rev Plant Physiol Plant Mol Biol 44:357–384

Risk BA, Spitzer WJ, Giddings MC (2013) Peppy: proteogenomic search software. J Proteome Res 12:3019–3025. https://doi.org/10.1021/pr400208w

RNAcentral Consortium (2017) RNAcentral: a comprehensive database of non-coding RNA sequences. Nucleic Acids Res 45:D128–D134. https://doi.org/10.1093/nar/gkw1008

Rosewick N, Durkin K, Momont M, Takeda H, Caiment F, Cleuter Y, Vernin C, Mortrex F, Wattel E, Burny A, Georges M, Van den Broeke A (2013) ST105 Deep sequencing reveals abundant Pol III retroviral microRNA cluster in Bovine Leukemia Virus induced leukemia. J Acquir Immune Defic Syndr 62:66. https://doi.org/10.1097/01.qai.0000429267.82844.b6

Schmidt R, Schippers JH, Welker A, Mieulet D, Guiderdoni E, Mueller-Roeber B (2012) Transcription factor OsHsfC1b regulates salt tolerance and development in *Oryza sativa ssp.* Japonica. AoB Plants 2012:pls011

Shah ZH, Rehman HM, Akhtar T, Daur I, Nawaz MA, Ahmad MQ, Rana IA, Atif RM, Yang SH, Chung G (2017) Redox and ionic homeostasis regulations against oxidative, salinity and drought stress in wheat (a systems biology approach). Front Genet 8:141. https://doi.org/10.3389/fgene.2017.00141

Shahbaz M, Ashraf M (2013) Improving salinity tolerance in cereals. Crit Rev Plant Sci 32:237–249

Shriram V, Kumar V, Devarumath RM, Khare T, Wani SH (2016) MicroRNAs as potent targets for abiotic stress tolerance in plants. Front Plant Sci 7:817. https://doi.org/10.3389/fpls.2016.00817

Srivastava A, Singh SS, Mishra AK (2014) Modulation in fatty acid composition influences salinity stress tolerance in Frankia strains. Ann Microbiol 64:1315–1323

Stanke M, Morgenstern B (2005) AUGUSTUS: a web server for gene prediction in eukaryotes that allows user-defined constraints. Nucleic Acids Res 33:W465–W467. https://doi.org/10.1093/nar/gki458

Stocks MB, Moxon S, Mapleson D, Woolfenden HC, Mohorianu I, Folkes L, Schwach F, Dalmay T, Moulton V (2012) The UEA sRNA workbench: a suite of tools for analysing and visualizing next generation sequencing microRNA and small RNA datasets. Bioinformatics 28:2059–2061. https://doi.org/10.1093/bioinformatics/bts311

Teige M, Scheikl E, Eulgem T, Doczi R, Ichimura K, Shinozaki K, Dangl JL, Hirt H (2004) The MKK2 pathway mediates cold and salt stress signaling in *Arabidopsis*. Mol Cell 15:141–152

The Arabidopsis Genome Initiative (2000) Analysis of the genome sequence of the flowering plant *Arabidopsis thaliana*. Nature 408:796–815
Thiery L, Leprince A, Lefebvre D, Ghars MA, Debabieux E, Savoure A (2004) Phospholipase D is a negative regulator of proline biosynthesis in *Arabidopsisthaliana*. J Biol Chem 279:14812–14818
Toubiana D, Fernie AR, Nikoloski Z, Fait A (2013) Network analysis: tackling complex data to study plant metabolism. Trends Biotechnol 31:29–36
Tripathi A, Goswami K, Sanan-Mishra N (2015) Role of bioinformatics in establishing microRNAs as modulators of abiotic stress responses: the new revolution. Front Physiol 6:286. https://doi.org/10.3389/fphys.2015.00286
Tuteja N (2007) Mechanisms of high salinity tolerance in plants. Methods Enzymol 428:419–438
Tuteja N, Sahoo RK, Garg B, Tuteja R (2013) *OsSUV3* dual helicase functions in salinity stress tolerance by maintaining photosynthesis and antioxidant machinery in rice (*Oryza sativa* L.cv. IR64). Plant J 76:115–127
Tyers M, Mann M (2003) From genomics to proteomics. Nature 422:193–197
Vashisht AA, Tuteja N (2006) Stress responsive DEAD-box helicases: a new pathway to engineer plant stress tolerance. J Phytochem Photobiol 84:150–160
Vinocur B, Altman A (2005) Recent advances in engineering plant tolerance to abiotic stress: achievements and limitations. Curr Opin Biotechnol 16:123–132
Walia H, Wilson A, Wahid A, Condamine P, Cui X et al (2006) Expression analysis of barley (*Hordeum vulgare* L.) during salinity stress. Funct Integr Genomics 6:143–156
Wang R, Fabregat A, Ríos D et al (2012) PRIDE Inspector: a tool to visualize and validate MS proteomics data. Nat Biotechnol 30:135–137. https://doi.org/10.1038/nbt.2112
White PJ, Broadley MR (2003) Calcium in plants. Ann Bot 92:487–511
Wilkinson S, Davies WJ (2010) Drought, ozone, ABA and ethylene: new insights from cell to plant to community. Plant Cell Environ 33:510–525. https://doi.org/10.1111/j.1365-3040.2009.02052.x
Yachdav G, Kloppmann E, Kajan L et al (2014) PredictProtein—an open resource for online prediction of protein structural and functional features. Nucleic Acids Res 42:W337–W343. https://doi.org/10.1093/nar/gku366
Yamaguchi T, Blumwald E (2005) Developing salt-tolerant crop plants: challenges and opportunities. Trends Plant Sci 10:615–620
Yi X, Zhang Z, Ling Y, Xu W, Su Z (2015) PNRD: a plant non-coding RNA database. Nucleic Acids Res 43:D982–D989. https://doi.org/10.1093/nar/gku1162
Yu J, Hu S, Wang J, Wong G, Li S et al (2002) A draft sequence of the rice genome (*Oryza sativa* L. ssp indica). Science 296:79–82
Yu J, Wang J, Lin W, Li S, Li H et al (2005) The genomes of *Oryza sativa*: a history of duplications. PLoS Biol 3:266–281
Zhang T, Liu Y, Yang T, Zhang L, Xu S, Xue L, An L (2006) Diverse signals converge at MAPK cascades in plant. Plant Physiol Biochem 44:274–283
Zhang C, Li G, Zhu S, Zhang S, Fang J (2014a) tasiRNAdb: a database of ta-siRNA regulatory pathways. Bioinformatics 30:1045–1046. https://doi.org/10.1093/bioinformatics/btt746
Zhang M, Lv D, Ge P et al (2014b) Phosphoproteome analysis reveals new drought response and defense mechanisms of seedling leaves in bread wheat (*Triticum aestivum* L.). J Proteome 109:290–308. https://doi.org/10.1016/j.jprot.2014.07.010
Zhang F, Zhu G, Du L, Shang X, Cheng C, Yang B et al (2016) Genetic regulation of salt stress tolerance revealed by RNA-Seq in cotton diploid wild species, Gossypium davidsonii. Sci Rep 6:20582. https://doi.org/10.1038/srep20582
Zhao Y, Dong W, Zhang N, Ai X, Wang M, Huang Z et al (2014) A wheat allene oxide cyclase gene enhances salinity tolerance via jasmonate signaling. Plant Physiol 164:1068–1076. https://doi.org/10.1104/pp.113.227595
Zhao Y, Li H, Fang S, Kang Y, Hao Y et al (2016) NONCODE 2016: an informative and valuable data source of long non-coding RNAs. Nucleic Acids Res 44:D203–D208. https://doi.org/10.1093/nar/gkv1252
Zhu JK (2003) Regulation of ion homeostasis under salt stress. Curr Opin Plant Biol 6:441–445

Printed in the United States
By Bookmasters